CHEMISTRY IN THE ENVIRONMENT

Readings from

**SCIENTIFIC
AMERICAN**

CHEMISTRY IN THE ENVIRONMENT

With Introductions by
Carole L. Hamilton
California Institute of Technology

W. H. Freeman and Company
San Francisco

Library of Congress Cataloging in Publication Data

Hamilton, Carole L.   comp.
    Chemistry in the environment.

    1. Chemistry, Technical—Environmental aspects.
2. Geochemistry.   3. Man—Influence on nature.
4. Power resources.   I. Scientific American.
II. Title.
TD194.H35   660   73-3172
ISBN 0-7167-0877-9 (pbk)
ISBN 0-7167-0878-7

Some of the SCIENTIFIC AMERICAN articles in
Chemistry in the Environment are available as
separate Offprints. For a complete list of more than
900 articles now available as Offprints, write to
W. H. Freeman and Company, 660 Market Street,
San Francisco, California 94104.

Printed in the United States of America

9 8 7 6 5 4 3 2 1

Cover by Andy Jurinko

Throughout history, mankind's progress has in large part been propelled by an increasing ability to manipulate materials. The originally unwitting practice of chemistry became more and more systematized, and in the past few decades our pursuit of the science has led to a phenomenal increase in our understanding of the behavior of atoms and molecules. Now, with building blocks at the atomic level, we have the ability to create almost at will substances to serve nearly any purpose that might occur to us. Meanwhile, the parallel growth of the other scientific and technical fields has assured that we will never be at a loss for ideas.

As our power to exploit our physical surroundings has increased, so have our numbers and the magnitude of the demands each of us makes to satisfy his needs and habits. Today's industrial societies function on a huge scale. Their consumption of some raw materials comprises measurable fractions of the global supply. The by-products of their activities can have massive impacts locally, and are even beginning to be significant in parts of the world far from where they originate. It has been pointed out many times in many contexts that man is but a part of his environment. We are now becoming aware that man has become a very large part of the environment—so large a part that he can no longer afford the luxury of thinking of himself and his activities as separate from it. We inhabit a finite region of the universe, depending for our survival on the dynamic balance of chemical and physical forces that have prevailed throughout our evolution. With the growing realization of our potential for disturbing that balance comes the need to evaluate our actions in the light of their possible effects on it.

One of industrial man's more noticeable effects on the integrity of the environment has been his intrusion into the orderly chemical processes of the biosphere. Since these are the processes that sustain life itself, the prospect of tampering with them blindly seems particularly threatening. No wonder, then, that there is growing interest in studying chemistry in the context of the outside world as well as within the limitations of isolated test tubes and reaction flasks. The thirty-seven articles contained in this volume were chosen to illustrate how the activities of industrial societies interact with the chemistry of the environment. They are by no means all-inclusive, either with respect to man's activities or to areas of environmental concern. Nor are they all of the articles dealing with such interactions that have appeared in *Scientific American* over the years. They are a sampling: but a sampling that has been carefully selected to make a coherent presentation of part of a very diffuse subject.

It may be surprising at first that so much of the material here is not "pure" chemistry—or even chemistry at all. Consider, however, just what it is we have set out to examine. The biosphere is a single reaction vessel, with contents that are far from homogeneous. The transformations that go on in that vessel occur in the gas phase, in solution, and in the solid phase, as well as at all the possible interfaces. Transport of reacting species is dependent on many phenomena that are normally dealt with separately in traditional academic disciplines— biology, geology, physics, fluid mechanics. In the laboratory we can pay attention to the reactants and their behavior alone; their containment can usually be more or less ignored. In the biosphere all the characteristics of the reaction system are important, and any treatise on chemistry in the environment has to consider them.

The first of this reader's five sections is intended to provide a general description of the sequences of processes in the biosphere into which life's chemistry is interwoven. The subsequent groups of articles relate to the ways in which the workings of industrial societies perturb those functions. Section II is devoted to the problem of feeding a population grown too large to survive on the fruits of simple cultivation. The technological support structure created by (and necessary for) that large population consumes tremendous amounts of energy, most of it stored in chemical bonds during the life cycles of ages past. The fueling of its operations, examined in Section III, might be defined roughly as the physical chemistry of society. Some of the direct consequences of our exploitation of energy are covered in Section IV, and the last illustrates the perils involved in the extensive chemical manipulation we use to satisfy our material demands.

If there is a lesson to be learned here, perhaps it is this: the scope of man's knowledge is vast, but his reach extends into a still more vast region about which he knows nothing. It has been amply demonstrated that what we don't know *can* hurt us. I hope that this volume serves to show its readers at least some of the directions our search must take for the knowledge on which our survival might depend.

April 1973   Carole L. Hamilton

CONTENTS

I THE STUFF OF LIFE

II OUR IMPACT ON THE LAND THAT FEEDS US

III ENERGY IN OUR SOCIETY

IV THE LEGACY OF ENERGY USE

V IMPLICATIONS OF MATERIAL WEALTH

Note on cross-references: References to articles included in this book are noted by the title of the article and the page on which it begins; references to articles that are available as Offprints, but are not included here, are noted by the article's title and Offprint number; references to articles published by SCIENTIFIC AMERICAN, but which are not available as Offprints, are noted by the title of the article and the month and year of its publication.

CHEMISTRY IN THE ENVIRONMENT

I

THE STUFF OF LIFE

At any one time, only a tiny fraction of this planet's atoms are involved in the fascinating set of processes we call life. What this small collection lacks in mass, however, it makes up in activity and complexity—and because we human animals are part of it, it assumes in our eyes *the* primary role in the "proper" functioning of the system Earth. Even from a viewpoint less prejudiced than ours, it is clear that the biomass exerts an influence on its surroundings out of all proportion to its physical size. The main purpose of the articles in this section is to place the activities of our special population of atoms in perspective, so we can see how the furious molecular busy-ness of living matter fits in with the more leisurely, grand inanimate processes to which it is irrevocably linked.

On a gross level, it is fairly easy to visualize our planet in terms of the concepts of basic chemistry. The system exists in three phases— solid (the lithosphere), liquid (the hydrosphere), and vapor (the atmosphere)—held together by gravity and surrounded by more-or-less empty space. It receives directional radiant energy from outside its boundaries, some of which is absorbed by each phase. Heat is transferred between phases by radiation, by conduction, and as the enthalpy of vaporization of water. The temperature distribution through the system is by no means uniform; thus all phases are in a state of constant agitation, and stirring of a sort is provided for the myriad of reacting species it contains. Material transfer occurs between phases also, on a scale depending on the appropriate distribution coefficients and rate constants.

Such a picture is really not detailed enough, however, to provide much understanding of how the chemistry of life's processes works into the scheme of things. For that we need to obtain more information about the composition of each phase and about those components that are used in the synthesis and sustenance of living matter. "Mineral Cycles," by Edward S. Deevey, Jr., provides an excellent introduction to that task. Starting with an examination of how the elements are distributed through the earth and its environs, he leads us through an amusing process of deducing the unique chemical nature of life. After identifying carbon, hydrogen, oxygen, and nitrogen as the major components of the biomass, he describes a conceptual model for keeping track of those elements as they circulate through the whole system. In the context of that model, Deevey discusses in more detail the paths of two elements, phosphorus and sulfur, that occur in relatively small amounts in living systems but are crucial for their functioning. That discussion brings up the important distinction between elements that can exist in compounds volatile enough to go into the atmosphere, allowing complete recycling through all phases in our system, and those that form no volatile compounds under biospheric conditions and so are restricted to a one-way trip from lithosphere to hydrosphere, unless there is some intervention by living matter. Biochemical processes, like all reactions, can proceed no faster than the rate at which their scarcest reactant becomes available to them. Nutrients (like the soluble, non-volatile phosphorus compounds) that exist in limited reserves with no mechanism for replacing losses therefore control the functioning of the biomass to a great extent; the overall balance of the system will be particularly sensitive to perturbations in the chemistry of these substances.

The second article, "The Chemical Elements of Life," deals in still greater detail with the makeup of living matter. In it, Earl Frieden goes all the way through the periodic table, showing the relationship between what is known about the biological functions of the twenty-four essential elements and their electronic structures. The long list of components that are used in very small amounts and the severe consequences resulting from their absence attest to the complexity and sensitivity of the links between biochemistry *in vivo* and the chemistry of the inanimate surroundings.

Fortunately, it is possible to construct a fairly comprehensive overview of the global interactions involving the major constituents of the biomass—carbon, hydrogen, oxygen, and nitrogen. We begin with water, the unique compound that forms the liquid phase of our planetary reaction system and comprises a significant fraction of the mass of living matter. It is the solvent in which biochemical reactions take place and provides the medium for transporting nonvolatile reactants from place to place in the biosphere, as well as being the source of the hydrogen incorporated into biochemical compounds. "The Water Cycle" by H. L. Penman describes briefly the chemical and physical properties of water and their importance to its function in the biosphere. The article discusses the distribution of the earth's water between solid, liquid, and vapor phases and its movement within and between phases. A portion of the article touches on the physical chemistry of water in soils. The remainder, dealing with the role of water in plant growth, is peripheral to the subject of chemistry in the environment but of interest in terms of overall perspective.

Water also serves as the source of the molecular oxygen released in photosynthesis, as pointed out in "The Oxygen Cycle," so must be the ultimate precursor of today's atmospheric oxygen. Cloud and Gibor outline here the basics of oxidative metabolism, the reactions that higher life forms use to fuel their activities, and trace the concurrent evolution of oxygen-processing organisms and appearance of molecular oxygen in the biosphere. Their description of the oxygen cycle through geologic time provides a striking illustration of the effect biological processes have had on the nature of our planet. Interaction is continuing, of course, because nothing can occur in a totally connected system of equilibria without ultimately affecting all of its components. The latest development has been the intrusion of man's activities into the complex of pathways followed by oxygen as it circulates through the biosphere. Cloud and Gibor speculate on the nature of possible adjustments the system might make in response to these perturbations, emphasizing that all we know for sure is that adjustments will indeed take place.

It is in the nitrogen cycle that industrial man's manipulations have the greatest impact. The world's reservoir of that element is the molecular nitrogen of the atmosphere, which is unusable to most living species. The present-day biomass, including man, relies in almost equal measure on a few species of microorganisms and on industrial chemistry to reduce atmospheric nitrogen to fulfill its metabolic needs. C. C. Delwiche, in "The Nitrogen Cycle," outlines the fixation processes, and shows how a variety of organisms are sustained by the energy differences between items on the rich menu of oxidation levels available to the element. In his analysis of the nitrogen cycle, the author dwells on the magnitude of the contribution of ammonia manufacture to total fixation. There is another massive input of human origin, the roughly fifty million tons of nitrogen oxides produced annually over the world as a consequence of combustion.

Man's burning of fossil fuels also interferes with the "natural" circulation of carbon. Bert Bolin, in "The Carbon Cycle," identifies the main pathway followed by that element on a global scale as the transformation of carbon dioxide into living matter and vice versa. Though the amount of carbon dioxide released in fossil fuel combustion is small compared with that produced by respiration of the planet's living creatures, it must create a continuing alteration in the dynamic equilibrium that exists between the world's major carbon dioxide reservoirs, since the reverse of the combustion reaction ($CO_2 \rightarrow$ coal, oil, or gas) occurs much more slowly than even the slowest circulation of CO_2-carrying media in the biosphere. Part of Bolin's discussion deals with the nature of the system's responses to the disturbance, and the time scale over which they will continue.

A complementary article is "The Mechanism of Photosynthesis," by R. P. Levine, who looks at one subsystem within the global system —the one that accomplishes the task of incorporating carbon dioxide into the molecules that make up living matter. This step in the carbon cycle is the one that taps our prime energy source, the sun, and turns its radiation into a form usable for driving biochemical reactions, thus providing the ultimate fuel source for the entire biomass. The exquisite complexity of the photosynthetic process is well illustrated by this article. Considering that photosynthesis is only one of many equally involved reaction sequences occurring as the essential elements cycle through the living component of the biosphere, it provides a strong reminder of just how limited our understanding of the whole system is.

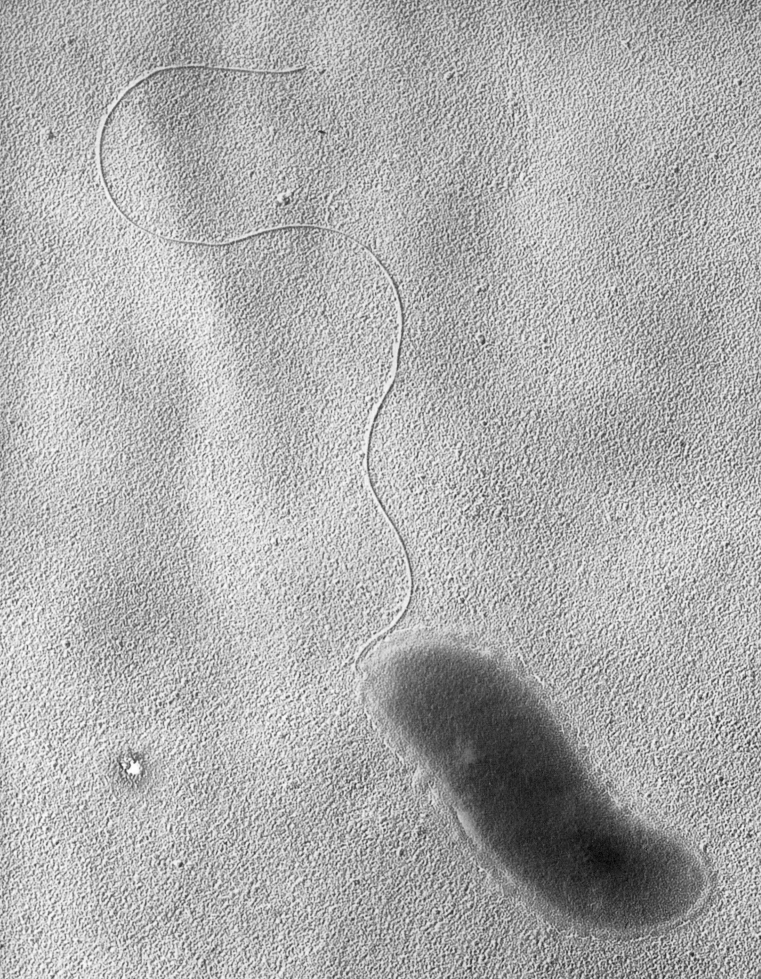

1

MINERAL CYCLES

EDWARD S. DEEVEY, JR.
September 1970

*Although the biosphere is mainly composed of
hydrogen, carbon, nitrogen and oxygen, other elements
are essential constitutents of living matter. Notable
among them are phosphorus and sulfur*

The periodic table lists more than 100 chemical elements. Yet ecologists have defined the biosphere as the locus of interaction of only four of them: hydrogen, carbon, nitrogen and oxygen. In the periodic table these four are numbered 1, 6, 7 and 8. This definition, although it deals handsomely with much of the chemistry of life, turns out to be a little too restrictive. But when we enlarge it to include phosphorus and sulfur, as we do here, we have gone no farther up the table than element No. 16. From this it should be apparent that no element lighter than sulfur can be ignored, either by ecologists or by anyone else. The fact is that most human problems—all environmental ones, anyway—arise from the exceptional reactivity of six of the 16 lightest elements.

Because our definition of the biosphere is based more on reactivity than on atomic number, it is a minimum definition. It is not intended to exclude heavier elements that react with the primary six. As a matter of empirical fact it is known that no element lighter than iron and cobalt, elements No. 26 and No. 27, is unimportant to the biosphere. Beyond copper, No. 29, there are a few conspicuously reactive elements such as the heavy halogens bromine and iodine. Most of the heavies are metals, such as gold, mercury and lead (Nos. 79, 80 and

82), however, and their main effect on the lightweight biosphere is to depress it. Toward the end of the periodic table are some famously overweight metals whose tendency to lighten themselves has disastrous effects on any light substances that get in the way.

In order to understand how it is that many elements interact with the essential six, one must briefly reflect on the biosphere as a whole. Because the biosphere is so reactive, its influence on the hydrosphere, the lithosphere and the atmosphere is inversely proportional to its mass. This mass is very small. An average square centimeter of the earth's surface supports a tiny amount of biosphere: 580 milligrams, less than the weight of two aspirin tablets. A roughly equivalent mass is found in the same area of hydrosphere a single centimeter deep, or in a paper-thin slice of lithosphere. Still, from a worm's-eye view the biosphere has real substance, particularly on land, where it amounts to 200 oven-dry tons on an average hectare.

A glance at a partial list of the elements that compose the biosphere shows why hydrogen, oxygen, carbon and nitrogen dominate conceptions of biosphere chemistry. Together these elements constitute all but a tiny fraction of the average terrestrial vegetation, which in turn constitutes more than 99 percent of the world's standing crop. The quantities are shown in the chart on the next page, based on a splendid compilation by L. E. Rodin and N. I. Basilevich. What I have done is to weight their chemical analyses in proportion to the kinds of land area they represent. The weighting factors, for desert, forest, tundra and so on, are the same ones I used to calculate the earth's production of carbon in an earlier article ["The Human Population," by Edward S. Deevey, Jr.; SCIENTIFIC AMERICAN Offprint 608]. Inciden-

tally, on the basis of this new calculation terrestrial carbon production comes out at 65×10^9 tons of carbon per year, about 15 percent more than the figure I computed before.

What chemical compounds do these elements form? The standard way to determine the chemical composition of an organic substance is to burn it and collect the products. The list of components that results from this destructive procedure expresses some obvious facts, such as the familiar one that the biosphere is mainly carbon dioxide and water. Nitrogen, a major constituent of protein, seems surprisingly scarce (about five parts per 1,000 by weight) until we remember that the biosphere is chiefly wood, that is, not protein but the carbohydrate cellulose.

The destructive procedure would also leave a smudge, about 12 parts per 1,000 of the total, loosely called ash. Its dominant elements calcium, potassium, silicon and magnesium have important biochemical functions. One atom of magnesium, for instance, lies at the center of every molecule of chlorophyll, and silicon, the stuff of sand, is obviously useful for building hard structures. Iron and manganese also play central roles in the biosphere, a fact that could not be guessed from their position in our chart. In biochemistry as in geochemistry the importance of these elements is in governing oxidation-reduction reactions, but the masses involved are small. As for the major cations—ions of such elements as calcium, potassium, magnesium and sodium—new insights have just begun to flood in with their discovery in rainwater.

There are many other metallic elements that appear in trace amounts. Not all of them are listed in the chart because some could be accidental con-

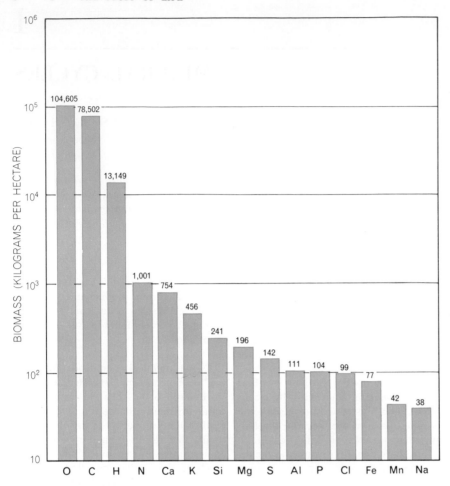

COMPOSITION OF THE BIOSPHERE is dominated by oxygen, carbon and hydrogen, as is indicated by the bars in this logarithmic chart. The units are kilograms per hectare of land surface. Key to the symbols for the chemical elements is at the bottom of the page.

charts on the opposite page can serve as a substitute, we can cast a Holmesian eye over the list. Our thinking will be more productive if we compare the composition of the biosphere with the composition of the lithosphere, the hydrosphere and the atmosphere. For this comparison all four of the "spheres" in the chart are converted from parts by weight to atoms per 100 atoms. (The masses of the four spheres being very different, these percentages will give no idea of the earth's mean or total composition.)

At first glance the four spheres do not seem to belong in the same universe. Not surprisingly, the lithosphere turns out to be a slightly metallic aluminum silicate. ("Here is no water but only rock/Rock and no water and the sandy road," as T. S. Eliot put it in "The Waste Land.") The biosphere, in sharp contrast, is both wet and carbonaceous. A single class of compounds, formaldehyde (CH_2O) and its polymers, including cellulose, could make up more than 98 percent of the total (by weight). Still, even when it is dried in an oven at 110 degrees Celsius, life is mainly hydrogen and oxygen, in close approximation to the proportions known as water. In other words, the biosphere is notably carboxylated: it is both more hydrated and chemically more reduced (hydrogenated) than is the lithosphere from which, in some sense, it came. Among the 10 most abundant elements of the lithosphere there is no obvious source for life's carbon. Hydrogen is also fairly far down the list for rock (and would be farther down if I had not copied some old figures from Frank W. Clarke's *The Data of Geochemistry*, which overweight the acidic rocks of continents).

Even the elementary Dr. Watson might conclude that life's hydrogen comes from some inorganic hydrate—water, for instance—and indeed the hydrosphere provides an ample and ready supply. This will not work for carbon, though, and in trying to account for carboxylation we can make a deduction that is truly elementary in the Holmesian, or nonobvious, sense. We begin

taminants. There remain two, sulfur and phosphorus, each amounting to more than 10 percent of the nitrogen, that do not look like contaminants. To ignore these elements as "traces" or even to think of them as "ash" or "inorganic" elements is to misconstrue the chemical architecture of the biosphere.

A listing of elements and compounds does not reveal that architecture. There is a big difference between a finished house and a pile of building materials. Nevertheless, a list is a useful point of departure. If it is made with care, it can protect ecologists from the kind of mistake that architects sometimes make, such as forgetting the plumbing.

When a list contains as much information as a shopping list—when it shows amounts as well as kinds of materials—some conclusions can be drawn from the relative proportions. (As a former bureaucrat I have learned that a "laundry list" contains even more ambiguous information than a "shopping list"; good bureaucrats keep both.) If a housewife's shopping list showed a pound of coffee, four pork chops and 100 pounds of

sugar, for example, we would know that madame is either hoarding or running a private business. If she also wants a ton of flour, she is evidently baking, not distilling. The inclusion of two dozen light bulbs would suggest that she works mainly at night, but the listing of 10 dozen light bulbs would point to a faulty generator.

As it happens, this kind of semiquantitative ratiocination was applied to ash, and to biogeochemistry, by the master of nonobvious deduction, Sherlock Holmes. Unfortunately no copy of his analytical results (the monograph on cigar ash, cited in Chapter 4 of "A Study in Scarlet") has yet come to light. If the

Al	ALUMINUM	Cl	CHLORINE	Mn	MANGANESE	P	PHOSPHORUS
Ar	ARGON	Fe	IRON	N	NITROGEN	S	SULFUR
B	BORON	H	HYDROGEN	Na	SODIUM	Si	SILICON
C	CARBON	K	POTASSIUM	Ne	NEON	Ti	TITANIUM
Ca	CALCIUM	Mg	MAGNESIUM	O	OXYGEN		

RELATIVE AMOUNTS OF ELEMENTS in the biosphere, the lithosphere, the hydrosphere and the atmosphere are presented in the charts on the opposite page. Here, however, amounts are given not as kilograms per hectare but as atoms per 100 atoms. Here again scale is logarithmic to show less abundant elements, which otherwise could not be compared.

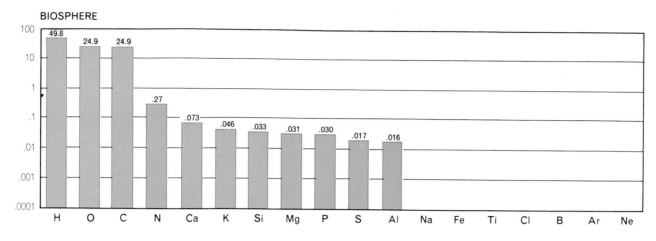

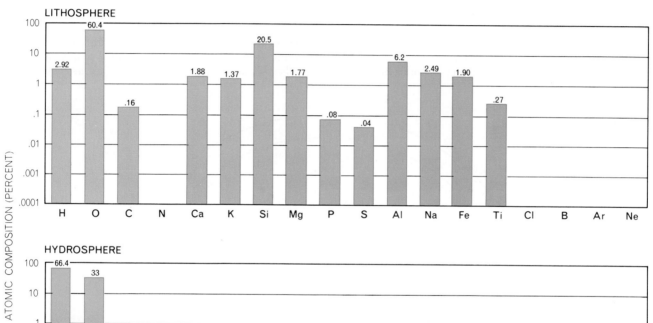

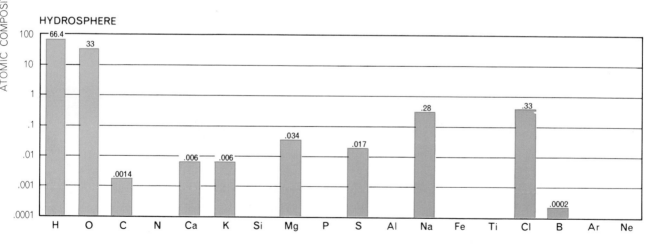

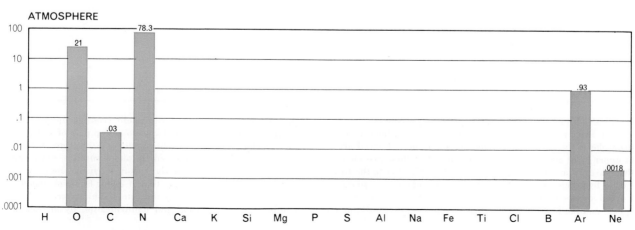

ATOMIC COMPOSITION (PERCENT)

again by noting that life is mainly aqueous, and also that it concentrates carbon in proportions far greater than those in any accessible source. Is it possible that these facts are related? If they are, what do we know about water that throws any light on this relation and on the behavior of carbon? (At this point a lesser detective might reach for the carbonated water and pause for a reply.)

Instead of guessing, Holmes would proceed with his review of the evidence. Water, of course, is continuously recycled near the earth's surface, by runoff, evaporation and condensation. That is, it flows in rivers from the lithosphere to the hydrosphere, and it returns to rewash the land by way of the atmosphere. Any water-soluble elements are certain to track this cycle at least partway, from land to sea, although they may find the sea to be a sink, as boron does. If they

are to get out, they can reach the land as part of an uplifted sea bottom, but that is a chancy mechanism. Recycling is both faster and surer if the element is volatile as well as soluble, so that one of its compounds can move landward through the atmosphere as water does.

In the biosphere there are at least three elements besides those of water—carbon, nitrogen and sulfur—that fall in this doubly mobile class. Among their airborne compounds are carbon dioxide (CO_2), methane (CH_4), free nitrogen (N_2), ammonia (NH_3), hydrogen sulfide (H_2S) and sulfur dioxide (SO_2). It is interesting that when carbon, nitrogen and sulfur are recycled, their valence changes. It may not be an accident that all three are more reduced in the biosphere than they are in the external world. Be that as it may, they all seem

to belong to the biosphere, which is otherwise mainly water. Hence all three must be recycled together, *along with the water* (said Holmes with an air of quiet triumph), if the earth is to sustain its most unusual hydrate. ("And what is that?" I asked. "Why, *carbohydrate*, of course," said Holmes.)

I call this deduction nonobvious, because in an obvious variant it has become so familiar as to inhibit thought. The outlines of the carbon cycle, in organisms at any rate, have been evident since Joseph Priestley's day. The critical step, "obviously," is the photosynthetic reduction of carbon dioxide. That reaction is a hydrogenation, yielding formaldehyde. Its source of hydrogen is the dehydrogenation of water, with the liberation of oxygen. The chemical energy thus captured, by a process unique to green plants, becomes available, inside

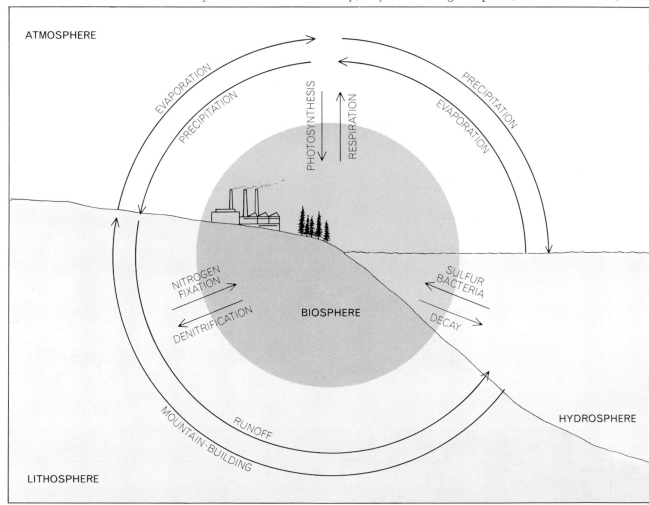

CARBOXYLATION CYCLE supplies the biosphere with carbon, oxygen, hydrogen, nitrogen and sulfur by carrying them from the lithosphere, the hydrosphere and the atmosphere. Curved arrows at upper right and left represent any or all of these five elements that travel from the atmosphere to the lithosphere or to the hydrosphere by precipitation, or back to the atmosphere by evaporation. Curved arrows at bottom indicate direct routes between the lithosphere and the hydrosphere such as runoff, mountain-building and the hydration of minerals. Biosphere (*color*) captures these elements by providing alternative routes. Top pair of straight arrows show exchange between the biosphere and the atmosphere, carbon, for example, being exchanged by photosynthesis and respiration. Pair of straight arrows at right show exchange between the biosphere and the hydrosphere, that of sulfur being mediated by bacteria. Pair of arrows at left indicate soil-biosphere exchanges including nitrogen fixation and denitrification by microorganisms.

the cell, for all other vital reactions (does it not?). After its utilization, which includes consumption by animals, the re-oxidized carbon dioxide can rejoin any geochemical cycles it likes.

All other vital reactions? Well, not quite all. The chemical reduction of nitrogen is one hydrogenation essential to green plants that they cannot perform for themselves. As one result, even elementary textbooks admit, the carbon and nitrogen cycles are necessarily interdependent. Without microorganisms that take nitrogen from the air and hydrogenate it (they can use carbon dioxide as a carbon source), all the nitrogen in the biosphere would soon appear in the atmosphere in stable, oxidized form. (The textbooks concede this point somewhat grudgingly, because much of the biological nitrogen cycle operates below the oxidation state of free nitrogen by the reversible reduction of nitrate and nitrite to amino acids and ammonia.)

If, as it turns out, sulfur too is recycled by way of the hydrologic cycle but independently of green plants, it becomes necessary to look beyond carbon and water for the clue to carboxylation. In other words, some biologists are not unlike architects who forget about the plumbing. In their preoccupation with carbon dioxide reduction as the starting point for cell biochemistry they tend to forget two other hydrogenations, those of sulfur and nitrogen, that are just as important.

A check is needed here, to be sure that these two elements are really intrinsic to the biosphere. In the case of sulfur the figures show it to be very scarce, and if it is a contaminant, the whole argument might be superfluous. Sulfur, however, is no contaminant; no protein can be made without it. In fact, sulfur is the "stiffening" in protein. A protein cannot perform its function unless it is folded and shaped in a particular way. This three-dimensional structure is maintained by bonds between sulfur atoms that link one segment of a protein molecule to another. Without these sulfur bonds a protein would coil randomly, like a carelessly dropped rope.

The reason for the apparent scarcity of sulfur is the low protein content of woody tissue; any animal body contains much more. Cod-meal protein, for example, with 2.26 percent of the sulfurous amino acid methionine, has the empirical formula $H_{555}C_{265}O_{174}N_{83}S$. Although other proteins differ in the proportions, the substance of the biosphere must always contain these five elements.

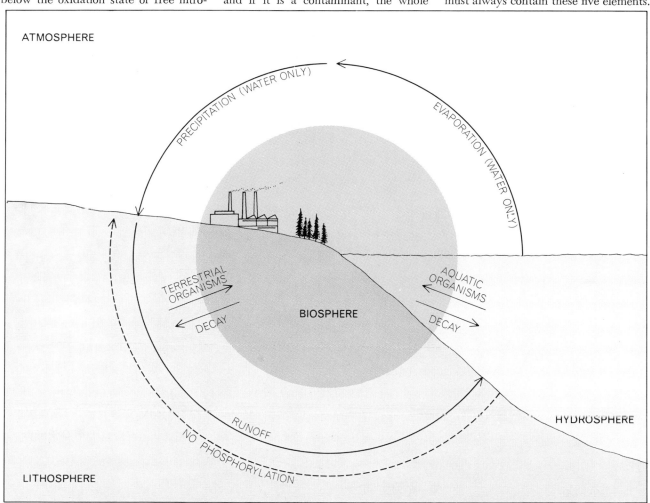

SOLUBLE-ELEMENT CYCLE is followed by minerals such as phosphorus that dissolve in water but are not volatile, that is, they are not carried into the air by evaporation (*curved arrow at right*). The curved arrow at bottom shows that phosphorus is washed from the lithosphere into the hydrosphere by runoff from rainfall (*curved arrow at top left*). The broken curved arrow at bottom indicates that phosphorus in the hydrosphere does not normally return to the lithosphere and that therefore the ocean would become a phosphorus sink. The upper straight arrows at right and left, however, show that the organisms of the biosphere impede this development by absorbing some phosphorus. The straight arrows pointing from the biosphere to the lithosphere and to the hydrosphere indicate the decay of organic matter. On land the soluble-element cycle is continued when decay returns phosphorus to the lithosphere. Without an atmospheric link from ocean to land, however, the cycle is actually a one-way flow with interruptions.

It has been known for many years that sulfur is recycled from the sea back to the land by way of the atmosphere. Calculations confirming this fact by Erik Eriksson of the International Meteorological Institution show that the world's rocks contain too little sulfur, by a factor of about three, to account for the sulfate delivered annually by the world's rivers. About three-quarters of the total budget (in 1940) is therefore inferred to have come from the atmosphere. Of this amount about a third, or a quarter of the total, can have come from industrial sources—better known these days as "sulfur dioxide pollution." The other two-thirds, or half the total budget as of 1940, must take some more natural route from the hydrosphere.

When Eriksson wrote, in 1959, the question was still open, whether the cycled sulfur reaches the land as an aerosol from sea spray or as hydrogen sulfide (H_2S). If the principal volatile compound is a sulfide, it must be made by sulfate-reducing bacteria, because no other "room temperature" source of sulfide is known. M. LeRoy Jensen and Noboyuki Nakai, then working at Yale University, settled this question in favor of the bacteria, by showing that atmospheric sulfur, although it falls in rain as sulfate, contains less of the heavy isotope sulfur 34 than seawater sulfate does. What the natural isotopic label shows is that the sulfate in rain entered the atmosphere not as sea spray but as sulfide, there to be oxidized to sulfur dioxide. After dissolution in rainwater, sulfate (and sulfuric acid) are formed.

The principle of the Jensen-Nakai demonstration is worth noticing, because it applies to the cycling of carbon as well as of sulfur, and barring some technical difficulties it could also apply to nitrogen. The route followed through oxidation-reduction reactions by ordinary sulfur (sulfur 32) is analogous to the route followed by ordinary carbon (carbon 12) in photosynthesis. These lighter, more mobile isotopes appear preferentially in reduced compounds such as hydrogen sulfide, methane and formaldehyde. At equilibrium in a closed system the oxidation products (carbon dioxide or sulfate) have correspondingly more of the heavier isotopes carbon 13 and sulfur 34 without change in the total mass. If, however, a reduced and isotopically light product escapes, as hydrogen sulfide does from the hydrosphere, equilibrium is not attained, and if the gaseous product is trapped and reoxidized in a separate system, the oxide (sulfur dioxide in this case) remains light.

Exactly where within the hydrosphere

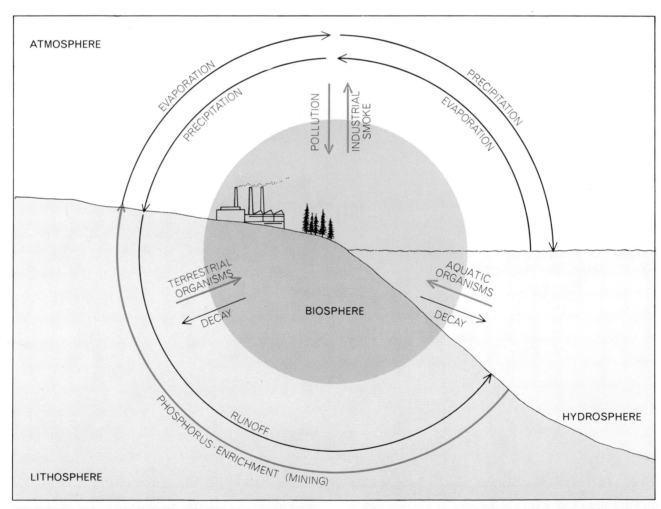

EUTROPHICATION OF THE BIOSPHERE is the intensive cycling of phosphorus, nitrogen and sulfur. Colored curved arrow at bottom represents beginning of the process: the human use of phosphorus as fertilizer, which returns phosphorus to the lithosphere, thereby reversing the phosphorus cycle. Colored straight arrows at left and right indicate that phosphorus added to the lithosphere (and to phosphorus already present) is then taken up by phytoplankton and other organisms as well as by crops. Other straight arrows at right and left show that phosphorus and other elements return to the lithosphere and hydrosphere by decay. Once phosphorus is plentiful, scarcity of nitrogen and sulfur may limit eutrophication. Arrows at top represent carbon dioxide, nitrate and sulfate from industrial activity rising into atmosphere and falling in rain. They may promote eutrophication of dry land since vegetation may reabsorb them from air and soil. Curved arrows indicate routes followed by elements that are both soluble and volatile.

most sulfate-reducers reside is still unclear. The known ones are obligate anaerobes, and their habitat is mud. Swamps, marshes and the floor of eutrophic lakes must all be important, and they may be quantitatively more important than the blue mud of estuaries and continental shelves. The sulfur metabolism of such large systems is not easy to study, even with isotopic tools. Minze Stuiver, now at the University of Washington, injected radioactively labeled sulfate ions into one eutrophic lake, Linsley Pond in Connecticut. Sulfate reduction proved to be intense, as had been expected. This lake, however, has quite a bit of ferrous iron in its deeper waters, and more in the mud itself. In the presence of the ferrous iron all the labeled sulfide was firmly held in the mud as ferrous sulfide, and no hydrogen sulfide escaped. At least for the duration of the radioactive label, with its half-life of 89 days, this mass of mud was not a source of atmospheric sulfur but a sink.

It follows from all of this that the cycling of sulfur in nature is no less relevant to carboxylation than the cycling of carbon and nitrogen. Without downgrading photosynthesis, we can say that carbon fixation is only one of at least three critical steps in the global synthesis of protein. All three are hydrogenations, achieved with the aid of enzymes, which are themselves proteins, and therefore occur only in the biosphere. Of the three reductions, however, only the reduction of carbon calls for green plants and sunlight. The other two, the reduction of nitrogen and of sulfur, are accomplished anaerobically, by microbes. Thus the locus of the nitrogen and sulfur reductions is, broadly speaking, oxygen-deficient soil and mud. Both loci are separated spatially from that airy, sunlit world where green plants (addicted, like human societies, to the external disposal of wastes) are thoughtlessly liberating oxygen.

With three critical steps for five elements, moving through four "spheres" of abstract space, one feels the need for a picture—a "systems model"—just to keep track of the relations. The two-dimensional analogue on page 8 is simple but adequate. Although it fails to specify fluxes, or any chemical quantities, it provides a mental framework for the movement of five elements: hydrogen, oxygen, carbon, nitrogen and sulfur, either alone or in combinations such as water, nitrate, the dioxides of carbon and sulfur, and carbohydrate. The synthetic output is the biosphere, with the

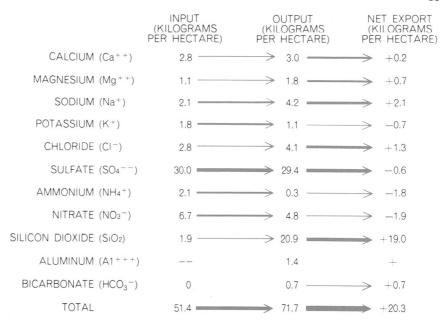

	INPUT (KILOGRAMS PER HECTARE)	OUTPUT (KILOGRAMS PER HECTARE)	NET EXPORT (KILOGRAMS PER HECTARE)
CALCIUM (Ca^{++})	2.8	3.0	+0.2
MAGNESIUM (Mg^{++})	1.1	1.8	+0.7
SODIUM (Na^+)	2.1	4.2	+2.1
POTASSIUM (K^+)	1.8	1.1	−0.7
CHLORIDE (Cl^-)	2.8	4.1	+1.3
SULFATE (SO_4^{--})	30.0	29.4	−0.6
AMMONIUM (NH_4^+)	2.1	0.3	−1.8
NITRATE (NO_3^-)	6.7	4.8	−1.9
SILICON DIOXIDE (SiO_2)	1.9	20.9	+19.0
ALUMINUM (Al^{+++})	−−	1.4	+
BICARBONATE (HCO_3^-)	0	0.7	+0.7
TOTAL	51.4	71.7	+20.3

EUTROPHICATION OF DRY LAND is indicated by the imbalance between the quantity of certain ions falling from the atmosphere on the forest at Watershed No. 6 at Hubbard Brook in New Hampshire and the output of these ions in the brook itself. Input (*smaller arrows at left*) of some elements such as calcium, magnesium and sodium is smaller than the output (*larger arrows at right*). The input of potassium, ammonium, sulfate and nitrate, however, is larger (*larger arrows at left*) than output of these substances (*smaller arrows*). The excess of input indicates that the forest is utilizing these four substances as it grows.

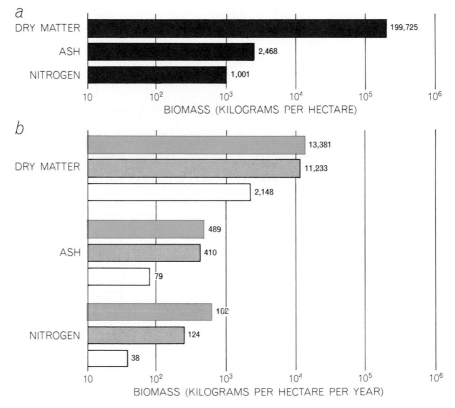

OTHER EVIDENCE FOR EUTROPHICATION is provided by studies of the earth's standing crop on dry land. In a biomass, or weighable dry matter that includes ash and nitrogen, totals about 200,000 kilograms per hectare. In b the first set of bars shows that the dry matter increases in an average year by 13,381 kilograms per hectare (*color*). About 11,000 kilograms is lost in the form of litter fall (*gray*) such as fallen leaves and branches, giving a "mean increment" of 2,148 kilograms per hectare (*open bar*). The second set of bars shows that ash increases by 489 kilograms per hectare (*color*) but is reduced by litter fall (*gray*) to 79 kilograms (*open bar*). The third set of bars shows that nitrogen increases by 162 kilograms (*color*), a gain that is reduced by litter fall (*gray*) to 38 kilograms per hectare.

empirical composition of protein. The central or regulatory position of the biosphere in this model follows from the fact that for all five elements it is both a source and a temporary sink. For any element that might be tempted to cycle around the edges of the model, the biosphere provides several high-energy alternatives. The most interesting of these are the reductions of carbon, nitrogen and sulfur, each concentrated at a different interface, two being out of immediate contact with air. Water, although it is able in principle to cycle independently, is the source of the hydrogen that energizes the biosphere, and cannot long avoid the biospheric loop as long as the biosphere functions.

It is easy to be bemused by so fascinating a model. Its function, however, is to clarify thought. If further thought disrupts the model, nothing is lost but a few lines on paper. More or less instantly, by reference to the table of biospheric composition, we can see that the model is incomplete. Phosphorus has been left out, along with calcium, potassium, silicon and magnesium, four elements that are commoner in the biosphere than sulfur is. Will any or all of these cycle tamely through the model, or will they disrupt it beyond repair?

For phosphorus, but not yet for the others, the answer is clear: With one significant modification, the model can accommodate phosphorus. First, let us be sure, as we made sure for sulfur, that phosphorus is necessary to the biosphere. It is not a constituent of protein, but no protein can be made without it. The "high-energy phosphate bond," reversibly moving between adenosine diphosphate (ADP) and adenosine triphosphate (ATP), is the universal fuel for all biochemical work within the cell. The photosynthetic fixation of carbon would be a fruitless tour de force if it were not followed by the phosphorylation of the sugar produced. Thus although neither ADP nor ATP contains much phosphorus, one phosphorus atom per molecule of adenosine is absolutely essential. No life (including microbial life) is possible without it.

UNIVERSAL FUEL of living matter is adenosine triphosphate (ATP). High-energy phosphate bonds of ATP ($\sim$) each store 12,000 calories and release 7,500 calories when broken.

PRODUCTION OF ATP, shown in generalized form, consists of two stages. The first stage begins as aldehyde reacts with an inorganic phosphate to produce hydrogen and acid phosphate. In second stage (*bottom*) acid phosphate (*shading*) reacts with ADP (adenosine diphosphate) to make an organic acid and ATP (*color*). R stands for radicals, or side groups.

Our provisional model of the biosphere has been constructed on two explicit assumptions: (1) the biosphere necessarily contains the five elements of protein and (2) all five are both soluble and volatile. If we now add phosphorus as a sixth necessary element, we can safely assume its solubility in water, and the crucial question concerns its volatility. Except as sea spray in coastal

GLUCOSE   GLUCOSE-1-PHOSPHATE

GLUCOSE-1-PHOSPHATE   FRUCTOSE   SUCROSE   INORGANIC PHOSPHATE

SYNTHESIS OF SUCROSE is an example of a reaction for which ATP (*color*) supplies energy. The reaction begins at upper left as the ATP molecule combines with glucose molecule, releasing 7,500 calories. The reaction produces ADP and glucose-1-phosphate. In a second stage of the reaction (*bottom*) glucose-1-phosphate combines with fructose, yielding sucrose and inorganic phosphate.

Chemical structure diagram of a segment of a bovine insulin molecule, with labels including: S, O, CH_2, CH, NH, C, R_{GLU}, R_{ALA}, R_{VAL}, R_{SER}, R_{HIS}, R_{LEU}, R_{GLY}, and disulfide (S—S) linkages between polypeptide chains.

BASIC FUNCTION OF SULFUR in living matter appears to be to provide a linkage between the polypeptide chains in a protein molecule. These linkages help the protein maintain its three-dimensional shape so that it can perform its function. In this segment of a bovine insulin molecule disulfide bonds (*color*) are formed between sulfur atoms, which are present in the amino acid cystine. Cystine is a subunit of both polypeptide chains. Because the molecule is displayed in two dimensions it is flattened. Therefore the bond between the top and bottom cystine groups on the upper chain appears broken. In the normal three-dimensional state, however, this chain is twisted and folded because of the disulfide bond in a way indicated by the colored line that joins the two sulfur atoms. (The other bond is one of two that links the chains.) The shape of the insulin molecule maintained by these bonds enables it to control the metabolism of sugar. The other amino acids in this molecular segment, whose side chains are indicated by the letter R, are glutamic acid (GLU), alanine (ALA), serine (SER), valine (VAL), histidine (HIS), leucine (LEU) and glycine (GLY).

zones, or as dust in the vicinity of exposed phosphate rock, phosphorus is unknown in the atmosphere; none of its ordinary compounds has any appreciable vapor pressure. It therefore tracks the hydrologic cycle only partway, from the lithosphere to the hydrosphere, and in a world uncomplicated by a biosphere the ocean would be its only sink. In terms of my model this amounts to uncoupling the atmospheric reservoir (except for water), omitting the half-arrow showing the return of phosphorus from the hydrosphere to the lithosphere and leaving the biosphere's phosphorus as a feedback loop, diverting some of the one-way flow from rock to ocean. Geometrically at least, the model is general enough to accommodate these changes, some version of which will be needed if any permanent sinks are discovered in the system.

For any soluble but nonvolatile element a closed natural cycle is possible only through the biosphere. The model hints at the reason why many elements—vanadium, cobalt, nickel and molybdenum among them—are best known in aquatic organisms and cycle mainly within the hydrosphere. Now, however, the model, nonquantitative though it is, suggests something else. If the biosphere demands such an element as phosphorus (and the cases of iron and manganese should be similar), two alternative inferences are permissible, depending on

the magnitudes of reservoirs and fluxes. If the lithosphere contains an ample supply of phosphorus, or if the flux to the hydrospheric sink is large, the biosphere can take off what it needs and waste the rest. It is commonly believed such elements as sodium and calcium are thus wasted by terrestrial vegetation, although ecologists are beginning to doubt it. On the other hand, if the quantity is scanty or the flux small, the element will be in critically short supply. And if short supply is chronic, the output of the entire system could be expected to be adjusted to the rate of exploitation of one critical element, much as the performance of a bureaucracy is closely geared to the supply of paper clips.

In undisturbed nature the chronic shortage of phosphorus is notorious; that is what most people mean by "soil infertility." In the lithosphere phosphorus is scarcer than carbon, and in the hydrosphere, because phosphorus falls in the parts-per-billion range, it fails to show up at all in the chart of constituents on page 3. Apart from its natural scarcity, phosphorus is freely soluble only in acid solution or under reducing conditions. On the surface of an alkaline and oxidized earth it tends to be immobilized as calcium phosphate or ferric phosphate. In lake waters, where the output of carbohydrate is thriftily attuned to phosphorus concentrations of the order of 50 micro-

grams per liter, doubling the phosphorus input commonly doubles the standing crop of plankton and pondweeds.

Under these conditions the situation in a lake changes drastically. If phosphorus is plentiful, nitrate may become the critically short nutrient for a crop that needs about 15 atoms of nitrogen for one of phosphorus. Blue-green algae may then take over the plankton because by reducing atmospheric nitrogen they escape the dependence that other algae have on nitrate. Meanwhile, judging from much recent experience, the phosphorus input will probably have doubled again—but the subject under discussion is no longer undisturbed nature. What started as "cottage eutrophication," by seepage from a few septic tanks, has been escalated into a noisome mess by "treated" sewage and polyphosphate detergents. To conserve biogeochemical parity the atmosphere has begun to deliver into lakes nitrate and sulfate from the combustion of fossil fuels.

It would be wrong to read too much into a systems model. "Conserving biogeochemical parity" is just a figure of speech, technically hyperbole, and ironical at that. After all, the pollution of the air by nitrate and sulfate is quite independent—technologically, spatially and politically—of the pollution of water by phosphorus. If the accelerated phosphorus cycle in lakes takes advantage of these added inputs, we dare not say that nitrate and sulfate have been drawn into the biosphere from the atmosphere, as U.S. power was drawn to a Canadian circuit breaker in the Northeast blackout of 1965. What the model tells us is that matters can look that way from the standpoint of the biosphere. If the lake segment of the phosphorus cycle is accelerated to the point where nitrogen and sulfur are as critical as phosphorus used to be, and if there is a new source of nitrate and sulfate in the atmosphere, the atmosphere is adequately coupled to all other subsystems to ensure the success of the newly accelerated loop. The loop, known as eutrophication, is thus amplified from a lacustrine nuisance to a systems problem, and around such lakes as Lake Erie it threatens to become a cancer in the global ecosystem.

The trouble started, of course, when the world's one-way phosphorus cycle was first reversed and then accelerated by human activity. Since bird guano was discovered on desert islands, later to be supplemented in fertilizers by phosphate rock, marine phosphate has been restored to the lithosphere in ever increasing amounts. As a device for growing people in ever increasing numbers the practice cannot be faulted, but if people

are to continue to flourish in the biosphere, they will have to pay more attention to scarce resources. Phosphorus is much too valuable to be thoughtlessly shared with blue-green algae.

The term eutrophication, which means enrichment, usually inadvertent, is not ordinarily applied to forests and deserts. I dare to extend it to the terrestrial biosphere because two new lines of evidence have suddenly appeared to suggest that the known pollution of air by nitrate and sulfate also encourages the bloom on dry land. The first line of evidence comes from Hubbard Brook, N.H., where F. Herbert Bormann of the Yale School of Forestry and Gene E. Likens of Cornell University have had six forested watersheds under close study since 1963. What interests us here is the difference, per hectare of ecosystem, between the input of ions in rainfall (plus dry fallout, if any) and the output as measured at a dam at the foot of each drainage basin.

Among the common ions entering and leaving Watershed No. 6 at Hubbard Brook, chloride and three positive ions—calcium, magnesium and sodium—show an excess of output over input, pre-

sumably derived from the local rocks and soil. These four ions conform, if only barely, to the idea that the biosphere wastes excess salts on their way to the sea. In contrast, potassium and ammonium (NH_4) and the two major negative ions, sulfate and nitrate, are avidly held by this segment of the biosphere, as is indicated by the fact that their input exceeds their output. In the case of potassium all but 700 grams per hectare is captured. Collectively the "nonvolatile" minerals, including silica, that fall from the clear New Hampshire sky amounted to some 13 kilograms per hectare in a typical year. With sulfate, nitrate and ammonium added, the total reached 51.4 kilograms per hectare.

The second line of evidence indicates that the biosphere as a whole is becoming larger. Ecologists expect to find growth in secondary forests, but climax vegetation should be in a steady state, with annual gains balancing losses. According to figures I have recompiled from Rodin and Basilevich, the mean world vegetation is not yet at climax. After the known quantity of dead leaves, branches and other litter is subtracted from the net production of new tissue, the difference is always positive, at an average 2,148 kilograms of new biomass per hectare of land per year. With ash making up 1.2 percent of this biomass, about 26 kilograms of ash is annually withdrawn from an average hectare to sustain the increment of carbohydrate. The input of airborne elements at Hubbard Brook could provide this ash twice over, with no contribution from the local lithosphere.

This comparison is impressionistic, and it may be misleading. Apart from industrial sulfate, which (as sulfuric acid) is perhaps as likely to corrode the biosphere as to nourish it, the world's vegetation may be in no danger of instant eutrophication. (If the biosphere is really becoming larger, the input of industrial carbon dioxide may constitute another major nutrient.) The modes of recycling discovered at Hubbard Brook are nonetheless astonishing. Added to what we know or can safely infer about other volatile elements, such studies underscore the necessity of a global view of biochemistry. What can be said with assurance is that there is a unique and nearly ubiquitous compound, with the empirical formula $H_{2960}O_{1480}C_{1480}N_{16}P_{1.8}S$, called living matter. Its synthesis, on an oxidized and uncarboxylated earth, is the most intricate feat of chemical engineering ever performed—and the most delicate operation that people have ever tampered with.

GUANO-COVERED ISLAND off the coast of Peru is a source of phosphate and nitrate for fertilizer. Guano has been deposited during many millenniums by generations of birds.

2

THE CHEMICAL ELEMENTS OF LIFE

EARL FRIEDEN
June 1972

Until recently it was believed that living matter incorporated 20 of the natural elements. Now it has been shown that a role is played by four others: fluorine, silicon, tin and vanadium

How many of the 90 naturally occurring elements are essential to life? After more than a century of increasingly refined investigation, the question still cannot be answered with certainty. Only a year or so ago the best answer would have been 20. Since then four more elements have been shown to be essential for the growth of young animals: fluorine, silicon, tin and vanadium. Nickel may soon be added to the list. In many cases the exact role played by these and other trace elements remains unknown or unclear. These gaps in knowledge could be critical during a period when the biosphere is being increasingly contaminated by synthetic chemicals and subjected to a potentially harmful redistribution of salts and metal ions. In addition, new and exotic chemical forms of metals (such as methyl mercury) are being discovered, and a complex series of competitive and synergistic relations among mineral salts has been encountered. We are led to the realization that we are ignorant of many basic facts about how our chemical milieu affects our biological fate.

Biologists and chemists have long been fascinated by the way evolution has selected certain elements as the building blocks of living organisms and has ignored others. The composition of the earth and its atmosphere obviously sets a limit on what elements are available. The earth itself is hardly a chip off the universe. The solar system, like the universe, seems to be 99 percent hydrogen and helium. In the earth's crust helium is essentially nonexistent (except in a few rare deposits) and hydrogen atoms constitute only about .22 percent of the total. Eight elements provide more than 98 percent of the atoms in the earth's crust: oxygen (47 percent), silicon (28 percent), aluminum (7.9 percent), iron (4.5 percent), calcium (3.5 percent), so-

dium (2.5 percent), potassium (2.5 percent) and magnesium (2.2 percent). Of these eight elements only five are among the 11 that account for more than 99.9 percent of the atoms in the human body. Not surprisingly nine of the 11 are also the nine most abundant elements in seawater [*see illustration on page 16*].

Two elements, hydrogen and oxygen, account for 88.5 percent of the atoms in the human body; hydrogen supplies 63 percent of the total and oxygen 25.5 percent. Carbon accounts for another 9.5 percent and nitrogen 1.4 percent. The remaining 20 elements now thought to be essential for mammalian life account for less than .7 percent of the body's atoms.

The Background of Selection

Three characteristics of the biosphere or of the elements themselves appear to have played a major part in directing the chemistry of living forms. First and foremost there is the ubiquity of water, the solvent base of all life on the earth. Water is a unique compound; its stability and boiling point are both unusually high for a molecule of its simple composition. Many of the other compounds essential for life derive their usefulness from their response to water: whether they are soluble or insoluble, whether or not (if they are soluble) they carry an electric charge in solution and, not least, what effect they have on the viscosity of water.

The second directing force involves the chemical properties of carbon, which evolution selected over silicon as the central building block for constructing giant molecules. Silicon is 146 times more plentiful than carbon in the earth's crust and exhibits many of the same properties. Silicon is directly below carbon in the periodic table of the elements;

like carbon, it has the capacity to gain four electrons and form four covalent bonds.

The crucial difference that led to the preference for carbon compounds over silicon compounds seems traceable to two chemical features: the unusual stability of carbon dioxide, which is readily soluble in water and always monomeric (it remains a single molecule), and the almost unique ability of carbon to form long chains and stable rings with five or six members. This versatility of the carbon atom is responsible for the millions of organic compounds found on the earth.

Silicon, in contrast, is insoluble in water and forms only relatively short chains with itself. It can enter into longer chains, however, by forming alternating bonds with oxygen, creating the compounds known as silicones ($-Si-O-Si-O-Si-$). Carbon-to-carbon bonds are more stable than silicon-to-silicon bonds, but not so stable as to be virtually immutable, as the silicon-oxygen polymers are. Nevertheless, silicon has recently been shown to be essential in a way as yet unknown for normal bone development and full growth in chicks.

The third force influencing the evolutionary selection of the elements essential for life is related to an atom's size and charge density. Obviously the heavy synthetic elements from neptunium (atomic number 93) to lawrencium (No. 103), along with two lighter synthetic elements, technetium (No. 43) and promethium (No. 61), were never available in nature. (The atomic number expresses the number of protons in the nucleus of an atom or the number of electrons around the nucleus.) The eight heavy elements in another group (Nos. 84 and 85 and Nos. 87 through 92) are too radioactive to be useful in living structures. Six more elements are inert gases

COMPOSITION OF UNIVERSE		COMPOSITION OF EARTH'S CRUST		COMPOSITION OF SEAWATER		COMPOSITION OF HUMAN BODY	
PERCENT OF TOTAL NUMBER OF ATOMS							
H	91	O	47	H	66	H	63
He	9.1	Si	28	O	33	O	25.5
O	.057	Al	7.9	Cl	.33	C	9.5
N	.042	Fe	4.5	Na	.28	N	1.4
C	.021	Ca	3.5	Mg	.033	Ca	.31
Si	.003	Na	2.5	S	.017	P	.22
Ne	.003	K	2.5	Ca	.006	Cl	.03
Mg	.002	Mg	2.2	K	.006	K	.06
Fe	.002	Ti	.46	C	.0014	S	.05
S	.001	H	.22	Br	.0005	Na	.03
		C	.19			Mg	.01
ALL OTHERS <.01		ALL OTHERS <.1		ALL OTHERS <.1		ALL OTHERS <.01	

Al ALUMINUM   C CARBON   Fe IRON   O OXYGEN   S SULFUR
B BORON   Cl CHLORINE   Mg MAGNESIUM   K POTASSIUM   Ti TITANIUM
Br BROMINE   He HELIUM   Ne NEON   Si SILICON
Ca CALCIUM   H HYDROGEN   N NITROGEN   Na SODIUM

CHEMICAL SELECTIVITY OF EVOLUTION can be demonstrated by comparing the composition of the human body with the approximate composition of seawater, the earth's crust and the universe at large. The percentages are based on the total number of atoms in each case; because of rounding the totals do not exactly equal 100. Elements in the colored boxes in the last column appear in one or more columns at the left. Thus one sees that phosphorus, the sixth most plentiful element in the body, is a rare element in inanimate nature. Carbon, the third most plentiful element, is also very scarce elsewhere.

with virtually no useful chemical reactivities: helium, neon, argon, krypton, xenon and radon. On various plausible grounds one can exclude another 24 elements, or a total of 38 natural elements, as being clearly unsatisfactory for incorporation in living organisms because of their relative unavailability (particularly the elements in the lanthanide and actinide series) or their high toxicity (for example mercury and lead). This leaves 52 of the 90 natural elements as being potentially useful.

Only three of the 24 elements known to be essential for animal life have an atomic number above 34. All three are needed only in trace amounts: molybdenum (No. 42), tin (No. 50) and iodine (No. 53). The four most abundant atoms in living organisms—hydrogen, carbon, oxygen and nitrogen—have atomic numbers of 1, 6, 7 and 8. Their preponderance seems attributable to their being the smallest and lightest elements that can achieve stable electronic configura-

tions by adding one to four electrons. The ability to add electrons by sharing them with other atoms is the first step in forming chemical bonds leading to stable molecules. The seven next most abundant elements in living organisms all have atomic numbers below 21. In the order of their abundance in mammals they are calcium (No. 20), phosphorus (No. 15), potassium (No. 19), sulfur (No. 16), sodium (No. 11), magnesium (No. 12) and chlorine (No. 17). The remaining 10 elements known to be present in either plants or animals are needed only in traces. With the exception of fluorine (No. 9) and silicon (No. 14), the remaining eight occupy positions between No. 23 and No. 34 in the periodic table [see illustration at right]. It is interesting that this interval embraces three elements for which evolution has evidently found no role: gallium, germanium and arsenic. None of the metals with properties similar to those of gallium (such as aluminum and indium) has

proved to be useful to living organisms. On the other hand, since silicon and tin, two elements with chemical activities similar to those of germanium, have just joined the list of essential elements, it seems possible that germanium too, in spite of its rarity, will turn out to have an essential role. Arsenic, of course, is a well-known poison.

Functions of Essential Elements

Some useful generalizations can be made about the role of the various elements. Six elements—carbon, nitrogen, hydrogen, oxygen, phosphorus and sulfur—make up the molecular building blocks of living matter: amino acids, sugars, fatty acids, purines, pyrimidines and nucleotides. These molecules not only have independent biochemical roles but also are the respective constituents of the following large molecules: proteins, glycogen, starch, lipids and nucleic acids. Several of the 20 amino acids contain sulfur in addition to carbon, hydrogen and oxygen. Phosphorus plays an important role in the nucleotides such as

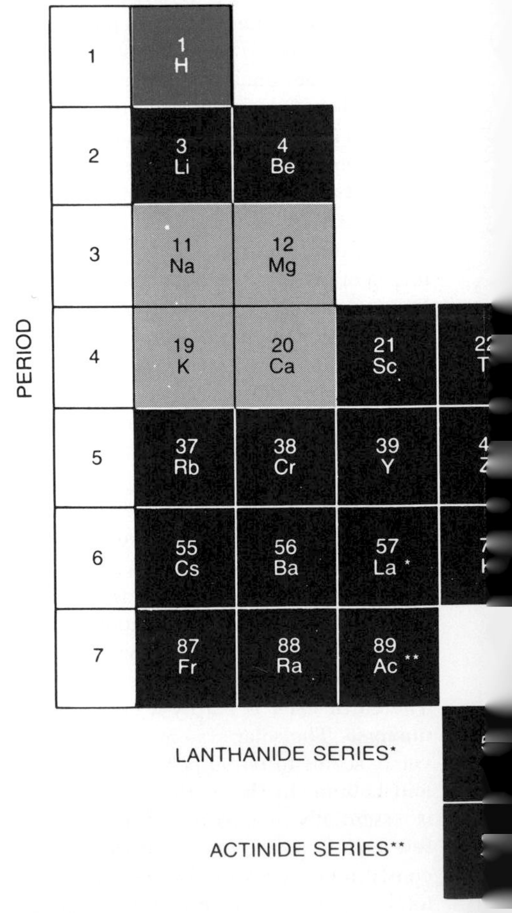

Period					
1	1 H				
2	3 Li	4 Be			
3	11 Na	12 Mg			
4	19 K	20 Ca	21 Sc	22 T
5	37 Rb	38 Cr	39 Y	
6	55 Cs	56 Ba	57 La *	
7	87 Fr	88 Ra	89 Ac **	

LANTHANIDE SERIES*

ACTINIDE SERIES**

ESSENTIAL LIFE ELEMENTS, 24 by the latest count, are clustered in the upper half of the periodic table. The elements are ar-

adenosine triphosphate (ATP), which is central to the energetics of the cell. ATP includes components that are also one of the four nucleotides needed to form the double helix of deoxyribonucleic acid (DNA), which incorporates the genetic blueprint of all plants and animals. Both sulfur and phosphorus are present in many of the small accessory molecules called coenzymes. In bony animals phosphorus and calcium help to create strong supporting structures.

The electrochemical properties of living matter depend critically on elements or combinations of elements that either gain or lose electrons when they are dissolved in water, thus forming ions. The principal cations (electron-deficient, or positively charged, ions) are provided by four metals: sodium, potassium, calcium and magnesium. The principal anions (ions with a negative charge because they have surplus electrons) are provided by the chloride ion and by sulfur and phosphorus in the form of sulfate ions and phosphate ions. These seven ions maintain the electrical neutrality of body fluids and cells and also play a part in maintaining the proper liquid volume of the blood and other fluid systems. Whereas the cell membrane serves as a physical barrier to the exchange of large molecules, it allows small molecules to pass freely. The electrochemical functions of the anions and cations serve to maintain the appropriate relation of osmotic pressure and charge distribution on the two sides of the cell membrane.

One of the striking features of the ion distribution is the specificity of these different ions. Cells are rich in potassium and magnesium, and the surrounding plasma is rich in sodium and calcium. It seems likely that the distribution of ions in the plasma of higher animals reflects the oceanic origin of their evolutionary antecedents. One would like to know how primitive cells learned to exclude the sodium and calcium ions in which they were bathed and to develop an internal milieu enriched in potassium and magnesium.

The third and last group of essential elements consists of the trace elements. The fact that they are required in extremely minute quantities in no way diminishes their great importance. In this sense they are comparable to the vitamins. We now know that the great majority of the trace elements, represented by metallic ions, serve chiefly as key components of essential enzyme systems or of proteins with vital functions (such as hemoglobin and myoglobin, which respectively transports oxygen in the blood and stores oxygen in muscle). The heaviest essential element, iodine, is an essential constituent of the thyroid hormones thyroxine and triiodothyronine, although its precise role in hormonal activity is still not understood.

The Trace Elements

To demonstrate that a particular element is essential to life becomes increasingly difficult as one lowers the threshold of the amount of a substance recognizable as a "trace." It has been known for more than 100 years, for example, that iron and iodine are essential to man. In a rapidly developing period of biochemistry between 1928 and 1935 four more elements, all metals, were shown to be

ranged according to their atomic number, which is equivalent to the number of protons in the atom's nucleus. The four most abundant elements that are found in living organisms (hydrogen, oxygen, carbon and nitrogen) are indicated by dark color. The seven next most common elements are in lighter color. The 13 elements that are shown in lightest color are needed only in traces.

ELEMENT	SYMBOL	ATOMIC NUMBER	COMMENTS
HYDROGEN	H	1	Required for water and organic compounds.
HELIUM	He	2	Inert and unused.
LITHIUM	Li	3	Probably unused.
BERYLLIUM	Be	4	Probably unused; toxic.
BORON	B	5	Essential in some plants; function unknown.
CARBON	C	6	Required for organic compounds.
NITROGEN	N	7	Required for many organic compounds.
OXYGEN	O	8	Required for water and organic compounds.
FLUORINE	F	9	Growth factor in rats; possible constituent of teeth and bone.
NEON	Ne	10	Inert and unused.
SODIUM	Na	11	Principal extracellular cation.
MAGNESIUM	Mg	12	Required for activity of many enzymes; in chlorophyll.
ALUMINUM	Al	13	Essentiality under study.
SILICON	Si	14	Possible structural unit of diatoms; recently shown to be essential in chicks.
PHOSPHORUS	P	15	Essential for biochemical synthesis and energy transfer.
SULFUR	S	16	Required for proteins and other biological compounds.
CHLORINE	Cl	17	Principal cellular and extracellular anion.
ARGON	A	18	Inert and unused.
POTASSIUM	K	19	Principal cellular cation.
CALCIUM	Ca	20	Major component of bone; required for some enzymes.
SCANDIUM	Sc	21	Probably unused.
TITANIUM	Ti	22	Probably unused.
VANADIUM	V	23	Essential in lower plants, certain marine animals and rats.
CHROMIUM	Cr	24	Essential in higher animals; related to action of insulin.
MANGANESE	Mn	25	Required for activity of several enzymes.
IRON	Fe	26	Most important transition metal ion; essential for hemoglobin and many enzymes.
COBALT	Co	27	Required for activity of several enzymes; in vitamin B_{12}.
NICKEL	Ni	28	Essentiality under study.
COPPER	Cu	29	Essential in oxidative and other enzymes and hemocyanin.
ZINC	Zn	30	Required for activity of many enzymes.
GALLIUM	Ga	31	Probably unused.
GERMANIUM	Ge	32	Probably unused.
ARSENIC	As	33	Probably unused; toxic.
SELENIUM	Se	34	Essential for liver function.
MOLYBDENUM	Mo	42	Required for activity of several enzymes.
TIN	Sn	50	Essential in rats; function unknown.
IODINE	I	53	Essential constituent of the thyroid hormones.

SOME TWO-THIRDS OF LIGHTEST ELEMENTS, or 21 out of the first 34 elements in the periodic table, are now known to be essential for animal life. These 21 plus molybdenum (No. 42), tin (No. 50) and iodine (No. 53) constitute the total list of the 24 essential elements, which are here enclosed in colored boxes. It is possible that still other light elements will turn out to be essential. The most likely candidates are aluminum, nickel and germanium. The element boron already appears to be essential for some plants.

essential: copper, manganese, zinc and cobalt. The demonstration can be credited chiefly to a group of investigators at the University of Wisconsin led by C. A. Elvehjem, E. B. Hart and W. R. Todd. At that time it seemed that these four metals might be the last of the essential trace elements. In the next 30 years, however, three more elements were shown to be essential: chromium, selenium and molybdenum. Fluorine, silicon, tin and vanadium have been added since 1970.

The essentiality of five of these last seven elements was discovered through the careful, painstaking efforts of Klaus Schwarz and his associates, initially located at the National Institutes of Health and now based at the Veterans Administration Hospital in Long Beach, Calif. For the past 15 years Schwarz's group has made a systematic study of the trace-element requirements of rats and other small animals. The animals are maintained from birth in a completely isolated sterile environment [*see illustration on page 21*].

The apparatus is constructed entirely of plastics to eliminate the stray contaminants contained in metal, glass and rubber. Although even plastics may contain some trace elements, they are so tightly bound in the structural lattice of the material that they cannot be leached out or be picked up by an animal even through contact. A typical isolator system houses 32 animals in individual acrylic cages. Highly efficient air filters remove all trace substances that might be present in the dust in the air. Thus the animals' only access to essential nutrients is through their diet. They receive chemically pure amino acids instead of natural proteins, and all other dietary ingredients are screened for metal contaminants.

Since the standards of purity employed in these experiments far exceed those for reagents normally regarded as analytically pure, Schwarz and his coworkers have had to develop many new analytical chemical methods. The most difficult problem turned out to be the purification of salt mixtures. Even the purest commercial reagents were contaminated with traces of metal ions. It was also found that trace elements could be passed from mothers to their offspring. To minimize this source of contamination animals are weaned as quickly as possible, usually from 18 to 20 days after birth.

With these precautions Schwarz and his colleagues have within the past several years been able to produce a new

deficiency disease in rats. The animals grow poorly, lose hair and muscle tone, develop shaggy fur and exhibit other detrimental changes [*see illustration on page 22*]. When standard laboratory food is given these animals, they regain their normal appearance. At first it was thought that all the symptoms were caused by the lack of one particular trace element. Eventually four different elements had to be supplied to complete the highly purified diets the animals had been receiving. The four elements proved to be fluorine, silicon, tin and vanadium. A convenient source of these

elements is yeast ash or liver preparations from a healthy animal. The animals on the deficiency diet grew less than half as fast as those on a normal or supplemented diet. Growth alone, however, may not tell the entire story. There is some evidence that even the addition of the four elements may not reverse the loss of hair and skin changes resulting from the deficiency diet.

Functions of Trace Elements

The addition of tin and vanadium to the list of essential trace metals brings

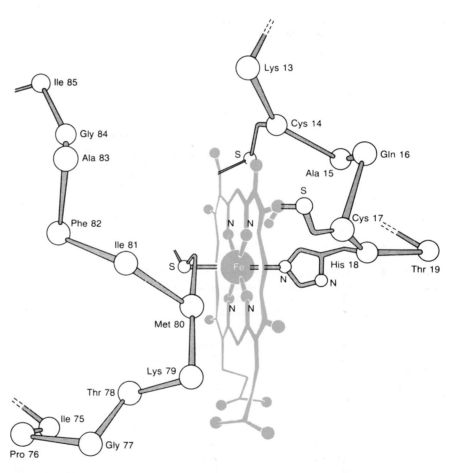

Ala	ALANINE	His	HISTIDINE	Phe	PHENYLALANINE
Cys	CYSTEINE	Ile	ISOLEUCINE	Pro	PROLINE
Gln	GLUTAMINE	Lys	LYSINE	Thr	THREONINE
Gly	GLYCINE	Met	METHIONINE		

THE METALLOENZYME CYTOCHROME *c* is typical of metal-protein complexes in which trace metals play a crucial role. Cytochrome *c* belongs to a family of enzymes that extract energy from food molecules. It consists of a protein chain of 104 amino acid units attached to a heme group (*color*), a rosette of atoms with an atom of iron at the center. This simplified molecular diagram shows only the heme group and several of the amino acid units closest to it. The iron atom has six coordination sites enabling it to form six bonds with neighboring atoms. Four bonds connect to nitrogen atoms in the heme group itself, and the remaining two bonds link up with amino acid units in the protein chain (histidine at site No. 18 and methionine at site No. 80). The illustration is based on the work of Richard E. Dickerson of the California Institute of Technology, in whose laboratory the complete structure of horse-heart cytochrome *c* was recently determined.

METAL	ENZYME	BIOLOGICAL FUNCTION
IRON	FERREDOXIN	Photosynthesis
	SUCCINATE DEHYDROGENASE	Aerobic oxidation of carbohydrates
IRON IN HEME	ALDEHYDE OXIDASE	Aldehyde oxidation
	CYTOCHROMES	Electron transfer
	CATALASE	Protection against hydrogen peroxide
	[HEMOGLOBIN]	Oxygen transport
COPPER	CERULOPLASMIN	Iron utilization
	CYTOCHROME OXIDASE	Principal terminal oxidase
	LYSINE OXIDASE	Elasticity of aortic walls
	TYROSINASE	Skin pigmentation
	PLASTOCYANIN	Photosynthesis
	[HEMOCYANIN]	Oxygen transport in invertebrates
ZINC	CARBONIC ANHYDRASE	CO_2 formation; regulation of acidity
	CARBOXYPEPTIDASE	Protein digestion
	ALCOHOL DEHYDROGENASE	Alcohol metabolism
MANGANESE	ARGINASE	Urea formation
	PYRUVATE CARBOXYLASE	Pyruvate metabolism
COBALT	RIBONUCLEOTIDE REDUCTASE	DNA biosynthesis
	GLUTAMATE MUTASE	Amino acid metabolism
MOLYBDENUM	XANTHINE OXIDASE	Purine metabolism
	NITRATE REDUCTASE	Nitrate utilization
CALCIUM	LIPASES	Lipid digestion
MAGNESIUM	HEXOKINASE	Phosphate transfer

WIDE VARIETY OF METALLOENZYMES is required for the successful functioning of living organisms. Some of the most important are given in this list. The giant oxygen-transporting molecules hemoglobin and hemocyanin are included in the list (in brackets) even though they are not strictly enzymes, that is, they do not act as biological catalysts.

to 10 the total number of trace metals needed by animals and plants. What role do these metals play? For six of the eight trace metals recognized from earlier studies (that is, for iron, zinc, copper, cobalt, manganese and molybdenum) we are reasonably sure of the answer. The six are constituents of a wide range of enzymes that participate in a variety of metabolic processes [see illustration above].

In addition to its role in hemoglobin and myoglobin, iron appears in succinate dehydrogenase, one of the enzymes needed for the utilization of energy from sugars and starches. Enzymes incorporating zinc help to control the formation of carbon dioxide and the digestion of proteins. Copper is present in more than a dozen enzymes, whose roles range from the utilization of iron to the pigmentation of the skin. Cobalt appears in enzymes involved in the synthesis of DNA and the metabolism of amino acids. Enzymes incorporating manga-

nese are involved in the formation of urea and the metabolism of pyruvate. Enzymes incorporating molybdenum participate in purine metabolism and the utilization of nitrogen.

These six metals belong to a group known as transition elements. They owe their uniqueness to their ability to form strong complexes with ligands, or molecular groups, of the type present in the side chains of proteins. Enzymes in which transition metals are tightly incorporated are called metalloenzymes, since the metal is usually embedded deep inside the structure of the protein. If the metal atom is removed, the protein usually loses its capacity to function as an enzyme. There is also a group of enzymes in which the metal ion is more loosely associated with the protein but is nonetheless essential for the enzyme's activity. Enzymes in this group are known as metal-ion-activated enzymes. In either group the role of the metal ion may be to maintain the proper confor-

mation of the protein, to bind the substrate (the molecule acted on) to the protein or to donate or accept electrons in reactions where the substrate is reduced or oxidized.

In 1968 the complete three-dimensional structure of the first metalloenzyme, cytochrome c, was published [see "The Structure and History of an Ancient Protein," by Richard E. Dickerson; SCIENTIFIC AMERICAN, April, 1972]. Cytochrome c, a red enzyme containing iron, is universally present in plants and animals. It is one of a series of enzymes, all called cytochromes, that extract energy from food molecules by the stepwise addition of oxygen.

The complete amino acid sequence of cytochrome c obtained from the human heart was determined some 10 years ago by a group led by Emil L. Smith of the University of California at Los Angeles and by Emanuel Margoliash of Northwestern University. The iron atom is partially complexed with an intricate organic molecule, protoporphyrin, to form a heme group similar to that in hemoglobin. Of the iron atom's six coordination sites, four are attached to the heme group through nitrogen atoms. The other two sites form bonds with the protein chain; one bond is through a nitrogen atom in the side chain of a histidine unit at site No. 18 in the protein sequence and the other bond is through a sulfur atom in the side chain of a methionine unit at site No. 80 [see illustration on preceding page].

Although the cytochrome c molecule is complicated, it is one of the simplest of the metalloenzymes. Cytochrome oxidase, probably the single most important enzyme in most cells, since it is responsible for transferring electrons to oxygen to form water, is far more complicated. Each molecule contains about 12 times as many atoms as cytochrome c, including two copper atoms and two heme groups, both of which participate in transferring the electrons.

More complicated yet is cysteamine oxygenase, which catalyzes the addition of oxygen to a molecule of cysteamine; it contains one atom each of three different metals: iron, copper and zinc. There are many other combinations of metal ions and unique molecular assemblies. An extreme example is xanthine oxidase, which contains eight iron atoms, two molybdenum atoms and two molecules incorporating riboflavin (one of the B vitamins) in a giant molecule more than 25 times the size of cytochrome c.

The metal-containing proteins of another group, the metalloproteins, closely resemble the metalloenzymes except that they lack an obvious catalytic function. Hemoglobin itself is an example. Others are hemocyanin, the copper-containing blue protein that carries oxygen in many invertebrates, metallothionein, a protein involved in the absorption and storage of zinc, and transferrin, a protein that transports iron in the bloodstream. There may be many more such compounds still unrecognized because their function has escaped detection.

The Newest Essential Elements

Much remains to be learned about the specific biochemical role of the most recently discovered essential elements. In 1957 Schwarz and Calvin M. Foltz, working at the National Institutes of Health, showed that selenium helped to prevent several serious deficiency diseases in different animals, including liver necrosis and muscular dystrophy. Rats were protected against death from liver necrosis by a diet containing one-tenth of a part per million of selenium. Comparably low doses reversed the white muscle disease observed in cattle and sheep that happen to graze in areas where selenium is scarce.

In April a group at the University of Wisconsin under J. T. Rotruck reported a direct biochemical role for selenium.

Oxidative damage to red blood cells was detected in rats kept on a selenium-deficient diet. This damage was related to reduced activity of an enzyme, glutathione peroxidase, that helps to protect hemoglobin against the injurious oxidative effects of hydrogen peroxide. The enzyme uses hydrogen peroxide to catalyze the oxidation of glutathione, thus keeping hydrogen peroxide from oxidizing the reduced state of iron in hemoglobin. Oxidized glutathione can readily be converted to reduced glutathione by a variety of intracellular mechanisms. There is some reason to believe glutathione peroxidase may even contain some form of selenium acting as an integral part of the functional enzyme molecule.

The physiological importance of chromium was established in 1959 by Schwarz and Walter Mertz. They found that chromium deficiency is characterized by impaired growth and reduced life-span, corneal lesions and a defect in sugar metabolism. When the diet is deficient in chromium, glucose is removed from the bloodstream only half as fast as it is normally. In rats the deficiency is relieved by a single administration of 20 micrograms of certain trivalent chromic salts. It now appears that the chromium ion works in conjunction with insulin, and that in at least some cases diabetes may reflect faulty chromium metabolism.

After developing the all-plastic trace-element isolator described above, Schwarz, David B. Milne and Elizabeth Vineyard discovered that tin, not previously suspected as being essential, was necessary for normal growth. Without one or two parts per million of tin in their diet, rats grow at only about two-thirds the normal rate.

The next element shown to be essential in mammals by the Schwarz group was vanadium, an element that had been detected earlier in certain marine invertebrates but whose essentiality had not been demonstrated. On a diet in which vanadium is totally excluded rats suffer a retardation of about 30 percent in growth rate. Schwarz and Milne found that normal growth is restored by adding one-tenth of a part per million of vanadium to the diet. At higher concentrations vanadium is known to have several biological effects, but its essential role in trace amounts remains to be established. A high dose of vanadium blocks the synthesis of cholesterol and reduces the amount of phospholipid and cholesterol in the blood. Vanadium also promotes the mineralization of teeth and is effective as a catalyst in the oxidation of many biological substances.

The third element most recently iden-

NUTRITIONAL NEEDS OF SMALL ANIMALS are studied in a trace-element isolator, a modification of the apparatus originally conceived to maintain animals in a germ-free environment. To prevent unwanted introduction of trace elements the isolator is built completely of plastics. It holds 32 animals in separate cages, individually supplied with food of precisely known composition. The system was designed by Klaus Schwarz and J. Cecil Smith of the Veterans Administration Hospital in Long Beach, Calif.

tified as being essential is fluorine. Even with tin and vanadium added to highly purified diets containing all other elements known to be essential, the animals in Schwarz's plastic cages still failed to grow at a normal rate. When up to half a part per million of potassium fluoride was added to the diet, the animals showed a 20 to 30 percent weight gain in four weeks. Although it had appeared that a trace amount of fluorine was essential for building sound teeth, Schwarz's study showed that fluorine's biochemical role was more fundamental than that. In any case fluoridated water provides more than enough fluorine to maintain a normal growth rate.

Although there were earlier clues that silicon might be an essential life element, firm proof of its essentiality, at least in young chicks, was reported only three months ago. Edith M. Carlisle of the School of Public Health at the University of California at Los Angeles finds that chicks kept on a silicon-free diet for only one or two weeks exhibit poor development of feathers and skeleton, including markedly thin leg bones. The addition of 30 parts per million of silicon to the diet increases the chicks' growth more than 35 percent and makes possible normal feathering and skeletal development. Considering that silicon is not only the second most abundant element in the earth's crust but is also similar to carbon in many of its chemical properties, it is hard to see how evolution could have totally excluded it from an essential biochemical role.

Nickel, nearly always associated with iron in natural substances, is another element receiving close attention. Also a transition element, it is particularly difficult to remove from the food used in special diets. Nickel seems to influence the growth of wing and tail feathers in chicks but more consistent data are needed to establish its essentiality. One incidental result of Schwarz's work has been the discovery of a previously unrecognized organic compound, which will undoubtedly prove to be a new vitamin.

Synergism and Antagonism

The interaction of the various essential metals can be extremely complicated. The absence of one metal in the diet can profoundly influence, either positively or negatively, the utilization of another metal that may be present. For example, it has been known for nearly 50 years that copper is essential for the proper metabolism of iron. An animal deprived of copper but not iron develops anemia because the biosynthetic machinery fails to incorporate iron in hemoglobin molecules. It has only recently been found in our laboratories at Florida State University that ceruloplasmin, the copper-containing protein of the blood, is a direct molecular link between the two metals. Ceruloplasmin promotes the release of iron from animal liver so that the iron-binding protein of the serum, transferrin, can complex with iron and transfer it to the developing red blood cells for direct utilization in the biosynthesis of hemoglobin. This represents a synergistic relation between copper and iron.

As an example of antagonism between elements one can cite the instance of copper and zinc. The ability of sheep or cattle to absorb copper is greatly reduced if too much zinc or molybdenum is present in their diet. Evidently either of the two metals can displace copper in an absorption process that probably involves competition for sites on a metal-binding protein in the intestines and liver.

The recent discoveries present many fresh challenges to biochemists. One can expect the discovery of previously unsuspected metalloenzymes containing vanadium, tin, chromium and selenium. New compounds or enzyme systems requiring fluorine and silicon may also be uncovered. The multiple and complex interdependencies of the elements suggest many hitherto unrecognized and important facts about the role and interrelations of metal ions in nutrition and in health and disease.

TRACE-ELEMENT DEFICIENCY developed when the rat at the top of this photograph was kept in the trace-element isolator for 20 days and fed a diet from which fluorine, tin and vanadium had been carefully excluded. The healthy animal at the bottom was fed the same diet but was kept under ordinary conditions. It was evidently able to obtain the necessary trace amounts of fluorine, tin and vanadium from dust and other contaminants.

3

THE WATER CYCLE

H. L. PENMAN
September 1970

Water is the medium of life processes and the source of their hydrogen. It flows through living matter mainly in the stream of transpiration: from the roots of a plant through its leaves

By far the most abundant single substance in the biosphere is the familiar but unusual inorganic compound called water. The earth's oceans, ice caps, glaciers, lakes, rivers, soils and atmosphere contain 1.5 billion cubic kilometers of water in one form or another. In nearly all its physical properties water is either unique or at the extreme end of the range of a property. Its extraordinary physical properties, in turn, endow it with a unique chemistry. From these physical and chemical characteristics flows the biological importance of water. It is the purpose of this article to describe some of water's principal qualities and their significance in the biosphere.

Water remains a liquid within the temperature range most suited to life processes, yet in due season there are occasions when liquid water exists in equilibrium with its solid and gaseous form, for example as ice on the top of a lake with water vapor in the air above it. Freezing starts at the surface of the water and proceeds downward; this follows from one of water's many peculiar attributes. Like everything else, ice included, liquid water contracts when it is cooled, but the shrinkage ceases before solidification, at about four degrees Celsius. From that temperature down to the freezing point the water expands, and because of its decreased density the cooler water floats on top of the warmer. Ice has a density of .92 with respect to the maximum density of water and hence an unconstrained block of ice will float in water with about an eleventh of its volume projecting above the surface. The biological significance of freezing from the surface downward, rather than from the bottom upward, is too well known to need repetition here.

Among its other thermal properties water has the greatest specific heat known among liquids (the ability to store heat energy for a given increase in temperature). The same is true of water's latent heat of vaporization: at 20 degrees C. (68 degrees Fahrenheit), 585 calories are required to evaporate one gram of water. Finally, with the exception of mercury, water has the greatest thermal conductivity of all liquids. Some consequences of water's large latent heat of evaporation, which is a major energizer of the atmosphere, will be considered below. Its great specific heat means that, for a given rate of energy input, the temperature of a given mass of water will rise more slowly than the temperature of any other material. Conversely, as energy is released its temperature will drop more slowly. This slow warming and cooling, together with other important factors, affects yearly, daily and even hourly changes in the temperature of oceans and lakes, which are quite different from the corresponding changes in the temperature of land. Among other things, this can lead to differences in the thermal regimes of soils that are of major importance in ecology. The type of soil, interacting with water, determines the earliness or lateness of plant growth at a given site; the interaction may also affect the local risk of frost.

In basic structure the water molecule has a small dipole moment and is feebly ionized. Water will dissolve almost anything to some extent (fortunately the extent is extremely small for many substances). The dissolved material tends to remain in solution because of another of water's exceptional attributes. The values given by the inverse-square law for the force that attracts separated positive and negative ions are determined by multiplying the square of the distance separating the ions by a constant that varies according to the nature of the separating medium. Known as the dielectric constant, this constant is greater for water than for any other substance. To get the same attractive force in water as in air, for example, the water separation has to be cut down to a ninth of the separation in air.

Because of its extreme dielectric constant liquid water in the biosphere is not chemically pure (unlike water vapor, which is always pure, or ice, which can be and often is pure). Instead liquid water is an ionic solution and one that always contains some hydrogen ions because the water itself can supply them. The concentration of hydrogen ions, expressed as a degree of dilution, gives the physical chemist a numerical index that describes the state of various water samples. The number is the logarithm (to the base 10) of the degree of dilution; the chemist labels it pH. For his tests he is armed with a pH meter, calibrated from zero to 14. Fourteen orders of magnitude is an enormous range for any terrestrial property, yet the water content of the soil may give a reading anywhere from pH 3 (very acid) to pH 10 (very alkaline), which is equivalent to a range of from one to 10 million. These are extremes, however, and most terrestrial plant growth—including much of the world's agriculture—proceeds in soil with a water content that ranges only a few units on each side of pH 6. The range for marine organisms is even more restrictive: coastal waters are about pH 9 and the general oceanic average is just over pH 8. Below pH 7.5 many marine animals die; eggs are particularly vulnerable. Below pH 7 the carbonate in seawater would remain in solution, rendering production of any kind of skeleton impossible.

Another method of describing the

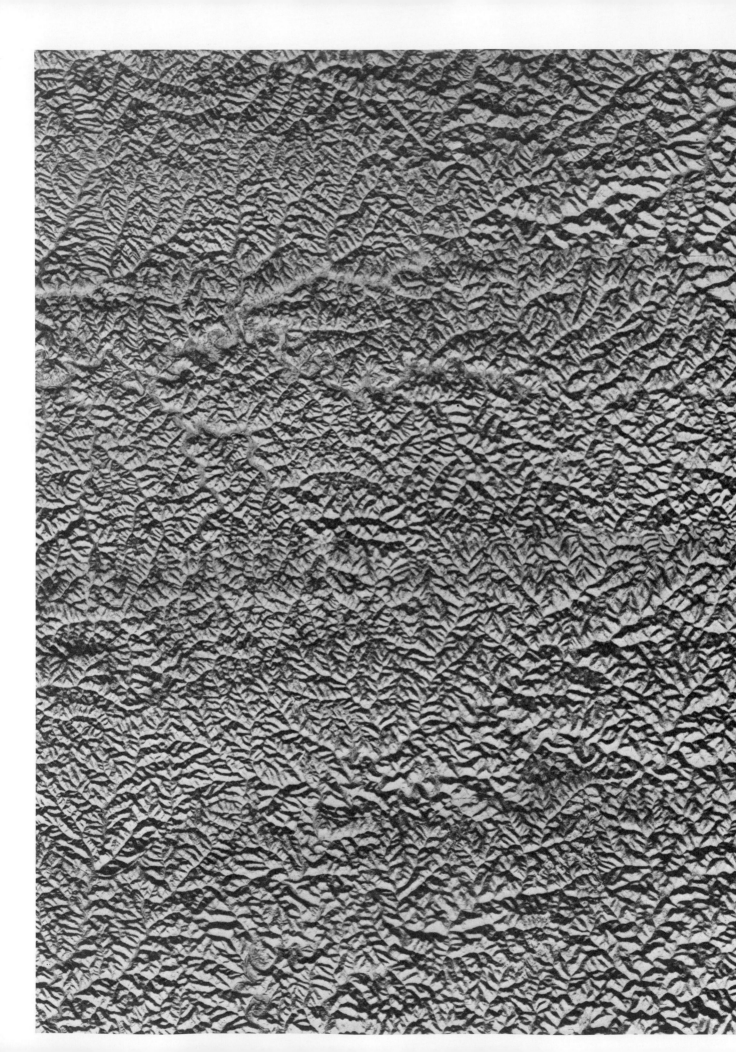

WATER AT WORK for millenniums in the form of rainfall and stream runoff has produced the dissected land surface seen in the side-looking radar image on the opposite page. The annual work of terrain modeling by rainfall and runoff has been estimated to equal the work of one horse-drawn scraper busy day and night on every 10 acres of land surface. This area, in the vicinity of Sandy Hook, Ky., is drained by tributaries of the Ohio River. Each inch equals 2.3 miles on the ground. The radar mosaic, made by the Autometric division of the Raytheon Company, is reproduced by the courtesy of the Army Topographic Command.

state of a given water sample is independent of hydrogen-ion content. Material in solution, whether it is ionized or not, disturbs the liquid structure of the water; in thermodynamic terms the presence of solutes decreases the free energy of the water. Many soil and plant workers find it convenient to use the symbol pF for such changes in free energy, with the steps between units also representing one order of magnitude. As with the pH range, the range of pF values is very great.

The quantity being measured in pF units is basically a potential, with the same dimensions as pressure. If all the water problems in soils, plants and animals were problems of solutions, it would be sufficient to describe the consequent variations in free energy as variations in osmotic potential, expressed in any of the conventional units of pressure. The free energy of water, however, can be decreased in other ways, notably in capillary systems. The energy to lift the water into a capillary tube (or in nature into the porous and cellular systems of soils and plants) comes out of the free energy of the water. Today this is known as "matric" potential, a term that has replaced the earlier "capillary" potential. In soils and plants the matric potential may be more than the osmotic potential. A comparison of numbers will give an idea of the pF scale and its ranges. The pressure is expressed as the height in centimeters of an equivalent column of water; thus one bar equals one atmosphere, which equals a 1,000-centimeter water column. This is equivalent to pF 3.

In a waterlogged soil, beginning to drain, the matric potential may be between pF 0 and pF 1; in a fully drained soil the potential may be near pF 1.7. In a soil that is as dry as plant uptake and the transpiration of water from leaves can make it, the matric potential will be about pF 4.2, which is close to 16 atmospheres of suction. The osmotic potential of seawater is near pF 4.5, which makes seawater too "dry" for plant roots; the salt content of plant cells might be anywhere in a range from less than pF 4 up to pF 4.5.

Here once again water is extreme. Associated with the matric potential in a capillary system there is a curved liquid-air interface; the value of the potential is found by doubling the known value of the liquid's surface tension and dividing the product by the radius of curvature. Water has the greatest surface tension of any liquid known, so that at any given matric potential the radius of curvature of a water meniscus will be greater than it could be for another liquid. The greater the radius of curvature, the greater the total water content. In a soil this means that more liquid can be retained as water just because it *is* water. In general, but not always, this is an advantage for plant growth.

The effects of a decrease in the free energy of water contained in porous soils or in the tissues of plants include a lowering of the fluid's freezing point and vapor pressure. If the source of the decrease is a matric potential, there is also negative pressure, or suction, that tends to pull all kinds of retaining walls together. The effect of freezing in soils and rocks is worth a brief aside. As the temperature falls the water in the larger soil pores freezes first and the free-energy gradient is such that water will be withdrawn from the smaller pores. As a result ice lenses form in the coarser pore spaces and the finer pore spaces are exposed to greater shrinkage forces. Because water expands on freezing, the ice lenses have a disruptive effect as they make room for themselves. In rock this is the beginning of one method of soil formation. There tends to be a preferred size for the rock fragments produced by ice disruption. This size is near the optimum for transport by wind and is the dominant size in many of the loess soils that have accumulated in areas near glaciers. In soil the ice disruption is the source of "frost tilth," which is sought by farmers when they leave land roughly plowed in the fall and hope for a sufficiently frosty winter.

There are still some uncertainties with respect to the world's water balance, but agreement was reached on probable values or ranges during an international symposium on the subject held in Britain this summer as one of the activities of the International Hydrological Decade. The figures that follow are taken from the proceedings of the symposium.

The world's water exists as liquid (salt and fresh), as solid (fresh) and as vapor (fresh). There is some uncertainty in the value of the total volume, but it is near 1,500 million cubic kilometers (in U.S. usage 1.5 billion). Estimates of the components are most easily expressed as

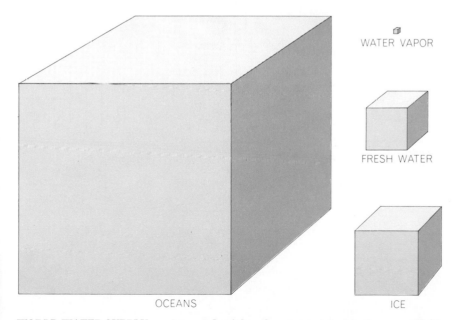

WORLD WATER SUPPLY consists mainly of the salt water contained in the oceans (*left*). The world's fresh water comprises only about 3 percent of the total supply; three-quarters of it is locked up in the world's polar ice caps and glaciers and most of the rest is found as ground water or in lakes. The very small amount of water in the atmosphere at any one time (*top right*) is nonetheless of vital importance as a major energizer of weather systems.

average depths per unit area of the entire surface of the earth, which has a total area of 510 million square kilometers. Oceans and seas—liquid salt water—make up about 97 percent of all water, with an equivalent depth of between 2,700 and 2,800 meters; the greater part is in the Southern Hemisphere. Of the remaining 3 percent, three-quarters is locked up as solid in the polar ice caps and in glaciers. Here measurement is quite difficult, and a spread in estimates is inevitable. The equivalent depth of ice and snow may be near 120 meters, but at the recent symposium a value of 50 meters was not challenged. The other large component of fresh liquid water is subject to similar uncertainty: the estimates for underground water may be near 45 meters, but again a value near 15 meters was not challenged. Estimates for surface water, mainly in the great lakes of the world,

ranged from .4 meter to one meter. There is general agreement on the average water-vapor content of the atmosphere, at an equivalent in liquid of .03 meter. Although this is a very small fraction of the total, size is no measure of importance. Without water in the atmosphere there would be no weather; Leonardo da Vinci's dictum, "Water is the driver of nature," is justified on meteorological grounds alone. A little detail at this point will be helpful as an introduction to another aspect of the world circulation of water.

The amount of water vapor is not the same everywhere, either geographically or seasonally. It is greatest at and near the Equator. If the air there were squeezed dry, it would yield about 44 millimeters of rainfall. In middle latitudes, say from 40 to 50 degrees, the summer yield would be near 20 millime-

ters and the winter yield near 10 millimeters, with large variations that depend on geography and weather patterns. In the polar regions the yield ranges from two millimeters in winter to as much as eight in summer.

Water vapor enters the atmosphere by evaporation (this term includes transpiration by vegetation), and the main oceanic sources are fairly identifiable. It leaves the atmosphere as rain or snow, and because the precipitation may take place close to the source or thousands of miles away, the residence time may vary from a few hours to a few weeks. A general average is nine or 10 days.

The general balance of evaporation and precipitation needs three sets of figures, one set for the entire earth, one for the oceans and one for the land surface. Here, within a few percent, there is almost complete agreement on values. For the entire earth, average evaporation

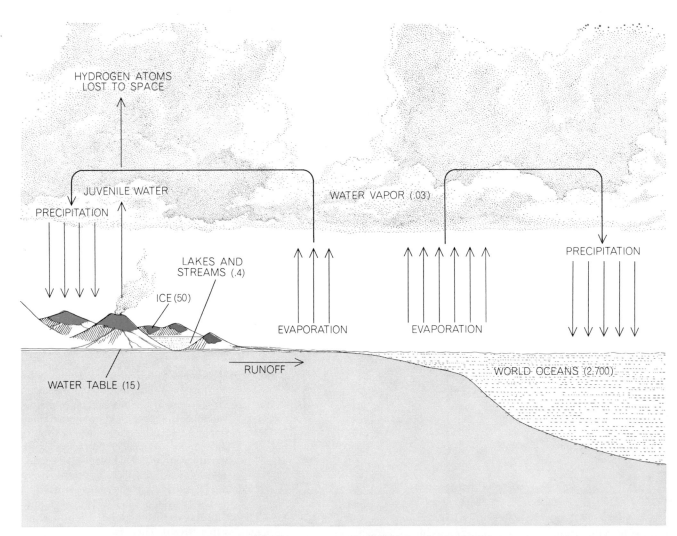

WATER CYCLE in the biosphere requires that worldwide evaporation and precipitation be equal; hydrogen losses to space are presumably replaced by juvenile water. Ocean evaporation, however, is greater than return precipitation; the reverse is true of the land. Excess land precipitation may end up in ice caps and glaciers that contain 75 percent of all fresh water, may replenish supplies taken from the water table by transpiring plants or may enter lakes and rivers, eventually returning to the sea as runoff. Numbers show minimum estimates of the amount of water present in each reserve, expressed as a depth in meters per unit area of the earth's surface.

and precipitation are equal—as they must be—at very nearly 100 centimeters per year. For the oceans, expressed as equivalent depths over the area of the oceans, the average annual precipitation is between 107 and 114 centimeters, the average annual evaporation is between 116 and 124 centimeters and balance is restored by river flow, with an annual value close to 10 centimeters in all estimates. For the land surface the average annual precipitation is near 71 centimeters, the average annual evaporation is near 47 centimeters and the average annual river discharge is near 24 centimeters. (The ocean figure of 10 centimeters corresponds to the 24-centimeter land figure.)

Because half of the land surface—ice caps, deserts, mountains, tundra—contributes little or nothing to evaporation, a better evaporation average would take into consideration only the land component of the biosphere where the availability of water is combined with the opportunity for evaporation. Here the average evaporation may total 100 centimeters per year. The evaporation in high latitudes would of course be far less than the evaporation nearer the Equator.

Available measurements support this conclusion. In Finland, at 65 degrees north latitude, the average evaporation is 20 centimeters per year; in southeastern England, at 50 degrees north, it is 50; in North Carolina, at 35 degrees north, it ranges from 80 to 120. On the Equator in the Congo basin the average is 120 centimeters per year; at the same latitude in Kenya it is 150. In the papyrus swamps of the Nile in the southern Sudan, 10 degrees north of the Equator, the average is 240 centimeters per year, but this is a special case. Here the river carries its water into the desert environment of the Sudd; evaporation rates are high not only because of the clear skies and intense sunshine overhead but also because the surrounding desert is a source of hot dry air that augments evaporation. This kind of advective augmentation operates in many places other than the Sudan, particularly in semiarid regions where irrigation is practiced, and not quite enough is known about it.

Once in the air, water vapor may circulate locally or become part of the general circulation of the atmosphere. The general circulation is one of the three important ways of moving water across the earth. Some indication of the worldwide volumes involved is given by the fact that the total annual precipitation over the U.S. comes to some 6,000 cubic kilometers, whereas the liquid equivalent of the water vapor that passes over the U.S. in a year owing to the general circulation of the atmosphere is 10 times that amount.

Of the two remaining important ways of moving water across the earth, the major ocean currents comprise one and the discharge of rivers comprises the other. Both have substantial effects on the biosphere. The ocean currents carry energy surpluses or deficits over great distances; one well-known instance accounts for the extreme contrast between the climates on the west and east sides of the Atlantic in the areas between 50 and 55 degrees north latitude. Without the Gulf Stream northwestern Europe would be a much less pleasant place in which to live and work; indeed, if the cold Labrador Current had replaced the Gulf Stream, the history of civilization would have been very different.

The rivers of the world not only are long-distance movers of water but also serve as conduits for dissolved and suspended material. Because of its chemical and physical properties, water is a very efficient erosive agent; erosion, transport and deposition have to be recognized as geological processes associated with water in the biosphere. They are the processes that have produced lands and soils, now densely populated and intensively cropped, where annual floods and silt deposition are regarded as the mainstay of life. Elsewhere, notably in the Americas, silt is an embarrassment in the deltas where it settles, and its production is equally unwelcome in river headwaters.

Two further points about river water deserve mention. First, the salt content of river water differs markedly in composition from that of the oceans. This suggests that the oceanic brine is not merely the accumulation of salts from aeons of land-surface leaching. Second, information about river discharge rates is scanty and not always reliable. As an example, it is only recently that a good estimate of the flow of the Amazon has been obtained. It proved to be twice the best previous estimate and indicates that almost a fifth of the world's river discharge comes from this one stream.

It is not possible to do more than guess at the average amount of water the world's plant and animal populations contain. Considered as the equivalent of rainfall, it may amount to about one millimeter over the entire surface of the earth. This is less by one order of magnitude than the amount of water vapor in the atmosphere, and its distribution is even more varied in space and time. For a fully grown good crop of corn in North America or of sugar beet in northwestern Europe the amount might come to the equivalent of five millimeters of rainfall, and its summer residence time would be two to three days. This is a measure of the rate of water supply needed to maintain optimum conditions for growth. Here, at the point of water uptake by the roots of plants, begins the problem with respect to water in the biosphere that makes all other water problems seem trifling.

With unimportant exceptions, the basis of all life on the earth is photosynthesis by green plants, a process that involves physics (in the fixation of solar energy) and chemistry (in the union of carbon dioxide and water to form carbohydrates and more complex biochemical compounds). Water comes into the story in two ways: in transit (as part of the transpiration stream) and in residence (as its hydrogen is chemically bound into the plant structure). The amount that is bound, however, may be less than a fifth of the amount in transit. To give scale to the argument that follows, here are some values based on a real crop in a real climate. In producing 20 fresh-weight tons of crop, 2,000 tons of water will pass into the plants at their roots. At harvest perhaps 15 tons of the water supply will be in transit, leaving the crop with a dry weight of five tons. To produce the five tons of dry matter three tons of water will have been fixed and transformed. The energy fixed in the dry matter will be 1 percent or less of the total solar energy received by the crop; nearly 40 percent of the energy will have been used to evaporate the transpired water.

The average value of 40 percent for the net solar radiation income retained by a green crop cover varies, of course, with season and climate. The first loss is to reflectivity: of the solar radiation reaching the crop about 30 percent is reflected. There is also an income of long-wave radiation from the sky, but this is outweighed by the outgo of long-wave radiation from the earth to the atmosphere. When the deficit is met by deducting it from the remaining balance of short-wave solar income, the net retained income is decreased to 40 percent of the initial input. As already noted, when the water is available, very nearly all this energy is used in evaporating water.

Here once again water stands at the extreme of a range of physical properties. The volume of water evaporated per unit of energy input is less than it would be for any other liquid. The relevant physical constant, the latent heat of vaporization, is somewhat less than 600 calories per gram at ordinary temperatures, but the rounded figure is adequate for the present purpose. If we let R_I represent the total radiant income in calories per square centimeter over a period of time, then the net radiation is about $.4R_I$ and the evaporation equivalent is near $R_I/$ 1,500 grams per square centimeter (or centimeters of water depth as the equivalent of rainfall). Consider some real midsummer values to show what this means. In a humid temperate climate the value of R_I is close to 450 calories per square centimeter per day. This works out to an evaporation equivalent of three millimeters per day, which is a good estimate for June in southeastern England. For many of the farming areas of the U.S. the R_I value is close to 650 calories per square centimeter, bringing the evaporation rate up to about 4.5 millimeters per day. The maximum rates known, which are found in irrigated areas, range from 4.5 to 7.5 millimeters per day. It is possible that the higher rates are influenced by advection from surrounding nonirrigated areas, as is the case in the papyrus swamps of the Nile.

The most important fact to be considered in connection with this wide range of evaporation rates is that there are only very small variations among the evaporation rates of different kinds of plants. Thus the governing factor in variation is almost exclusively a climatic one. This fact and much other evidence suggest that the supposed water "need" of a crop is dictated not by the plants but by the weather. In this connection the concept of "potential transpiration," which came into use simultaneously and independently in at least two parts of the world, is of great value both in research and in the practical aspects of soil water management. It is worthwhile seeing how potential transpiration is linked with elementary plant physiology and with some of the physics of soil water already considered.

A growing plant takes in water at the roots and, in the absence of immediate replenishment, the process dries the soil so that more and more energy is required for further extraction. The energy requirement is very small, however, compared with the amount of energy needed to evaporate the same quantity of water from the plant's leaves. There can be no serious error in assuming, as Frank J. Veihmeyer of the University of California at Davis does, that all soil water is equally available for transpiration up to the stage marked by the onset of wilting. The purpose of well-managed irrigation, of course, is to make sure that plants never get to the wilting stage. For maximum growth irrigation may have to consist of frequent small applications of water rather than occasional large ones.

Given an adequate supply of water, the chain of consequences is simple. There are maximum values for each of several factors: water content in the plant, hydrostatic pressure in the plant and leaf turgidity. When neither the intensity of the light nor the concentration of carbon dioxide constitutes a limiting factor, maximum leaf turgidity permits

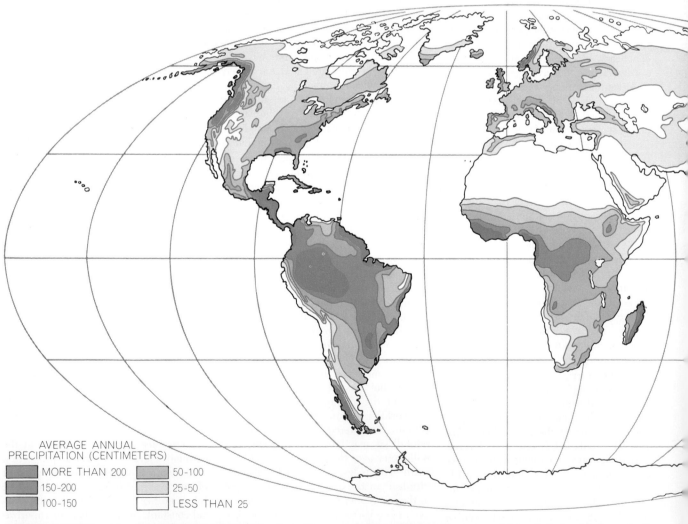

AVERAGE ANNUAL
PRECIPITATION (CENTIMETERS)

MORE THAN 200   50-100
150-200   25-50
100-150   LESS THAN 25

PRECIPITATION reaches the land areas of the world principally in the form of rainfall, which is heaviest at and near the Equator and along some western coasts at higher latitudes (*darker colors*). Variations in precipitation are the result of atmospheric circula-

maximum opening of the stomatal apertures in the leaf surface, thus affording the best possible opportunity for movement of carbon dioxide into the leaf. The state of the stomatal opening that allows easy inflow of carbon dioxide, however, also allows equally easy outflow of water vapor. By far the greater part of the water need of plants is actually a "leakage" process that has to be kept going to ensure continued growth. Given a sufficiently wet soil around the plants' roots, the rate of leakage is dictated not by plant physiology but by the physical factors of temperature, humidity and ventilation. The sole constraint is imposed by the law of energy conservation. In its last stages the transpiration stream undergoes a change of state from liquid to vapor, and the rate of change depends on the rate at which energy can reach the system to supply the necessary latent heat of vaporization.

So much for the physics of the process. When the supply of water in the soil approaches exhaustion, plant physiology rather than physics begins to predominate. Plant type, root structure, phase of plant development, soil type, soil depth—these become the important factors. What is available for utilization has more significance than the weather has, particularly in semiarid zones.

Because agriculture is most active in the more humid zones of the biosphere, it is useful to estimate how much reserve soil water is available on the average in these zones. Factors already described prevent any exact answer to this question. Nonetheless, a cautious estimate, advanced with considerable reservation, would be about 10 centimeters of rainfall equivalent. Three examples will suffice to show the need for caution. There are large agricultural areas of North Carolina and neighboring states where an inert subsoil is covered by no more than 20 centimeters of useful topsoil. Here the entire water reserve available to the agricultural cycle cannot exceed a rainfall equivalent of five centimeters. This is one extreme; the deep volcanic soils of East Africa are at the other. In those soils the roots of many plants go down as much as six meters below the surface. The available water in a profile that deep is equivalent to nearly 50 centimeters of rain, and the plant can extract water throughout a long dry season at something very close to the potential transpiration rate. An example from France falls somewhere in between. There the drying of the soil was observed while a crop of sugar beet transpired at the full potential rate throughout a dry summer. At the driest stage the crop had withdrawn from the soil available water equivalent to 27 centimeters of rainfall.

Soil water and ground water are closely related, but whereas soil water is always biologically important, the importance of ground water may range from being trivial to being all that matters. The soil is a kind of buffer between rainfall and ground water. In general any deficit in soil moisture that has built up in a dry period must be completely restored by rain before there is any water surplus available to move down to ground water. This is an important consideration for the water engineer, who may be drawing a water supply from a stream (permanent streams are sustained by ground water) or may be tapping an aquifer directly by means of a well. In the first instance the engineer will presumably have to work within legal constraints on how much river water he can divert. In the second, if he is to choose a safe aquifer pumping rate, the engineer must (or should!) have some awareness of the current soil-moisture deficit and of the likely rates of rainfall in the months ahead. The river engineer can use the same information for another purpose: the soil-moisture deficit will enable him to estimate the risk of flooding in the event of a heavy storm.

In some countries the control of ground water is a major outlet for engineering skill, and its exploitation is the basis of farming technique. One need only think on the one hand of the Netherlands and on the other of such semiarid regions as Iran, where deep tunnels tap the buried aquifers and carry ground water to valley bottoms. In many semiarid regions the vegetation along transient streams maintains its luxuriance because the ground-water level there is close to the surface and within reach of plant roots. The plants' effective reach depends both on the soil and on the kind of plant, but in general it is seldom more than a few meters. The movement of any water table deeper than that is unaffect-

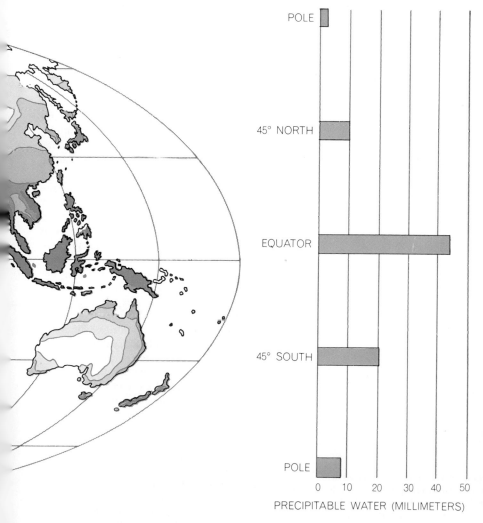

PRECIPITABLE WATER (MILLIMETERS)

tion patterns and also reflect the amount of precipitable water vapor present. This is greatest at the Equator, least at the poles and more in summer than in winter (*graph at right*).

ed by the plant growth or the evaporation processes taking place above it, and its ground water contributes nothing to the biological activity at the surface.

What has been said about water so far has involved terms that are generally accepted, and the concepts themselves are supported by good reasoning, good evidence or both. The remarks that follow, although also based on reasoning and evidence, are more speculative and personal. If the biosphere is taken to be the place where water and energy interact, can the interaction be expressed quantitatively in terms of biological productivity? The answer has to be no. There are too many variables. All the same, by rearranging some of the water quantities and energy quantities that are known, a suggestive relation can be obtained.

Start with the fact that, for a good crop, 1 percent or less of the incoming solar radiation is fixed as dry matter (here and in what follows the 1 percent refers to the total botanical yield, irrespective of economic value). We shall give this percentage the symbol ε, and express it numerically as 100 per 10,000. This degree of efficiency is achieved only by an experiment station or by an extremely competent commercial farmer. Based on the statistics of world cereal production, including straw as well as grain, the average achievement in highly mechanized industrial farming shows an efficiency of only about 35 per 10,000. This decreases to roughly 17 per 10,000 in North America, and in tropical Africa and Asia subsistence farming rarely shows an efficiency better than 8, even when the possibility of two crops per year is allowed for. There is obviously room for improvement everywhere and the question in the present context is: Where does water come in, and how?

Some evidence is now being accumulated suggesting that, when there are no limitations on water supply, the total crop growth is proportional to the total of potential transpiration over the period of growth. The factor of proportionality depends on many things: plant variety, management, kinds and quantities of fertilizer, pest and disease controls and the like. Hence the negative answer to the earlier question. Still, something can be done with ratios. There is reason to believe potential transpiration is a fairly constant fraction of the solar radiation income. Combining this fact with the relation between potential transpiration and total crop growth, it is possible to derive a connection between growth rate and utilization of water, assuming that unlimited water is available (by inference, there would be a similar response to timely irrigation). In what may seem too precise a form, the answer is that the increase in yield (t) equals .39ε. Here ε represents efficiency and t can be read, according to preference, as metric tons per hectare per centimeter of water applied or as tons per acre per inch of water. Considering the fivefold (or perhaps tenfold) world variation in efficiency, some uncertainty in the multiplying factor is unimportant; others may prefer, or find, a different value. Taking the illustration already given, suppose the area involved is one acre and the efficiency is exactly or almost 100 per 10,000. The predicted increase in yield in response to water, *applied when it is needed,* is .39 ton of dry matter per acre per inch of water. In terms of fresh weight the gain is about 1.5 tons per acre per inch. This is the kind of response obtained in experiments with irrigated potatoes in Britain.

There are some countries where the value of ε is small because of lack of water, but there are many, including several of the rice-growing nations, where the small value of ε is more truly a measure of the inefficiency of the farming system itself. To get the most out of water, whether it comes from irrigation or from rainfall, the standard of performance elsewhere in the system must be improved: better varieties, better soil management, better crop husbandry, better plant hygiene and better pest control. Then water may be the driver of nature in agriculture as well as in the atmosphere.

SOLAR ENERGY

DRY WEIGHT (5 TONS)

WATER FIXED (3 TONS)

FRESH WEIGHT (20 TONS)

WATER IN TRANSIT (15 TONS)

WATER (2,000 TONS)

ROLE OF WATER in photosynthesis is quantitatively minor compared with its role in transpiration, as this crop-water graph indicates. To produce 20 fresh-weight tons of crop in a season, some 2,000 tons of water will be drawn from the soil. At the harvest, water in transit will account for some 15 tons of the crop's fresh weight. Drying reduces the crop's weight to five tons. Of these, three tons, or .15 percent of the water used in the season, comprise hydrogen atoms from water molecules, photosynthetically bound to carbon atoms.

4

THE OXYGEN CYCLE

PRESTON CLOUD AND AHARON GIBOR
September 1970

The oxygen in the atmosphere was originally put there by plants. Hence the early plants made possible the evolution of the higher plants and animals that require free oxygen for their metabolism

The history of our planet, as recorded in its rocks and fossils, is reflected in the composition and the biochemical peculiarities of its present biosphere. With a little imagination one can reconstruct from that evidence the appearance and subsequent evolution of gaseous oxygen in the earth's air and water, and the changing pathways of oxygen in the metabolism of living things.

Differentiated multicellular life (consisting of tissues and organs) evolved only after free oxygen appeared in the atmosphere. The cells of animals that are truly multicellular in this sense, the Metazoa, obtain their energy by breaking down fuel (produced originally by photosynthesis) in the presence of oxygen in the process called respiration. The evolution of advanced forms of animal life would probably not have been possible without the high levels of energy release that are characteristic of oxidative metabolism. At the same time free oxygen is potentially destructive to all forms of carbon-based life (and we know no other kind of life). Most organisms have therefore had to "learn" to conduct their oxidations anaerobically, primarily by removing hydrogen from foodstuff rather than by adding oxygen. Indeed, the anaerobic process called fermentation is still the fundamental way of life, underlying other forms of metabolism.

Oxygen in the free state thus plays a role in the evolution and present functioning of the biosphere that is both pervasive and ambivalent. The origin of life and its subsequent evolution was contingent on the development of systems that shielded it from, or provided chemical defenses against, ordinary molecular oxygen (O_2), ozone (O_3) and atomic oxygen (O). Yet the energy requirements of higher life forms can be met only by

oxidative metabolism. The oxidation of the simple sugar glucose, for example, yields 686 kilocalories per mole; the fermentation of glucose yields only 50 kilocalories per mole.

Free oxygen not only supports life; it arises from life. The oxygen now in the atmosphere is probably mainly, if not wholly, of biological origin. Some of it is converted to ozone, causing certain high-energy wavelengths to be filtered out of the radiation that reaches the surface of the earth. Oxygen also combines with a wide range of other elements in the earth's crust. The result of these and other processes is an intimate evolutionary interaction among the biosphere, the atmosphere, the hydrosphere and the lithosphere.

Consider where the oxygen comes from to support the high rates of energy release observed in multicellular organisms and what happens to it and to the carbon dioxide that is respired [*see illustration on page 35*]. The oxygen, of course, comes from the air, of which it constitutes roughly 21 percent. Ultimately, however, it originates with the decomposition of water molecules by light energy in photosynthesis. The 1.5 billion cubic kilometers of water on the earth are split by photosynthesis and reconstituted by respiration once every two million years or so. Photosynthetically generated oxygen temporarily enters the atmospheric bank, whence it is itself recycled once every 2,000 years or so (at current rates). The carbon dioxide that is respired joins the small amount (.03 percent) already in the atmosphere, which is in balance with the carbon dioxide in the oceans and other parts of the hydrosphere. Through other interactions it may be removed from circulation as a part of the carbonate ion (CO_3^{--}) in calcium carbonate precipitated from solu-

tion. Carbon dioxide thus sequestered may eventually be returned to the atmosphere when limestone, formed by the consolidation of calcium carbonate sediments, emerges from under the sea and is dissolved by some future rainfall.

Thus do sea, air, rock and life interact and exchange components. Before taking up these interactions in somewhat greater detail let us examine the function oxygen serves within individual organisms.

Oxygen plays a fundamental role as a building block of practically all vital molecules, accounting for about a fourth of the atoms in living matter. Practically all organic matter in the present biosphere originates in the process of photosynthesis, whereby plants utilize light energy to react carbon dioxide with water and synthesize organic substances. Since carbohydrates (such as sugar), with the general formula $(CH_2O)_n$, are the common fuels that are stored by plants, the essential reaction of photosynthesis can be written as $CO_2 + H_2O +$ light energy $\rightarrow CH_2O + O_2$. It is not immediately obvious from this formulation which of the reactants serves as the source of oxygen atoms in the carbohydrates and which is the source of free molecular oxygen. In 1941 Samuel Ruben and Martin D. Kamen of the University of California at Berkeley used the heavy oxygen isotope oxygen 18 as a tracer to demonstrate that the molecular oxygen is derived from the splitting of the water molecule. This observation also suggested that carbon dioxide is the source of the oxygen atoms of the synthesized organic molecules.

The primary products of photosynthesis undergo a vast number of chemical transformations in plant cells and subsequently in the cells of the animals that

RED BEDS rich in the oxidized (ferric) form of iron mark the advent of oxygen in the atmosphere. The earliest continental red beds are less than two billion years old; the red sandstones and shales of the Nankoweap Formation in the Grand Canyon (*opposite page*) are about 1.3 billion years old. The appearance of oxygen in the atmosphere, the result of photosynthesis, led in time to the evolution of cells that could survive its toxic effects and eventually to cells that could capitalize on the high energy levels of oxidative metabolism.

feed on plants. During these processes changes of course take place in the atomic composition and energy content of the organic molecules. Such transformations can result in carbon compounds that are either more "reduced" or more "oxidized" than carbohydrates. The oxidation-reduction reactions between these compounds are the essence of biological energy supply and demand. A more reduced compound has more hydrogen atoms and fewer oxygen atoms per carbon atom; a more oxidized compound has fewer hydrogen atoms and more oxygen atoms per carbon atom. The combustion of a reduced compound liberates more energy than the combustion of a more oxidized one. An example of a molecule more reduced than a carbohydrate is the familiar alcohol ethanol (C_2H_6O); a more oxidized molecule is pyruvic acid ($C_3H_4O_3$).

Differences in the relative abundance of hydrogen and oxygen atoms in organic molecules result primarily from one of the following reactions: (1) the removal (dehydrogenation) or addition (hydrogenation) of hydrogen atoms, (2) the addition of water (hydration), followed by dehydrogenation; (3) the direct addition of oxygen (oxygenation). The second and third of these processes introduce into organic matter additional oxygen atoms either from water or from molecular oxygen. On decomposition the oxygen atoms of organic molecules are released as carbon dioxide and water. The biological oxidation of molecules such as carbohydrates can be written as the reverse of photosynthesis: $CH_2O + O_2 \rightarrow CO_2 + H_2O + energy$. The oxygen atom of the organic molecule appears in the carbon dioxide and the molecular oxygen acts as the acceptor for the hydrogen atoms.

The three major nonliving sources of oxygen atoms are therefore carbon dioxide, water and molecular oxygen, and since these molecules exchange oxygen atoms, they can be considered as a common pool. Common mineral oxides such as nitrate ions and sulfate ions are also oxygen sources for living organisms, which reduce them to ammonia (NH_3) and hydrogen sulfide (H_2S). They are subsequently reoxidized, and so as the oxides circulate through the biosphere their oxygen atoms are exchanged with water.

The dynamic role of molecular oxygen is as an electron sink, or hydrogen acceptor, in biological oxidations. The biological oxidation of organic molecules proceeds primarily by dehydrogenation: enzymes remove hydrogen atoms from the substrate molecule and transfer them to specialized molecules that function as hydrogen carriers [*see top illustration on pages 36 and 37*]. If these carriers become saturated with hydrogen, no further oxidation can take place until some other acceptor becomes available. In the anaerobic process of fermentation organic molecules serve as the hydrogen acceptor. Fermentation therefore results in the oxidation of some organic compounds and the simultaneous reduction of others, as in the fermentation of glucose by yeast: part of the sugar molecule is oxidized to carbon dioxide and other parts are reduced to ethanol.

In aerobic respiration oxygen serves as the hydrogen acceptor and water is produced. The transfer of hydrogen atoms (which is to say of electrons and protons) to oxygen is channeled through an array of catalysts and cofactors. Prominent among the cofactors are the iron-containing pigmented molecules called cytochromes, of which there are several kinds that differ in their affinity for electrons. This affinity is expressed as the oxidation-reduction, or "redox," potential of the molecule; the more positive the potential, the greater the affinity of the oxidized molecule for electrons. For example, the redox potential of cytochrome b is .12 volt, the potential of cytochrome c is .22 volt and the potential of cytochrome a is .29 volt. The redox potential for the reduction of oxygen to water is .8 volt. The passage of electrons from one cytochrome to another down a potential gradient, from cytochrome b to cytochrome c to the cytochrome a complex and on to oxygen, results in the alternate reduction and oxidation of these cofactors. Energy liberated in such oxidation-reduction reactions is coupled to the synthesis of high-energy phosphate compounds such as adenosine triphosphate (ATP). The special copper-containing enzyme cytochrome oxidase mediates the ultimate transfer of electrons from the cytochrome a complex to oxygen. This activation and binding of oxygen is seen as the fundamental step, and possibly the original primitive step, in the evolution of oxidative metabolism.

In cells of higher organisms the oxidative system of enzymes and electron carriers is located in the special organelles called mitochondria. These organelles can be regarded as efficient low-temperature furnaces where organic molecules are burned with oxygen. Most of the released energy is converted into the high-energy bonds of ATP.

Molecular oxygen reacts spontaneously with organic compounds and other reduced substances. This reactivity explains the toxic effects of oxygen above tolerable concentrations. Louis Pasteur discovered that very sensitive organisms such as obligate anaerobes cannot tolerate oxygen concentrations above about 1 percent of the present atmospheric level. Recently the cells of higher organisms have been found to contain organelles called peroxisomes, whose major function is thought to be the protection of cells from oxygen. The peroxisomes contain enzymes that catalyze the direct reduction of oxygen molecules through the oxidation of metabolites such as amino acids and other organic acids. Hydrogen peroxide (H_2O_2) is one of the products of such oxidation. Another of the peroxisome enzymes, catalase, utilizes the hydrogen peroxide as a hydrogen acceptor in the oxidation of substrates such as ethanol or lactic acid. The rate of reduction of oxygen by the peroxisomes increases proportionately with an increase in oxygen concentration, so that an excessive amount of oxygen in the cell increases the rate of its reduction by peroxisomes.

Christian de Duve of Rockefeller University has suggested that the peroxisomes represent a primitive enzyme system that evolved to cope with oxygen when it first appeared in the atmosphere. The peroxisome enzymes enabled the first oxidatively metabolizing cells to use oxygen as a hydrogen acceptor and so reoxidize the reduced products of fermentation. In some respects this process is similar to the oxidative reactions of the mitochondria. Both make further dehydrogenation possible by liberating oxidized hydrogen carriers. The basic difference between the mitochondrial oxidation reactions and those of peroxisomes is that in peroxisomes the steps of oxidation are not coupled to the synthesis of ATP. The energy released in the peroxisomes is thus lost to the cell; the function of the organelle is primarily to protect against the destructive effects of free molecular oxygen.

Oxygen dissolved in water can diffuse

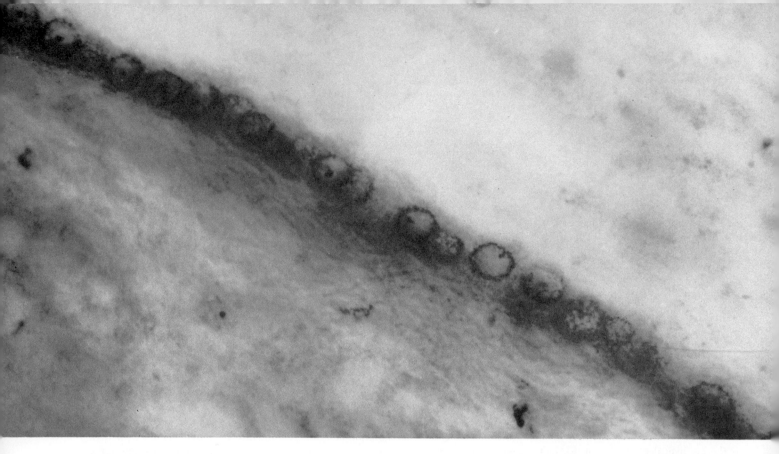

EUCARYOTIC CELLS, which contain a nucleus and divide by mitosis, were, like oxygen, a necessary precondition for the evolution of higher life forms. The oldest eucaryotes known were found in the Beck Spring Dolomite of eastern California by Cloud and his colleagues. The photomicrograph above shows eucaryotic cells with an average diameter of 14 microns, probably green algae. The regular occurrence and position of the dark spots suggest they may be remnants of nuclei or other organelles. Other cell forms, which do not appear in the picture, show branching and large filament diameters that also indicate the eucaryotic level of evolution.

PROCARYOTIC CELLS, which lack a nucleus and divide by simple fission, were a more primitive form of life than the eucaryotes and persist today in the bacteria and blue-green algae. Procaryotes were found in the Beck Spring Dolomite in association with the primitive eucaryotes such as those in the photograph at the top of the page. A mat of threadlike procaryotic blue-green algae, each thread of which is about 3.5 microns in diameter, is seen in the photomicrograph below. It was made, like the one at top of page, by Gerald R. Licari. Cells of this kind, among others, presumably produced photosynthetic oxygen before eucaryotes appeared.

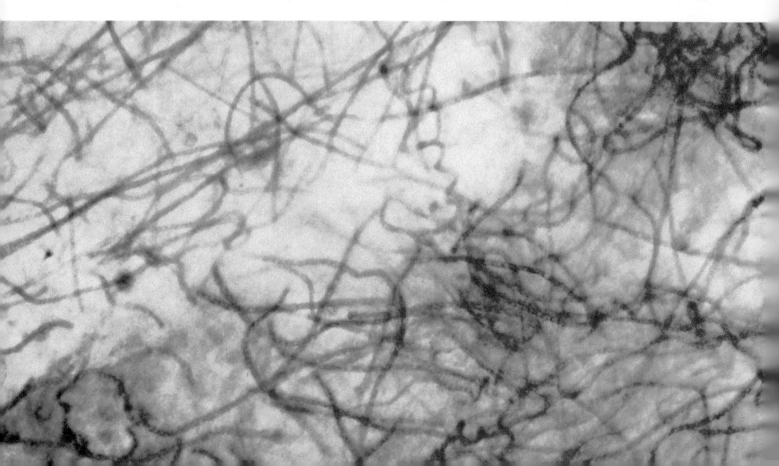

across both the inner and the outer membranes of the cell, and the supply of oxygen by diffusion is adequate for single cells and for organisms consisting of small colonies of cells. Differentiated multicellular organisms, however, require more efficient modes of supplying oxygen to tissues and organs. Since all higher organisms depend primarily on mitochondrial aerobic oxidation to generate the energy that maintains their active mode of life, they have evolved elaborate systems to ensure their tissues an adequate supply of oxygen, the gas that once was lethal (and still is, in excess). Two basic devices serve this purpose: special chemical carriers that increase the oxygen capacity of body fluids, and anatomical structures that provide relatively large surfaces for the rapid exchange of gases. The typical properties of an oxygen carrier are exemplified by those of hemoglobin and of myoglobin, or muscle hemoglobin. Hemoglobin in blood readily absorbs oxygen to near-saturation at oxygen pressures such as those found in the lung. When the blood is exposed to lower oxygen pressures as it moves from the lungs to other tissues, the hemoglobin discharges most of its bound oxygen. Myoglobin, which acts as a reservoir to meet the sharp demand for oxygen in muscle contraction, gives up its oxygen more rapidly. Such reversible bonding of oxygen in response to changes in oxygen pressure is an essential property of biochemical oxygen carriers.

Lungs and gills are examples of anatomical structures in which large wet areas of thin membranous tissue come in contact with oxygen. Body fluids are pumped over one side of these membranes and air, or water containing oxygen, over the other side. This ensures a rapid gas exchange between large volumes of body fluid and the environment.

How did the relations between organisms and gaseous oxygen happen to evolve in such a curiously complicated manner? The atmosphere under which life arose on the earth was almost certainly devoid of free oxygen. The low concentration of noble gases such as neon and krypton in the terrestrial atmosphere compared with their cosmic abundance, together with other geochemical evidence, indicates that the terrestrial atmosphere had a secondary origin in volcanic outgassing from the earth's interior. Oxygen is not known among the gases so released, nor is it found as inclusions in igneous rocks. The chemistry of rocks older than about two billion years is also inconsistent with the presence of more than trivial quantities

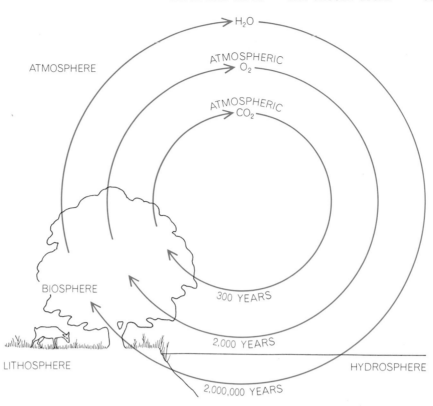

BIOSPHERE EXCHANGES water vapor, oxygen and carbon dioxide with the atmosphere and hydrosphere in a continuing cycle, shown here in simplified form. All the earth's water is split by plant cells and reconstituted by animal and plant cells about every two million years. Oxygen generated in the process enters the atmosphere and is recycled in about 2,000 years. Carbon dioxide respired by animal and plant cells enters the atmosphere and is fixed again by plant cells after an average atmospheric residence time of about 300 years.

of free atmospheric oxygen before that time. Moreover, it would not have been possible for the essential chemical precursors of life—or life itself—to have originated and persisted in the presence of free oxygen before the evolution of suitable oxygen-mediating enzymes.

On such grounds we conclude that the first living organism must have depended on fermentation for its livelihood. Organic substances that originated in nonvital reactions served as substrates for these primordial fermentations. The first organism, therefore, was not only an anaerobe; it was also a heterotroph, dependent on a preexisting organic food supply and incapable of manufacturing its own food by photosynthesis or other autotrophic processes.

The emergence of an autotroph was an essential step in the onward march of biological evolution. This evolutionary step left its mark in the rocks as well as on all living forms. Some fated eobiont, as we may call these early life forms whose properties we can as yet only imagine, evolved and became an autotroph, an organism capable of manufacturing its own food. Biogeological evidence suggests that this critical event may have occurred more than three bil-

lion years ago.

If, as seems inescapable, the first autotrophic eobiont was also anaerobic, it would have encountered difficulty when it first learned to split water and release free oxygen. John M. Olson of the Brookhaven National Laboratory recently suggested biochemical arguments to support the idea that primitive photosynthesis may have obtained electrons from substances other than water. He argues that large-scale splitting of water and release of oxygen may have been delayed until the evolution of appropriate enzymes to detoxify this reactive substance.

We nevertheless find a long record of oxidized marine sediments of a peculiar type that precedes the first evidence of atmospheric oxygen in rocks about 1.8 billion years old; we do not find them in significant amounts in more recent strata. These oxidized marine sediments, known as banded iron formations, are alternately iron-rich and iron-poor chemical sediments that were laid down in open bodies of water. Much of the iron in them is ferric (the oxidized form, Fe^{+++}) rather than ferrous (the reduced form, Fe^{++}), implying that there was a source of oxygen in the column of water above them. Considering the

a

OH OH

H H OH OH

HO OH H

H OH

MYO-INOSITOL

INOSITOL OXYGENASE

O_2 H_2O

COOH

H H O H

HO OH H OH

OH·

D-GLUCURONIC ACID

b

COOH

H—C—H

H—C—H

COOH

SUCCINIC ACID

SUCCINIC DEHYDROGENASE

FAD   FADH$_2$

COOH

H—C

C—H

COOH

FUMARIC ACID

OXIDATION involves a decrease in the number of hydrogen atoms in a molecule or an increase in the number of oxygen atoms. It may be accomplished in several ways. In oxygenation (*a*) oxygen is added directly. In dehydrogenation (*b*) hydrogen is re-

problems that would face a water-splitting photosynthesizer before the evolution of advanced oxygen-mediating enzymes such as oxidases and catalases, one can visualize how the biological oxygen cycle may have interacted with ions in solution in bodies of water during that time. The first oxygen-releasing photoautotrophs may have used ferrous compounds in solution as oxygen acceptors—oxygen for them being merely a toxic waste product. This would have precipitated iron in the ferric form ($4FeO + O_2 \rightarrow 2Fe_2O_3$) or in the ferro-ferric form (Fe_3O_4). A recurrent imbalance of supply and demand might then account for the cyclic nature and differing types of the banded iron formations.

Once advanced oxygen-mediating enzymes arose, oxygen generated by increasing populations of photoautotrophs containing these enzymes would build up in the oceans and begin to escape into the atmosphere. There the ultraviolet component of the sun's radiation would dissociate some of the molecular oxygen into highly reactive atomic oxygen and also give rise to equally reactive ozone. Atmospheric oxygen and its reactive derivatives (even in small quantities) would lead to the oxidation of iron in sediments produced by the weathering of rocks, to the greatly reduced solubility of iron in surface waters (now oxygenated), to the termination of the banded iron formations as an important sedimentary type and to the extensive formation of continental red beds rich in ferric iron [*see illustration on page 32*]. The record of the rocks supports this succession of events: red beds are essentially restricted to rocks younger than about 1.8 billion years, whereas banded iron formation is found only in older rocks.

So far we have assumed that oxygen accumulated in the atmosphere as a consequence of photosynthesis by green plants. How could this happen if the entire process of photosynthesis and respiration is cyclic, representable by the reversible equation $CO_2 + H_2O +$ energy

$\rightleftharpoons CH_2O + O_2$? Except to the extent that carbon or its compounds are somehow sequestered, carbohydrates produced by photosynthesis will be reoxidized back to carbon dioxide and water, and no significant quantity of free oxygen will accumulate. The carbon that is sequestered in the earth as graphite in the oldest rocks and as coal, oil, gas and other carbonaceous compounds in the younger ones, and in the living and dead bodies of plants and animals, is the

equivalent of the oxygen in oxidized sediments and in the earth's atmosphere! In attempting to strike a carbon-oxygen balance we must find enough carbon to account not only for the oxygen in the present atmosphere but also for the "fossil" oxygen that went into the conversion of ferrous oxides to ferric oxides, sulfides to sulfates, carbon monoxide to carbon dioxide and so on.

Interestingly, rough estimates made some years ago by William W. Rubey,

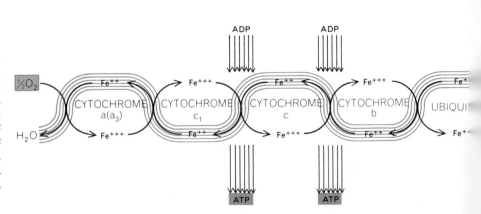

ATP

GLUCOSE

GLUCOSE-6-PHOSPHATE

ATP

FRUCTOSE-6-PHOSPHATE

FRUCTOS 1,6-DIPHOSP

● CARBON   ○ OXYGEN   ∘ HYDROGEN   Ⓟ PHOSPHATE

ADP   ADP

½O_2 Fe++   Fe+++   Fe++   Fe+++   Fe++

CYTOCHROME a(a$_3$)   CYTOCHROME c$_1$ CYTOCHROME c   CYTOCHROME b   UBIQUI

H_2O Fe+++   Fe++   Fe+++   Fe++   Fe++

ATP   ATP

OXIDATIVE METABOLISM provides the energy that powers all higher forms of life. It proceeds in two phases: glycolysis (*top*), an anaerobic phase that does not require oxygen, and aerobic respiration (*bottom*), which requires oxygen. In glycolysis (or fermentation, the anaerobic process by which organisms such as yeast derive their energy) a molecule of the six-carbon sugar glucose is broken down into two molecules of the three-carbon compound pyruvic acid with a net gain of two molecules of adenosine triphosphate, the cellular

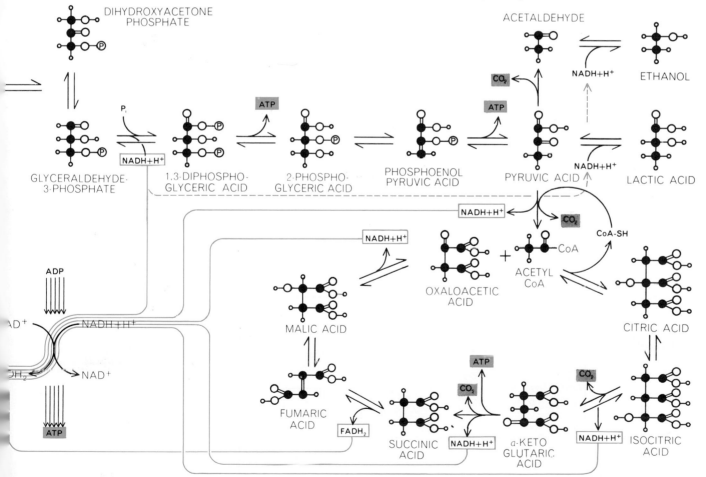

C

FUMARIC ACID

L-MALIC ACID

OXALOACETIC ACID

FUMARATE HYDROTASE

MALIC ACID DEHYDROGENASE

moved. In hydration-dehydrogenation (*c*) water is added and hydrogen is removed. Oxygenation does not occur in respiration, in which oxygen serves only as a hydrogen acceptor.

now of the University of California at Los Angeles, do imply an approximate balance between the chemical combining equivalents of carbon and oxygen in sediments, the atmosphere, the hydrosphere and the biosphere [*see bottom illustration on page 38*]. The relatively small excess of carbon in Rubey's estimates could be accounted for by the oxygen used in converting carbon monoxide to carbon dioxide. Or it might be due to an underestimate of the quantities of sulfate ion or ferric oxide in sediments. (Rubey's estimates could not include large iron formations recently discovered in western Australia and elsewhere.) The carbon dioxide in carbonate rocks does not need to be accounted for, but the oxygen involved in converting it to carbonate ion does. The recycling of sediments through metamorphism, mountain-building and the movement of ocean-floor plates under the continents is a variable of unknown dimensions, but

it probably does not affect the approximate balance observed in view of the fact that the overwhelmingly large pools to be balanced are all in the lithosphere and that carbon and oxygen losses would be roughly equivalent. The small amounts of oxygen dissolved in water are not included in this balance.

Nonetheless, water does enter the picture. Another possible source of oxygen in our atmosphere is photolysis, the ultraviolet dissociation of water vapor in the outer atmosphere followed by the escape of the hydrogen from the earth's gravitational field. This has usually been regarded as a trivial source, however. Although R. T. Brinkmann of the California Institute of Technology has recently argued that nonbiological photolysis may be a major source of atmospheric oxygen, the carbon-oxygen balance sheet does not support that belief, which also runs into other difficulties.

When free oxygen began to accumulate in the atmosphere some 1.8 billion years ago, life was still restricted to sites

energy carrier. The pyruvic acid is converted into lactic acid in animal cells deprived of oxygen and into some other compound, such as ethanol, in fermentation. In aerobic cells in the presence of oxygen, however, pyruvic acid is completely oxidized to produce carbon dioxide and water. In the process hydrogen ions are removed. The electrons of these hydrogens (and of two removed in

glycolysis) are passed along by two electron carriers, nicotinamide adenine dinucleotide (NAD) and flavin adenine dinucleotide (FAD), to a chain of respiratory enzymes, ubiquinone and the cytochromes, which are alternately reduced and oxidized. Energy released in the reactions is coupled to synthesis of ATP, 38 molecules of which are produced for every molecule of glucose consumed.

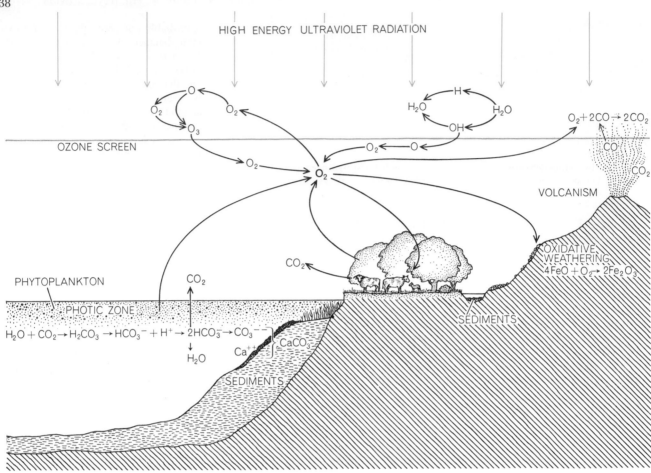

HIGH ENERGY ULTRAVIOLET RADIATION

OZONE SCREEN

VOLCANISM

$O_2 + 2CO \rightarrow 2CO_2$

OXIDATIVE WEATHERING
$4FeO + O_2 \rightarrow 2Fe_2O_3$

PHYTOPLANKTON

PHOTIC ZONE

$H_2O + CO_2 \rightarrow H_2CO_3 \rightarrow HCO_3^- + H^+ \rightarrow 2HCO_3^- \rightarrow CO_3^{--}$

Ca^{++} } $CaCO_3$

SEDIMENTS

SEDIMENTS

OXYGEN CYCLE is complicated because oxygen appears in so many chemical forms and combinations, primarily as molecular oxygen (O_2), in water and in organic and inorganic compounds. Some global pathways of oxygen are shown here in simplified form.

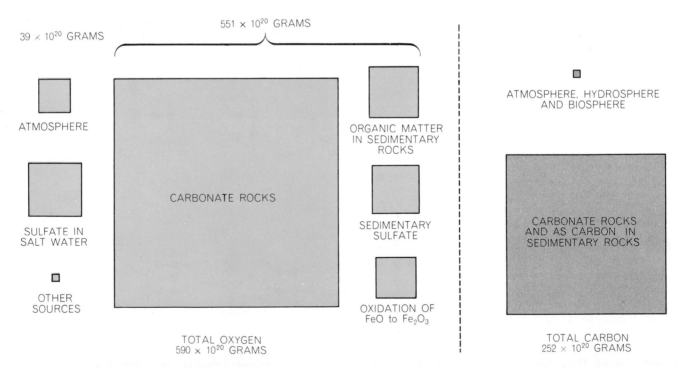

39×10^{20} GRAMS

551×10^{20} GRAMS

ATMOSPHERE

SULFATE IN SALT WATER

OTHER SOURCES

CARBONATE ROCKS

ORGANIC MATTER IN SEDIMENTARY ROCKS

SEDIMENTARY SULFATE

OXIDATION OF FeO to Fe_2O_3

TOTAL OXYGEN
590×10^{20} GRAMS

ATMOSPHERE, HYDROSPHERE AND BIOSPHERE

CARBONATE ROCKS AND AS CARBON IN SEDIMENTARY ROCKS

TOTAL CARBON
252×10^{20} GRAMS

OXYGEN-CARBON BALANCE SHEET suggests that photosynthesis can account not only for all the oxygen in the atmosphere but also for the much larger amount of "fossil" oxygen, mostly in compounds in sediments. The diagram, based on estimates made by William W. Rubey, indicates that the elements are present in about the proportion, 12/32, that would account for their derivation through photosynthesis from carbon dioxide (one atom of carbon, molecular weight 12, to two of oxygen, molecular weight 16).

shielded from destructive ultraviolet radiation by sufficient depths of water or by screens of sediment. In time enough oxygen built up in the atmosphere for ozone, a strong absorber in the ultraviolet, to form a shield against incoming ultraviolet radiation. The late Lloyd V. Berkner and Lauriston C. Marshall of the Graduate Research Center of the Southwest in Dallas calculated that only 1 percent of the present atmospheric level of oxygen would give rise to a sufficient level of ozone to screen out the most deleterious wavelengths of the ultraviolet radiation. This also happens to be the level of oxygen at which Pasteur found that certain microorganisms switch over from a fermentative type of metabolism to an oxidative one. Berkner and Marshall therefore jumped to the conclusion (reasonably enough on the evidence they considered) that this was the stage at which oxidative metabolism arose. They related this stage to the first appearance of metazoan life somewhat more than 600 million years ago.

The geological record has long made it plain, however, that free molecular oxygen existed in the atmosphere well before that relatively late date in geo-logic time. Moreover, recent evidence is consistent with the origin of oxidative metabolism at least twice as long ago. Eucaryotic cells—cells with organized nuclei and other organelles—have been identified in rocks in eastern California that are believed to be about 1.3 billion years old [*see top illustration on page 34*]. Since all living eucaryotes depend on oxidative metabolism, it seems likely that these ancestral forms did too. The oxygen level may nonetheless have still been quite low at this stage. Simple diffusion would suffice to move enough oxygen across cell boundaries and within the cell, even at very low concentrations, to supply the early oxidative metabolizers. A higher order of organization and of atmospheric oxygen was required, however, for advanced oxidative metabolism. Perhaps that is why, although the eucaryotic cell existed at least 1.2 billion years ago, we have no unequivocal fossils of metazoan organisms from rocks older than perhaps 640 million years.

In other words, perhaps Berkner and Marshall were mistaken only in trying to make the appearance of the Metazoa coincide with the onset of oxidative metabolism. Once the level of atmospheric oxygen was high enough to generate an effective ozone screen, photosynthetic organisms would have been able to spread throughout the surface waters of the sea, greatly accelerating the rate of oxygen production. The plausible episodes in geological history to correlate with this development are the secondary oxidation of the banded iron formations and the appearance of sedimentary calcium sulfate (gypsum and anhydrite) on a large scale. These events occurred just as or just before the Metazoa first appeared in early Paleozoic time. The attainment of a suitable level of atmospheric oxygen may thus be correlated with the emergence of metazoan root stocks from premetazoan ancestors beginning about 640 million years ago. The fact that oxygen could accumulate no faster than carbon (or hydrogen) was removed argues against the likelihood of a rapid early buildup of oxygen.

That subsequent biospheric and atmospheric evolution were closely interlinked can now be taken for granted. What is not known are the details. Did oxygen levels in the atmosphere increase steadily throughout geologic time, marking regular stages of biological evolution such as the emergence of land plants, of

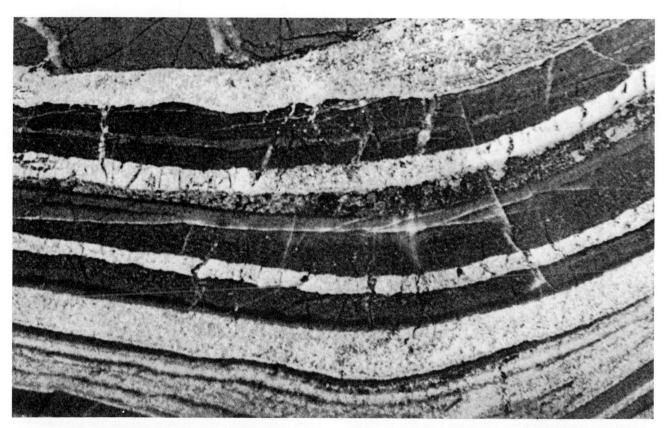

BANDED IRON FORMATION provides the first geological evidence of free oxygen in the hydrosphere. The layers in this polished cross section result from an alternation of iron-rich and iron-poor depositions. This sample from the Soudan Iron Formation in Minnesota is more than 2.7 billion years old. The layers, originally horizontal, were deformed while soft and later metamorphosed.

insects, of the various vertebrate groups and of flowering plants, as Berkner and Marshall suggested? Or were there wide swings in the oxygen level? Did oxygen decrease during great volcanic episodes, as a result of the oxidation of newly emitted carbon monoxide to carbon dioxide, or during times of sedimentary sulfate precipitation? Did oxygen increase when carbon was being sequestered during times of coal and petroleum formation? May there have been fluctuations in both directions as a result of plant and animal evolution, of phytoplankton eruptions and extinctions and of the extent and type of terrestrial plant cover? Such processes and events are now being seriously studied, but the answers are as yet far from clear.

What one can say with confidence is that success in understanding the oxy-

YEARS BEFORE PRESENT	LITHOSPHERE	BIOSPHERE	HYDROSPHERE	ATMOSPHERE
20 MILLION	GLACIATION	MAMMALS DIVERSIFY GRASSES APPEAR		OXYGEN APPROACHES PRESENT LEVEL
50 MILLION	COAL FORMATION VOLCANISM			
100 MILLION		SOCIAL INSECTS, FLOWERING PLANTS — MAMMALS		ATMOSPHERIC OXYGEN INCREASES AT FLUCTUATING RATE
200 MILLION	GREAT VOLCANISM — COAL FORMATION		OCEANS CONTINUE TO INCREASE IN VOLUME	
		INSECTS APPEAR LAND PLANTS APPEAR		
500 MILLION		METAZOA APPEAR RAPID INCREASE IN PHYTOPLANKTON		OXYGEN AT 3-10 PERCENT OF PRESENT ATMOSPHERIC LEVEL
	GLACIATION SEDIMENTARY CALCIUM SULFATE		SURFACE WATERS OPENED TO PHYTOPLANKTON	OXYGEN AT 1 PERCENT OF PRESENT ATMOSPHERIC LEVEL, OZONE SCREEN EFFECTIVE
1 BILLION	VOLCANISM	EUCARYOTES		OXYGEN INCREASING, CARBON DIOXIDE DECREASING
2 BILLION	RED BEDS	ADVANCED OXYGEN-MEDIATING ENZYMES	OXYGEN DIFFUSES INTO ATMOSPHERE	OXYGEN IN ATMOSPHERE
	GLACIATION — BANDED IRON FORMATIONS OLDEST SEDIMENTS OLDEST EARTH ROCKS	FIRST OXYGEN-GENERATING PHOTOSYNTHETIC CELLS PROCARYOTES ABIOGENIC EVOLUTION	START OF OXYGEN GENERATION WITH FERROUS IRON AS OXYGEN SINK	
5 BILLION	(ORIGIN OF SOLAR SYSTEM)			NO FREE OXYGEN

CHRONOLOGY that interrelates the evolutions of atmosphere and biosphere is gradually being established from evidence in the geological record and in fossils. According to calculations by Lloyd V. Berkner and Lauriston C. Marshall, when oxygen in the atmosphere reached 1 percent of the present atmospheric level, it provided enough ozone to filter out the most damaging high-energy ultraviolet radiation so that phytoplankton could survive everywhere in the upper, sunlit layers of the seas. The result may have been a geometric increase in the amount of photosynthesis in the oceans that, if accompanied by equivalent sequestration of carbon, might have resulted in a rapid buildup of atmospheric oxygen, leading in time to the evolution of differentiated multicelled animals.

gen cycle of the biosphere in truly broad terms will depend on how good we are at weaving together the related strands of biospheric, atmospheric, hydrospheric and lithospheric evolution throughout geologic time. Whatever we may conjecture about any one of these processes must be consistent with what is known about the others. Whereas any one line of evidence may be weak in itself, a number of lines of evidence, taken together and found to be consistent, reinforce one another exponentially. This synergistic effect enhances our confidence in the proposed time scale linking the evolution of oxygen in the atmosphere and the management of the gaseous oxygen budget within the biosphere [see illustration on opposite page].

The most recent factor affecting the oxygen cycle of the biosphere and the oxygen budget of the earth is man himself. In addition to inhaling oxygen and exhaling carbon dioxide as a well-behaved animal does, man decreases the oxygen level and increases the carbon dioxide level by burning fossil fuels and paving formerly green land. He is also engaged in a vast but unplanned experiment to see what effects oil spills and an array of pesticides will have on the world's phytoplankton. The increase in the albedo, or reflectivity, of the earth as a result of covering its waters with a molecule-thick film of oil could also affect plant growth by lowering the temperature and in other unforeseen ways. Reductions in the length of growing seasons and in green areas would limit terrestrial plant growth in the middle latitudes. (This might normally be counterbalanced by increased rainfall in the lower latitudes, but a film of oil would also reduce evaporation and therefore rainfall.) Counteracting such effects, man moves the earth's fresh water around to increase plant growth and photosynthesis in arid and semiarid regions. Some of this activity, however, involves the mining of ground water, thereby favoring processes that cause water to be returned to the sea at a faster rate than evaporation brings it to the land.

He who is willing to say what the final effects of such processes will be is wiser or braver than we are. Perhaps the effects will be self-limiting and self-correcting, although experience should warn us not to gamble on that. Oxygen in the atmosphere might be reduced several percent below the present level without adverse effects. A modest increase in the carbon dioxide level might enhance plant growth and lead to a cor-

responding increase in the amount of oxygen. Will a further increase in carbon dioxide also have (or renew) a "greenhouse effect," leading to an increase in temperature (and thus to a rising sea level)? Or will such effects be counterbalanced or swamped by the cooling effects of particulate matter in the air or by increased albedo due to oil films? It is anyone's guess. (Perhaps we should be more alarmed about a possible decrease of atmospheric carbon dioxide, on which all forms of life ultimately depend, but the sea contains such vast amounts that it can presumably keep carbon dioxide in the atmosphere balanced at about the present level for a long time to come.) The net effect of the burning of fossil fuels may in the long run be nothing more than a slight increase (or decrease?) in the amount of limestone deposited. In any event the recoverable fossil fuels whose combustion releases carbon dioxide are headed for depletion in a few more centuries, and then man will have other problems to contend with.

What we want to stress is the indivisibility and complexity of the environment. For example, the earth's atmosphere is so thoroughly mixed and so

rapidly recycled through the biosphere that the next breath you inhale will contain atoms exhaled by Jesus at Gethsemane and by Adolf Hitler at Munich. It will also contain atoms of radioactive strontium 90 and iodine 131 from atomic explosions and gases from the chimneys and exhaust pipes of the world. Present environmental problems stand as a grim monument to the cumulatively adverse effects of actions that in themselves were reasonable enough but that were taken without sufficient thought to their consequences. If we want to ensure that the biosphere continues to exist over the long term and to have an oxygen cycle, each new action must be matched with an effort to foresee its consequences throughout the ecosystem and to determine how they can be managed favorably or avoided. Understanding also is needed, and we are woefully short on that commodity. This means that we must continue to probe all aspects of the indivisible global ecosystem and its past, present and potential interactions. That is called basic research, and basic research at this critical point in history is gravely endangered by new crosscurrents of anti-intellectualism.

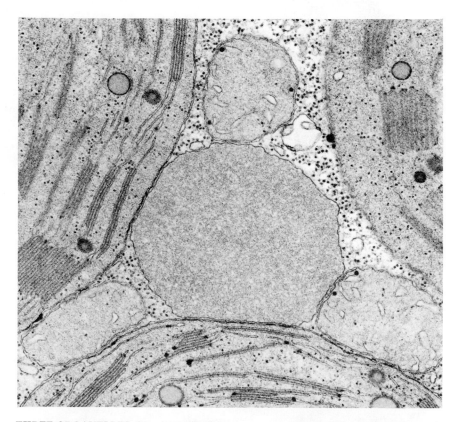

THREE ORGANELLES that are involved in oxygen metabolism in the living cell are enlarged 40,000 diameters in an electron micrograph of a tobacco leaf cell made by Sue Ellen Frederick in the laboratory of Eldon H. Newcomb at the University of Wisconsin. A peroxisome (center) is surrounded by three mitochondria and three chloroplasts. Oxygen is produced in the grana (layered objects) in the chloroplasts and is utilized in aerobic respiration in the mitochondria. Peroxisomes contain enzymes involved in oxygen metabolism.

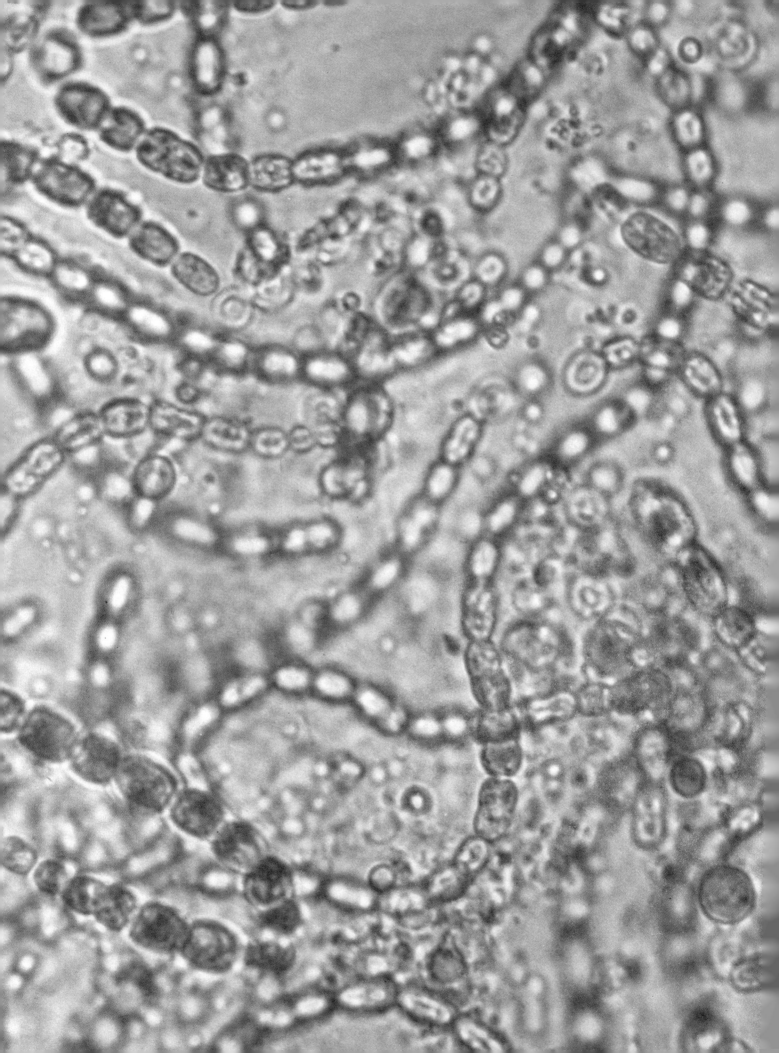

5

THE NITROGEN CYCLE

C. C. DELWICHE
September 1970

Nitrogen is 79 percent of the atmosphere, but it cannot be used directly by the large majority of living things. It must first be "fixed" by specialized organisms or by industrial processes

Although men and other land animals live in an ocean of air that is 79 percent nitrogen, their supply of food is limited more by the availability of fixed nitrogen than by that of any other plant nutrient. By "fixed" is meant nitrogen incorporated in a chemical compound that can be utilized by plants and animals. As it exists in the atmosphere nitrogen is an inert gas except to the comparatively few organisms that have the ability to convert the element to a combined form. A smaller but still significant amount of atmospheric nitrogen is fixed by ionizing phenomena such as cosmic radiation, meteor trails and lightning, which momentarily provide the high energy needed for nitrogen to react with oxygen or the hydrogen of water. Nitrogen is also fixed by marine organisms, but the largest single natural source of fixed nitrogen is probably terrestrial microorganisms and associations between such microorganisms and plants.

Of all man's recent interventions in the cycles of nature the industrial fixation of nitrogen far exceeds all the others in magnitude. Since 1950 the amount of nitrogen annually fixed for the production of fertilizer has increased approximately fivefold, until it now equals the amount that was fixed by all terrestrial ecosystems before the advent of modern agriculture. In 1968 the world's annual output of industrially fixed nitrogen amounted to about 30 million tons of nitrogen; by the year 2000 the industrial fixation of nitrogen may well exceed 100 million tons.

Before the large-scale manufacture of synthetic fertilizers and the wide cultivation of the nitrogen-fixing legumes one could say with some confidence that the amount of nitrogen removed from the atmosphere by natural fixation processes was closely balanced by the amount returned to the atmosphere by organisms that convert organic nitrates to gaseous nitrogen. Now one cannot be sure that the denitrifying processes are keeping pace with the fixation processes. Nor can one predict all the consequences if nitrogen fixation were to exceed denitrification over an extended period. We do know that excessive runoff of nitrogen compounds in streams and rivers can result in "blooms" of algae and intensified biological activity that deplete the available oxygen and destroy fish and other oxygen-dependent organisms. The rapid eutrophication of Lake Erie is perhaps the most familiar example.

To appreciate the intricate web of nitrogen flow in the biosphere let us trace the course of nitrogen atoms from the atmosphere into the cells of microorganisms, and then into the soil as fixed nitrogen, where it is available to higher plants and ultimately to animals. Plants and animals die and return the fixed nitrogen to the soil, at which point the nitrogen may simply be recycled through a new generation of plants and animals

or it may be broken down into elemental nitrogen and returned to the atmosphere [*see illustration on next two pages*].

Because much of the terminology used to describe steps in the nitrogen cycle evolved in previous centuries it has an archaic quality. Antoine Laurent Lavoisier, who clarified the composition of air, gave nitrogen the name azote, meaning without life. The term is still found in the family name of an important nitrogen-fixing bacterium: the Azotobacteraceae. One might think that fixation would merely be termed nitrification, to indicate the addition of nitrogen to some other substance, but nitrification is reserved for a specialized series of reactions in which a few species of microorganisms oxidize the ammonium ion (NH_4^+) to nitrite (NO_2^-) or nitrite to nitrate (NO_3^-). When nitrites or nitrates are reduced to gaseous compounds such as molecular nitrogen (N_2) or nitrous oxide (N_2O), the process is termed denitrification. "Ammonification" describes the process by which the nitrogen of organic compounds (chiefly amino acids) is converted to ammonium ion. The process operates when microorganisms decompose the remains of dead plants and animals. Finally, a word should be said about the terms oxidation and reduction, which have come to mean more than just the addition of oxygen or its removal. Oxidation is any process that removes electrons from a substance. Reduction is the reverse process: the addition of electrons. Since electrons can neither be created nor destroyed in a chemical reaction, the oxidation of one substance always implies the reduction of another.

One may wonder how it is that some organisms find it profitable to oxidize

BLUE-GREEN ALGAE, magnified 4,200 diameters on the opposite page, are among the few free-living organisms capable of combining nitrogen with hydrogen. Until this primary fixation process is accomplished, the nitrogen in the air (or dissolved in water) cannot be assimilated by the overwhelming majority of plants or by any animal. A few bacteria are also free-living nitrogen fixers. The remaining nitrogen-fixing microorganisms live symbiotically with higher plants. This micrograph, which shows blue-green algae of the genus *Nostoc,* was made by Herman S. Forest of the State University of New York at Geneseo.

nitrogen compounds whereas other organisms—even organisms in the same environment—owe their survival to their ability to reduce nitrogen compounds. Apart from photosynthetic organisms, which obtain their energy from radiation, all living forms depend for their energy on chemical transformations.

These transformations normally involve the oxidation of one compound and the reduction of another, although in some cases the compound being oxidized and the compound being reduced are different molecules of the same substance, and in other cases the reactants are fragments of a single molecular species. Ni-

trogen can be cycled because the reduced inorganic compounds of nitrogen can be oxidized by atmospheric oxygen with a yield of useful energy. Under anaerobic conditions the oxidized compounds of nitrogen can act as oxidizing agents for the burning of organic compounds (and a few inorganic com-

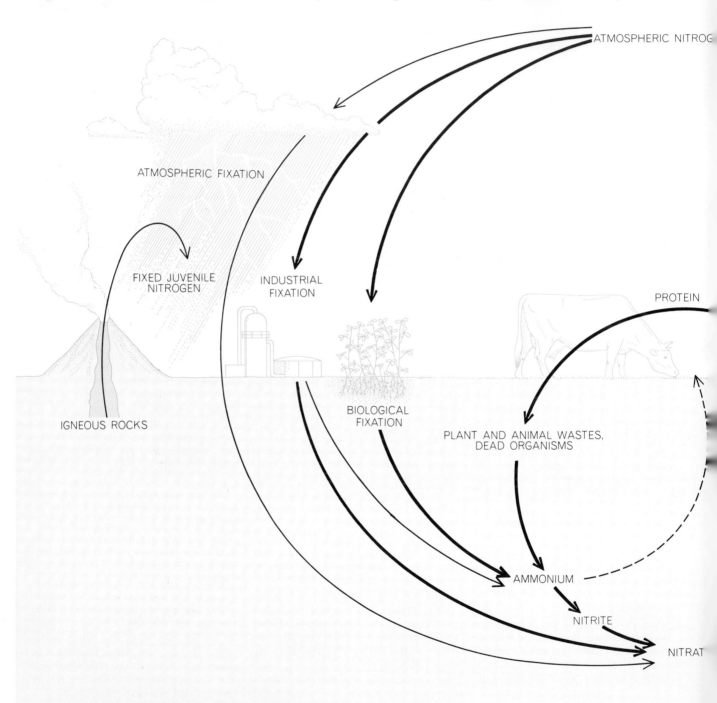

ATMOSPHERIC NITROG

ATMOSPHERIC FIXATION

FIXED JUVENILE NITROGEN

INDUSTRIAL FIXATION

PROTEIN

BIOLOGICAL FIXATION

IGNEOUS ROCKS

PLANT AND ANIMAL WASTES, DEAD ORGANISMS

AMMONIUM

NITRITE

NITRAT

NITROGEN CYCLE, like the water, oxygen and carbon cycles, involves all regions of the biosphere. Although the supply of nitrogen in the atmosphere is virtually inexhaustible, it must be combined with hydrogen or oxygen before it can be assimilated by higher plants, which in turn are consumed by animals. Man has intervened in the historical nitrogen cycle by the large-scale cultivation of nitrogen-fixing legumes and by the industrial fixation of nitrogen. The amount of nitrogen fixed annually by these two expedients now exceeds by perhaps 10 percent the amount of nitrogen fixed by terrestrial ecosystems before the advent of agriculture.

pounds), again with a yield of useful energy.

Nitrogen is able to play its complicated role in life processes because it has an unusual number of oxidation levels, or valences [see *illustration on page 47*]. An oxidation level indicates the number of electrons that an atom in a

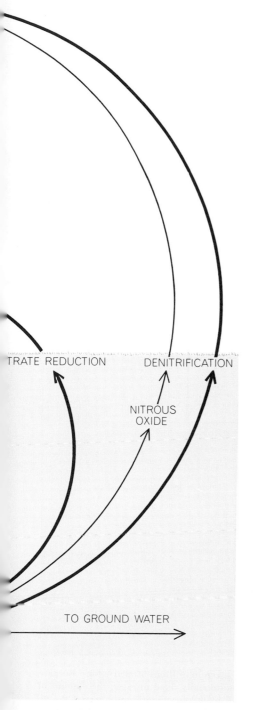

particular compound has "accepted" or "donated." In plants and animals most nitrogen exists either in the form of the ammonium ion or of amino ($-NH_2$) compounds. In either case it is highly reduced; it has acquired three electrons by its association with three other atoms and thus is said to have a valence of minus 3. At the other extreme, when nitrogen is in the highly oxidized form of the nitrate ion (the principal form it takes in the soil), it shares five of its electrons with oxygen atoms and so has a valence of plus 5. To convert nitrogen as it is found in the ammonium ion or amino acids to nitrogen as it exists in soil nitrates involves a total valence change of eight, or the removal of eight electrons. Conversely, to convert nitrate nitrogen into amino nitrogen requires the addition of eight electrons.

By and large the soil reactions that reduce nitrogen, or add electrons to it, release considerably more energy than the reactions that oxidize nitrogen, or remove electrons from it. The illustration on page 8 lists some of the principal reactions involved in the nitrogen cycle, together with the energy released (or required) by each. As a generalization one can say that for almost every reaction in nature where the conversion of one compound to another yields an energy of at least 15 kilocalories per mole (the equivalent in grams of a compound's molecular weight), some organism or group of organisms has arisen that can exploit this energy to survive.

The fixation of nitrogen requires an investment of energy. Before nitrogen can be fixed it must be "activated," which means that molecular nitrogen must be split into two atoms of free nitrogen. This step requires at least 160 kilocalories for each mole of nitrogen (equivalent to 28 grams). The actual fixation step, in which two atoms of nitrogen combine with three molecules of hydrogen to form two molecules of ammonia (NH_3), releases about 13 kilocalories. Thus the two steps together require a net input of at least 147 kilocalories. Whether nitrogen-fixing organisms actually invest this much energy, however, is not known. Reactions catalyzed by enzymes involve the penetration of activation barriers and not a simple change in energy between a set of initial reactants and their end products.

Once ammonia or the ammonium ion has appeared in the soil, it can be absorbed by the roots of plants and the nitrogen can be incorporated into amino acids and then into proteins. If the plant is subsequently eaten by an animal, the nitrogen may be incorporated into a new protein. In either case the protein ultimately returns to the soil, where it is decomposed (usually with bacterial help) into its component amino acids. Assuming that conditions are aerobic, meaning that an adequate supply of oxygen is present, the soil will contain many microorganisms capable of oxidizing amino acids to carbon dioxide, water and ammonia. If the amino acid happens to be glycine, the reaction will yield 176 kilocalories per mole.

A few microorganisms represented by the genus *Nitrosomonas* employ nitrification of the ammonium ion as their sole source of energy. In the presence of oxygen, ammonia is converted to nitrite ion (NO_2^-) plus water, with an energy yield of about 65 kilocalories per mole, which is quite adequate for a comfortable existence. *Nitrosomonas* belongs to the group of microorganisms termed autotrophs, which get along without an organic source of energy. Photoautotrophs obtain their energy from light; chemoautotrophs (such as *Nitrosomonas*) obtain energy from inorganic compounds.

There is another specialized group of microorganisms, represented by *Nitrobacter*, that are capable of extracting additional energy from the nitrite generated by *Nitrosomonas*. The result is the oxidation of a nitrite ion to a nitrate ion with the release of about 17 kilocalories per mole, which is just enough to support the existence of *Nitrobacter*.

In the soil there are numerous kinds of denitrifying bacteria (for example *Pseudomonas denitrificans*) that, if obliged to exist in the absence of oxygen, are able to use the nitrate or nitrite ion as electron acceptors for the oxidation of organic compounds. In these reactions the energy yield is nearly as large as it would be if pure oxygen were the oxidizing agent. When glucose reacts with oxygen, the energy yield is 686 kilocalories per mole of glucose. In microorganisms living under anaerobic conditions the reaction of glucose with nitrate ion yields about 545 kilocalories per mole of glucose if the nitrogen is reduced to nitrous oxide, and 570 kilocalories if the nitrogen is reduced all the way to its elemental gaseous state.

The comparative value of ammonium and nitrate ions as a source of nitrogen for plants has been the subject of a number of investigations. One might think that the question would be readily resolved in favor of the ammonium ion: its valence is minus 3, the same as the valence of nitrogen in amino acids, whereas the valence of the nitrate ion is plus 5.

TRATE REDUCTION   DENITRIFICATION

NITROUS OXIDE

TO GROUND WATER

A cycle similar to the one illustrated also operates in the ocean, but its characteristics and transfer rates are less well understood. A global nitrogen flow chart, using the author's estimates, appears on the next page.

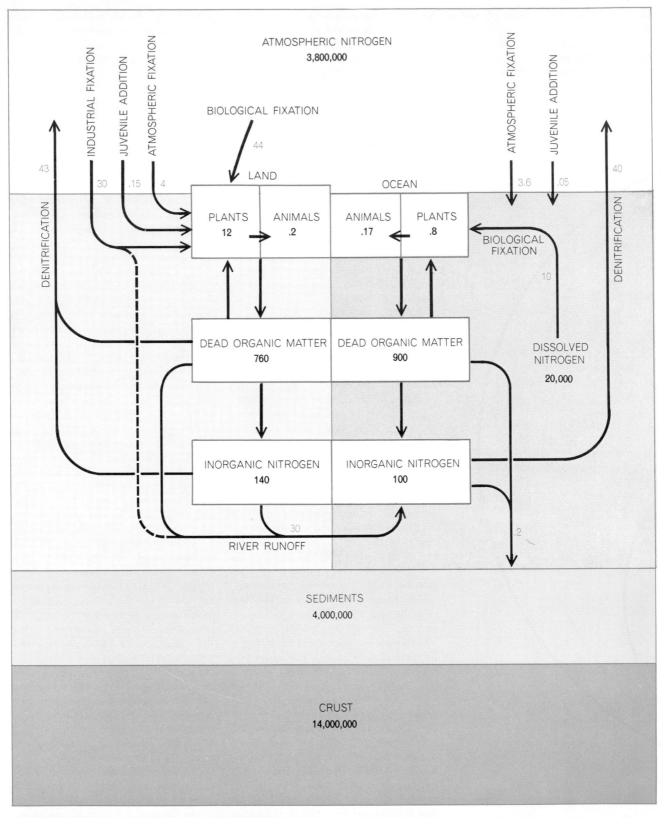

DISTRIBUTION OF NITROGEN in the biosphere and annual transfer rates can be estimated only within broad limits. The two quantities known with high confidence are the amount of nitrogen in the atmosphere and the rate of industrial fixation. The apparent precision in the other figures shown here reflects chiefly an effort to preserve indicated or probable ratios among different inventories. Thus the figures for atmospheric fixation and biological fixation in the oceans could well be off by a factor of 10. The figures for inventories are given in billions of metric tons; the figures for transfer rates (color) are given in millions of metric tons. Because of the extensive use of industrially fixed nitrogen the amount of nitrogen available to land plants may significantly exceed the nitrogen returned to the atmosphere by denitrifying bacteria in the soil. A portion of this excess fixed nitrogen is ultimately washed into the sea but it is not included in the figure shown for river runoff. Similarly, the value for oceanic denitrification is no more than a rough estimate that is based on the assumption that the nitrogen cycle was in overall balance before man's intervention.

On this basis plants must expend energy to reduce nitrogen from a valence of plus 5 to one of minus 3. The fact is, however, that there are complicating factors; the preferred form of nitrogen depends on other variables. Because the ammonium ion has a positive charge it tends to be trapped on clay particles near the point where it is formed (or where it is introduced artificially) until it has been oxidized. The nitrate ion, being negatively charged, moves freely through the soil and thus is more readily carried downward into the root zone. Although the demand for fertilizer in solid form (such as ammonium nitrate and urea) remains high, anhydrous ammonia and liquid ammoniacal fertilizers are now widely applied. The quantity of nitrogen per unit weight of ammonia is much greater than it is per unit of nitrate; moreover, liquids are easier to handle than solids.

Until the end of the 19th century little was known about the soil organisms that fix nitrogen. In fact, at that time there was some concern among scientists that the denitrifying bacteria, which had just been discovered, would eventually deplete the reserve of fixed nitrogen in the soil and cripple farm productivity. In an address before the Royal Society of London, Sir William Crookes painted a bleak picture for world food production unless artificial means of fixing nitrogen were soon developed. This was a period when Chilean nitrate reserves were the main source of fixed nitrogen for both fertilizer and explosives. As it turned out, the demand for explosives provided the chief incentive for the invention of the catalytic fixation process by Fritz Haber and Karl Bosch of Germany in 1914. In this process atmospheric nitrogen and hydrogen are passed over a catalyst (usually nickel) at a temperature of about 500 degrees Celsius and a pressure of several hundred atmospheres. In a French version of the process, developed by Georges Claude, nitrogen was obtained by the fractional liquefaction of air. In current versions of the Haber process the source of hydrogen is often the methane in natural gas [see illustration on page 49].

As the biological fixation of nitrogen and the entire nitrogen cycle became better understood, the role of the denitrifying bacteria fell into place. Without such bacteria to return nitrogen to the atmosphere most of the atmospheric nitrogen would now be in the oceans or locked up in sediments. Actually, of course, there is not enough oxygen in the

VALENCE	COMPOUND	FORMULA	VALENCE ELECTRONS
+ 5	NITRATE ION	NO_3^-	
+ 3	NITRITE ION	NO_2^-	
+ 1	NITROXYL	[HNO]	
0	NITROGEN GAS	N_2	
− 1	HYDROXYLAMINE	$HONH_2$	
− 3	AMMONIA	NH_3	

NITROGEN'S VARIETY OF OXIDATION LEVELS, or valence states, explains its ability to combine with hydrogen, oxygen and other atoms to form a great variety of biological compounds. Six of its valence states are listed with schematic diagrams (right) showing the disposition of electrons in the atom's outer (valence) shell. The ions are shown combined with potassium (K). In the oxidized (+) states nitrogen's outer electrons complete the outer shells of other atoms. In the reduced (−) states the two electrons needed to complete the outer shell of nitrogen are supplied by other atoms. Actually the outer electrons of two bound atoms spend some time in the shells of both atoms, contributing to the electrostatic attraction between them. Electrons of nitrogen (N) are in color; those of other atoms are black dots or open circles. The nitroxyl radical, HNO, is placed in brackets because it is not stable. It can exist in its dimeric form, hyponitrous acid (HONNOH).

atmosphere today to convert all the free nitrogen into nitrates. One can imagine, however, that if a one-way process were to develop in the absence of denitrifying bacteria, the addition of nitrates to the ocean would make seawater slightly more acidic and start the release of carbon dioxide from carbonate rocks. Eventually the carbon dioxide would be taken up by plants, and if the carbon were then deposited as coal or other hydrocarbons, the remaining oxygen would be available in the atmosphere to be combined with nitrogen. Because of the large number of variables involved it is difficult to predict how the world would look without the denitrification reaction, but it would certainly not be the world we know.

The full story of the biological fixation of nitrogen has not yet been written. One would like to know how the activating enzyme (nitrogenase) used by nitrogen-fixing bacteria can accomplish at ordinary temperatures and pressures what takes hundreds of degrees and thousands of pounds of pressure in a synthetic-ammonia reactor. The total amount of nitrogenase in the world is probably no more than a few kilograms.

The nitrogen-fixing microorganisms are divided into two broad classes: those that are "free-living" and those that live in symbiotic association with higher plants. This distinction, however, is not as sharp as it was once thought to be, because the interaction of plants and microorganisms has varying degrees of intimacy. The symbionts depend directly on the plants for their energy supply and probably for special nutrients as well. The free-living nitrogen fixers are indirectly dependent on plants for their energy or, as in the case of the blue-green algae and photosynthetic bacteria, obtain energy directly from sunlight.

Although the nitrogen-fixation reaction is associated with only a few dozen species of higher plants, these species are widely distributed in the plant kingdom. Among the more primitive plants whose symbionts can fix nitrogen are the cycads and the ginkgos, which can be traced back to the Carboniferous period of some 300 million years ago [see bottom illustration on page 51]. It is probable that the primitive atmosphere of the earth contained ammonia, in which case the necessity for nitrogen fixation did not arise for hundreds of millions of years.

Various kinds of bacteria, particularly the Azotobacteraceae, are evidently the chief suppliers of fixed nitrogen in grasslands and other ecosystems where plants with nitrogen-fixing symbionts are absent. Good quantitative information on the rate of nitrogen fixation in such ecosystems is hard to obtain. Most investigations indicate a nitrogen-fixation rate of only two or three kilograms per hectare per year, with a maximum of perhaps five or six kilograms. Blue-green algae seem to be an important source of fixed nitrogen under conditions that favor their development [see illustration on page 42]. They may be a significant source in rice paddies and other environments favoring their growth. In natural ecosystems with mixed vegetation the symbiotic associations involving such plant genera as Alnus (the alders) and Ceanothus (the buckthorns) are important suppliers of fixed nitrogen.

For the earth as a whole, however, the greatest natural source of fixed nitrogen is probably the legumes. They are certainly the most important from an agronomic standpoint and have therefore been the most closely studied. The input of nitrogen from the microbial symbionts of alfalfa and other leguminous crops can easily amount to 350 kilograms per hectare, or roughly 100 times the annual rate of fixation attainable by nonsymbiotic organisms in a natural ecosystem.

Recommendations for increasing the world's food supply usually emphasize increasing the cultivation of legumes not only to enrich the soil in nitrogen but also because legumes (for example peas and beans) are themselves a food crop containing a good nutritional balance of amino acids. There are, however, several obstacles to carrying out such recommendations. The first is custom and taste. Many societies with no tradition of growing and eating legumes are reluctant to adopt them as a basic food.

For the farmer legumes can create a more immediate problem: the increased yields made possible by the extra nitrogen lead to the increased consumption of other essential elements, notably potassium and phosphorus. As a consequence farmers often say that legumes are "hard on the soil." What this really means is that the large yield of such crops places

REACTION	ENERGY YIELD (KILOCALORIES)
DENITRIFICATION	
1 $C_6H_{12}O_6 + 6KNO_3 \longrightarrow 6CO_2 + 3H_2O + 6KOH + 3N_2O$ GLUCOSE POTASSIUM NITRATE POTASSIUM HYDROXIDE NITROUS OXIDE	545
2 $5C_6H_{12}O_6 + 24KNO_3 \longrightarrow 30CO_2 + 18H_2O + 24KOH + 12N_2$ NITROGEN	570 (PER MOLE OF GLUCOSE)
3 $5S + 6KNO_3 + 2CaCO_3 \longrightarrow 3K_2SO_4 + 2CaSO_4 + 2CO_2 + 3N_2$ SULFUR POTASSIUM SULFATE CALCIUM SULFATE	132 (PER MOLE OF SULFUR)
RESPIRATION	
4 $C_6H_{12}O_6 + 6O_2 \longrightarrow 6CO_2 + 6H_2O$ CARBON DIOXIDE WATER	686
AMMONIFICATION	
5 $CH_2NH_2COOH + 1\frac{1}{2}O_2 \longrightarrow 2CO_2 + H_2O + NH_3$ GLYCINE OXYGEN AMMONIA	176
NITRIFICATION	
6 $NH_3 + 1\frac{1}{2}O_2 \longrightarrow HNO_2 + H_2O$ NITROUS ACID	66
7 $KNO_2 + \frac{1}{2}O_2 \longrightarrow KNO_3$ POTASSIUM NITRITE	17.5
NITROGEN FIXATION	
8 $N_2 \longrightarrow 2N$ "ACTIVATION" OF NITROGEN	− 160
9 $2N + 3H_2 \longrightarrow 2NH_3$	12.8

ENERGY YIELDS OF REACTIONS important in the nitrogen cycle show the various means by which organisms can obtain energy and thereby keep the cycle going. The most profitable are the denitrification reactions, which add electrons to nitrate nitrogen, whose valence is plus 5, and shift it either to plus 1 (as in N_2O) or zero (as in N_2). In the process glucose (or sulfur) is oxidized. Reactions No. 1 and No. 2 release nearly as much energy as conventional respiration (No. 4), in which the agent for oxidizing glucose is oxygen itself. The ammonification reaction (No. 5) is one of many that release ammonium for nitrification. The least energy of all, but still enough to provide the sole energetic support for certain bacteria, is released by the nitrification reactions (No. 6 and No. 7), which oxidize nitrogen. Only nitrogen fixation, which is accomplished in two steps, calls for an input of energy. The true energy cost of nitrogen fixation to an organism is unknown, however.

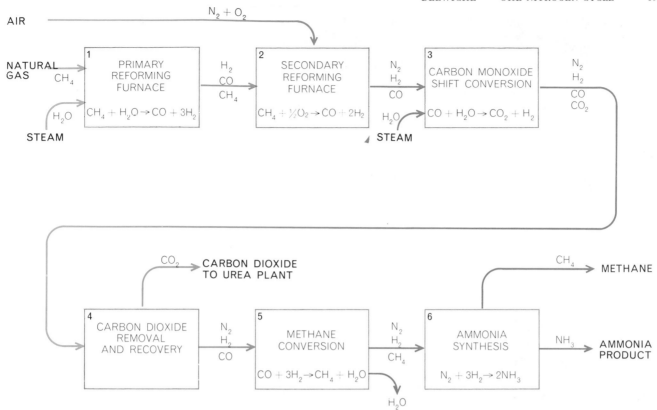

INDUSTRIAL AMMONIA PROCESS is based on the high-pressure catalytic fixation method invented in 1914 by Fritz Haber and Karl Bosch, which supplied Germany with nitrates for explosives in World War I. This flow diagram is based on the process developed by the M. W. Kellogg Company. As in most modern plants, the hydrogen for the basic reaction is obtained from methane, the chief constituent of natural gas, but any hydrocarbon source will do. In Step 1 methane and steam react to produce a gas rich in hydrogen. In Step 2 atmospheric nitrogen is introduced; the oxygen accompanying it is converted to carbon monoxide by partial combustion with methane. The carbon monoxide reacts with steam in Step 3. The carbon dioxide is removed in Step 4 and can be used elsewhere to convert some of the ammonia to urea, which has the formula $CO(NH_2)_2$. The last traces of carbon monoxide are converted to methane in Step 5. In Step 6 nitrogen and hydrogen combine at elevated temperature and pressure, in the presence of a catalyst, to form ammonia. A portion of the ammonia product can readily be converted to nitric acid by reacting it with oxygen. Nitric acid and ammonia can then be combined to produce ammonium nitrate, which, like urea, is another widely used fertilizer.

a high demand on all minerals, and unless the minerals are supplied the full benefit of the crop is not realized.

Symbiotic nitrogen fixers have a greater need for some micronutrients (for example molybdenum) than most plants do. It is now known that molybdenum is directly incorporated in the nitrogen-fixing enzyme nitrogenase. In Australia there were large areas where legumes refused to grow at all until it was discovered that the land could be made fertile by the addition of as little as two ounces of molybdenum per acre. Cobalt turns out to be another essential micronutrient for the fixation of nitrogen. The addition of only 10 parts per trillion of cobalt in a culture solution can make the difference between plants that are stunted and obviously in need of nitrogen and plants that are healthy and growing vigorously.

Although legumes and their symbionts are energetic fixers of nitrogen, there are indications that the yield of a legume crop can be increased still further by direct application of fertilizer instead of depending on the plant to supply all its own needs for fixed nitrogen. Additional experiments are needed to determine just how much the yield can be increased and how this increase compares with the industrial fixation of nitrogen in terms of energy investment. Industrial processes call for some 6,000 kilocalories per kilogram of nitrogen fixed, which is very little more than the theoretical minimum. The few controlled studies with which I am familiar suggest that the increase in crop yield achieved by the addition of a kilogram of nitrogen amounts to about the same number of calories. This comparison suggests that one can exchange the calories put into industrial fixation of nitrogen for the calories contained in food. In actuality this trade-off applies to the entire agricultural enterprise. The energy required for preparing, tilling and harvesting a field and for processing and distributing the product is only slightly less than the energy contained in the harvested crop.

Having examined the principal reactions that propel the nitrogen cycle, we are now in a position to view the process as a whole and to interpret some of its broad implications. One must be cautious in trying to present a worldwide inventory of a particular element in the biosphere and in indicating annual flows from one part of a cycle to another. The balance sheet for nitrogen [see top illustration on page 51] is particularly crude because we do not have enough information to assign accurate estimates to the amounts of nitrogen that are fixed and subsequently returned to the atmosphere by biological processes.

Another source of uncertainty involves the amount of nitrogen fixed by ionizing phenomena in the atmosphere. Although one can measure the amount of fixed nitrogen in rainfall, one is forced to guess how much is produced by ionization and how much represents nitrogen that has

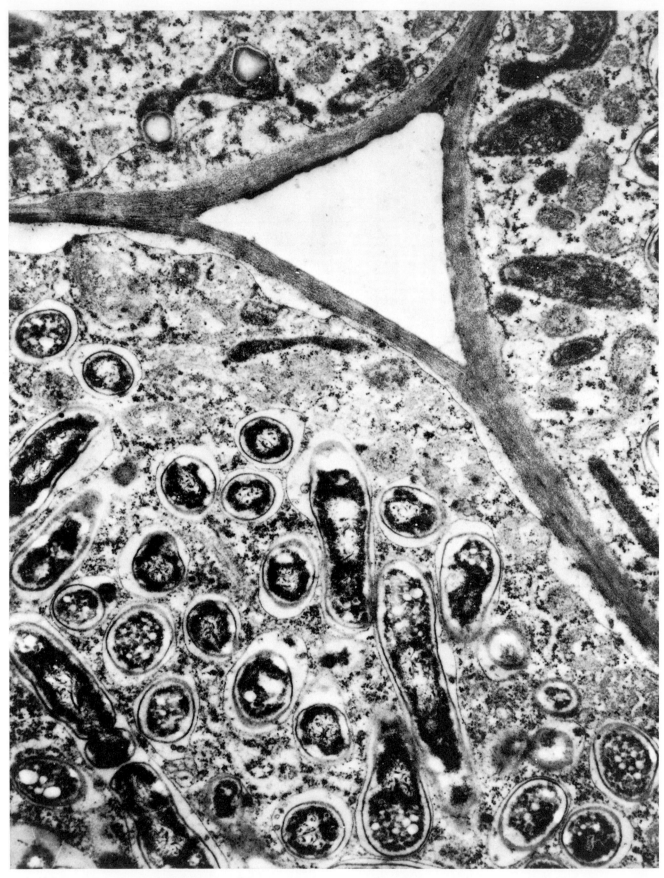

CROSS SECTION OF SOYBEAN ROOT NODULE, enlarged 22,-000 diameters, shows portions of three cells that have been infected by the nitrogen-fixing bacterium *Rhizobium japonicum*. More than two dozen bacteria are visible, each surrounded by a membrane. After the bacteria have divided, within a few days, each membrane will contain four to six "bacteroids." This electron micrograph was made by D. J. Goodchild and F. J. Bergersen of the Commonwealth Scientific and Industrial Research Organization in Australia.

entered the atmosphere from the land or the sea, either as ammonia or as oxides of nitrogen. Because the ocean is slightly alkaline it could release ammonia at a low rate, but that rate is almost impossible to estimate. Land areas are a more likely source of nitrogen oxides, and some reasonable estimates of the rate of loss are possible. One can say that the total amount of fixed nitrogen delivered to the earth by rainfall is of the order of 25 million metric tons per year. My own estimate is that 70 percent of this total is previously fixed nitrogen cycling through the biosphere, and that only 30 percent is freshly fixed by lightning and other atmospheric phenomena.

Another factor that is difficult to estimate is the small but steady loss of nitrogen from the biosphere to sedimentary rocks. Conversely, there is a continuous delivery of new nitrogen to the system by the weathering of igneous rocks in the crust of the earth. The average nitrogen content of igneous rocks, however, is considerably lower than that of sedimentary rocks, and since the quantities of the two kinds of rock are roughly equal, one would expect a net loss of nitrogen from the biosphere through geologic time. Conceivably this loss is just about balanced by the delivery of "juvenile" nitrogen to the atmosphere by volcanic action. The amount of fixed nitrogen reintroduced in this way probably does not exceed two or three million tons per year.

Whereas late-19th-century scientists worried that denitrifying bacteria were exhausting the nitrogen in the soil, we must be concerned today that denitrification may not be keeping pace with nitrogen fixation, considering the large amounts of fixed nitrogen that are being introduced in the biosphere by industrial fixation and the cultivation of legumes. It has become urgent to learn much more about exactly where and under what circumstances denitrification takes place.

We know first of all that denitrification does not normally proceed to any great extent under aerobic conditions. Whenever free oxygen is available, it is energetically advantageous for an organism to use it to oxidize organic compounds rather than to use the oxygen bound in nitrate salts. One can conclude that there must be large areas in the biosphere where conditions are sufficiently anaerobic to strongly favor the denitrification reaction. Such conditions exist wherever the input of organic materials exceeds the input of oxygen for their degradation. Typical areas where the deni-

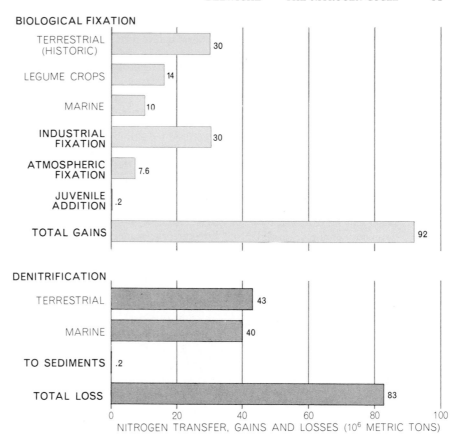

BALANCE SHEET FOR NITROGEN CYCLE, based on the author's estimates, indicates that nitrogen is now being introduced into the biosphere in fixed form at the rate of some 92 million metric tons per year (*colored bars*), whereas the total amount being denitrified and returned to the atmosphere is only about 83 million tons per year. The difference of some nine million tons may represent the rate at which fixed nitrogen is building up in the biosphere: in the soil, in ground-water reservoirs, in rivers and lakes and in the ocean.

ASSOCIATIONS OF TREES AND BACTERIA are important fixers of nitrogen in natural ecosystems. The ginkgo tree (*left*), a gymnosperm, has shown little outward change in millions of years. The alder (*right*), an angiosperm, is common in many parts of the world.

trification process operates close to the surface are the arctic tundra, swamps and similar places where oxygen input is limited. In many other areas where the input of organic material is sizable, however, denitrification is likely to be proceeding at some point below the surface, probably close to the level of the water table.

There are even greater uncertainties regarding the nitrogen cycle in the ocean. It is known that some marine organisms do fix nitrogen, but quantitative information is scanty. A minimum rate of denitrification can be deduced by estimating the amount of nitrate carried into the ocean by rivers. A reasonable estimate is 10 million metric tons per year in the form of nitrates and perhaps twice that amount in the form of organic material, a total of about 30 million tons. Since the transfer of nitrogen into sedi-

ments is slight, one can conclude that, at least before man's intervention in the nitrogen cycle, the ocean was probably capable of denitrifying that amount of fixed nitrogen.

The many blanks in our knowledge of the nitrogen cycle are disturbing when one considers that the amount of nitrogen fixed industrially has been doubling about every six years. If we add to this extra nitrogen the amounts fixed by the cultivation of legumes, it already exceeds (by perhaps 10 percent) the amount of nitrogen fixed in nature. Unless fertilizers and nitrogenous wastes are carefully managed, rivers and lakes can become loaded with the nitrogen carried in runoff waters. In such waterways and in neighboring ground-water systems the nitrogen concentration could, and in some cases already does, exceed the levels acceptable for human

consumption. Under some circumstances bacterial denitrification can be exploited to control the buildup of fixed nitrogen, but much work has to be done to develop successful management techniques.

The problem of nitrogen disposal is aggravated by the nitrogen contained in the organic wastes of a steadily increasing human and domestic-animal population. Ideally this waste nitrogen should be recycled back to the soil, but efficient and acceptable means for doing so remain to be developed. At present it is economically sounder for the farmer to keep adding industrial fertilizers to his crops. The ingenuity that has been used to feed a growing world population will have to be matched quickly by an effort to keep the nitrogen cycle in reasonable balance.

6

THE CARBON CYCLE

BERT BOLIN
September 1970

The main cycle is from carbon dioxide to living matter and back to carbon dioxide. Some of the carbon, however, is removed by a slow epicycle that stores huge inventories in sedimentary rocks

The biosphere contains a complex mixture of carbon compounds in a continuous state of creation, transformation and decomposition. This dynamic state is maintained through the ability of phytoplankton in the sea and plants on land to capture the energy of sunlight and utilize it to transform carbon dioxide (and water) into organic molecules of precise architecture and rich diversity. Chemists and molecular biologists have unraveled many of the intricate processes needed to create the microworld of the living cell. Equally fundamental and no less interesting is the effort to grasp the overall balance and flow of material in the worldwide community of plants and animals that has developed in the few billion years since life began. This is ecology in the broadest sense of the word: the complex interplay between, on the one hand, communities of plants and animals and, on the other, both kinds of community and their nonliving environment.

We now know that the biosphere has not developed in a static inorganic environment. Rather the living world has profoundly altered the primitive lifeless earth, gradually changing the composition of the atmosphere, the sea and the top layers of the solid crust, both on land and under the ocean. Thus a study of the carbon cycle in the biosphere is fundamentally a study of the overall global interactions of living organisms and their physical and chemical environment. To bring order into this world of complex interactions biologists must combine their knowledge with the information available to students of geology, oceanography and meteorology.

The engine for the organic processes that reconstructed the primitive earth is photosynthesis. Regardless of whether it takes place on land or in the sea, it can be summarized by a single reaction: $CO_2 + 2H_2A + light \rightarrow CH_2O + H_2O + 2A + energy$. The formaldehyde molecule CH_2O symbolizes the simplest organic compound; the term "energy" indicates that the reaction stores energy in chemical form. H_2A is commonly water (H_2O), in which case 2A symbolizes the release of free oxygen (O_2). There are, however, bacteria that can use compounds in which A stands for sulfur, for some organic radical or for nothing at all.

Organisms that are able to use carbon dioxide as their sole source of carbon are known as autotrophs. Those that use light energy for reducing carbon dioxide are called phototrophic, and those that use the energy stored in inorganic chemical bonds (for example the bonds of nitrates and sulfates) are called chemolithotrophic. Most organisms, however, require preformed organic molecules for growth; hence they are known as heterotrophs. The nonsulfur bacteria are an unusual group that is both photosynthetic and heterotrophic. Chemoheterotrophic organisms, for example animals, obtain their energy from organic compounds without need for light. An organism may be either aerobic or anaerobic regardless of its source of carbon or energy. Thus some anaerobic chemoheterotrophs can survive in the deep ocean and deep lakes in the total absence of light or free oxygen.

There is more to plant life than the creation of organic compounds by photosynthesis. Plant growth involves a series of chemical processes and transformations that require energy. This energy is obtained by reactions that use the oxygen in the surrounding water and air to unlock the energy that has been stored by photosynthesis. The process, which releases carbon dioxide, is termed respiration. It is a continuous process and is therefore dominant at night, when photosynthesis is shut down.

If one measures the carbon dioxide at various levels above the ground in a forest, one can observe pronounced changes in concentration over a 24-hour period [*see top illustration on page 56*]. The average concentration of carbon dioxide in the atmosphere is about 320 parts per million. When the sun rises, photosynthesis begins and leads to a rapid decrease in the carbon dioxide concentration as leaves (and the needles of conifers) convert carbon dioxide into organic compounds. Toward noon, as the temperature increases and the humidity decreases, the rate of respiration rises and the net consumption of carbon dioxide slowly declines. Minimum values of carbon dioxide 10 to 15 parts per million below the daily average are reached around noon at treetop level. At sunset photosynthesis ceases while respiration continues, with the result that the carbon dioxide concentration close to the ground may exceed 400 parts per million. This high value reflects partly the release of carbon dioxide from the decomposition of organic matter in the soil and partly the tendency of air to stagnate near the ground at night, when there is no solar heating to produce convection currents.

The net productivity, or net rate of fixation, of carbon dioxide varies greatly from one type of vegetation to another. Rapidly growing tropical rain forests annually fix between one kilogram and two kilograms of carbon (in the form of carbon dioxide) per square meter of land surface, which is roughly equal to the amount of carbon dioxide in a column of air extending from the same area of the earth's surface to the top of the atmosphere. The arctic tundra and the nearly barren regions of the desert may fix as little as 1 percent of that amount. The

CARBON LOCKED IN COAL and oil exceeds by a factor of about 50 the amount of carbon in all living organisms. The estimated world reserves of coal alone are on the order of 7,500 billion tons. The photograph on the opposite page shows a sequence of lignite coal seams being strip-mined in Stanton, N.D., by the Western Division of the Consolidation Coal Company. The seam, about two feet thick, is of low quality and is discarded. The second seam from the top, about three feet thick, is marketable, as is the third seam, 10 feet farther down. This seam is really two seams separated by about 10 inches of gray clay. The upper is some 3½ feet thick; the lower is about two feet thick. Twenty-four feet below the bottom of this seam is still another seam (*not shown*) eight feet thick, which is also mined.

forests and cultivated fields of the middle latitudes assimilate between .2 and .4 kilogram per square meter. For the earth as a whole the areas of high productivity are small. A fair estimate is that the land areas of the earth fix into organic compounds 20 to 30 billion net metric tons of carbon per year. There is considerable uncertainty in this figure; published estimates range from 10 to 100 billion tons.

The amount of carbon in the form of carbon dioxide consumed annually by phytoplankton in the oceans is perhaps 40 billion tons, or roughly the same as the gross assimilation of carbon dioxide by land vegetation. Both the carbon dioxide consumed and the oxygen released are largely in the form of gas dissolved near the ocean surface. Therefore most of the carbon cycle in the sea is self-con-

tained: the released oxygen is consumed by sea animals, and their ultimate decomposition releases carbon dioxide back into solution. As we shall see, however, there is a dynamic exchange of carbon dioxide (and oxygen) between the atmosphere and the sea, brought about by the action of the wind and waves. At any given moment the amount of carbon dioxide dissolved in the surface layers of the sea is in close equilibrium with the concentration of carbon dioxide in the atmosphere as a whole.

The carbon fixed by photosynthesis on land is sooner or later returned to the atmosphere by the decomposition of dead organic matter. Leaves and litter fall to the ground and are oxidized by a series of complicated processes in the soil. We can get an approximate idea of the rate at which organic matter in the soil is

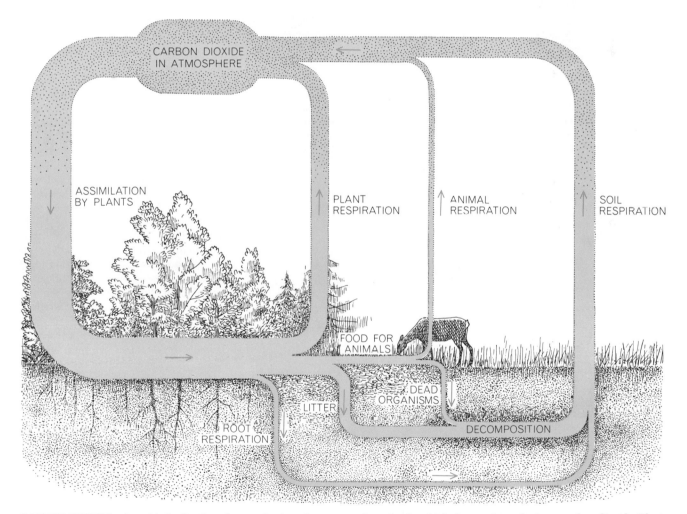

CARBON CYCLE begins with the fixation of atmospheric carbon dioxide by the process of photosynthesis, conducted by plants and certain microorganisms. In this process carbon dioxide and water react to form carbohydrates, with the simultaneous release of free oxygen, which enters the atmosphere. Some of the carbohydrate is directly consumed to supply the plant with energy; the carbon dioxide so generated is released either through the plant's leaves or through its roots. Part of the carbon fixed by plants is consumed by animals, which also respire and release carbon dioxide. Plants and animals die and are ultimately decomposed by microorganisms in the soil; the carbon in their tissues is oxidized to carbon dioxide and returns to the atmosphere. The widths of the pathways are roughly proportional to the quantities involved. A similar carbon cycle takes place within the sea. There is still no general agreement as to which of the two cycles is larger. The author's estimates of the quantities involved appear in the flow chart on page 59.

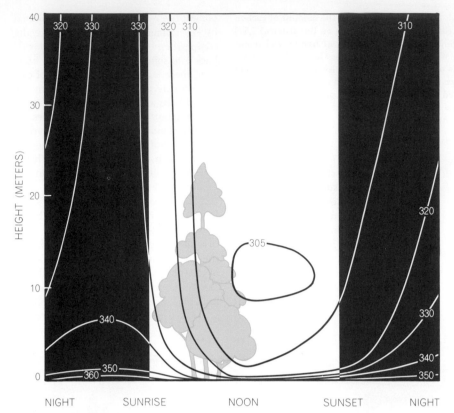

VERTICAL DISTRIBUTION OF CARBON DIOXIDE in the air around a forest varies with time of day. At night, when photosynthesis is shut off, respiration from the soil can raise the carbon dioxide at ground level to as much as 400 parts per million (ppm). By noon, owing to photosynthetic uptake, the concentration at treetop level can drop to 305 ppm.

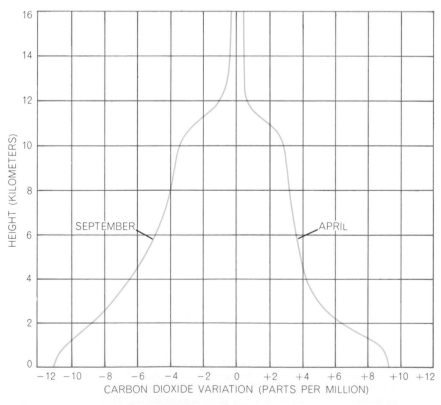

SEASONAL VARIATIONS in the carbon dioxide content of the atmosphere reach a maximum in September and April for the region north of 30 degrees north latitude. The departure from a mean value of about 320 ppm varies with altitude as shown by these two curves.

being transformed by measuring its content of the radioactive isotope carbon 14. At the time carbon is fixed by photosynthesis its ratio of carbon 14 to the nonradioactive isotope carbon 12 is the same as the ratio in the atmosphere (except for a constant fractionation factor), but thereafter the carbon 14 decays and becomes less abundant with respect to the carbon 12. Measurements of this ratio yield rates for the oxidation of organic matter in the soil ranging from decades in tropical soils to several hundred years in boreal forests.

In addition to the daily variations of carbon dioxide in the air there is a marked annual variation, at least in the Northern Hemisphere. As spring comes to northern regions the consumption of carbon dioxide by plants greatly exceeds the return from the soil. The increased withdrawal of carbon dioxide can be measured all the way up to the lower stratosphere. A marked decrease in the atmospheric content of carbon dioxide occurs during the spring. From April to September the atmosphere north of 30 degrees north latitude loses nearly 3 percent of its carbon dioxide content, which is equivalent to about four billion tons of carbon [see bottom illustration at left]. Since the decay processes in the soil go on simultaneously, the net withdrawal of four billion tons implies an annual gross fixation of carbon in these latitudes of at least five or six billion tons. This amounts to about a fourth of the annual terrestrial productivity referred to above (20 to 30 billion tons), which was based on a survey of carbon fixation. In this global survey the estimated contribution from the Northern Hemisphere, where plant growth shows a marked seasonal variation, constituted about 25 percent of the total tonnage. Thus two independent estimates of worldwide carbon fixation on land show a quite satisfactory agreement.

The forests of the world not only are the main carbon dioxide consumers on land; they also represent the main reservoir of biologically fixed carbon (except for fossil fuels, which have been largely removed from the carbon cycle save for the amount reintroduced by man's burning of it). The forests contain between 400 and 500 billion tons of carbon, or roughly two-thirds of the amount present as carbon dioxide in the atmosphere (700 billion tons). The figure for forests can be estimated only approximately. The average age of a tree can be assumed to be about 30 years, which implies that about 15 billion tons of carbon in the form of carbon dioxide is annually transformed into wood, which seems reasonable in comparison with a total annual assimilation of 20 to 30 billion tons.

The pattern of carbon circulation in the sea is quite different from the pattern on land. The productivity of the soil is mostly limited by the availability of fresh water and phosphorus, and only to a degree by the availability of other nutrients in the soil. In the oceans the overriding limitation is the availability of inorganic substances. The phytoplankton require not only plentiful supplies of phosphorus and nitrogen but also trace amounts of various metals, notably iron.

The competition for food in the sea is so keen that organisms have gradually developed the ability to absorb essential minerals even when these nutrients are available only in very low concentration. As a result high concentrations of nutrients are rarely found in surface waters, where solar radiation makes it possible for photosynthetic organisms to exist. If an ocean area is uncommonly productive, one can be sure that nutrients are supplied from deeper layers. (In limited areas they are supplied by the wastes of human activities.) The most productive waters in the world are therefore near the Antarctic continent, where the deep waters of the surrounding oceans well up and mix with the surface layers. There are similar upwellings along the coast of Chile, in the vicinity of Japan and in the Gulf Stream. In such regions fish are abundant and the maximum annual fixation of carbon approaches .3 kilogram per square meter. In the "desert" areas of the oceans, such as the open seas of subtropical latitudes, the fixation rate may be less than a tenth of that value. In the Tropics warm surface layers are usually effective in blocking the vertical water exchange needed to carry nutrients up from below.

Phytoplankton, the primary fixers of carbon dioxide in the sea, are eaten by the zooplankton and other tiny animals. These organisms in turn provide food for the larger animals. The major part of the oceanic biomass, however, consists of microorganisms. Since the lifetime of such organisms is measured in weeks, or at most in months, their total mass can never accumulate appreciably. When microorganisms die, they quickly disintegrate as they sink to deeper layers. Soon most of what was once living tissue has become dissolved organic matter.

A small fraction of the organic particulate matter escapes oxidation and settles into the ocean depths. There it

profoundly influences the abundance of chemical substances because (except in special regions) the deep layers exchange water with the surface layers very slowly. The enrichment of the deep layers goes hand in hand with a depletion of oxygen. There also appears to be an increase in carbon dioxide (in the form of carbonate and bicarbonate ions) in the ocean depths. The overall distribution of carbon dioxide, oxygen and various minor constituents in the sea reflects a balance between the marine life and its chemical milieu in the surface layers and the slow transport of substances by the general circulation of the ocean. The net effect is to prevent the ocean from becoming saturated with oxygen and to enrich the deeper strata with carbonate and bicarbonate ions.

The particular state in which we find the oceans today could well be quite different if the mechanisms for the exchange of water between the surface layers and the deep ones were either more intense or less so. The present state is determined primarily by the sinking of cold water in the polar regions, particularly the Antarctic. In these regions the water is also slightly saltier, and therefore still denser, because some of it has been frozen out in floating ice. If the climate of the earth were different, the distribution of carbon dioxide, oxygen and minerals might also be quite different. If the difference were large enough, oxygen might completely vanish from the ocean depths, leaving them to be populated only by chemibarotrophic bacteria. (This is now the case in the depths of the Black Sea.)

The time required to establish a new equilibrium in the ocean is determined

by the slowest link in the chain of processes that has been described. This link is the oceanic circulation; it seems to take at least 1,000 years for the water in the deepest basins to be completely replaced. One can imagine other conditions of circulation in which the oceans would interact differently with sediments and rocks, producing a balance of substances that one can only guess at.

So far we have been concerned only with the basic biological and ecological processes that provide the mechanisms for circulating carbon through living organisms. Plants on land, with lifetimes measured in years, and phytoplankton in the sea, with lifetimes measured in weeks, are merely the innermost wheels in a biogeochemical machine that embraces the entire earth and that retains important characteristics over much longer time periods. In order to understand such interactions we shall need some rough estimates of the size of the various carbon reservoirs involved and the nature of their contents [*see illustration on page 59*]. In the context of the present argument the large uncertainties in such estimates are of little significance.

Only a few tenths of a percent of the immense mass of carbon at or near the surface of the earth (on the order of 20×10^{15} tons) is in rapid circulation in the biosphere, which includes the atmosphere, the hydrosphere, the upper portions of the earth's crust and the biomass itself. The overwhelming bulk of near-surface carbon consists of inorganic deposits (chiefly carbonates) and organic fossil deposits (chiefly oil shale, coal and petroleum) that required hundreds of

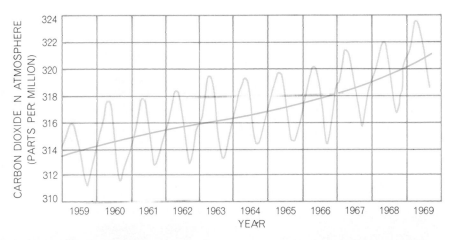

LONG-TERM VARIATIONS in the carbon dioxide content of the atmosphere have been followed at the Mauna Loa Observatory in Hawaii by the Scripps Institution of Oceanography. The sawtooth curve indicates the month-to-month change in concentration since January, 1959. The oscillations reflect seasonal variations in the rate of photosynthesis, as depicted in the bottom illustration on the opposite page. The smooth curve shows the trend.

OIL SHALE is one of the principal sedimentary forms in which carbon has been deposited over geologic time. This photograph, taken at Anvil Points, Colo., shows a section of the Green River Formation, which extends through Colorado, Utah and Wyoming. The formation is estimated to contain the equivalent of more than a trillion barrels of oil in seams containing more than 10 barrels of oil per ton of rock. Of this some 80 billion barrels is considered recoverable. The shale seams are up to 130 feet thick.

WHITE CLIFFS OF DOVER consist of almost pure calcium carbonate, representing the skeletons of phytoplankton that settled to the bottom of the sea over a period of millions of years more than 70 million years ago. The worldwide deposits of limestone, oil shale and other carbon-containing sediments are by far the largest repository of carbon: an estimated 20 quadrillion (10^{15}) tons.

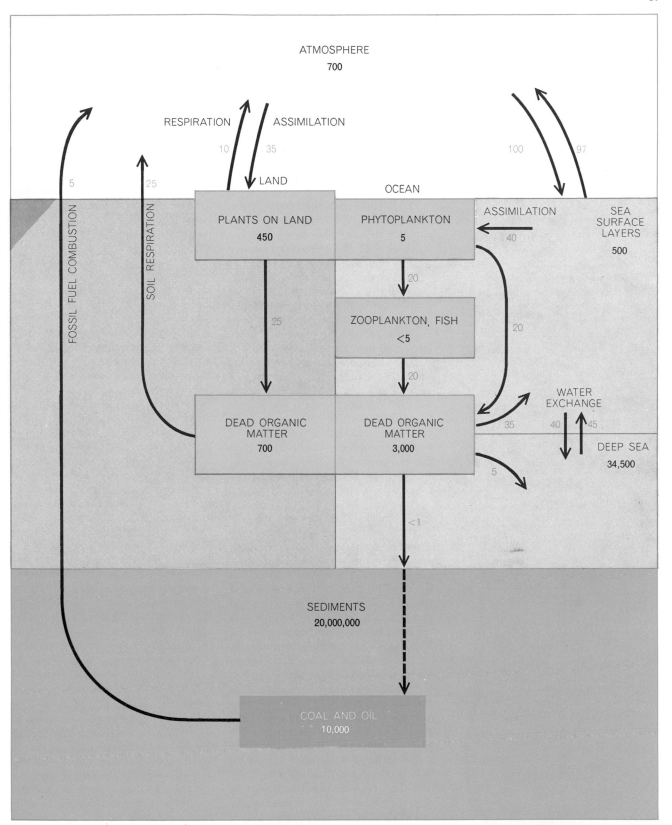

ATMOSPHERE
700

RESPIRATION   ASSIMILATION

10   35   100   97

LAND   OCEAN   ASSIMILATION   SEA
SURFACE
LAYERS

PLANTS ON LAND   PHYTOPLANKTON   40
450   5   500

FOSSIL FUEL COMBUSTION   5   SOIL RESPIRATION   25

20   20

ZOOPLANKTON, FISH
<5

25   20

WATER
EXCHANGE

DEAD ORGANIC
MATTER   DEAD ORGANIC
MATTER   35   40   45
700   3,000   DEEP SEA
34,500

5

<1

SEDIMENTS
20,000,000

COAL AND OIL
10,000

CARBON CIRCULATION IN BIOSPHERE involves two quite distinct cycles, one on land and one in the sea, that are dynamically connected at the interface between the ocean and the atmosphere. The carbon cycle in the sea is essentially self-contained in that phytoplankton assimilate the carbon dioxide dissolved in seawater and release oxygen back into solution. Zooplankton and fish consume the carbon fixed by the phytoplankton, using the dissolved oxygen for respiration. Eventually the decomposition of organic matter replaces the carbon dioxide assimilated by the phytoplank- ton. All quantities are in billions of metric tons. It will be seen that the combustion of fossil fuels at the rate of about five billion tons per year is sufficient to increase the carbon dioxide in the atmosphere by about .7 percent, equivalent to adding some two parts per million to the existing 320 ppm. Since the observed an- nual increase is only about .7 ppm, it appears that two-thirds of the carbon dioxide released from fossil fuels is quickly removed from the atmosphere, going either into the oceans or adding to the total mass of terrestrial plants. The estimated tonnages are the author's.

millions of years to reach their present magnitude. Over time intervals as brief as those of which we have been speaking—up to 1,000 years for the deep-ocean circulation—the accretion of such deposits is negligible. We may therefore consider the life processes on land and in the sea as the inner wheels that spin at comparatively high velocity in the carbon-circulating machine. They are coupled by a very low gear to more majestic processes that account for the overall circulation of carbon in its various geologic and oceanic forms.

We now know that the two great systems, the atmosphere and the ocean, are closely coupled to each other through the transfer of carbon dioxide across the surface of the oceans. The rate of exchange has recently been estimated by measuring the rate at which the radio-

GIANT FERN of the genus *Pecopteris*, which fixed atmospheric carbon dioxide 300 million years ago, left the imprint of this frond in a thin layer of shale just above a coal seam in Illinois. The specimen is in the collection of the Smithsonian Institution.

active isotope carbon 14 produced by the testing of nuclear weapons has disappeared from the atmosphere. The neutrons released in such tests form carbon 14 by reacting with the nitrogen 14 of the atmosphere. In this reaction a nitrogen atom ($_7N^{14}$) captures a neutron and subsequently releases a proton, yielding $_6C^{14}$. (The subscript before the letter represents the number of protons in the nucleus; the superscript after the letter indicates the sum of protons and neutrons.)

The last major atmospheric tests were conducted in 1963. Sampling at various altitudes and latitudes shows that the constituents of the atmosphere became rather well mixed over a period of a few years. The decline of carbon 14, however, was found to be rapid; it can be explained only by assuming an exchange of atmospheric carbon dioxide, enriched in carbon 14, with the reservoir of much less radioactive carbon dioxide in the sea. The measurements indicate that the characteristic time for the residence of carbon dioxide in the atmosphere before the gas is dissolved in the sea is between five and 10 years. In other words, every year something like 100 billion tons of atmospheric carbon dioxide dissolves in the sea and is replaced by a nearly equivalent amount of oceanic carbon dioxide.

Since around 1850 man has inadvertently been conducting a global geochemical experiment by burning large amounts of fossil fuel and thereby returning to the atmosphere carbon that was fixed by photosynthesis millions of years ago. Currently between five and six billion tons of fossil carbon per year are being released into the atmosphere. This would be enough to increase the amount of carbon dioxide in the air by 2.3 parts per million per year if the carbon dioxide were uniformly distributed and not removed. Within the past century the carbon dioxide content of the atmosphere has risen from some 290 parts per million to 320, with more than a fifth of the rise occurring in just the past decade [*see illustration on page 57*]. The total increase accounts for only slightly more than a third of the carbon dioxide (some 200 billion tons in all) released from fossil fuels. Although most of the remaining two-thirds has presumably gone into the oceans, a significant fraction may well have increased the total amount of vegetation on land. Laboratory studies show that plants grow faster when the surrounding air is enriched in carbon dioxide. Thus it is possible that man is fertilizing fields and

forests by burning coal, oil and natural gas. The biomass on land may have increased by as much as 15 billion tons in the past century. There is, however, little concrete evidence for such an increase.

Man has of course been changing his environment in other ways. Over the past century large areas covered with forest have been cleared and turned to agriculture. In such areas the character of soil respiration has undoubtedly changed, producing effects that might have been detectable in the atmospheric content of carbon dioxide if it had not been for the simultaneous increase in the burning of fossil fuels. In any case the dynamic equilibrium among the major carbon dioxide reservoirs in the biomass, the atmosphere, the hydrosphere and the soil has been disturbed, and it can be said that they are in a period of transition. Since even the most rapid processes of adjustment among the reservoirs take decades, new equilibriums are far from being established. Gradually the deep oceans become involved; their turnover time of about 1,000 years and their rate of exchange with bottom sediments control the ultimate partitioning of carbon.

Meanwhile human activities continue to change explosively. The acceleration in the consumption of fossil fuels implies that the amount of carbon dioxide in the atmosphere will keep climbing from its present value of 320 parts per million to between 375 and 400 parts per million by the year 2000, in spite of anticipated large removals of carbon dioxide by land vegetation and the ocean reservoir [*see illustrations on next page*]. A fundamental question is: What will happen over the next 100 or 1,000 years? Clearly the exponential changes cannot continue.

If we extend the time scale with which we are viewing the carbon cycle by several orders of magnitude, to hundreds of thousands or millions of years, we can anticipate large-scale exchanges between organic carbon on land and carbonates of biological origin in the sea. We do know that there have been massive exchanges in the remote past. Any discussion of these past events and their implications for the future, however, must necessarily be qualitative and uncertain.

Although the plants on land have probably played an important role in the deposition of organic compounds in the soil, the oceans have undoubtedly acted as the main regulator. The amount of carbon dioxide in the atmosphere is essentially determined by the partial pressure of carbon dioxide dissolved in the sea. Over a period of, say, 100,000 years the leaching of calcium carbonates from land areas tends to increase the amount of carbon dioxide in the sea, but at the same time a converse mechanism—the precipitation and deposition of oceanic carbonates—tends to reduce the amount of carbon dioxide in solution. Thus the two mechanisms tend to cancel each other.

Over still longer periods of time—millions or tens of millions of years—the concentrations of carbonate and bicarbonate ions in the sea are probably buffered still further by reactions involving potassium, silicon and aluminum, which are slowly weathered from rocks and carried into the sea. The net effect is to stabilize the carbon dioxide content of the oceans and hence the carbon dioxide content of the atmosphere. Therefore it appears that the carbon dioxide environment, on which the biosphere fundamentally depends, may have been fairly constant right up to the time, barely a moment ago geologically speaking, when man's consumption of fossil fuels began to change the carbon dioxide content of the atmosphere.

The illustration on page 59 represents an attempt to synthetize into a single picture the circulation of carbon in nature, particularly in the biosphere. In addition to the values for inventories and transfers already mentioned, the flow chart contains other quantities for which the evidence is still meager. They have been included not only to balance the books but also to suggest where further investigation might be profitable. This may be the principal value of such an exercise. Such a flow chart also provides a semiquantitative model that enables one to begin to discuss how the global carbon system reacts to disturbances. A good model should of course include inventories and pathways for all the elements that play a significant role in biological processes.

The greatest disturbances of which we are aware are those now being introduced by man himself. Since his tampering with the biological and geochemical balances may ultimately prove injurious —even fatal—to himself, he must understand them much better than he does today. The story of the circulation of carbon in nature teaches us that we cannot control the global balances. Therefore we had better leave them close to the natural state that existed until the beginning of the Industrial Revolution. Out of a simple realization of this necessity may come a new industrial revolution.

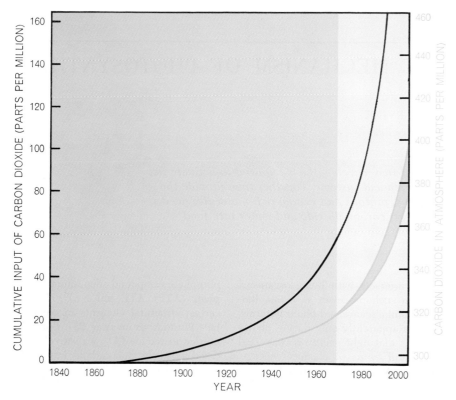

INCREASE IN ATMOSPHERIC CARBON DIOXIDE since 1860 is shown by the lower curve, with a projection to the year 2000. The upper curve shows the cumulative input of carbon dioxide. The difference between the two curves represents the amount of carbon dioxide removed by the ocean or by additions to the total biomass of vegetation on land.

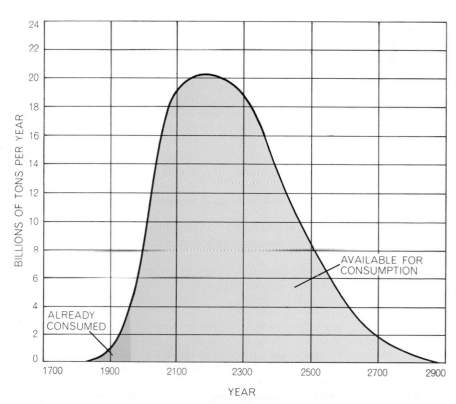

POSSIBLE CONSUMPTION PATTERN OF FOSSIL FUELS was projected by Harrison Brown in the mid-1950's. Here the fuel consumed is updated to 1960. If a third of the carbon dioxide produced by burning it all were to remain in the atmosphere, the carbon dioxide level would rise from 320 ppm today to about 1,500 ppm over the next several centuries.

7

THE MECHANISM OF PHOTOSYNTHESIS

R. P. LEVINE

December 1969

Light of two wavelengths is required to activate two photochemical systems. Together they provide the electrons, protons and energy-rich molecules needed to convert carbon dioxide and water into food

Light interacts with living organisms in such processes as vision, bioluminescence and photosynthesis, but it is apparently only during photosynthesis that light energy is converted into useful forms of chemical energy. That energy, in turn, is used to build up complex molecules—notably carbohydrates—that animals require as food. Photosynthetic organisms also provide most of the oxygen in the atmosphere, and the evolution of animals was certainly dependent on the existence of oxygen-evolving microorganisms in the primitive seas.

The study of photosynthesis spans the disciplines of photophysics, photochemistry, biochemistry and physiology. In recent decades such studies have revealed many remarkable aspects of the photosynthetic process. The simple equation that summarizes the process has been known since the 19th century: water (H_2O) plus carbon dioxide (CO_2) yields some form of carbohydrate (represented by CH_2O) and oxygen (O_2). The reaction is driven by light energy. Photons are first absorbed by chlorophyll and other photosynthetic pigments. The red and blue-green algae, for example, contain pigments called phycobilins in addition to chlorophyll [*see illustration on page 64*].

Once the light energy is absorbed, it is used for two purposes. First, it is used to generate what a chemist calls "reducing power." Reduction involves the addition of electrons or the removal of protons, or both. Molecules that are rich in reducing power can transfer electrons to more oxidized molecules. The reducing agent produced by photosynthesis is NADPH, the reduced form of nicotinamide adenine dinucleotide diphosphate (NADP). Second, the light energy becomes converted into the energy-rich phosphate compound adenosine triphosphate (ATP). ATP and NADPH have certain structural elements in common [*see illustrations on page 65*]. Both are needed to reduce CO_2, a relatively oxidized molecule, into carbohydrate. The overall balance sheet for photosynthesis shows that three molecules of ATP and two molecules of NADPH are required for each molecule of CO_2 reduced.

In most algae and in higher plants photosynthesis occurs in the intricate, membrane-filled structure known as the chloroplast [*see illustration on opposite page*]. Within the chloroplast light energy is trapped and the rapid photophysical and photochemical reactions take place that generate the ATP and NADPH to be used in the more leisurely biochemical process of reducing carbon dioxide to carbohydrate. Once ATP and NADPH have been formed they are released into the nonmembranous, or soluble, phase of the chloroplast; there the fixation of carbon dioxide can proceed in the absence of light with the assistance of a number of soluble enzymes.

The Absorption of Light Energy

When chlorophyll or one of the other photosynthetic pigments absorbs photons, the pigment passes from its lowest energy state, or ground state, to a higher energy state. The excited state is not stable, and the pigment can return to the ground state within 10^{-9} second. If in that brief period the energy is not used for the generation of ATP or NADPH, it can be dissipated as fluorescent light. Since 100 percent efficiency is never achieved in biological systems, fluorescence is always observed during photosynthesis; the fluorescence is at a longer —hence less energetic—wavelength than the wavelength originally absorbed. As we shall see, fluorescence has been useful to the investigator of photosynthesis.

Each of the photosynthetic pigments has its characteristic absorption spectrum: it absorbs more or less light at different wavelengths depending on its molecular structure. For example, the two chlorophylls designated *a* and *b* have major and distinctive absorption bands in the blue and red regions of the spectrum [*see top illustration on page 66*]. When studied in isolation outside the chloroplast, each pigment also has a characteristic fluorescence spectrum. Inside the chloroplast, however, chlorophyll *b* fluorescence is never detected, even when the incident light is of a wavelength known to be absorbed by that chlorophyll. Similarly, fluorescence is never observed from the carotenoid pigments and the phycobilins. Only one of the pigments fluoresces naturally inside the chloroplast: chlorophyll *a*.

This surprising phenomenon has now been explained, largely through the work of Louis N. M. Duysens of the Netherlands. He showed that chlorophyll *b*, the carotenoids and the phycobilins do not participate directly in photosynthesis but rather act only as "antennas" to help gather light energy. When they absorb energy and become excited, they transfer their excitation energy to chlorophyll *a*. Only chlorophyll *a* is actively involved in the subsequent reactions of photosynthesis; when its energy cannot be used for photosynthesis, it dissipates its excitation energy as fluorescence.

The mechanism for the transfer of excitation energy between pigment molecules in photosynthesis is not clearly understood, but a process called inductive resonance is one possibility. If an excited molecule is close enough to an unexcited one (say within 30 angstroms), it can dissipate its energy by inducing an excited state in the neighboring molecule. In this way energy can pass from chlorophyll *b* to chlorophyll *a*. The reverse process is not possible, however, because to become excited chlorophyll *b* requires more excitation energy than

an excited chlorophyll *a* can provide.

Ultimately the energy of excitation reaches a photosynthetic reaction center, where it is transferred to a special long-wavelength form of chlorophyll *a*. Because this pigment absorbs at a longer wavelength, and hence at a lower ener-

gy, than the surrounding pigment molecule, it can be considered a kind of energy sink. The transfer of excitation energy from an excited molecule of normal chlorophyll *a* to such a special chlorophyll *a* molecule probably takes place within 10^{-12} second, which is 1,000 times

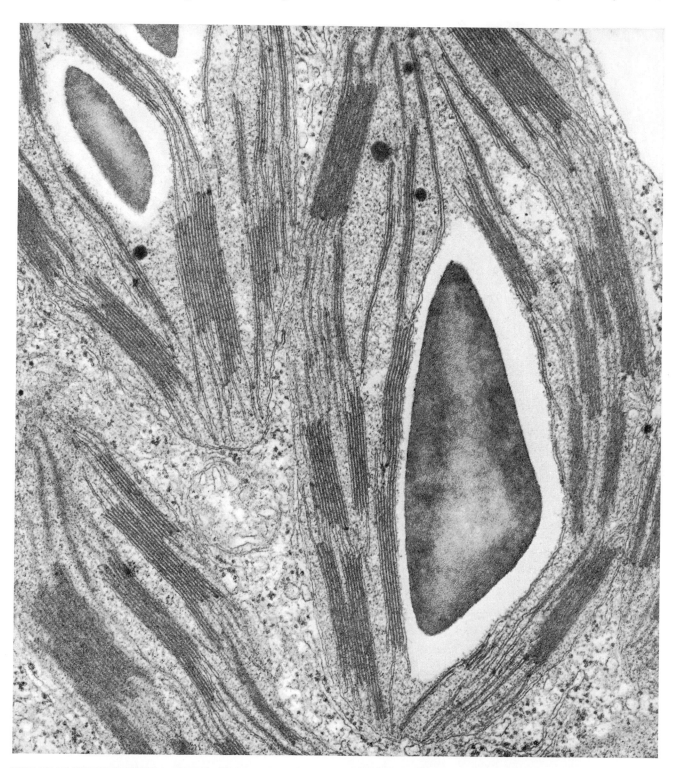

SITE OF PHOTOSYNTHESIS is the organelle known as a chloroplast, present in the cells of all higher plants and most algae. This electron micrograph made by Peter Hepler of Harvard University shows portions of three chloroplasts. Photosynthesis takes place inside the dark membranes that lie in long parallel bundles. The large triangular object in the chloroplast at the right is a kernel of starch, produced by photosynthesis. The chloroplast at the upper left contains two kernels. The magnification is 58,000 diameters.

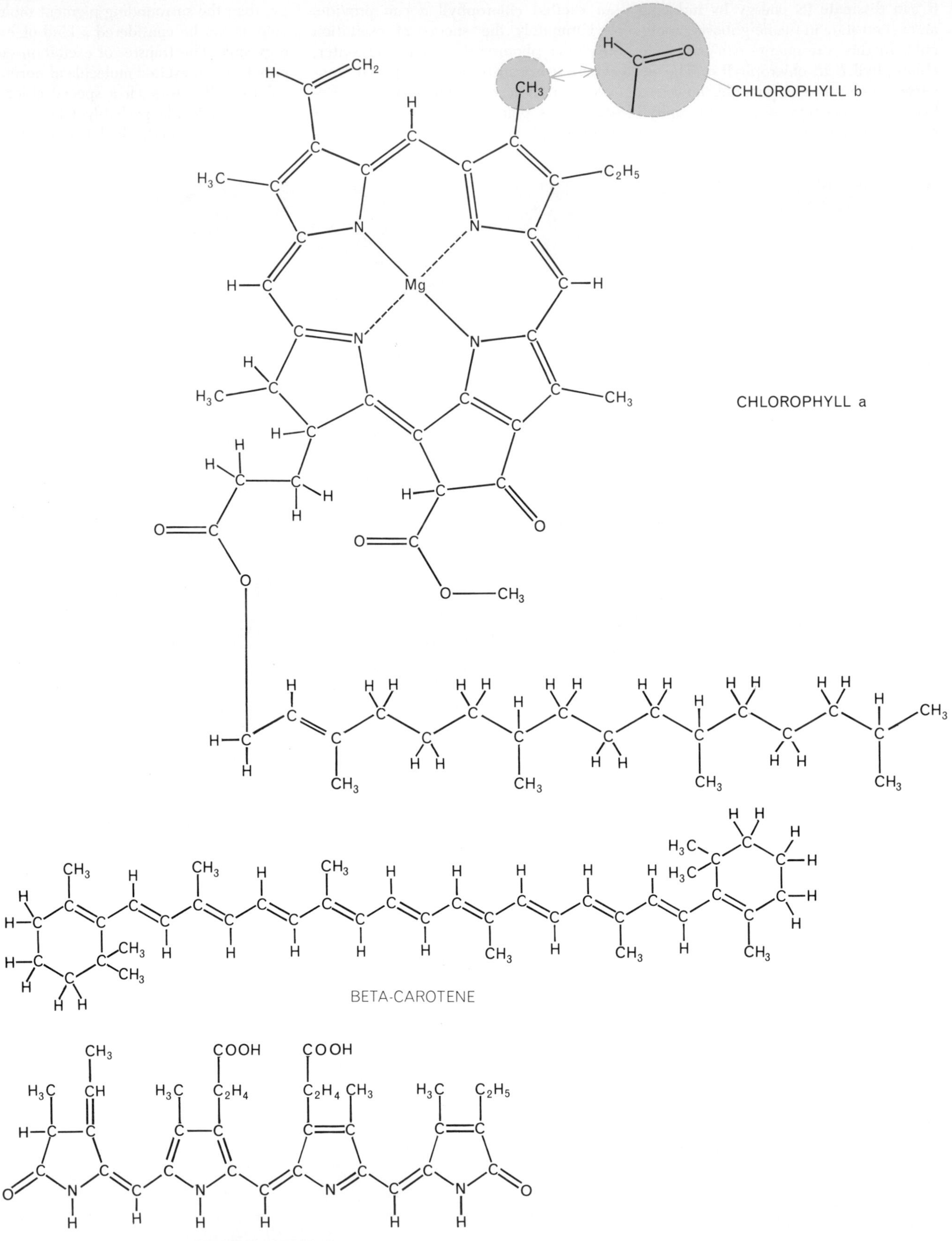

CHLOROPHYLL b

CHLOROPHYLL a

BETA-CAROTENE

PHYCOCYANOBILIN

PHOTOSYNTHETIC PIGMENTS have the ability to capture photons and convert their energy into molecular excitation energy. Chlorophyll *a* is the pigment found in algae and in the leaves of higher plants. Chlorophyll *b* has the same structure except that a –CHO group replaces a –CH₃ group in one corner of the porphyrin ring. Beta-carotene is another photosynthetic pigment present in many higher plants. Red and blue-green algae contain still a third class of pigments, the phycobilins, of which phycocyanobilin is one.

faster than the time taken for the "waste" energy of chlorophyll *a* to emerge as fluorescence. Thus there is ample time for an excited chlorophyll molecule to disperse its energy in a chemically useful way.

The First Chemical Steps

Once light energy has been relayed to the special chlorophyll *a* molecule, the energy sink in the reaction center, the chemistry begins: the excitation energy must be used to form an oxidant and a reductant. The oxidant must be capable of oxidizing water, that is, capable of splitting the water molecule into free oxygen, protons and electrons. (Actually two molecules of H_2O are split into one molecule of O_2 plus four electrons and four protons.) The reductant must accept the reducing equivalents (electrons and protons) that arise from the oxidation of water. Ultimately these equivalents will be used in the reduction of carbon dioxide. The oxidant and the reductant must be formed within the very short lifetime of the excited state of chlorophyll *a*. How this comes about constitutes one of the biggest gaps in our knowledge of photosynthesis, and a great deal of what follows must be conjecture.

One simple way to visualize the initiation of the first chemical steps of photosynthesis is to imagine the existence of an electron-donor molecule *D* and an electron-acceptor molecule *A*. The donor in the oxidized form (D^+) will oxidize water and the acceptor in reduced form (A^-) will ultimately transfer its reducing equivalents to NADP, converting it to NADPH.

A simple model of the primary reaction sequence involving the donor, the acceptor and an excited molecule of chlorophyll *a* is shown in the top illustration on page 67. In this sequence the chlorophyll *a* in the reaction center is raised to an excited state by receiving excitation energy from surrounding pigment molecules. Each reaction-center chlorophyll molecule is in close association with the donor and acceptor molecules in the membrane of the chloroplast. When the chlorophyll returns to the ground state, the release of the excitation energy is sufficient to extract an electron from the donor molecule, thereby oxidizing it to D^+, and to transfer this electron to the acceptor, thus reducing it to A^-. Such charge-transfer processes are known to operate in nonbiological systems where the organic molecules involved have properties similar to those involved in photosynthesis.

ADENOSINE TRIPHOSPHATE, or ATP, is produced from adenosine diphosphate (ADP), with the energy collected by the photosynthetic pigments. The wavy lines are links to energy-rich phosphate groups. If the last group (*color*) is removed, ATP becomes ADP. In the process ATP supplies energy for converting carbon dioxide into carbohydrates.

NICOTINAMIDE ADENINE DINUCLEOTIDE DIPHOSPHATE, or NADP, is reduced to NADPH during photosynthesis. NADP becomes NADPH by the addition of two hydrogen atoms. One binds directly to the molecule while the other loses its electron and is released as a proton (H^+). NADPH supplies "reducing power" for fixation of carbon dioxide.

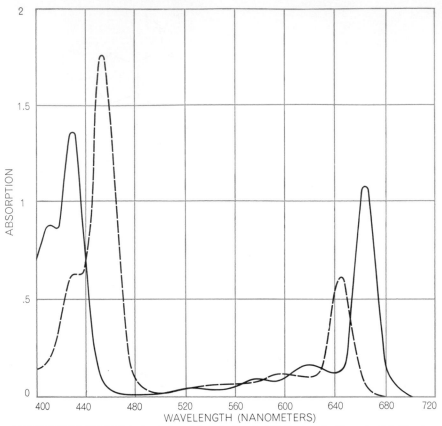

ABSORPTION SPECTRA show that chlorophyll *a* (*solid line*) and chlorophyll *b* (*broken line*) strongly absorb blue and far-red light. The green, yellow and orange wavelengths lying between the peaks are reflected and give both pigments their familiar green color.

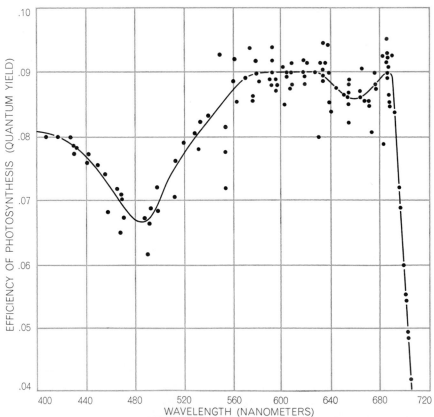

EFFICIENCY OF LIGHT ABSORPTION in the alga chlorella was determined by Robert Emerson of the University of Illinois. Efficiency of photosynthesis falls off sharply in the far-red region beyond 680 nanometers even though chlorophyll *a* still absorbs at that wavelength. If light of shorter wavelength is added to far-red light, efficiency rises sharply.

Regardless of how the charge transfer may be accomplished in the chloroplast, its net effect is to separate oxidizing and reducing equivalents. Very little is known about how the hypothetical D^+ participates in the oxidation of water with the concomitant evolution of oxygen. Much more is known about the transfer of reducing equivalents from the hypothetical A^- to NADP. With this step the mechanism of photosynthesis moves from a high-speed photochemical phase to a slower biochemical phase in which electrons are transported through a series of reactions, ultimately to yield NADPH and ATP.

The Biochemical Phase

Our understanding of the biochemical phase of photosynthesis owes much to investigations showing that two light re-actions (and not one, as has been tacitly assumed so far) take place in the photo-synthetic process used by algae and higher plants. Two experiments set the stage for this discovery. The first pro-vided measurements of the rate of pho-tosynthesis at different wavelengths of light over the range absorbed by the photosynthetic pigments. The result is a curve showing how the quantum effi-ciency of the process varies at different wavelengths [*see bottom illustration at left*]. The curve reveals a curious fact: in the far-red region, beyond a wavelength of 680 nanometers, the efficiency of photosynthesis falls rapidly to zero even though the pigments still absorb light.

This surprising result led to the second set of experiments, reported in 1956 by Robert Emerson and his colleagues at the University of Illinois. They found that although photosynthesis is very in-efficient at wavelengths greater than 680 nanometers, it can be enhanced by add-ing light of a shorter wavelength, 650 nanometers for example. Moreover, the rate of photosynthesis in the presence of both wavelengths is greater than the sum of the rates obtained when the two wavelengths are supplied separately. This phenomenon, now known as the Emerson enhancement effect, can be ex-plained if photosynthesis is assumed to require two light-driven reactions, both of which can be driven by light of less than 680 nanometers but only one by light of longer wavelength.

These two sets of experiments marked the beginning of an exciting period in the effort to understand the mechanism of photosynthesis. They gave rise to a provocative hypothesis and to several revealing lines of research. The hypothesis was provided by Robert Hill and Fay Bendall of the University of Cambridge. They proposed a scheme showing how electrons could be transported along a biochemical chain in which two separate reactions are triggered by light. Before describing the Hill-Bendall scheme I should briefly touch on some characteristics of electron transport.

One must distinguish first between a transfer in which electrons go *with* an electrochemical gradient (the easy direction) and a transfer in which the electrons go *against* that gradient (the hard direction). Electron-donor and electron-acceptor molecules can be characterized by the quantity called oxidation-reduction potential, which can be positive or negative and is usually expressed in volts. Electrons can be transferred from donors that have a more negative potential to acceptors that have a more positive potential without any input of energy. In fact, when electron transfer takes place along this gradient, energy is released; the greater the gap in potential between donor and acceptor, the greater the yield of energy. To transfer electrons against the electrochemical gradient, on the other hand, requires an input of energy. The greater the gap between the donor and the acceptor, the greater the energy required.

In the mitochondria of both plant and animal cells energy, in the form of ATP, is generated as a consequence of electron transport down an electrochemical gradient between a series of electron-donors and -acceptors called cytochromes. At least some of the electron-transport steps in photosynthesis, however, must go against an electrochemical gradient because the oxidation-reduction potential of water (the primary electron-donor) is +.8 volt whereas that of NADP (the terminal acceptor) is −.3 volt.

The Two Photochemical Systems

Hill and his co-workers had earlier identified and characterized a number of cytochromes found in the chloroplast; Hill and Bendall saw that ATP might be generated in photosynthesis if advantage were taken of the difference in oxidation-reduction potential between two of these cytochromes. One of them, a *b*-type cytochrome, has a potential close to zero and the other, a *c*-type cytochrome, has a potential of about +.35 volt. In the Hill-Bendall scheme, therefore, two light reactions provide the energy to go *against* the electrochemical

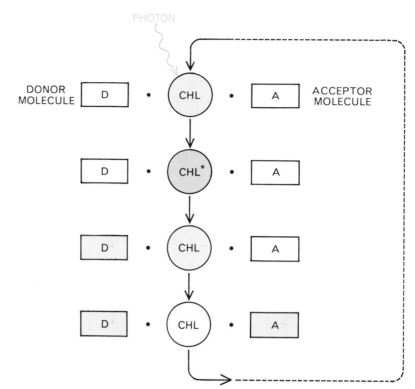

FIRST CHEMICAL STEPS in photosynthesis involve an electron-donor molecule (D) and an electron-acceptor molecule (A) in close association with a special chlorophyll (Chl) in the reaction center. An incoming photon can raise the chlorophyll to an excited state (Chl^*). When the excited chlorophyll returns to the ground state in less than 10^{-9} second, the energy released extracts an electron from D, oxidizing it to D^+, and transfers the electron to A, reducing it to A^-. Later D^+ oxidizes water and A^- reduces NADP to NADPH.

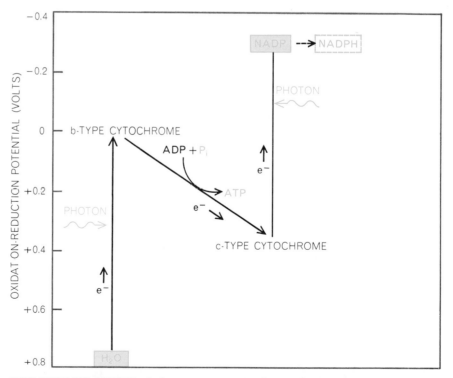

HILL-BENDALL MODEL of electron transport in photosynthesis suggests how electrons (e^-) removed from water can be boosted against an electrochemical gradient, finally reaching NADP. With the aid of protons, also provided by water, NADP is converted to NADPH. A change in an upward direction requires an input of energy, supplied by photons; a change in a downward direction yields energy. ADP is presumably converted to ATP on the downslope between the two cytochromes, which act as electron-acceptors and -donors.

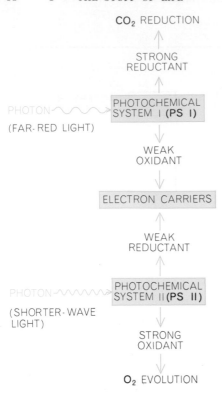

CO₂ REDUCTION

↑

STRONG
REDUCTANT

↑

PHOTOCHEMICAL
SYSTEM I (PS I)

↑ PHOTON

(FAR-RED LIGHT)

WEAK
OXIDANT

↓

ELECTRON CARRIERS

↑

WEAK
REDUCTANT

↑

PHOTON

(SHORTER-WAVE
LIGHT)

PHOTOCHEMICAL
SYSTEM II (PS II)

↓

STRONG
OXIDANT

↓

O₂ EVOLUTION

PS I AND PS II, two photochemical systems, cooperate in the fixation of carbon dioxide in algae and higher plants. Each system has its own reaction center containing a photosynthetic pigment. The pigment in PS I is a species of chlorophyll *a* known as P-700 because its maximum absorption is at a wavelength of 700 nanometers. The strong oxidant of PS II is able to oxidize water. The strong reductant of PS I has the power to reduce NADP to NADPH. The reactions driven by the photochemical systems are shown in the equations at the bottom of the page.

gradient while electron transport between the two cytochromes goes *with* the gradient [see *bottom illustration on preceding page*].

The Hill-Bendall formulation indicated that the two light reactions occur in two different photochemical systems [see *illustration above*]. Each system has a reaction center within which an oxi-

dant and a reductant are formed. Photochemical system II (PS II) sensitizes a reaction that results in the oxidation of water and in the formation of a weak reductant. The chlorophyll in the reaction center of PS II has not yet been identified, but it is presumed to be some form of chlorophyll *a*. Photochemical system I (PS I) sensitizes a reaction that yields a weak oxidant and a strong reductant. The chlorophyll in the reaction center of PS I has been identified as a species of chlorophyll *a* whose absorption peak is at 700 nanometers and is therefore known as P-700. The two photochemical systems are linked in series by electron-carriers, so that the weak reductant produced in PS II is oxidized by the weak oxidant produced in PS I.

Duysens and his co-workers provided some of the early evidence for this model of two photochemical systems acting in series when they showed that the *c*-type cytochrome in the chloroplast is reduced by the shorter-wavelength light absorbed by PS II and oxidized by the longer-wavelength light absorbed by PS I. Such "antagonistic" effects indicate that the cytochrome lies in the path of electron flow between the two systems. Other investigators have since demonstrated similar antagonistic effects on the *b*-type cytochrome of the chloroplast and on P-700.

Additional evidence for the series model has involved the use of the potent weed killer DCMU (dichlorophenyldimethyl urea), which owes its effectiveness to its ability to inhibit the flow of electrons from water to NADP. In its presence both the *c*-type and the *b*-type cytochromes can be oxidized by PS I but they cannot be reduced by PS II. One can therefore assume that the DCMU acts at a site somewhere between PS II and the cytochromes [see *illustration on page 69*]. The photoreduction of NADP is thus blocked by DCMU, but it can be restored if an artificial electron-donor (such as a reduced

indophenol dye) is introduced into the chloroplast. In the presence of the dye NADP can once again be photoreduced but now light absorbed by PS I alone is sufficient. The effect of DCMU indicates the existence of two light reactions coupled by a system of electron-donors and -acceptors.

Electron Path from PS II

The portion of the photosynthetic electron-transport chain that carries electrons from water to PS II is known as the oxidizing "side" of PS II. As mentioned above, the oxidation of water is effected by the oxidized form of a hypothetical donor molecule, D^+. Experimental evidence for electron transport between water and PS II has recently been provided by Takashi Yamashita and Warren L. Butler of the University of California at San Diego, but the nature of the electron-carrier (or carriers) involved, and its relation to D, has not yet been determined.

The reducing "side" of PS II is the portion of the electron-transport chain between PS II and its electron-acceptor A. The fluorescence properties of the chloroplast have provided information on the nature of A. If chloroplasts are irradiated with short-wavelength light, the yield of fluorescence is high, but if they are illuminated with the longer wavelength of light that can be absorbed by PS I, the fluorescence yield decreases. From these observations Duysens and his associates have inferred that when A is reduced by PS II, fluorescence is high, but when it is oxidized by PS I, fluorescence is low. They called the acceptor component Q rather than A, Q standing for quencher of fluorescence. Q in the oxidized form quenches fluorescence, whereas Q in the reduced form does not and therefore the fluorescence yield increases. The yield increases even more in the presence of DCMU, suggesting that the weed killer acts at a site between Q and PS I in the photosynthetic electron-transport chain.

The chemical nature of Q has not been determined with certainty. Norman I. Bishop of Oregon State University has obtained evidence suggesting that it may be a compound known as plastoquinone. Regardless of its chemical nature, Q is probably the electron-acceptor of PS II.

Electron Path from Q to P-700

Proceeding along the electron-transport chain from PS II to PS I, one finds

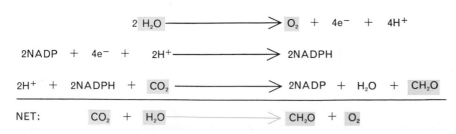

$$2\ H_2O \longrightarrow O_2 + 4e^- + 4H^+$$

$$2NADP + 4e^- + 2H^+ \longrightarrow 2NADPH$$

$$2H^+ + 2NADPH + CO_2 \longrightarrow 2NADP + H_2O + CH_2O$$

NET: $$CO_2 + H_2O \longrightarrow CH_2O + O_2$$

CHEMISTRY OF PHOTOSYNTHESIS is summarized by these four equations. Water is oxidized, releasing free oxygen, electrons and protons. The electrons and two protons reduce NADP to NADPH. NADPH plus two protons and carbon dioxide yield NADP, water and carbohydrate (CH₂O). Thus carbon dioxide and water yield carbohydrate and oxygen.

that there is a cytochrome on the downhill slope from Q to PS I. This is a b-type cytochrome. It is followed by a c-type cytochrome. The reduction of both cytochromes is sensitized by PS II; their oxidation is sensitized by PS I. As mentioned above, this differential oxidation and reduction pattern localizes the cytochromes between the two photochemical systems.

Between the b-type and the c-type cytochromes there is at least one more component, which is not yet identified. Evidence for its existence comes from experiments that Donald Gorman and I have conducted in the Biological Laboratories of Harvard University, using mutant strains of a unicellular green alga that had lost the capacity to carry out normal photosynthetic electron transport. Of the mutant strains, one lacked the b-type cytochrome, another lacked the c-type cytochrome and a third lacked an unknown component.

The third mutant strain proved to possess both cytochromes (b-type and c-type), but when it was illuminated with long-wavelength light of the kind absorbed by PS I, only the c-type cytochrome was oxidized. When light of the kind absorbed by PS II was used, only the b-type cytochrome was reduced. Since the first kind of light normally oxidizes both cytochromes and the second kind of light normally reduces both cytochromes, it was clear that some component was missing from the mutant strain that ordinarily acts as an electron-donor and -acceptor between the two cytochromes. For want of a specific identification we have designated it M.

Another component in the electron-transport chain is the copper-containing protein plastocyanin. Although this protein can act as an electron-acceptor and -donor, its role is not clearly understood. At present some experiments indicate that plastocyanin lies between the c-type cytochrome and PS I, and other experiments give equally convincing evidence that it lies on the uphill side of the c-type cytochrome.

We now come to P-700, the chlorophyll that absorbs far-red light in the reaction center of PS I. Its discoverers, Bessel Kok of the Research Institute for Advanced Study in Baltimore and George E. Hoch of the University of Rochester, showed that P-700 is oxidized by light absorbed by PS I and reduced by light absorbed by PS II. On being oxidized it transfers its electron to its electron-acceptor. It can then be reduced again by electrons coming from water and passed along the transport

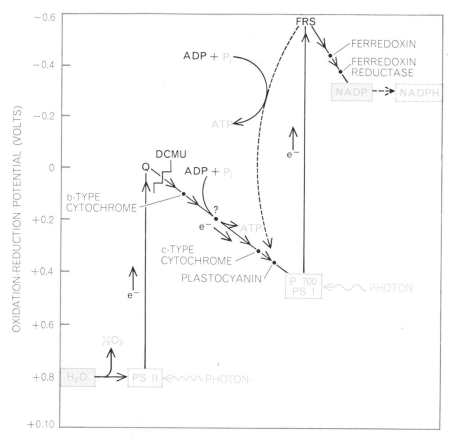

ELECTRON-TRANSPORT CHAIN in photosynthesis follows the Hill-Bendall model (*see bottom illustration on page 67*). It shows how electrons removed from water are passed along by various acceptors and donors until they finally reach NADP and participate in its reduction to NADPH. Along the chain, at one or more places not yet clearly identified, energy is extracted to form ATP from ADP and inorganic phosphate (P_i). The energy for boosting electrons against the electrochemical gradient is supplied by photons that excite chlorophyll molecules in the two photochemical systems, PS I and PS II. Electrons leaving PS II are evidently accepted directly by a substance called Q (for quencher of fluorescence) and electrons leaving PS I by the ferredoxin-reducing substance (FRS). The downhill path from Q to PS I contains a number of donors and acceptors, which are discussed in the text. The path that transports electrons from FRS to NADP appears to be less complicated; the end result is NADPH. Many details of the electron-transport chain have been clarified with the help of a weed killer called DCMU, which interrupts electron flow to the right of Q.

chain from PS II. The immediate electron-donor to P-700 is either the c-type cytochrome or plastocyanin or both.

The Path between P-700 and NADP

Moving along the electron-transport chain, we see that the electron donated by PS I, which has a potential of about +.45 volt, must travel against a large electrochemical gradient before it can reach NADP, which has a potential of −.3 volt. The electron-acceptor of PS I has been much debated ever since it was proposed a few years ago that the acceptor might be ferredoxin, an electron-acceptor and -donor molecule whose negative potential (−.43 volt) is even higher than that of NADP. It was clear that if light energy could boost an electron against the potential gradient from

P-700 to ferredoxin, the final step to NADP would be downhill. The difficulty was that several investigators found that PS I can boost electrons against potentials even more negative than that of ferredoxin. It did not seem economical of nature to provide a greater boosting capacity than is actually required, and this cast doubt on ferredoxin's being the primary acceptor of electrons from PS I.

Recently Charles Yocum and Anthony San Pietro of Indiana University and Achim Trebst of the University of Göttingen have discovered what is apparently the true acceptor. A substance with a potential of about −.6 volt, it has been given the tentative name ferredoxin-reducing substance, or FRS. Its chemical nature is now under study. Its absorption spectrum suggests that it will turn out to have a complex structure consisting of more than one molecular species.

We have now nearly reached the end of the photosynthetic electron-transport chain. The FRS transfers its electron to ferredoxin, and NADP is reduced to NADPH in the presence of an enzyme called ferredoxin-NADP reductase.

ATP Formation and CO_2 Fixation

Much current research is focused on how the production of ATP is coupled to photosynthetic electron transport. Theoretically, as Hill and Bendall originally suggested, sufficient energy is available in the downhill flow of electrons between the b-type and the c-type cytochromes to phosphorylate a molecule of ADP, converting it into a molecule of ATP. And indeed there is evidence for a site of ATP formation between the two cytochromes. There is also evidence for a cyclic flow of electrons around PS I (most likely involving FRS), and ATP formation is coupled to this electron flow.

In spite of extensive investigation there is uncertainty regarding the mechanism of the coupling of ATP formation and electron flow not only in the chloroplast but also in mitochondria. To review current opinions regarding the mechanism would require an article in itself.

We have now reached the last phase of the photosynthetic process: the reduction of carbon dioxide to carbohydrate. Much of our knowledge of this final phase is due to the work of Melvin Calvin, James A. Bassham and Andrew A. Benson of the University of California [see "The Path of Carbon in Photosynthesis," by J. A. Bassham; SCIENTIFIC AMERICAN Offprint 122]. In this cycle one molecule of ribulose diphosphate and one molecule of carbon dioxide react, with the aid of suitable enzymes, to form two molecules of phosphoglyceric acid (PGA). The two molecules of PGA are converted to two molecules of glyceraldehyde phosphate in a reaction that requires two molecules of NADPH and two of ATP. One other step requires ATP (the production of ribulose diphosphate from the monophosphate), so that the overall requirement is three molecules of ATP and two of NADPH for each molecule of carbon dioxide reduced to carbohydrate [see illustration below]. This sequence is thought to represent the pathway of carbon dioxide fixation in higher plants, algae and photosynthetic bacteria.

Quite recently, however, M. D. Hatch and C. R. Slack in Australia have shown that there is a different kind of pathway in certain species of tropical grasses. The first step of CO_2 fixation in these grasses involves the carboxylation of phosphopyruvic acid (rather than of ribulose diphosphate), yielding oxaloacetic acid, which then serves as a precursor of PGA.

We have now followed the mechanism of photosynthesis from the initial trapping of the electromagnetic energy of light, through the conversion of energy into chemical energy, then through the electron-transport steps that lead to the generation of NADPH and ATP and finally to the terminal events of carbon dioxide fixation. We have seen that some parts of the process are much better understood than others. The most enigmatic part is the one associated with events at photochemical system II. The means by which four electrons and four protons are extracted from water with the concomitant evolution of a molecule of oxygen is one of the most fascinating problems still to be solved.

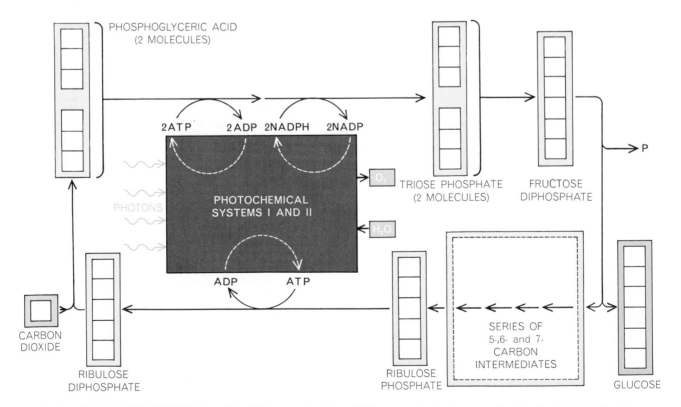

FIXATION OF CARBON DIOXIDE is achieved by a cycle of chemical reactions powered by photons that are trapped by the two photochemical systems. These systems package part of the energy in the form of ATP and remove electrons and protons from water, releasing oxygen. The electrons and protons enter the cycle in the form of NADPH. Two molecules of NADPH and three of ATP are required to fix one molecule of carbon dioxide, shown entering the cycle at the lower left. In the cycle each white square represents a carbon atom. The carbon atoms from CO_2 can be incorporated into a variety of compounds and removed at various points in the cycle. Here six atoms of carbon supplied by CO_2 are shown leaving the cycle as glucose, $C_6H_{12}O_6$, a simple carbohydrate.

II

OUR IMPACT ON THE LAND THAT FEEDS US

Our species, like all in the biosphere, depends for survival on the ability to compete with other creatures for the elements necessary for growth and reproduction. Ultimately, this reduces to competition for available energy. Wherever man maintains a more-or-less advanced state of civilization, he does so by exploiting the energy content of as many components of his environment as he can manage to appropriate for his own purposes. In our present state of development, the maintenance of the species requires extensive conscious cultivation of organisms that we can use as food. This practice, in effect, amounts to pirating incident solar energy to sustain our own metabolisms at the expense of whatever less palatable life forms would otherwise populate the areas we have diverted to food production. The tactics we use to maximize our share of the planet's photosynthetic production have direct and widespread impact on the landscape and its inhabitants.

In order to understand the extent to which our agricultural activities affect the environment, we should be familiar with the basic energetics of the biosphere. This does not mean the detailed energy transfer that takes place in the chemical reactions of photosynthesis or metabolism, but rather the gross characteristics of the way the energy in the solar flux is distributed among members of the biomass. "The Flow of Energy in the Biosphere," by David M. Gates, provides some excellent background in the subject, tracing what happens to the solar radiation that strikes the earth, with special emphasis on the small fraction that is converted by photosynthesis to vegetation. The author contrasts the low net productivity of natural plant communities with the high storage capability of agricultural crops, and discusses the factors (sunlight, temperature, and nutrients) that place an upper limit on possible photosynthetic yield. His concluding paragraphs analyze the efficiencies of various pathways for providing the energy required for human metabolism.

This section's second article, "Human Food Production as a Process in the Biosphere," presents an overview of the factors that must be considered in feeding the people of the world. In it, Lester R. Brown points out the correlation between availability of food and population size. His main theme deals with the stress on the biosphere created by our attempts to increase food supplies in response to the exponential growth of the population. We have fostered extensive alterations in the relative abundance and geographic distribution of species to satisfy our tastes and digestive capabilities. In addition, the biosphere must absorb the consequences of the technological components of modern agriculture: Brown identifies and discusses four.

Each of the four technologies can be viewed as a tool for overcoming one limiting factor or another. Mechanization, by increasing the surface area that man can bring under cultivation, increases the number of quanta available for photosynthetic conversion by edible plants. Irrigation can not only raise production levels in regions with moderate water supplies, it can also allow use of areas formerly too dry for cultivation. Fertilizers remove limitations imposed by the scarcity of nutrients in the soil, and pesticides eliminate competition from organisms that could limit yield either by diverting soil nutrients to themselves or by eating a product directly. Fertilizers and pesticides are decidedly chemical technologies, and are examined in more depth in the remaining articles in this section.

Christopher J. Pratt's article, "Chemical Fertilizers," comprises an interesting and informative discourse on the cheapest, most effective tool at our disposal for boosting crop productivity. The author discusses the need for providing nutrients to cultivated land, and describes the compounds used for fertilizer and how they are made. A section is devoted to some of the factors (plant physiology, soil chemistry) that must be considered if fertilizers are to be used most effectively. Written before public awareness of environmental problems became really widespread, the presentation emphasizes the benefits derived from application of chemical fertilizers and touches only lightly on the potential hazards attendant on dosing the soil with millions of tons of extra nutrients each year.

One observation made by Pratt is of fundamental importance, and deserves a little amplification here. That is that crop yields do not increase linearly with the amount of fertilizer applied. Rather, the incremental yield diminishes with increasing nutrient doses until finally they have essentially no additional beneficial effect. This phenomenon is one manifestation of a universal characteristic of complicated systems, one that has its basis in the principles of chemical kinetics. The overall rate of a set of linked kinetic processes is controlled by the rate of the slowest one. If that process is slow because one of the reactants required for it is available only in small amounts, then providing more of that substance will increase the net rate of the set, because it speeds up the most sluggish step. The rate of that step can only increase so much, however, before it is no longer the slowest, and another process in the set will assume control over the overall rate. Any other effort to accelerate what is at any given time the slowest step, such as changing the rate constant, will have the same effect. This behavior holds an important lesson for anyone concerned with the responses of any multicomponent system.

Pesticides are the weapons man uses to maintain an advantage over those species that reduce either his food supply or, by carrying disease organisms, his numbers. The articles that follow in this section provide a somewhat historical look at the ramifications of our attempts at chemical control of competitiors. "Insects v. Insecticides," written by Robert L. Metcalf in 1952, outlines the beginnings of the race between chemical poisons and the development of resistance to them on the part of their targets. It is interesting to note how soon resistant insects appeared after the initial use of DDT in the 1940s, though not surprising in the light of the research elucidating the ways in which mutant insects could detoxify the chemical and how the mutation was passed on. Even in 1952 the problem of DDT resistance and its generalization to other chlorinated hydrocarbons was fairly well understood, as was the fact that pyrethrins and organophosphorus insecticides remained effective over many generations. Why did it take twenty years to read the handwriting on the wall and acknowledge that our first miracle insecticide was not really suitable for general use? Several possible reasons come to mind. DDT and its relatives showed such spectacular initial success that it would have been difficult for the developers to abandon them in the face of what then seemed to be minor difficulties. The alternatives had problems of their own. The chlorinated hydrocarbons were cheap and easy to synthesize, but pyrethrins had to be isolated from chrysanthemums, their structure, when finally determined, proving to be too complex for economical large-scale production. Organophosphorus compounds are highly toxic to all organisms; DDT had not been observed to threaten anything but insects — yet.

The outcome of chronic exposure to chlorinated hydrocarbons is outlined in "Pesticides and the Reproduction of Birds" by David B.

Peakall. Instances of large-scale reproductive failure in colonies of predatory birds were observed to correlate in time and geographic distribution with extensive use of the persistent pesticides, which by now are known to become highly concentrated in the tissues of creatures at the top of a food chain. The author discusses experiments implicating DDE (the now ubiquitous detoxification product of DDT) and related compounds in the disruption of calcium metabolism in birds and in upsetting hormone balance in many other vertebrates, including squirrel monkeys.

The whole situation poses a disturbing dilemma. Our highly specialized agricultural ecosystems are vulnerable to plant pests, and need protection. Dense human populations are vulnerable to disease, and need protection. Yet the chemicals most used to provide that protection hold the potential for causing as much trouble as they prevent. Carroll M. Williams, in "Third-Generation Pesticides," discusses one possible strategy for circumventing the problem. He describes the action of juvenile hormones, the properly timed application of which makes insect development go fatally awry. Some excellent natural-product chemistry is involved in the isolation and structure determination of some of these compounds, and in the synthesis of similar substances with juvenile hormone activity. The chemicals tested kill only insects, but show no specificity within that order. Williams talks about the possibility of attaining greater selectivity, but when the article was written, in 1967, only one such pesticide specific for one insect was known. The logistic problems involved in finding, isolating, and producing on a large scale a natural product to poison each of our insect enemies still seems overwhelming.

There is a class of natural products, however, that has intriguing potential for use in pest control, at least for some applications. In "Pheromones," Edward O. Wilson describes some of the remarkable substances that certain organisms release into the environment in order to influence the behavior of other organisms. Response to these compounds is highly specific, with regard to both the creature affected and the behavior elicited. The problems of structure elucidation and synthesis are no less difficult than those posed by juvenile hormones, but pheromones offer a distinct advantage in their phenomenal potency. A tiny amount of insect sex attractant can exert its effect over a large area; one can visualize use of such lures to bring large numbers of a pest into contact with small, contained applications of a nonspecific poison.

Taken together, the last four articles present a picture of the past, the present, and the speculative future that covers just part of mankind's activities in chemical pest control. (Another result of these activities, mercury-containing pesticides, is mentioned in Section V, in "Mercury in the Environment.") That picture should help to provide some perspective in thinking about environmental questions: although chemistry has caused real harm in this area, it will also be chemistry that, one way or another, will solve the problem. If there is a real villain here, it is the human animal, with his short-sightedness, ignorance, and lack of caution in exploiting the fruits of his ingenuity.

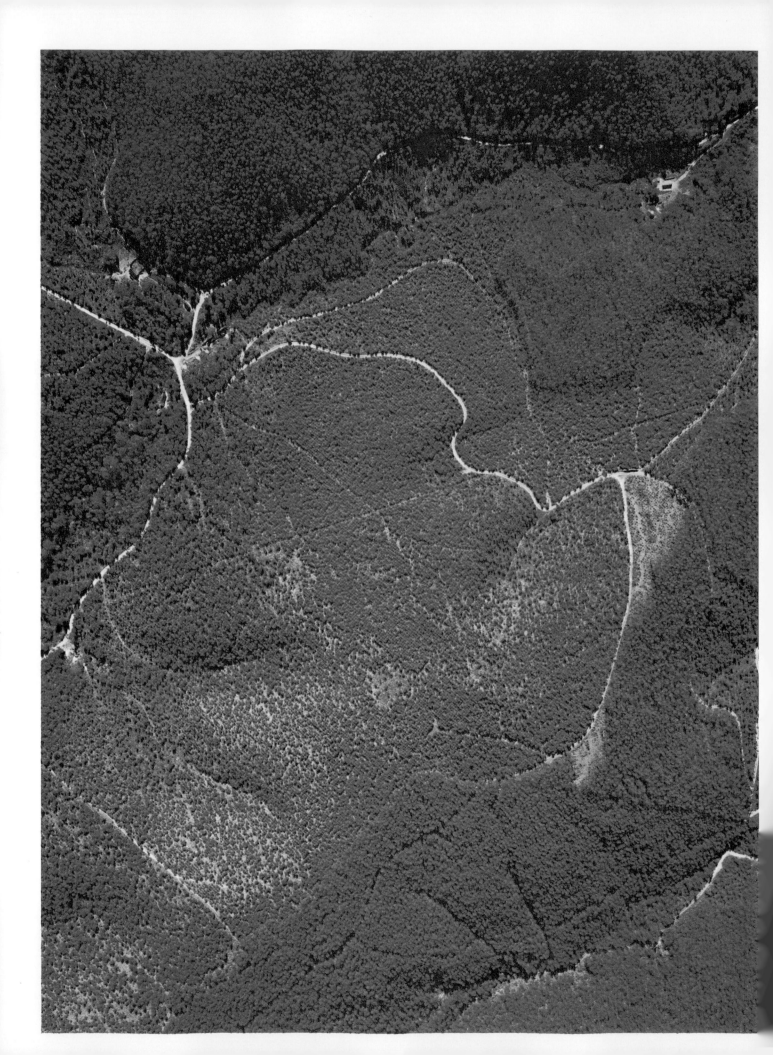

8

THE FLOW OF ENERGY IN THE BIOSPHERE

DAVID M. GATES

September 1971

The solar energy that falls on the earth warms the surface and is ultimately radiated back into space. The tiny fraction of it that is absorbed by photosynthetic plants maintains all living matter

The radiant energy that bathes the earth builds order from disorder through the processes of life. Most events in the universe proceed toward increasing entropy, but life postpones the effect of this basic law by using the stream of sunlight to build highly complex assemblages of proteins, carbohydrates, lipids and other biological molecules. The aim of this article is to trace the radiant energy as it sustains the remarkable diversity of living organisms.

A living organism can be viewed as a chemical system designed to maintain and replicate itself by utilizing energy that originates with the sun. Life cannot be sustained merely by an adequate quantity of radiation; the light must also be of a suitable spectral quality. The flux of solar radiation received at the ground is highly variable in both quantity and quality because of the variable transmissivity of the atmosphere and the changing degree of cloudiness. The earth's atmosphere filters sunlight by absorbing most of the ultraviolet wavelengths and some of the infrared.

Light entering a chemical system can be utilized in several ways. It can be absorbed and then simply dissipated as heat through the increased motion of the molecules in the system. It can be reradiated at the resonant frequencies of the molecules or as fluorescence or phosphorescence. It can be utilized to accelerate a chemical reaction that either increases or decreases the free energy of the participating molecules.

Light consists of the bundles of energy called quanta. The energy content of a quantum is proportional to the frequency of the light: the shorter the wavelength, the higher the frequency and the greater the energy content. A mole of any substance (a weight in grams equal to the molecular weight of the substance) contains 6×10^{23} molecules; that is the universal constant known as Avogadro's number. In discussing the interaction of light and matter it is convenient to use one mole of a substance. The energy content of a molecular bond can then be multiplied by the number of molecules per mole (6×10^{23}) to get the bond energy of the substance per mole. One can also regard one mole as containing 6×10^{23} quanta and can multiply that by the energy per quantum in order to get the mole equivalent energy of radiation.

The mole equivalent energy of blue light at a wavelength of 450 nanometers (a nanometer is a billionth of a meter) is 64 kilocalories per mole; of infrared radiation at 900 nanometers, 32 kilocalories per mole, and of ultraviolet radiation at 225 nanometers, 128 kilocalories per mole. The strength of molecular bonds (or the energy required to break them) can be expressed in kilocalories per mole. A single bond between two carbon atoms can be broken with only 82.6 kilocalories per mole; a double bond between the two atoms requires 145.8 kilocalories per mole and a triple bond 199.6 kilocalories per mole.

It is evident from these numbers that ultraviolet radiation has the energy per mole necessary to break bonds. It is also clear that visible light has relatively little potential for breaking or forming bonds and that infrared radiation has even less. Light absorbed by a molecule kicks one of the electrons associated with the molecule into an excited energy state, thereby making the electron available for pairing with an electron from a neighboring atom or molecule in an electron-pair bond. By this photochemical process new molecules are formed.

The most fundamental photochemical reaction of life is photosynthesis in plants. Photosynthesis combines molecules of carbon dioxide and water to form carbohydrate and oxygen; the energy converted in the process is 112 kilocalories per mole. It is known that photosynthesis proceeds by means of blue light and red light. From the relation of energy and wavelength I have described it is clear that neither blue nor red light can directly provide enough energy for photosynthesis.

It turns out that photosynthesis is a complicated stepwise process. Light is absorbed by the chlorophyll molecule (and by other pigments in the plant) and is transferred to electrons in such a way as to create strong oxidants and reductants, that is, molecules that readily remove electrons from other molecules (oxidize them) or readily supply electrons to other molecules (reduce them). In photosynthesis the oxidants and reductants assist with the storage of energy in chemical bonds, notably those of carbohydrate and of adenosine triphosphate (ATP), the basic energy currency of all living cells.

Animals, by eating plants, are able to release the energy stored in them by

ABSORPTION AND REFLECTION characteristics of vegetation are partly indicated by the aerial photograph on the opposite page. The vegetation, which is in a forested area northwest of São Paulo in Brazil, is red because the photograph was made with a special emulsion that is sensitive in the near-infrared. Green plants absorb about 92 percent of the blue and red light that energizes the process of photosynthesis. They absorb some 60 percent of the near-infrared; the rest is reflected. Thus a photograph of vegetation in the near-infrared shows considerably more intense reflection than one in visible region of spectrum.

means of the various oxidative reactions of metabolic processes. ATP interacts with the carbohydrate glucose to prepare it, through glycolysis, for a long series of complex reactions in the metabolic sequence known as the citric acid cycle. The energy released is employed to do muscular work, to generate nerve impulses and to synthesize proteins and other molecules for the building of new cells. The entire chain of life proceeds in this way as energy cascades through the communities of plants and animals.

Living systems must be protected against an excess of bond-breaking radiation. The primordial atmosphere of the earth contained no free oxygen and was highly transparent to ultraviolet radiation. Once photosynthesis began (with microorganisms in the ocean) oxygen was released to the atmosphere. Since the metabolic processes of the primitive organisms were primarily anaerobic, the oxygen in the atmosphere built up. As the oxygen molecules (O_2) diffused upward they were decomposed by ultraviolet radiation into oxygen atoms (O), some of which formed ozone (O_3). Ozone strongly absorbs ultraviolet radiation, and as it built up in the stratosphere it acted as a filter. In this way the earth's surface was shielded against the energetic ultraviolet radiation of the sun but remained transparent to visible light.

The interaction of life and the atmosphere has many more components than the ozone shield against ultraviolet. For example, the atmosphere contains .032 percent, or 320 parts per million, of carbon dioxide, which is essential to the part of photosynthesis that assimilates carbon into carbohydrates. Carbon dioxide, which at visible wavelengths is a clear transparent gas, strongly absorbs the radiation in certain infrared bands of the spectrum. The earth's surface radiates heat into space entirely at infrared wavelengths. If there were no atmosphere, or if the atmosphere were fully transparent, the temperature of the ground at night would be considerably colder than it is.

What happens is that the absorption bands of atmospheric carbon dioxide capture some of the infrared radiation headed toward space from the earth. The captured energy is then reradiated in two directions: back toward the ground and into outer space. Therefore the ground is exposed not to the cosmic cold of outer space but to a warm flow of radiation emitted by atmospheric carbon dioxide.

Clouds and water vapor in the sky also absorb and emit infrared radiation. When the sky is clear, the water vapor and the carbon dioxide only partly

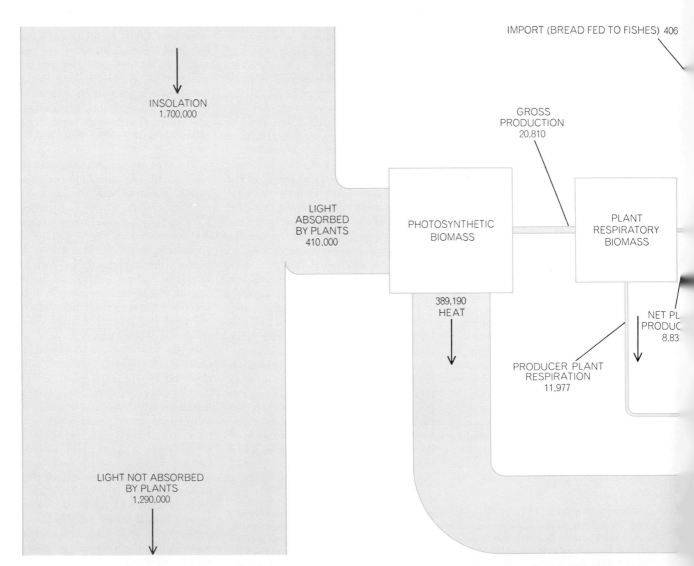

IMPORT (BREAD FED TO FISHES) 406

INSOLATION
1,700,000

LIGHT ABSORBED
BY PLANTS
410,000

LIGHT NOT ABSORBED
BY PLANTS
1,290,000

PHOTOSYNTHETIC
BIOMASS

389,190
HEAT

PRODUCER PLANT
RESPIRATION
11,977

GROSS
PRODUCTION
20,810

PLANT
RESPIRATORY
BIOMASS

NET PL
PRODUC
8,83

ENERGY IN A NATURAL SYSTEM flows as indicated in this diagram of the ecosystem at Silver Springs, Fla., which consists of a clear, spring-fed stream with vegetation covering the bottom and numerous species of animals living in or near the water. The nu-

shield the ground from the cold of space. When the sky is overcast, the cloud cover serves as an opaque thermal blanket. At such times the radiation from the earth to space originates at the top of the cloud deck rather than at the ground.

Thus green plants not only get the benefit of carbon dioxide but also are warmed by the radiant flux returned to the ground from the atmosphere. The atmosphere's window on space is transparent to visible light but is closed at the ultraviolet end by ozone absorption and at the infrared end by absorption in carbon dioxide and water vapor. This grand-scale synergy of green plants and the atmosphere is the result of millions of years in evolution of life and of the atmosphere, which are therefore closely interdependent. Life depends on both the clarity and the opaqueness of the atmospheric window, and the waste

products of man that are discharged into the atmosphere dirty the window, on whose clarity all life depends.

The growth of green vegetation depends simultaneously on the amount of sunlight reaching the ground, the temperature near the surface and the amount of water available. If any of these conditions is inadequate, growth is reduced. Much sunlight and little water produce a desert. Much sunlight and low temperature produce tundra. Much water and little sunlight make for a stunted rain forest.

The annual productivity of green vegetation is limited by the seasonal distribution of sunlight, temperature and moisture. The solar radiation reaching the atmosphere is partly absorbed by ozone, carbon dioxide, water vapor, nitrogen, oxygen, dust and aerosols. By the time it reaches the ground it is weakened in intensity and modified in

spectral quality. Solar radiation at the ground—direct sunlight plus skylight—varies from a maximum of between 200 and 220 kilocalories per square centimeter per year in desert areas to 70 kilocalories per square centimeter per year in polar regions. Tropical rain forests receive from 120 to 160 kilocalories; much of Europe, 80 to 120. The solar radiation at the Equator varies relatively little during the year except as it is affected by cloudiness. Polar regions experience the midnight sun of summer and perpetual darkness during the winter.

Now that we have traced the solar flux down through the atmosphere to the surface of the ground, let us see how it is partitioned and what it does. Clearly if it strikes bare rock or soil, it will be partly reflected and partly absorbed, and the rock or soil will gain energy. If the surface bears vegetation,

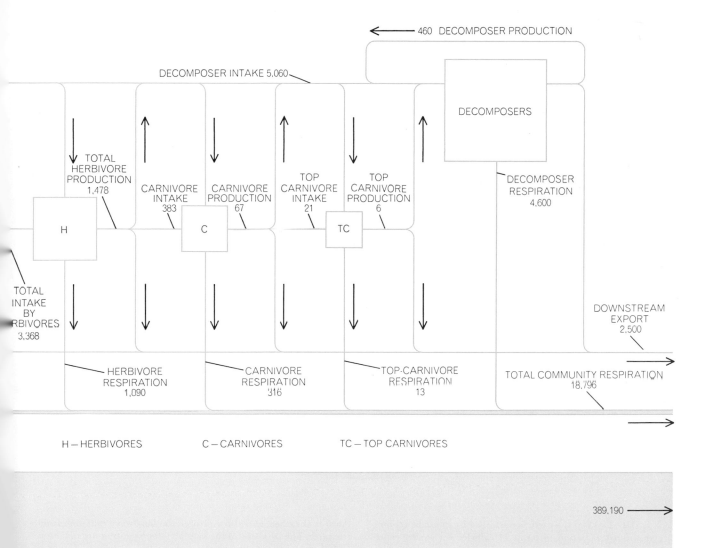

H — HERBIVORES   C — CARNIVORES   TC — TOP CARNIVORES

merals give inputs and outputs in kilocalories per square meter per year, with relative energies indicated by the widths of the bands and lines. The data were obtained by Howard T. Odum of the University of Florida. Top carnivores are at top of food chain.

some of the incident solar radiation is utilized in photosynthesis. Green fields reflect from 10 to 15 percent of visible light; dark green coniferous forests may reflect only 5 to 10 percent.

Of the total amount of solar energy entering the earth's atmosphere only about 53 percent is available at the ground after all scattering, absorbing and reflecting processes are taken into account. The ground exchanges energy by radiation, by the evaporation and condensation of water, by the exchange of sensible heat between the surface and the air and by conduction into or out of the soil. All the energy flowing to or from the ground must be accounted for in the energy budget relating to the surface.

During the day the surface has a net influx of radiation; during the night it loses a net quantity of radiation. During the day, when the ground is warmer than the air, heat is transferred from ground to air by convection. At night the air is usually warmer than the ground, so that the convectional transfer of heat is from air to ground.

Evaporation of water away from the surface requires both a moisture gradient away from the surface and energy sufficient to supply the latent heat of

vaporization. The amount of energy required is about 580 calories per gram at a temperature of 30 degrees Celsius. Evaporation and evapotranspiration through the leaves of plants are almost always taking place in daylight, except when it is raining; sometimes they proceed at night as well.

If the ground is quite dry, the net radiation input during the day will go into convection and conduction. The environment will be turbulent and windy, as is typical of deserts. If the ground is moist or the vegetation is well watered, evapotranspiration will consume the major fraction of net radiation and the atmosphere will be more quiescent. These are basically physical processes related to the thermodynamics of the earth's surface and to the conditions of climate. Let us now leave them in order to trace out the pathway of light—the visible wavelengths of radiation—as it affects primary productivity (the growth of vegetation) and the food chain of life.

Of the total amount of sunlight reaching the ground only about 25 percent is of wavelengths that stimulate photosynthesis, and only a fraction of the 25 percent is actually used by green plants. Most plants in the open are using

light at their maximum rate during most of the hours of daylight. A forest or a field receives, on a typical summer day in the U.S., from 500 to 700 calories of solar energy per square centimeter per day. Assuming that plants are receiving such an input and that they are using as much sunlight as they can, one can estimate the productivity of growing things.

Robert S. Loomis and William A. Williams of the University of California at Davis have made such estimates. Considering 500 calories per square centimeter per day a typical daily input of energy during the growing season, they found that potential net plant production (gross production minus respiration) is about 71 grams per square meter per day. Assuming that this net productivity represents as storage in carbon compounds about 3,740 calories per gram, one finds that 26.6 calories per square centimeter per day of solar radiation ends up in biomass. This represents 5.3 percent of the total incident solar radiation and about 12 percent of the energy received as visible light, which is 222 calories per square centimeter per day.

Plants reflect about 8 percent of photosynthetically active wavelengths. The

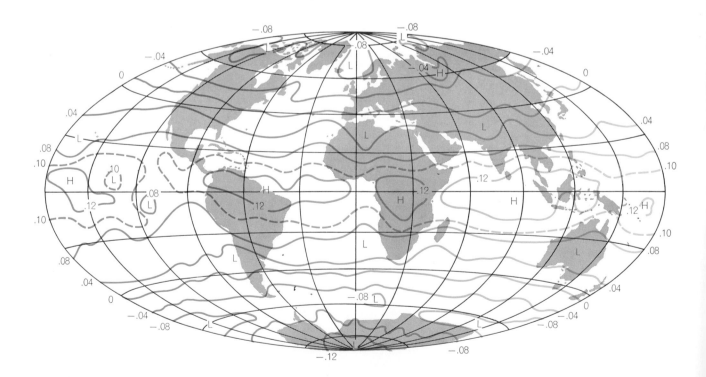

MEAN RADIATION of the earth is portrayed by isopleths (*color*) that give the net radiation in terms of calories per square centimeter per minute. Areas marked *H* and *L* are respectively high or low compared with their surroundings. The data were obtained by satellites that measured the earth's albedo, which indicates how much of the solar radiation reaching the earth is reflected and how

much is absorbed, and also measured the long-wave radiation from the earth. The isopleths of the map give the resulting net radiation and thereby provide information about the exchange of energy between the earth and space. The work was done in the department of meteorology of the University of Wisconsin by Thomas H. Vonder Haar, now at Colorado State University, and Verner E. Suomi.

percentage is considerably higher for the portion of such wavelengths in the near-infrared region of the spectrum; there individual leaves have reflectivities of 40 percent or more. An individual leaf will absorb some 60 percent of the total incident sunlight. A dense stand of vegetation, however, will absorb considerably more. For example, a dense corn crop may reflect 17 percent of the total incident solar radiation, transmit to the soil about 13 percent and absorb in the leaves about 70 percent.

About 10 quanta of light are required to reduce one molecule of carbon dioxide to carbohydrate. Respiration consumes from 20 to 40 percent of gross photosynthesis; the value used by Loomis and Williams for their estimates of productivity was 33 percent. Gross productivity was 107 grams per square meter per day and respiration burned up 36 grams per meter per day. Of the 500 calories per square centimeter per day incident on the crop 375 calories were absorbed by the crop and the soil, converted to heat and then transferred by radiation to the atmosphere, by evapotranspiration, by convection to the air and by conduction into the soil. The evapotranspiration component may account for as much as 200 calories per square centimeter per day.

A crop of Sudan grass at Davis produced 51 grams per square meter per day during a 35-day period when the incident solar radiation averaged 690 calories per square centimeter per day. It is estimated that the maximum potential production by this crop was 104 grams per square meter per day, so that the actual production was 49 percent of the potential yield. At 51 grams per square meter per day and with a caloric content of 4,000 calories per gram, the crop put into storage 3 percent of the total incident solar radiation and 6.7 percent of the visible radiation. Barley has been observed to convert as much as 14 percent of the incident visible light to carbon compounds. In general the productivity of crops is much lower than these maximum values.

Jen-hu Chang of the University of Hawaii has made careful estimates of potential photosynthesis and crop productivity for various regions of the world. He based his estimates on the intensity and duration of sunshine and the mean monthly temperatures over the period under consideration, and he assumed a well-watered crop. He made estimates of potential net photosynthesis, expressed in terms of crop productivity in grams per square meter per

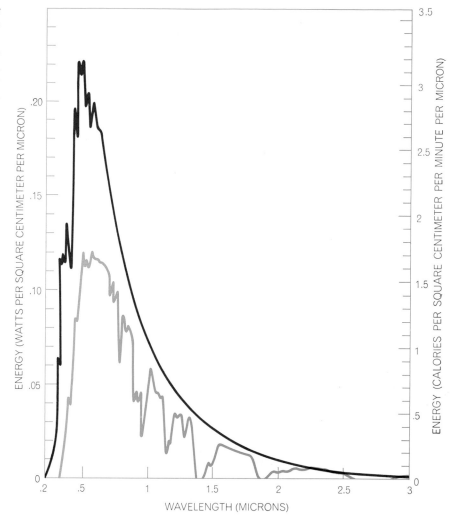

SPECTRAL DISTRIBUTION of solar radiation reaching the earth is given for the top of the atmosphere (*black*) and the ground (*color*). Curve for ground takes into account the absorbing effects of water vapor, carbon dioxide, oxygen, nitrogen, ozone and particles of dust. Data are based on a solar constant of 1.95 calories per square centimeter per minute.

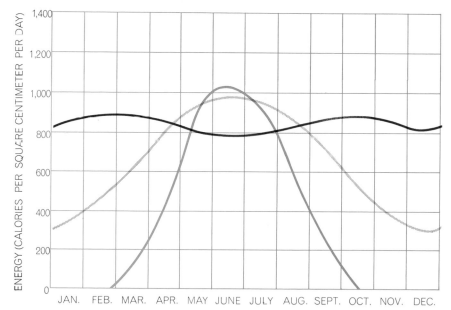

SEASONAL VARIATION of solar radiation on a horizontal surface outside earth's atmosphere is given for Equator (*black*) and latitudes 40 degrees (*gray*) and 80 degrees (*color*).

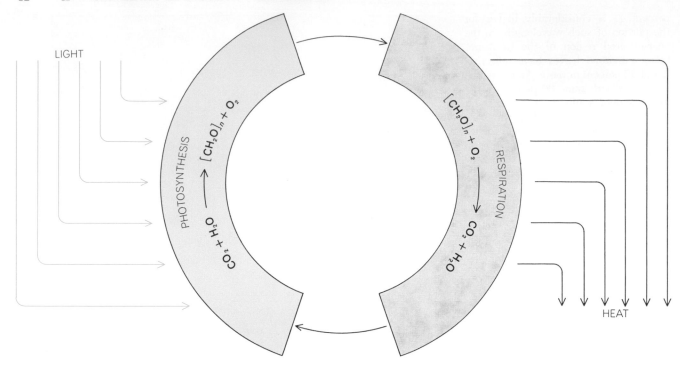

PHOTOSYNTHESIS AND RESPIRATION, which are the basic metabolic processes of most living plants, obtain their energy from sunlight. In photosynthesis energy from light is used to remove carbon dioxide and water from the environment; for each mole-cule of CO_2 and H_2O removed, part of a molecule of carbohydrate (CH_2O) is produced and one molecule of oxygen (O_2) is returned to the environment. In respiration by a plant or an animal com-bustion of carbohydrates and oxygen yields energy, CO_2 and H_2O.

day, for a four-month summer period and an eight-month period centered on the summer; he also calculated an annual mean [*see illustration below*]. His results are of much interest and should be compared with the highest levels of plant production discussed above.

During the four-month summer pe-riod the lowest potential photosynthesis is in the Tropics, and in particular at 10 degrees north latitude, which is the heat equator of the earth. There the potential photosynthesis is 25 percent lower than it is in temperate regions. Highlands in the Tropics have a higher yield than hot, humid lowlands. Southern Alaska, the upper Mackenzie River region in northwest Canada, southern Scandi-navia and Iceland have the highest po-tential photosynthesis, exceeding 37.5 grams per square meter per day. The reason is that these regions receive many hours of sunlight per day during the summer. Farther north the effect of low temperature drops productivity even though the summer day is still longer.

Across the central U.S. the potential net photosynthesis is about 30 grams per square meter per day for both the four-month period and the eight-month period. At the Canadian border the

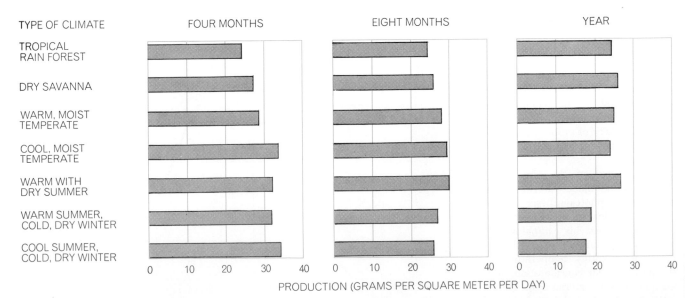

POTENTIAL NET PHOTOSYNTHESIS is expressed in terms of plant productivity in grams per square meter per day for various climates. Calculations are made for a four-month period, repre-senting the normal growing season for most plants, for an eight-month period centered on the summer and for a year. Net produc-tion is found by subtracting respiration from gross production.

four-month value is 36 grams per square meter per day, the eight-month value 27.5 grams per square meter per day and the annual value 17.5 grams per square meter per day. The annual value through the central U.S. is from 20 to 22.5 grams per square meter per day. The same levels are found throughout much of Europe, except in Spain, where the level is above 25 grams per square meter per day. Northern Europe has enormous four-month potential photosynthesis, with values of up to 38 grams per square meter per day. Clearly these high-latitude regions are best suited to crops with a short growing season. It is significant in this context that the part of western Europe between 50 and 60 degrees north latitude leads the world in wheat production.

Chang has made a number of highly interesting assessments of actual yields compared with potential photosynthesis in various countries. He finds that all the developed countries have a four-month potential photosynthesis in excess of 27.5 grams per square meter per day, whereas all the underdeveloped countries have much lower values. What the difference means is that countries such as the Philippines can expect to increase their yields by 30 percent through improved agricultural methods, but no matter what they do they cannot improve 500 percent or more in order to reach the yields of such countries as Spain. In other words, the underdeveloped countries are climatically limited. Chang has compared rice yields with the four-month potential photosynthesis, cotton yields with the eight-month potential (since cotton is planted in early spring and harvested in late fall) and sugarcane yields with the annual potential values (because sugarcane has a long growing season). The situation is the same throughout: the underdeveloped countries are the climatically deprived countries.

It is useful to compare the maximum rates of photosynthesis by agricultural crops, which are in principle plants selected and cultivated for high productivity, with the levels of productivity achieved by natural plant communities. On an annual average the net productivity in grams per square meter per day was nine for spartina grass in a salt marsh in Georgia; six for a pine forest in England during the years of most rapid growth; three for a deciduous forest in England; 1.22 for tall-grass prairies in Oklahoma and Nebraska; .19 for a short-grass prairie in Wyoming, and .11 for a desert in Nevada with five

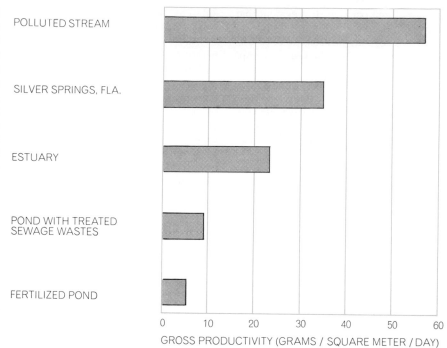

GROSS PRODUCTIVITY in several environments during short periods of time favorable for growing is expressed in terms of grams of dry matter produced per square meter per day. The polluted stream was in Indiana, the estuary was one of several in Texas, the pond with treated sewage wastes was in Denmark and the fertilized pond was in North Carolina.

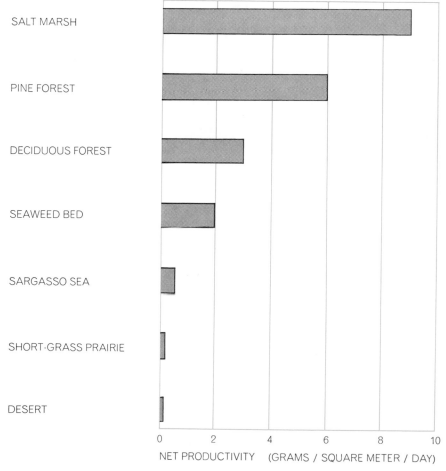

NET PRODUCTIVITY of seven environments is depicted. The salt marsh was in Georgia, forests were in England, prairie and desert in the U.S. and seaweed beds in Nova Scotia.

inches of rain per year. It is evident that the net productivity of most natural stands of vegetation is considerably lower than the levels achieved for crops managed by man.

Since each gram of dry matter produced has a caloric value of about 4,000 calories, the productivities can be converted into percentages of solar radiation utilized. For example, the salt marsh in Georgia converted nine times 4,000 calories per square meter per day into dry matter from the incident solar radiation of about 4.5 million calories per square meter per day for a conversion factor of .8 percent. The conversion by a forest is about .5 percent, by a tall-grass prairie about .1 percent and by a desert .05 percent or less. The average cornfield utilizes about 1 percent of the incident solar radiation and seldom exceeds 2 percent.

Many factors can limit primary productivity. If only one of them is greatly different from the optimum, productivity decreases. Among the important factors are sunlight, carbon dioxide, temperature, water, nitrogen, phosphorus and trace amounts of several minerals. A tropical forest may have an enormous standing mass of vegetation, but the rate of growth is not as high as in other regions because of high temperatures and limitations of soil nutrients. The primary productivity of deserts and grasslands is severely limited for lack of water. Alpine and arctic tundras, which are very wet because of low evaporation and the presence of permafrost not far below the surface, are limited in productivity by low temperatures.

All vegetation is limited in growth by the concentration of carbon dioxide in the atmosphere or in bodies of water. Plants will increase photosynthesis with increasing concentration of carbon dioxide to at least three times the normal concentration of 12.5 nanomoles per cubic centimeter (.03 percent by volume).

All life on the earth depends in one way or another on primary productivity, that is, on the growth of vegetation. Herbivores feed on the carbohydrates and proteins generated by photosynthesis, and carnivores get their energy by feeding on the herbivores. Decomposing organisms feed on both plants and animals, so that the material ingredients of life are returned to the soil and the cycle continues.

Evolutionary events over the past three billion years have created on the earth some two million species of insects, perhaps a million species of plants, 20,000 species of fishes, 8,700 species of birds and almost numberless kinds of microorganisms. Together they form a continuum of life over the surface of the earth. The organisms of the world are all interdependent, forming a vast web of protoplasm through which matter cycles and energy flows.

AGE (YEARS)			1	2	3 TO 20	
TYPE OF GROWTH	BARE FIELD		GRASS		GRASS AND SHRUBS	

CRABGRASS

HORSEWEED

ASTER

BROOM SEDGE

GROWTH SUCCESSION in the piedmont region of the southeastern U.S. progresses from grass to forest over a period of about 150 years. This is the succession that follows the abandonment of land once used for crops. Energy flows from a less mature ecosystem to a

The diversity of species within habitats varies enormously, from the incredible numbers of plant and animal species in a tropical rain forest to the severely limited varieties of life in a desert or a tundra. The diversity of species diminishes along gradients of water from moist to dry, of temperature from warm to cold and of light from bright to dark. By far the most intense speciation is found in a tropical rain forest.

Such a forest possesses an enormous biomass, consisting mostly of large trees laced with vines and lianas of various kinds struggling skyward from the dark interior to compete for light in the canopy. Tremendous numbers of insects and birds coexist in the canopy, and a constant rain of detritus falls to the forest floor, where decay returns the nutrients to the soil. Every conceivable niche of this biome is filled by a plant or an animal.

Biologists have often debated the reasons for the great diversity and stability of life in a rain forest. There is no single reason; there are several. The temperature is favorable to life and moisture is abundant. Physiological processes in animals proceed faster in warm climates than in cold ones. As a result life cycles are shorter and are repeated more often, so that genetic mutations are more likely to arise.

Perhaps the most significant feature is the relative constancy of the climate and the relative absence of fluctuation of environmental factors. For most animals there is a dependable supply throughout the year of the fruit, flower or seed that a particular species needs. The species can specialize in one or two kinds of food because it does not have to confront a lack of food at any time.

Mature and stabilized ecosystems tend to have less productivity than immature or transitional ones within the same general environment. Energy flows from the less mature one to the more mature one. The total amount of energy required to maintain a diverse and complex community of organisms is considerable, but per unit of biomass the amount of energy is smaller than it is in less complex communities.

A case in point is a grass meadow next to a forest. Left to itself the meadow will eventually become forested through a succession of plant communities beginning with grass and extending through herbs and shrubs to trees [see illustration below]. Compared with the forest, the meadow has high instability, low diversity of species and high productivity per unit of biomass. Excess energy is available. It is tapped by the plant succession, each member of which requires more energy than the preceding one, and by the animals of the forest, which feed on the plants and animals of the changing meadow. The natural

| 25 TO 100 | MORE THAN 150 |
| PINE FOREST | OAK-HICKORY FOREST CLIMAX |

SHRUBS   PINE   HARDWOOD UNDERSTORY   OAK   HICKORY

more mature one, which is to say that a mature, stabilized system is likely to have less productivity than an immature, transitional one.

The succession typical of abandoned farmland in the Southeast was ascertained by Eugene P. Odum of the University of Georgia.

trend in succession within communities is toward a decreasing flow of energy per unit of biomass and toward increasing organization.

In 1957 Howard T. Odum of the University of Florida published a thorough analysis of the flow of energy through a river in Florida: the famous tourist attraction of Silver Springs. It is of interest to follow the energy flow through this system [see illustration on pages 4 and 5]. During one year each square meter of the surface received 1.7 million kilocalories of solar energy. The green plants of the stream fixed 20,810 kilocalories per year in gross productivity, which represented an efficiency of 1.2 percent of the incident sunlight and 5.1 percent of the sunlight actually absorbed by the green plants. Respiration by the plants accounted for 11,977 kilocalories per square meter per year, so that the net productivity was 8,833 kilocalories per square meter per year. The herbivores converted 1,478 kilocalories per square meter per year into tissue while respiring 1,890 kilocalories. The carnivores had a net productivity of 73 kilocalories per square meter per year and respired 329. The efficiency of conversion to net productivity from the primary level to the secondary level, represented by the herbivores, was 18 percent; to the tertiary level, represented by the carnivores, it was 5 percent. The energy stored in the bodies of the carnivores was only one part in 23,300 parts of incident sunlight—a very small fraction indeed. Hence when man derives energy from a wild animal, he converts only a fraction of 1 percent of solar energy into body tissue. By domesticating plants and animals, however, he has considerably shortened the food chain and increased the efficiency of the total system.

Considering the relation in which man eats beef and beef animals eat corn (a food chain far commoner in the U.S. than in other countries), one can estimate the efficiency of a domesticated system. For simplicity the numbers that follow are only approximate. The cornfield can be considered to convert about 1 percent of solar energy. The beef animal will convert to body tissue about 10 percent of the energy stored in corn, and man will utilize about 10 percent of the energy stored in the tissue of the animal. Hence man derives at best about .01 percent of the incident solar energy through the food chain.

Man's basal metabolism is between 65 and 85 watts, depending on body size, or about .062 calorie per square centimeter per minute. An active adult walking slowly has a metabolic rate of 200 watts; if he walks rapidly, the rate rises to as much as 400 watts. On a daily basis man's minimum energy requirement is about 1,320 kilocalories if he is wholly sedentary. With moderate activity the need approaches 2,400 kilocalories per day. In cold climates the requirement rises to 3,900 kilocalories per day.

Assuming that the normal adult in our society needs 3,000 kilocalories per day, the requirement is equivalent to 30,000 kilocalories per day of beef, which in turn requires 300,000 kilocalories per day of energy in corn, which means 30 million kilocalories per day of sunshine. If the corn is produced in a region with an incident solar radiation of 500 calories per square centimeter per day, one can compute that it takes a cornfield with an area of 60 million square centimeters, or approximately 1.5 acres, to feed one person for one day by means of this food chain. The amount of land needed might be reduced somewhat by improved productivity, but at best it would not be less than one acre per day per person. How does this compare with the amount of land used to feed the world's population today?

Some 3.5 billion (3.5×10^9) acres are under cultivation, and some five billion more acres are used for grazing. That amounts to one acre of cultivated land and 1.5 acres of grazing land for each of the 3.5 billion people now living. The total of potential arable land is estimated at eight billion acres and of grazing land at an additional eight billion acres. Clearly in order to deliver 3,000 kilocalories per day to each person it would be possible to support only about 2.5 times the present population. Either we must increase the productivity per acre or substantially reduce the input per person. Many peoples of the world subsist on about 2,000 kilocalories per day (the global average is 2,350), but human beings at that level cannot be very energetic and cannot function effectively in a complex industrialized society.

This analysis is undoubtedly too simplistic. Even allowing for considerable improvement in productivity, higher levels of protein production, increased land use and extensive use of the oceans, however, it is unlikely that the earth could support more than 10 to 12 billion people reasonably well. If the global population never exceeds eight billion, the chances of feeding them well are much higher and the risks are greatly reduced.

9

HUMAN FOOD PRODUCTION
AS A PROCESS IN THE BIOSPHERE

LESTER R. BROWN
September 1970

Human population growth is mainly the result of increases in food production. This relation raises the question: How many people can the biosphere support without impairment of its overall operation?

Throughout most of man's existence his numbers have been limited by the supply of food. For the first two million years or so he lived as a predator, a herbivore and a scavenger. Under such circumstances the biosphere could not support a human population of more than 10 million, a population smaller than that of London or Afghanistan today. Then, with his domestication of plants and animals some 10,000 years ago, man began to shape the biosphere to his own ends.

As primitive techniques of crop production and animal husbandry became more efficient the earth's food-producing capacity expanded, permitting increases in man's numbers. Population growth in turn exerted pressure on food supply, compelling man to further alter the biosphere in order to meet his food needs. Population growth and advances in food production have thus tended to be mutually reinforcing.

It took two million years for the human population to reach the one-billion mark, but the fourth billion now being added will require only 15 years: from 1960 to 1975. The enormous increase in the demand for food that is generated by this expansion in man's numbers, together with rising incomes, is beginning to have disturbing consequences. New signs of stress on the biosphere are reported almost daily. The continuing expansion of land under the plow and the evolution of a chemically oriented modern agriculture are producing ominous alterations in the biosphere not just on a local scale but, for the first time in history, on a global scale as well. The natural cycles of energy and the chemical elements are clearly being affected by man's efforts to expand his food supply.

Given the steadily advancing demand for food, further intervention in the biosphere for the expansion of the food supply is inevitable. Such intervention, however, can no longer be undertaken by an individual or a nation without consideration of the impact on the biosphere as a whole. The decision by a government to dam a river, by a farmer to use DDT on his crops or by a married couple to have another child, thereby increasing the demand for food, has repercussions for all mankind.

The revolutionary change in man's role from hunter and gatherer to tiller and herdsman took place in circumstances that are not well known, but some of the earliest evidence of agriculture is found in the hills and grassy plains of the Fertile Crescent in western Asia. The cultivation of food plants and the domestication of animals were aided there by the presence of wild wheat, barley, sheep, goats, pigs, cattle and horses. From the beginnings of agriculture man naturally favored above all other species those plants and animals that had been most useful to him in the wild. As a result of this favoritism he has altered the composition of the earth's plant and animal populations. Today his crops, replacing the original cover of grass or forest, occupy some three billion acres. This amounts to about 10 percent of the earth's total land surface and a considerably larger fraction of the land capable of supporting vegetation, that is, the area excluding deserts, polar regions and higher elevations. Two-thirds of the cultivated cropland is planted to cereals. The area planted to wheat alone is 600 million acres—nearly a million square miles, or an area equivalent to the U.S.

east of the Mississippi. As for the influence of animal husbandry on the earth's animal populations, Hereford and Black Angus cattle roam the Great Plains, once the home of an estimated 30 to 40 million buffalo; in Australia the kangaroo has given way to European cattle; in Asia the domesticated water buffalo has multiplied in the major river valleys.

Clearly the food-producing enterprise has altered not only the relative abundance of plant and animal species but also their global distribution. The linkage of the Old and the New World in the 15th century set in motion an exchange of crops among various parts of the world that continues today. This exchange greatly increased the earth's capacity to sustain human populations, partly because some of the crops transported elsewhere turned out to be better suited there than to their area of origin. Perhaps the classic example is the introduction of the potato from South America into northern Europe, where it greatly augmented the food supply, permitting marked increases in population. This was most clearly apparent in Ireland, where the population increased rapidly for several decades on the strength of the food supply represented by the potato. Only when the potato-blight organism (*Phytophthora infestans*) devastated the potato crop was population growth checked in Ireland.

The soybean, now the leading source of vegetable oil and principal farm export of the U.S., was introduced from China several decades ago. Grain sorghum, the second-ranking feed grain in the U.S. (after corn), came from Africa as a food store in the early slave ships. In the U.S.S.R. today the principal source of vegetable oil is the sunflower, a plant that originated on the southern Great Plains of the U.S. Corn, unknown in the Old World before Columbus, is now grown on every continent. On the other hand, North America is indebted to the Old World for all its livestock and poultry species with the exception of the turkey.

To man's accomplishments in exploiting the plants and animals that natural evolution has provided, and in improving them through selective breeding over the millenniums, he has added in this century the creation of remarkably productive new breeds, thanks to the discoveries of genetics. Genetics has made possible the development of cereals and other plant species that are more tolerant to cold, more resistant to drought, less susceptible to disease, more responsive to fertilizer, higher in yield and richer in protein. The story of hybrid corn is only one of many spectacular examples. The breeding of short-season corn varieties has extended the northern limit of this crop some 500 miles.

Plant breeders recently achieved a historic breakthrough in the development of new high-yielding varieties of wheat and rice for tropical and subtropical regions. These wheats and rices, bred by Rockefeller Foundation and Ford Foundation scientists in Mexico and the Philippines, are distinguished by several characteristics. Most important, they are short-statured and stiff-strawed, and are highly responsive to chemical fertilizer. They also mature earlier. The first of the high-yielding rices, IR-8, matures in 120 days as against 150 to 180 days for other varieties.

Another significant advance incorporated into the new strains is the reduced sensitivity of their seed to photoperiod (length of day). This is partly the result of their cosmopolitan ancestry: they were developed from seed collections all over the world. The biological clocks of traditional varieties of cereals were keyed to specific seasonal cycles, and these cereals could be planted only at a certain time of the year, in the case of rice say at the onset of the monsoon season. The new wheats, which are quite flexible in terms of both seasonal and latitudinal variations in length of day, are now being grown in developing countries as far north as Turkey and as far south as Paraguay.

The combination of earlier maturity and reduced sensitivity to day length creates new opportunities for multiple cropping in tropical and subtropical re-

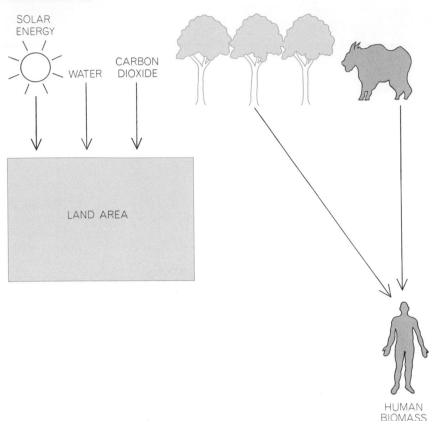

IMPACT OF THE AGRICULTURAL REVOLUTION on the human population is outlined in these two diagrams. The diagram at left shows the state of affairs before the invention of agriculture: the plants and animals supported by photosynthesis on the total land area could support a human population of only about 10 million. The diagram at right shows

gions where water supplies are adequate, enabling farmers to harvest two, three and occasionally even four crops per year. Workers at the International Rice Research Institute in the Philippines regularly harvest three crops of rice per year. Each acre they plant yields six tons annually, roughly three times the average yield of corn, the highest-yielding cereal in the U.S. Thousands of farmers in northern India are now alternating a crop of early-maturing winter wheat with a summer crop of rice, greatly increasing the productivity of their land. These new opportunities for farming land more intensively lessen the pressure for bringing marginal land under cultivation, thus helping to conserve precious topsoil. At the same time they increase the use of agricultural chemicals, creating environmental stresses more akin to those in the advanced countries.

The new dwarf wheats and rices are far more efficient than the traditional varieties in their use of land, water, fertilizer and labor. The new opportunities for multiple cropping permit conversion of far more of the available solar energy into food. The new strains are not the

solution to the food problem, but they are removing the threat of massive famine in the short run. They are buying time for the stabilization of population, which is ultimately the only solution to the food crisis. This "green revolution" may affect the well-being of more people in a shorter period of time than any technological advance in history.

The progress of man's expansion of food production is reflected in the way crop yields have traditionally been calculated. Today the output of cereals is expressed in yield per acre, but in early civilizations it was calculated as a ratio of the grain produced to that required for seed. On this basis the current ratio is perhaps highest in the U.S. corn belt, where farmers realize a four-hundred-fold return on the hybrid corn seed they plant. The ratio for rice is also quite high, but the ratio for wheat, the third of the principal cereals, is much lower, possibly 30 to one on a global basis.

The results of man's efforts to increase the productivity of domestic animals are equally impressive. When the ancestors of our present chickens were domesticated, they laid a clutch of about 15 eggs once a year. Hens in the U.S. today

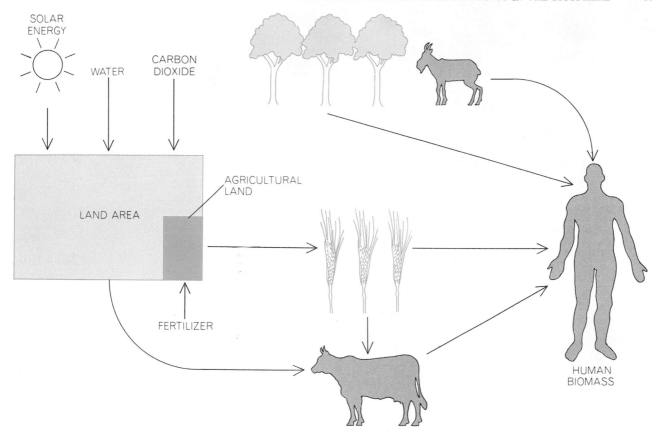

the state of affairs after the invention of agriculture. The 10 percent of the land now under the plow, watered and fertilized by man, is the primary support for a human population of 3.5 billion. Some of the agricultural produce is consumed directly by man; some is consumed indirectly by first being fed to domestic animals. Some of the food for domestic animals, however, comes from land not under the plow (*curved arrow at bottom left*). Man also obtains some food from sources other than agriculture, such as fishing.

average 220 eggs per year, and the figure is rising steadily as a result of continuing advances in breeding and feeding. When cattle were originally domesticated, they probably did not produce more than 600 pounds of milk per year, barely enough for a calf. (It is roughly the average amount produced by cows in India today.) The 13 million dairy cows in the U.S. today average 9,000 pounds of milk yearly, outproducing their ancestors 15 to one.

Most such advances in the productivity of plant and animal species are recent. Throughout most of history man's efforts to meet his food needs have been directed primarily toward bringing more land under cultivation, spreading agriculture from valley to valley and continent to continent. He has also, however, invented techniques to raise the productivity of land already under cultivation, particularly in this century, when the decreasing availability of new lands for expansion has compelled him to turn to a more intensive agriculture. These techniques involve altering the biosphere's cycles of energy, water, nitrogen and minerals.

Modern agriculture depends heavily on four technologies: mechanization, irrigation, fertilization and the chemical control of weeds and insects. Each of these technologies has made an important contribution to the earth's increased capacity for sustaining human populations, and each has perturbed the cycles of the biosphere.

At least as early as 3000 B.C. the farmers of the Middle East learned to harness draft animals to help them till the soil. Harnessing animals much stronger than himself enabled man to greatly augment his own limited muscle power. It also enabled him to convert roughage (indigestible by humans) into a usable form of energy and thus to free some of his energy for pursuits other than the quest for food. The invention of the internal-combustion engine and the tractor 5,000 years later provided a much greater breakthrough. It now became possible to substitute petroleum (the product of the photosynthesis of aeons ago) for oats, corn and hay grown as feed for draft animals. The replacement of horses by the tractor not only provided the farmer with several times as much power but

also released 70 million acres in the U.S. that had been devoted to raising feed for horses.

In the highly mechanized agriculture of today the expenditure of fossil fuel energy per acre is often substantially greater than the energy yield embodied in the food produced. This deficit in the output is of no immediate consequence, because the system is drawing on energy in the bank. When fossil fuels become scarcer, man will have to turn to some other source of motive energy for agriculture: perhaps nuclear energy or some means, other than photosynthesis, of harnessing solar energy. For the present and for the purposes of agriculture the energy budget of the biosphere is still favorable: the supply of solar energy—both the energy stored in fossil fuels and that taken up daily and converted into food energy by crops—enables an advanced nation to be fed with only 5 percent of the population directly employed in agriculture.

The combination of draft animals and mechanical power has given man an enormous capacity for altering the earth's surface by bringing additional

EXPERIMENTAL FARM in Brazil, one of thousands around the world where improvements in agricultural technology are pioneered, is seen as an image on an infrared-sensitive film in the aerial photograph on the opposite page. The reflectance of vegetation at near-infrared wavelengths of .7 to .9 micron registers on the film in false shades of red that are proportional to the intensity of the energy. The most reflective, and reddest, areas (bottom) are land still uncleared of forest cover. Most of the tilled fields, although irregular in shape, are contour-plowed. Regular patterns (left and bottom right) are citrus-orchard rows. The photograph was taken by a National Aeronautics and Space Administration mission in cooperation with the Brazilian government in a joint study of the assessment of agricultural resources by remote sensing. The farm is some 80 miles northwest of São Paulo.

land under the plow (not all of it suited for cultivation). In addition, in the poorer countries his expanding need for fuel has forced him to cut forests far in excess of their ability to renew themselves. The areas largely stripped of forest include mainland China and the subcontinent of India and Pakistan, where much of the population must now use cow dung for fuel. Although statistics are not available, the proportion of mankind using cow dung as fuel to prepare meals may far exceed the proportion using natural gas. Livestock populations providing draft power, food and fuel tend to increase along with human populations, and in many poor countries the needs of livestock for forage far exceed its self-renewal, gradually denuding the countryside of grass cover.

As population pressure builds, not only is more land brought under the plow but also the land remaining is less suited to cultivation. Once valleys are filled, farmers begin to move up hillsides, creating serious soil-erosion problems. As the natural cover that retards runoff is reduced and soil structure deteriorates, floods and droughts become more severe.

Over most of the earth the thin layer of topsoil producing most of man's food is measured in inches. Denuding the land of its year-round natural cover of grass or forest exposes the thin mantle of life-sustaining soil to rapid erosion by wind and water. Much of the soil ultimately washes into the sea, and some of it is lifted into the atmosphere. Man's actions are causing the topsoil to be removed faster than it is formed. This unstable relationship between man and the land from which he derives his subsistence obviously cannot continue indefinitely.

Robert R. Brooks of Williams College, an economist who spent several years in India, gives a wry description of the process occurring in the state of Rajasthan, where tens of thousands of acres of rural land are being abandoned yearly because of the loss of topsoil: "Overgrazing by goats destroys the desert plants which might otherwise hold the soil in place. Goatherds equipped with sickles attached to 20-foot poles strip the leaves of trees to float downward into the waiting mouths of famished goats and sheep. The trees die and the soil blows away 200 miles to New Delhi, where it comes to rest in the lungs of its inhabitants and on the shiny cars of foreign diplomats."

Soil erosion not only results in a loss of soil but also impairs irrigation systems. This is illustrated in the Mangla irrigation reservoir, recently built in the foothills of the Himalayas in West Pakistan as part of the Indus River irrigation system. On the basis of feasibility studies indicating that the reservoir could be expected to have a lifetime of at least 100 years, $600 million was invested in the construction of the reservoir. Denuding and erosion of the soil in the watershed, however, accompanying a rapid growth of population in the area, has already washed so much soil into the reservoir that it is now expected to be completely filled with silt within 50 years.

A historic example of the effects of man's abuse of the soil is all too plainly visible in North Africa, which once was the fertile granary of the Roman Empire and now is largely a desert or near-desert whose people are fed with the aid of food imports from the U.S. In the U.S. itself the "dust bowl" experience of the 1930's remains a vivid lesson on the folly of overplowing. More recently the U.S.S.R. repeated this error, bringing 100 million acres of virgin soil under the plow only to discover that the region's rainfall was too scanty to sustain continuous cultivation. Once moisture reserves in the soil were depleted the soil began to blow.

Soil erosion is one of the most pressing and most difficult problems threatening the future of the biosphere. Each year it is forcing the abandonment of millions of acres of cropland in Asia, the Middle East, North Africa and Central America. Nature's geological cycle continuously produces topsoil, but its pace is far too slow to be useful to man. Someone once defined soil as rock on its way to the sea. Soil is produced by the weathering of rock and the process takes several centuries to form an inch of topsoil. Man is managing to destroy the topsoil

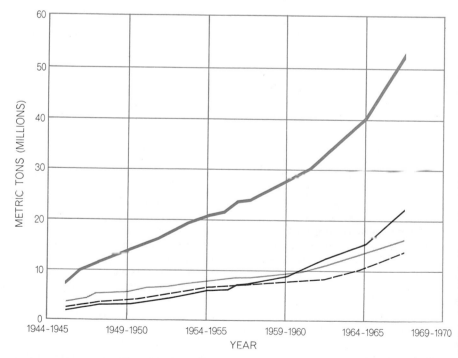

FERTILIZER CONSUMPTION has increased more than fivefold since the end of World War II. The top line in the graph (color) shows the tonnage of all kinds of fertilizers combined. The lines below show the tonnages of the three major types: nitrogen (black), now the leader, phosphate (gray) and potash (broken line). Figures, from the most recent report by the UN Food and Agriculture Organization, omit fertilizer consumption in China.

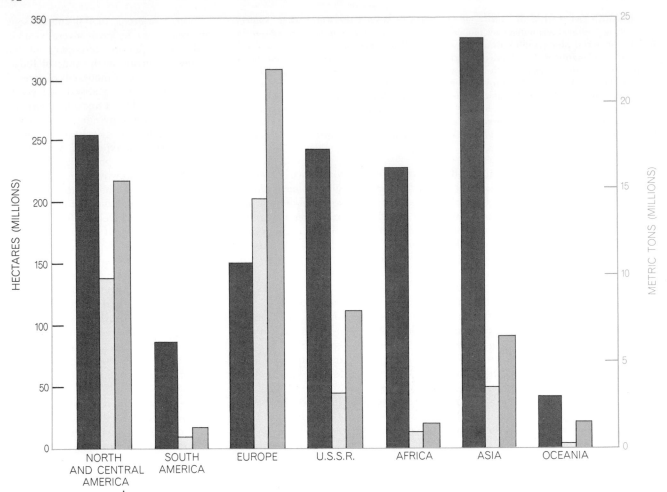

TONS OF FERTILIZER used in seven world areas are compared with the amount of agricultural land in each area. Two tonnages are shown in each instance: the amount used in 1962–1963 (*light color*) and the amount used in 1967–1968 (*solid color*). The great-est use of fertilizer occurs in Europe, the least fertilized area is Africa and the greatest percentage increase in the period was in Australia and New Zealand. Figures, from the Food and Agriculture Organization, omit China, North Korea and North Vietnam.

in some areas of the world in a fraction of this time. The only possible remedy is to find ways to conserve the topsoil more effectively.

The dust-bowl era in the U.S. ended with the widespread adoption of conservation practices by farmers. Twenty million acres were fallowed to accumulate moisture and thousands of miles of windbreaks were planted across the Great Plains. Fallow land was alternated with strips of wheat ("strip-cropping") to reduce the blowing of soil while the land was idle. The densely populated countries of Asia, however, are in no position to adopt such tactics. Their food needs are so pressing that they cannot afford to take large areas out of cultivation; moreover, they do not yet have the financial resources or the technical skills for the immense projects in reforestation, controlled grazing of cattle, terracing, contour farming and systematic management of watersheds that would be required to preserve their soil.

The significance of wind erosion goes

far beyond the mere loss of topsoil. As other authors in this issue have observed, a continuing increase in particulate matter in the atmosphere could affect the earth's climate by reducing the amount of incoming solar energy. This particulate matter comes not only from the technological activities of the richer countries but also from wind erosion in the poorer countries. The poorer countries do not have the resources for undertaking the necessary effort to arrest and reverse this trend. Should it be established that an increasing amount of particulate matter in the atmosphere is changing the climate, the richer countries would have still another reason to provide massive capital and technical assistance to the poor countries, joining with them to confront this common threat to mankind.

Irrigation, which agricultural man began to practice at least as early as 6,000 years ago, even earlier than he harnessed animal power, has played its

great role in increasing food production by bringing into profitable cultivation vast areas that would otherwise be unusable or only marginally productive. Most of the world's irrigated land is in Asia, where it is devoted primarily to the production of rice. In Africa the Volta River of Ghana and the Nile are dammed for irrigation and power purposes. The Colorado River system of the U.S. is used extensively for irrigation in the Southwest, as are scores of rivers elsewhere. Still to be exploited for irrigation are the Mekong of southeastern Asia and the Amazon.

During the past few years there has been an important new irrigation development in Asia: the widespread installation of small-scale irrigation systems on individual farms. In Pakistan and India, where in many places the water table is close to the surface, hundreds of thousands of tube wells with pumps have been installed in recent years. Interestingly, this development came about partly as an answer to a problem that had been presented by irrigation itself.

Like many of man's other interventions in the biosphere, his reshaping of the hydrologic cycle has had unwanted side effects. One of them is the raising of the water table by the diversion of river water onto the land. Over a period of time the percolation of irrigation water downward and the accumulation of this water underground may gradually raise the water table until it is within a few feet or even a few inches of the surface. This not only inhibits the growth of plant roots by waterlogging but also results in the surface soil's becoming salty as water evaporates through it, leaving a concentrated deposit of salts in the upper few inches. Such a situation developed in West Pakistan after its fertile plain had been irrigated with water from the Indus for a century. During a visit by President Ayub to Washington in 1961 he appealed to President Kennedy for help: West Pakistan was losing 60,-000 acres of fertile cropland per year because of waterlogging and salinity as its population was expanding 2.5 percent yearly.

This same sequence, the diversion of river water into land for irrigation, followed eventually by waterlogging and salinity and the abandonment of land, had been repeated many times throughout history. The result was invariably the decline, and sometimes the disappearance, of the civilizations thus intervening in the hydrologic cycle. The remains of civilizations buried in the deserts of the Middle East attest to early experiences similar to those of contemporary Pakistan. These civilizations, however, had no one to turn to for foreign aid. An interdisciplinary U.S. team led by Roger Revelle, then Science Adviser to the Secretary of the Interior, studied the problem and proposed among other things a system of tube wells that would lower the water table by tapping the ground water for intensive irrigation. Discharging this water on the surface, the wells would also wash the soil's salt downward. The stratagem worked, and the salty, waterlogged land of Pakistan is steadily being reclaimed.

Other side effects of river irrigation are not so easily remedied. Such irrigation has brought about a great increase in the incidence of schistosomiasis, a disease that is particularly prevalent in the river valleys of Africa and Asia. The disease is produced by the parasitic larva of a blood fluke, which is harbored by aquatic snails and burrows into the flesh of people standing in water or in water-soaked fields. The Chinese call schistosomiasis "snail fever"; it might also be called the poor man's emphysema, because, like emphysema, this extremely debilitating disease is environmentally induced through conditions created by man. The snails and the fluke thrive in perennial irrigation systems, where they are in close proximity to large human populations. The incidence of the disease is rising rapidly as the world's large rivers are harnessed for irrigation, and today schistosomiasis is estimated to afflict 250 million people. It now surpasses malaria, the incidence of which is declining, as the world's most prevalent infectious disease.

As a necessity for food production water is of course becoming an increasingly crucial commodity. The projected increases in population and in food requirements will call for more and more water, forcing man to consider still more massive and complex interventions in the biosphere. The desalting of seawater for irrigation purposes is only one major departure from traditional practices. Another is a Russian plan to reverse the flow of four rivers currently flowing northward and emptying into the Arctic Ocean. These rivers would be diverted southward into the semiarid lands of southern Russia, greatly enlarging the irrigated area of the U.S.S.R. Some climatologists are concerned, however, that the shutting off of the flow of relatively warm water from these four rivers would have far-reaching implications for not only the climate of the Arctic but also the climatic system of the entire earth.

The growing competition for scarce water supplies among states and among various uses in the western U.S. is also forcing consideration of heroic plans. For example, a detailed engineering proposal exists for the diversion of the Yukon River in Alaska southward across Canada into the western U.S. to meet the growing need for water for both agricultural and industrial purposes. The effort would cost an estimated $100 billion.

Representing an even greater intervention in the biosphere is the prospect that man may one day consciously alter the earth's climatic patterns, shifting some of the rain now falling on the oceans to the land. Among the steps needed for the realization of such a scheme are the construction of a comprehensive model of the earth's climatic system and the development of a computational facility capable of simulating

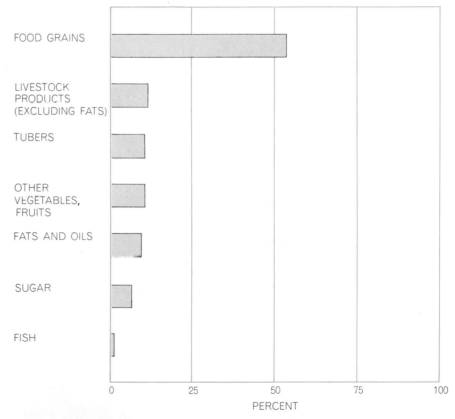

WORLDWIDE FOOD ENERGY comes in different amounts from different products. Cereals outstrip other foodstuffs; wheat and rice each supply a fifth of mankind's food energy.

and manipulating the model. The required information includes data on temperatures, humidity, precipitation, the movement of air masses, ocean currents and many other factors that enter into the weather. Earth-orbiting satellites will doubtless be able to collect much of this information, and the present generation of advanced computers appears to be capable of carrying out the necessary experiments on the model. For the implementation of the findings, that is, for the useful control of rainfall, there will of course be a further requirement: the project will have to be managed by a global and supranational agency if it is not to lead to weather wars among nations working at cross purposes. Some commercial firms are already in the business of rainmaking, and they are operating on an international basis.

The third great technology that man has introduced to increase food production is the use of chemical fertilizers. We owe the foundation for this development to Justus von Liebig of Germany, who early in the 19th century determined the specific requirements of nitrogen, phosphorus, potassium and other nutrients for plant growth. Chemical fertilizers did not come into widespread use, however, until this century, when the pressure of population and the dis-

appearance of new frontiers compelled farmers to substitute fertilizer for the expansion of cropland to meet growing food needs. One of the first countries to intensify its agriculture, largely by the use of fertilizers, was Japan, whose output of food per acre has steadily risen (except for wartime interruptions) since the turn of the century. The output per acre of a few other countries, including the Netherlands, Denmark and Sweden, began to rise at about the same time. The U.S., richly endowed with vast farmlands, did not turn to the heavy use of fertilizer and other intensive measures until about 1940. Since then its yields per acre, assisted by new varieties of grain highly responsive to fertilizer, have also shown remarkable gains. Yields of corn, the production of which exceeds that of all other cereals combined in the U.S., have nearly tripled over the past three decades.

Experience has demonstrated that in areas of high rainfall the application of chemical fertilizers in conjunction with other inputs and practices can double, triple or even quadruple the productivity of intensively farmed soils. Such levels of productivity are achieved in Japan and the Netherlands, where farmers apply up to 300 pounds of plant nutrients per acre per year. The use of chemical fertilizers is estimated to account for at

least a fourth of man's total food supply. The world's farmers are currently applying 60 million metric tons of plant nutrients per year, an average of nearly 45 pounds per acre for the three billion acres of cropland. Such application, however, is unevenly distributed. Some poor countries do not yet benefit from the use of fertilizer in any significant amounts. If global projections of population and income growth materialize, the production of fertilizer over the remaining three decades of this century must almost triple to satisfy food demands.

Can the projected demand for fertilizer be met? The key ingredient is nitrogen, and fortunately man has learned how to speed up the fixation phase of the nitrogen cycle [see the article "The Nitrogen Cycle," by C. C. Delwiche, beginning on page 43]. In nature the nitrogen of the air is fixed in the soil by certain microorganisms, such as those present in the root nodules of leguminous plants. Chemists have now devised various ways of incorporating nitrogen from the air into inorganic compounds and making it available in the form of nitrogen fertilizers. These chemical processes produce the fertilizer much more rapidly and economically than the growing of leguminous-plant sources such as clover, alfalfa or soybeans. More than 25 million tons of nitrogen fertilizer is now being synthesized and added to the earth's soil annually.

The other principal ingredients of chemical fertilizer are the minerals potassium and phosphorus. Unlike nitrogen, these elements are not replenished by comparatively fast natural cycles. Potassium presents no immediate problem; the rich potash fields of Canada alone are estimated to contain enough potassium to supply mankind's needs for centuries to come. The reserves of phosphorus, however, are not nearly so plentiful as those of potassium. Every year 3.5 million tons of phosphorus washes into the sea, where it remains as sediment on the ocean floor. Eventually it will be thrust above the ocean surface again by geologic uplift, but man cannot wait that long. Phosphorus may be one of the first necessities that will prompt man to begin to mine the ocean bed.

The great expansion of the use of fertilizers in this century has benefited mankind enormously, but the benefits are not unalloyed. The runoff of chemical fertilizers into rivers, lakes and underground waters creates two important hazards. One is the chemical pollution

EXPERIMENTAL PLANTINGS at the International Rice Research Institute in the Philippine Republic are seen in an aerial photograph. IR-8, a high-yield rice, was bred here.

RUINED FARM in the "dust bowl" area of the U.S. in the 1930's is seen in an aerial photograph. The farm is near Union in Terry County, Tex. The wind has eroded the powdery, drought-parched topsoil and formed drifts among the buildings and across the fields.

of drinking water. In certain areas in Illinois and California the nitrate content of well water has risen to a toxic level. Excessive nitrate can cause the physiological disorder methemoglobinemia, which reduces the blood's oxygen-carrying capacity and can be particularly dangerous to children under five. This hazard is of only local dimensions and can be countered by finding alternative sources of drinking water. A much more extensive hazard, profound in its effects on the biosphere, is the now well-known phenomenon called eutrophication.

Inorganic nitrates and phosphates discharged into lakes and other bodies of fresh water provide a rich medium for the growth of algae; the massive growth of the algae in turn depletes the water of oxygen and thus kills off the fish life. In the end the eutrophication, or overfertilization, of the lake slowly brings about its death as a body of fresh water, converting it into a swamp. Lake Erie is a prime example of this process now under way.

How much of the now widespread eutrophication of fresh waters is attributable to agricultural fertilization and how much to other causes remains an open question. Undoubtedly the runoff of nitrates and phosphates from farmlands plays a large part. There are also other important contributors, however. Considerable amounts of phosphate, coming mainly from detergents, are discharged into rivers and lakes from sewers carrying municipal and industrial

wastes. And there is reason to believe that in some rivers and lakes most of the nitrate may come not from fertilizers but from the internal-combustion engine. It is estimated that in the state of New Jersey, which has heavy automobile traffic, nitrous oxide products of gasoline combustion, picked up and deposited by rainfall, contribute as much as 20 pounds of nitrogen per acre per year to the land. Some of this nitrogen washes into the many rivers and lakes of New Jersey and its adjoining states. A way must be found to deal with the eutrophication problem because even in the short run it can have damaging effects, affecting as it does the supply of potable water, the cycles of aquatic life and consequently man's food supply.

Recent findings have presented us with a related problem in connection with the fourth technology supporting man's present high level of food production: the chemical control of diseases, insects and weeds. It is now clear that the use of DDT and other chlorinated hydrocarbons as pesticides and herbicides is beginning to threaten many species of animal life, possibly including man. DDT today is found in the tissues of animals over a global range of life forms and geography from penguins in Antarctica to children in the villages of Thailand. There is strong evidence that it is actually on the way to extinguishing some animal species, notably predatory birds such as the bald eagle and the peregrine falcon, whose capacity for using calcium is so impaired by DDT

that the shells of their eggs are too thin to avoid breakage in the nest before the fledglings hatch. Carnivores are particularly likely to concentrate DDT in their tissues because they feed on herbivores that have already concentrated it from large quantities of vegetation. Concentrations of DDT in mothers' milk in the U.S. now exceed the tolerance levels established for foodstuffs by the Food and Drug Administration.

It is ironic that less than a generation after 1948, when Paul Hermann Müller of Switzerland received a Nobel prize for the discovery of DDT, the use of the insecticide is being banned by law in many countries. This illustrates how little man knows about the effects of his intervening in the biosphere. Up to now he has been using the biosphere as a laboratory, sometimes with unhappy results.

Several new approaches to the problem of controlling pests are now being explored. Chemists are searching for pesticides that will be degradable, instead of long-lasting, after being deposited on vegetation or in the soil, and that will be aimed at specific pests rather than acting as broad-spectrum poisons for many forms of life. Much hope is placed in techniques of biological control, such as are exemplified in the mass sterilization (by irradiation) of male screwworm flies, a pest of cattle that used to cost U.S. livestock producers $100 million per year. The release of 125 million irradiated male screwworm flies weekly in the U.S. and in adjoining areas of Mexico (in a cooperative effort with the Mexican government) is holding the fly population to a negligible level. Efforts are now under way to get rid of the Mexican fruit fly and the pink cotton bollworm in California by the same method.

Successes are also being achieved in breeding resistance to insect pests in various crops. A strain of wheat has been developed that is resistant to the Hessian fly; resistance to the corn borer and the corn earworm has been bred into strains of corn, and work is in progress on a strain of alfalfa that resists aphids and leafhoppers. Another promising approach, which already has a considerable history, is the development of insect parasites, ranging from bacteria and viruses to wasps that lay their eggs in other insects. The fact remains, however, that the biological control of pests is still in its infancy.

I have here briefly reviewed the major agricultural technologies evolved to meet man's increasing food needs, the problems arising from them and some possible solutions. What is the present balance sheet on the satisfaction of human food needs? Although man's food supply has expanded several hundredfold since the invention of agriculture, two-thirds of mankind is still hungry and malnourished much of the time. On the credit side a third of mankind, living largely in North America, Europe, Australia and Japan, has achieved an adequate food supply, and for the remaining two-thirds the threat of large-scale famine has recently been removed, at least for the immediate future. In spite of rapid population growth in the developing countries since World War II, their peoples have been spared from massive famine (except in Biafra in 1969–1970) by huge exports of food from the developed countries. As a result of two consecutive monsoon failures in India, a fifth of the total U.S. wheat crop was shipped to India in both 1966 and 1967, feeding 60 million Indians for two years.

Although the threat of outright famine has been more or less eliminated for the time being, human nutrition on the global scale is still in a sorry state. Malnutrition, particularly protein deficiency, exacts an enormous toll from the physical and mental development of the young in the poorer countries. This was dramatically illustrated when India held tryouts in 1968 to select a team to represent it in the Olympic games that year. Not a single Indian athlete, male or female, met the minimum standards for qualifying to compete in any of the 36

track and field events in Mexico City. No doubt this was partly due to the lack of support for athletics in India, but poor nutrition was certainly also a large factor. The young people of Japan today are visible examples of what a change can be brought about by improvement in nutrition. Well-nourished from infancy, Japanese teen-agers are on the average some two inches taller than their elders.

Protein is as crucial for children's mental development as for their physical development. This was strikingly shown in a recent study extending over several years in Mexico: children who had been severely undernourished before the age of five were found to average 13 points lower in I.Q. than a carefully selected control group. Unfortunately no amount of feeding or education in later life can repair the setbacks to development caused by undernourishment in the early years. Protein shortages in the poor countries today are depreciating human resources for at least a generation to come.

Protein constitutes the main key to human health and vigor, and the key to the protein diet at present is held by grain consumed either directly or indirectly (in the form of meat, milk and eggs). Cereals, occupying more than 70 percent of the world's cropland, provide 52 percent of man's direct energy intake. Eleven percent is supplied by livestock products such as meat, milk and eggs, 10 percent by potatoes and other tubers, 10 percent by fruits and vegetables, 9 percent by animal fats and vegetable oils, 7 percent by sugar and 1 percent by fish. As in the case of the total quantity of the individual diet, however, the composition of the diet varies greatly around the world. The difference is most marked in the per capita use of grain consumed directly and indirectly.

The two billion people living in the poor countries consume an average of about 360 pounds of grain per year, or about a pound per day. With only one pound per day, nearly all must be consumed directly to meet minimal energy requirements; little remains for feeding to livestock, which may convert only a tenth of their feed intake into meat or other edible human food. The average American, in contrast, consumes more than 1,600 pounds of grain per year. He eats only about 150 pounds of this directly in the form of bread, breakfast cereal and so on; the rest is consumed indirectly in the form of meat, milk and eggs. In short, he enjoys the luxury of the highly inefficient animal conversion

of grain into tastier and somewhat more nutritious proteins.

Thus the average North American currently makes about four times as great a demand on the earth's agricultural ecosystem as someone living in one of the poor countries. As the income levels in these countries rise, so will their demand for a richer diet of animal products. For the increasing world population at the end of the century, which is expected to be twice the 3.5 billion of today, the world production of grain would have to be doubled merely to maintain present consumption levels. This increase, combined with the projected improvement in diet associated with gains in income over the next three decades, could nearly triple the demand for grain, requiring that the food supply increase more over the next three decades than it has in the 10,000 years since agriculture began.

There are ways in which this pressure can be eased somewhat. One is the breeding of higher protein content in grains and other crops, making them nutritionally more acceptable as alternatives to livestock products. Another is the development of vegetable substitutes for animal products, such as are already available in the form of oleomargarine, soybean oil, imitation meats and other replacements (about 65 percent of the whipped toppings and 35 percent of the coffee whiteners now sold in U.S. supermarkets are nondairy products). Pressures on the agricultural ecosystem would thus drive high-income man one step down in the food chain to a level of more efficient consumption of what could be produced by agriculture.

What is clearly needed today is a cooperative effort—more specifically, a world environmental agency—to monitor, investigate and regulate man's interventions in the environment, including those made in his quest for more food. Since many of his efforts to enlarge his food supply have a global impact, they can only be dealt with in the context of a global institution. The health of the biosphere can no longer be separated from our modes of political organization. Whatever measures are taken, there is growing doubt that the agricultural ecosystem will be able to accommodate both the anticipated increase of the human population to seven billion by the end of the century and the universal desire of the world's hungry for a better diet. The central question is no longer "Can we produce enough food?" but "What are the environmental consequences of attempting to do so?"

10

CHEMICAL FERTILIZERS

CHRISTOPHER J. PRATT
June 1965

In a world that is obliged to produce more food, enriching the soil with elements that are needed for plant growth is a major concern. What are these elements and how can they be artificially supplied?

Whatever estimate one accepts of the increase of the human population in the finite future, or whatever estimate of how long it will take to bring this increase under control, it is clear that the present rate of increase is alarmingly high. Three centuries ago the number of people in the world was probably about 500 million; now it is more than three billion, and if the current rate of increase holds, it will be six billion by the end of the century and millennium. In some underdeveloped areas, where the rate is highest, the Malthusian prediction that population would eventually outrun food supplies seems close to reality.

Clearly mankind faces a formidable problem in making certain that future populations have enough to eat. Doubtless a partial solution lies in improved technology, which has already done so much to keep the food supply abreast of population, and in the spread of existing technology from the developed to the underdeveloped countries. It should also be possible to bring some new areas under cultivation or grazing, but the opportunities in that direction appear to be limited. Even though only about 2.4 billion acres, or approximately 7 percent of the earth's land area, are used for crop production in any one year, most of the unused land is too dry or too cold for agriculture or is in some other way unsuitable. Neither extensive clearing of forests nor large-scale cultivation of tropical lands offers as much promise as one might think, because much of the soil in such regions is lateritic and turns hard as the result of an oxidizing effect when it is put to the plow [see "Lateritic Soils," by Mary McNeil; SCIENTIFIC AMERICAN Offprint 870].

With huge amounts of capital and carefully planned projects it would be possible to create much new cropland by vast undertakings of irrigation, drainage and other kinds of reclamation. Even if such projects were launched, however, they would take decades to complete. It seems more feasible to look to shorter-range ventures, particularly in those developing areas where famine is an imminent threat.

Of all the short-range factors capable of increasing agricultural production readily—factors including pesticides, improved plant varieties and mechanization—the largest yields and the most substantial returns on invested capital come from chemical fertilizers. The application of these substances to underfertilized soils can have dramatic results. In a typical situation the ratio of the extra weight of grain produced per unit weight of nutrients applied can be as high as 10 to 1. To put it another way, an investment of this kind alone can quickly produce increases in crop yields of 100 to 200 percent.

Today some 30 million tons of the so-called primary nutrients—nitrogen, phosphorus and potassium—are annually supplied to world agriculture by chemical fertilizers. This amount is hardly adequate, for reasons I shall discuss. Moreover, crop yields diminish in proportion to the amount of fertilizer applied. Therefore it can be estimated that a population of six billion in the year 2000 will require at least 120 million tons of primary nutrients. An increase of 90 million tons of nutrients for three billion more people means that 60 pounds of primary nutrients will be needed to help sustain each additional person for a year. This is equivalent to about one 100-pound bag of modern high-analysis chemical fertilizer.

Stated in such a way, the amount of effort required to supply the additional fertilizer may seem modest. Actually the expansion of capacity required is enormous; achieving it may well become a major preoccupation of technology. Fortunately processes for manufacturing the needed substances are already well established on a large scale and are capable of rapid expansion, provided that enough capital is made available and the necessary priorities are given. Considering all these factors, it is appropriate to review briefly the fertilizer situation: how plants utilize nutrients, how chemical fertilizers came into use, how they are manufactured, how they are best applied and how the increasing demand for them can be met by chemical technology.

Plants and Nutrients

A growing plant requires most or all of 16 nutrients, nine in large amounts and seven in small. The former are sometimes called macronutrients, the latter micronutrients. Most plants obtain three of the macronutrients—carbon, hydrogen and oxygen—from the air and all the other nutrients from the soil. (A few species, such as clover, are able to fill their nitrogen needs from the air.) The primary soil macronutrients—nitrogen, phosphorus and potassium—are the N, P and K often seen on bags of fertilizer; they are also the substances represented by the set of three figures, such as 10-12-8, that normally designates the nutrient content of a fertilizer. Usually these figures respectively denote the percentage in the fertilizer of total nitrogen (N), of phosphorus pentoxide (P_2O_5, often called phosphoric acid or phosphate) in a form available for use by plants and of water-soluble potassium oxide (K_2O, usually called potash).

The three other soil macronutrients—calcium, magnesium and sulfur—are often called secondary. Agricultural lime, limestone and dolomite, which are used to correct soil acidity, also serve as sources of calcium and magnesium. Sulfur deficiencies can be remedied by certain commercial fertilizers. The seven micronutrients, which are sometimes added in traces to fertilizers providing one or more of the primary nutrients, are boron, copper, iron, manganese, zinc, molybdenum and chlorine.

The growth of plants is a highly complicated process that is far from fully understood. For the purposes of this article it is enough to say that the usual path of mineral nutrients from the soil to the plant is from the solid particles of soil to the water in the soil and thence into the root. The actual transfer of nutrients from soil to root involves the movement of mineral ions. These ions are contained mostly in the soil water, but some of them are adsorbed on solid soil particles.

It follows that nutrients must be in ionic form or capable of transformation to ionic form by soil processes if they are to be of any value to the plant. Hence it is not necessarily a lack of minerals in a soil that causes plants to show signs of nutrient deficiency; the problem can also be that the nutrients are not in a form readily available to the plant. For example, it is quite possible for crops to starve in soils that are amply supplied with phosphorus and potassium if these nutrients are insoluble in water or plant juices. Essentially what the chemical fertilizer industry does, in addition to converting inert nitrogen from the air into soluble salts, is employ processes to "open" the molecules containing the vital nutrients so that these molecules form soluble salts that plants can assimilate readily.

One can best grasp the need for mineral nutrients in agriculture by taking account of the nutrients that are removed from the soil by cropping and grazing. A ton of wheat grain is equivalent to about 40 pounds of nitrogen, eight pounds of phosphorus and nine pounds of potassium. If the straw, husks, roots and other agricultural wastes of such a crop are not returned to the soil, they represent additional large losses of nutrients. A ton of fat cattle corresponds to a depletion of about 54 pounds of nitrogen, 15 pounds of phosphorus, three pounds of potassium and 26 pounds of calcium. Such rates of removal will quickly exhaust a typical soil unless the losses are made up by regular additions of suitable fertilizer.

Equivalent additions of fertilizer, however, are not really enough. There are other factors to be taken into account, and they explain why the present consumption of fertilizers is barely adequate. Nutrients are leached from soils by the flow of water; moreover, they are fixed in forms not readily available to plants. As a result of such losses the proportion of soil nitrogen and phosphorus utilized by a crop is rarely more

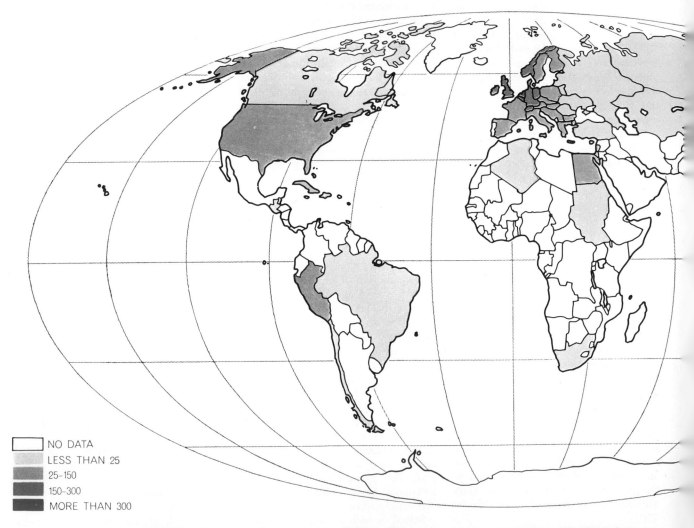

NO DATA
LESS THAN 25
25–150
150–300
MORE THAN 300

USE OF CHEMICAL FERTILIZERS is shown according to data assembled by the Food and Agriculture Organization of the United Nations. Figures give average consumption of fertilizer in metric tons per 1,000 hectares of arable land, defined as land planted to

than 75 percent. In some instances the utilization of phosphorus is as low as 10 percent.

Even allowing for losses, the increased crop value resulting from the proper application of fertilizer can be substantial. On a poor soil the gain can approach 10 times the cost of the material applied. Where the soil is good and the crop yields are high the gain from fertilizing is more likely to be three to five times the cost of the fertilizer. Because of this diminishing return there eventually comes a point at which the additional yield no longer justifies the cost of the corresponding extra fertilizer. There is also an agronomic reason for avoiding the overapplication of fertilizer: ultimately a point can be reached at which the high concentration of nutrient salts in the soil can damage the plants.

Long before men began to write history they knew about the effects of organic, or natural, fertilizers on grow-

crops, in temporary use as meadow for mowing or pasture, or temporarily lying fallow.

ing plants. The effects must surely have been evident in the relatively lush growth induced by animal droppings and carcasses. Eventually farmers began to collect dung and apply it to crops. The first English settlers in North America reported that the Indians substantially increased their yield of maize by burying a fish with each seed they planted. In medieval times farmers in Europe had commonly undertaken to grow nitrogen-converting legumes such as clover and to rotate crops in order to maintain soil fertility. By the early 19th century the use of farm manure, blood, bones, animal wastes and Peruvian guano became widespread, particularly in England, where the Industrial Revolution had brought about a rapid expansion of population and a simultaneous movement of workers from the land to the manufacturing towns. For a time it appeared that the limited supply of organic fertilizer would be insufficient to meet the rising demand for food in the industrializing countries; it is said that even human bones from the battlefields of Europe were recovered, crushed and used as plant foods.

Although the use of organic fertilizers was well established by the 19th century, the basic reasons for their effectiveness were not understood. This lack of knowledge hampered the discovery of alternative substances that could relieve the pressure on the limited supply of organic matter. Another obstacle was the passionate belief held by many that organic materials had special fertilizing properties not shared by inorganic substances. Even after the Swiss chemist Nicolas de Saussure demonstrated in 1804 that plants can grow luxuriantly on carbon and oxygen from the air and mineral nutrients from the soil, strong feelings about organic fertilizers persisted. (Today the view is still sometimes expressed that organic fertilizers possess inexplicable virtues unrelated to their content of primary nutrient. Such materials—manure, sewage sludge, compost and the like are indeed valuable as conditioners of soil and as minor contributors of plant nutrients, but there is not nearly enough organic material to meet present needs, let alone those of the future.)

Gradually the advances of chemistry revealed the processes of plant nutrition and pointed the way toward the substitution of chemical fertilizers for organic fertilizers. In some instances the process was very slow. Nitrogen, for example, was recognized as an important plant nutrient in manures and other organic matter long before anybody un-

derstood the complex cycle by which unreactive atmospheric nitrogen is converted by legumes and soil bacteria into ammonia and soluble nitrogen salts. By the time the process was understood, early in this century, conditions were ripe for a rapid evolution of industrial replacements for organic nitrogen in fertilizer. For one thing, ammonia in the form of ammonium sulfate had become available as a by-product of coal-gas works. For another, mine operators in Chile had begun large-scale production and export of sodium nitrate for use in explosives and other chemicals. As a result of their availability these salts rapidly overtook organic nitrogen as an ingredient of fertilizer. The speed of the transformation is indicated by the fact that the proportion of organic nitrogen materials in fertilizers used in the U.S. fell from 91 percent in 1900 to 40 percent in 1913.

Chilean nitrate was not to hold its position for long. A prolonged effort to synthesize ammonia by combining nitrogen with hydrogen succeeded at last in 1910, when the German chemist Fritz Haber found that the reaction would proceed at high pressure (at least 3,000 pounds per square inch) and in the presence of osmium as a catalyst. The achievement gave rise to a revolution in chemical fertilizer technology. In 1913 Haber and Karl Bosch, having worked out many difficult engineering problems, designed a commercial plant that soon produced 20 tons of ammonia a day. The requirements of the two world wars made ammonia available on a large scale, together with such derivatives as ammonium nitrate and urea. These compounds in time largely replaced Chilean nitrate as a source of nitrogen and also reduced the proportion of fertilizers containing organic nitrogen to a few percent of total fertilizer consumption.

Phosphorus moved from the organic to the chemical stage in fertilizer sooner than nitrogen but by a similarly slow process. The first association of phosphorus with bones was made by the Swedish mineralogist and chemist Johan Gottlieb Gahn in 1769. It took until 1840, however, for chemistry to advance to the stage where it was possible to recognize that phosphorus was the key ingredient in the bone manure that had come into wide use. In that same year the great German chemist Justus von Liebig, who is regarded by many scholars as the founder of agricultural chemistry, put forward the thesis that the action of sulfuric acid on bones would make the phosphorus in the bones more readily available to plants.

CHEMICAL FERTILIZER is meticulously placed on a field in Oklahoma by a spreader 34 feet wide. The spreader is of the "drill" type, meaning that it lays fertilizer in precise rows instead of broadcasting it generally over the field as would be done by other types of spreaders.

TREATMENT PLANT of the V-C Chemical Company in Florida removes organic material and some carbon dioxide from phosphate rock to provide a raw material for making phosphoric acid, which is used in the manufacture of such chemical fertilizers as triple superphosphate and diammonium phosphate. Piles of phosphate rock as brought from the mine are at top right. Horizontal tube in foreground is a calcine kiln in which rock is given thermal treatment. Treated rock is stored in tanks behind kiln until shipped.

This idea was promptly developed in England, where the need for additional sources of fertilizer was acute. In 1842 John Bennet Lawes, a wealthy farmer and industrialist who spent many years conducting agricultural experiments on his estate at Rothamsted, obtained a patent covering the treatment of bones and bone ash with sulfuric acid to make an improved phosphorus-containing fertilizer. Significantly he included "other phosphoritic substances" in his patent, indicating that he foresaw the role of minerals as sources of phosphate. With-

in 20 years the production in Britain of "chemical manures" made from sulfuric acid, local coprolites (fossil manures) and various phosphatic minerals had risen to a level of 200,000 tons a year. The phosphate fertilizer industry, thus firmly established, spread rapidly to other countries. Toward the end of the 19th century slag removed during the production of iron and steel from high-phosphate ores became another major source of phosphorus for agricultural purposes in Britain and Europe, where even now several million tons of "basic

slag" are used annually as a phosphate fertilizer.

As for potassium, the benefits of adding wood ashes ("pot ash") to the soil must have been recognized in ancient times. By early in the 19th century the progress of chemistry was sufficient for a start to be made in the use of potassium chloride deposits in Germany and France as sources of potassium in fertilizer. The first factory producing potash from these deposits was built in 1861. Germany and France continued to be the principal sources of potash until

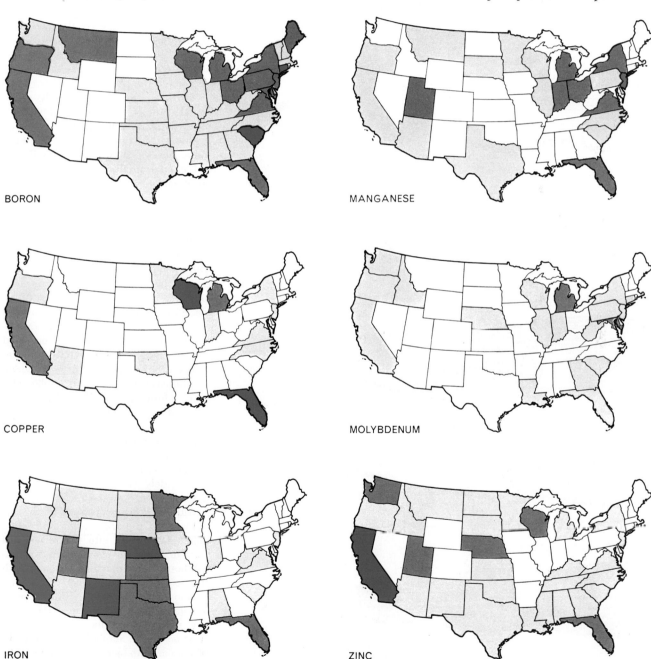

BORON

MANGANESE

COPPER

MOLYBDENUM

IRON

ZINC

NUTRIENT DEFICIENCIES appearing in various parts of the U.S. mainland are indicated. The findings, based on work done by K. C. Berger of the University of Wisconsin, pertain to several "micronutrients," meaning minerals needed by plants in small but important amounts. "Macronutrients" such as nitrogen, phosphorus and potassium are needed by plants in large amounts and usually must be supplied wherever commercial crops are grown. The colors indicate the degree of deficiency from modest (*light*) to severe (*dark*) as based on reports of the number of crops affected. The absence of color means that the state has not reported a deficiency.

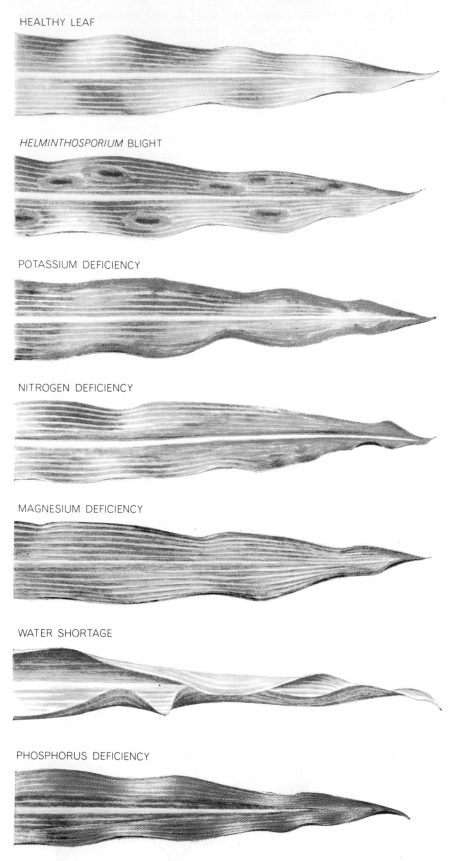

HEALTHY LEAF

HELMINTHOSPORIUM BLIGHT

POTASSIUM DEFICIENCY

NITROGEN DEFICIENCY

MAGNESIUM DEFICIENCY

WATER SHORTAGE

PHOSPHORUS DEFICIENCY

CORN-LEAF VARIATIONS directly or indirectly related to the amount of nutrients and water available to the plant are depicted. Gray represents green; the other colors are approximately as they appear in nature. *Helminthosporium* blight is a common fungus disease to which poorly nourished plants are vulnerable. Signs of potassium deficiency usually appear at the tips and along the edges of the lower leaves; of nitrogen deficiency, at the leaf tip, and of phosphorus, on young plants. Water shortage makes leaves a grayish-green.

rather recently, when major deposits were developed in the U.S., Israel, the U.S.S.R. and Canada.

Modern Fertilizer Production

Today a farmer can buy a wide variety of chemical fertilizers. If he wants only one nutrient, he can find a fertilizer that provides it; he can also find fertilizers that contain almost any combination of nitrogen, phosphorus, potassium and the micronutrients. The industry that produces them is enormous, having a worldwide output, according to a recent estimate by the Food and Agriculture Organization, of more than 33 million tons a year. I shall briefly describe the processes now involved in producing the primary nutrients.

Synthetic ammonia is firmly established as the principal source of nitrogen in fertilizer. Ammonia synthesis remains unchanged in principle from the technique developed by Haber and Bosch. Large-scale production often presents additional problems, however, because of the need to obtain huge supplies of pure gaseous nitrogen and hydrogen at low cost. Pure nitrogen can be produced in quantity with relative ease by removing oxygen and other gases from air through liquefaction or combustion. Hydrogen is another matter. Some early ammonia plants used hydrogen made by electrolysis, but the prohibitive cost led to a search for cheaper sources. Methods for producing hydrogen from solid fuels such as coal and lignite were developed in Europe. In the U.S., where natural gas is plentiful, the simpler catalytic re-forming of methane has proved an ideal way of making hydrogen. More recently the catalytic re-forming of light petroleum fractions such as naphtha with the aid of steam and the partial oxidation of heavy oil with oxygen have been widely used in countries that lack natural gas.

Although there is a strong trend, particularly in the U.S., toward injecting ammonia directly into the soil in the form of anhydrous ammonia or aqueous solutions, most agricultural ammonia is still converted into solid derivatives. Ammonium nitrate is a form popular among manufacturers, since the nitric acid needed to produce it is also made from ammonia. Similarly, large amounts of urea are produced by combining ammonia with carbon dioxide derived from oxidation of the raw material used to produce the hydrogen. Ammonium sulfate is also made on a large scale by reacting ammonia with sulfuric acid. In the Far East substantial quantities of

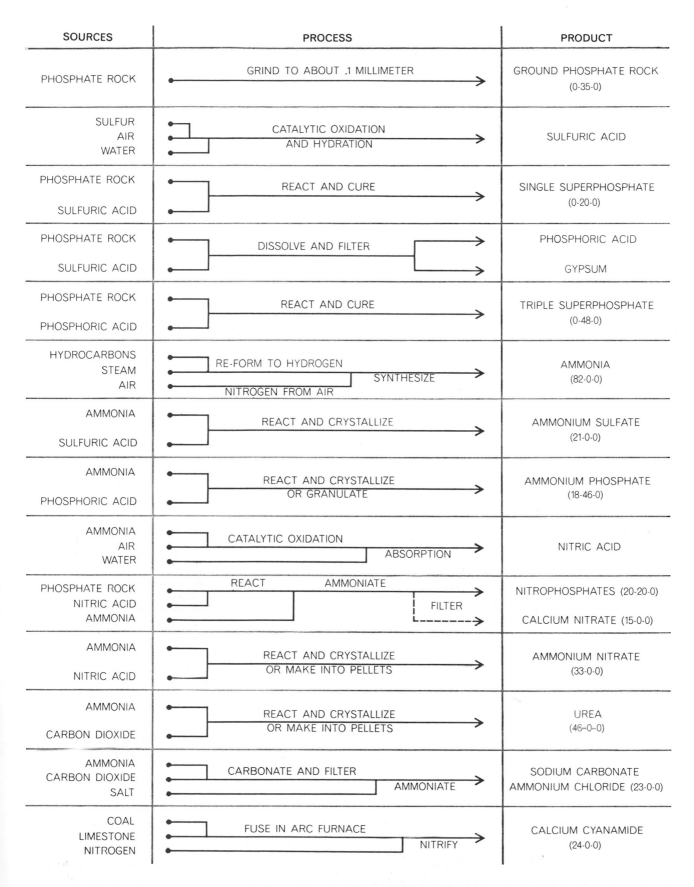

SOURCES	PROCESS	PRODUCT
PHOSPHATE ROCK	GRIND TO ABOUT .1 MILLIMETER	GROUND PHOSPHATE ROCK (0-35-0)
SULFUR AIR WATER	CATALYTIC OXIDATION AND HYDRATION	SULFURIC ACID
PHOSPHATE ROCK SULFURIC ACID	REACT AND CURE	SINGLE SUPERPHOSPHATE (0-20-0)
PHOSPHATE ROCK SULFURIC ACID	DISSOLVE AND FILTER	PHOSPHORIC ACID GYPSUM
PHOSPHATE ROCK PHOSPHORIC ACID	REACT AND CURE	TRIPLE SUPERPHOSPHATE (0-48-0)
HYDROCARBONS STEAM AIR	RE-FORM TO HYDROGEN SYNTHESIZE NITROGEN FROM AIR	AMMONIA (82-0-0)
AMMONIA SULFURIC ACID	REACT AND CRYSTALLIZE	AMMONIUM SULFATE (21-0-0)
AMMONIA PHOSPHORIC ACID	REACT AND CRYSTALLIZE OR GRANULATE	AMMONIUM PHOSPHATE (18-46-0)
AMMONIA AIR WATER	CATALYTIC OXIDATION ABSORPTION	NITRIC ACID
PHOSPHATE ROCK NITRIC ACID AMMONIA	REACT AMMONIATE FILTER	NITROPHOSPHATES (20-20-0) CALCIUM NITRATE (15-0-0)
AMMONIA NITRIC ACID	REACT AND CRYSTALLIZE OR MAKE INTO PELLETS	AMMONIUM NITRATE (33-0-0)
AMMONIA CARBON DIOXIDE	REACT AND CRYSTALLIZE OR MAKE INTO PELLETS	UREA (46-0-0)
AMMONIA CARBON DIOXIDE SALT	CARBONATE AND FILTER AMMONIATE	SODIUM CARBONATE AMMONIUM CHLORIDE (23-0-0)
COAL LIMESTONE NITROGEN	FUSE IN ARC FURNACE NITRIFY	CALCIUM CYANAMIDE (24-0-0)

BASIC PROCESSES used in manufacturing the major kinds of chemical fertilizers are charted. Each horizontal line shows the flow of the ingredient listed at left opposite the line. A vertical line shows a combining of ingredients. Numbers in parentheses show respectively the typical percentage of nitrogen, phosphorus and potassium materials used as fertilizer. For example, 0-35-0 means no nitrogen, 35 percent phosphorus pentoxide and no potassium oxide. Figures thus show amounts of primary nutrients.

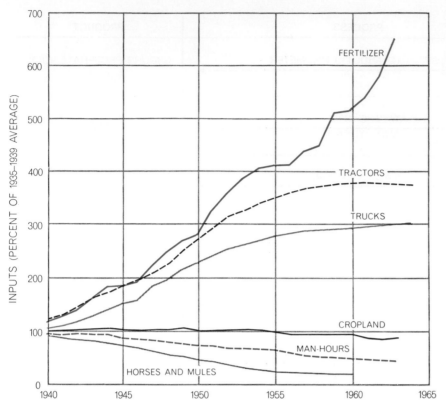

CHANGED TECHNOLOGY of U.S. agriculture over the past 30 years is reflected in a comparison of current inputs with those of 1935 through 1939. The changes are expressed as percentages of the average input in each category for the five-year base period. Concurrent with these changes of input has been a steady rise in the nation's agricultural output.

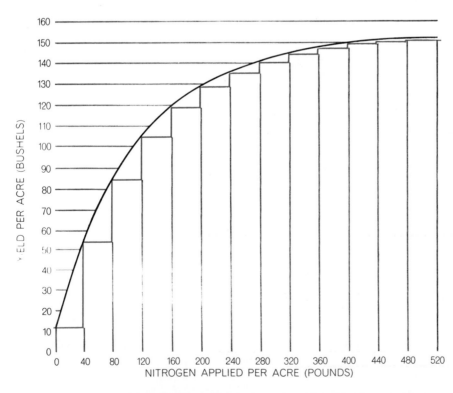

FERTILIZER ECONOMICS are indicated by the curve of yield compared with applications of fertilizer. This curve is based on a crop of irrigated corn grown in the state of Washington. With nitrogen as with other fertilizers, crop yields diminish with increasing applications of nutrients. In time additional increments of fertilizer become uneconomical.

ammonium chloride are made from ammonia and salt or hydrochloric acid. Ammonium phosphates and nitrophosphates are additional fertilizers derived from ammonia. A principal advantage of most solid forms of ammonia is the ease with which they can be transported and applied to the soil. The high nitrogen content of urea (46 percent) and ammonium nitrate (33.5 percent) make them particularly advantageous.

Most phosphate fertilizers now come from mineral deposits, chiefly those in Florida, the western U.S., North Africa and parts of the U.S.S.R. Although both igneous and sedimentary phosphate deposits exist, about 90 percent of the world's fertilizer needs are supplied from the sedimentary sources because they are more plentiful than the igneous minerals and also easier to mine and process. The origin of sedimentary phosphates has generated much speculation among geologists. Some of them believe that the minerals were precipitated from seawater after it had been saturated with phosphate and fluorine ions derived from the contact of the water with igneous rocks and gases. It is also possible that these phosphates resulted to some extent from the replacement of calcium carbonate with calcium phosphate in particles of the mineral aragonite on the ocean floor, a slow process that may still be taking place. Marine deposits of this nature may well become future sources of phosphate.

In any event, most of the primary deposits of sedimentary phosphate were laid down on ocean floor that subsequently became dry land. In time the weathering of such areas removed cementing substances such as calcium carbonate and magnesium carbonate, leaving extensive deposits of phosphate in the form of small pellets. Some of these deposits were later moved by surface water and redeposited elsewhere. Because of this extensive redeposition, and because pellet phosphates are insoluble in water, few minerals are found more widely scattered. By the same token, few have been formed over a longer span of time; phosphate minerals were laid down over the 400 million years from the Ordovician period to the Tertiary period and even later.

Often the phosphate pellets are covered by several feet of sand, clay or leached ore that must be taken off by scrapers or draglines before the phosphate matrix can be removed. In the extensive operations in Florida the matrix is excavated, dropped into sumps, slurried with powerful jets of water and then pumped to the processing plants. The material thus obtained may be only about 15 percent phosphate because of the large amounts of sand and clay in the matrix. Much of the sand and clay is removed by various processes to yield concentrates containing 30 to 36 percent phosphate. These concentrates are then blended and dried before further processing or shipment. Somewhat different methods are used in North Africa; there large tonnages of high-grade phosphate rock are mined by underground methods. Often they are only crushed, screened and dried before shipment.

Several types of fertilizer are made from the phosphate rock processed by the methods I have described. The simplest type consists of high-grade rock ground to particles less than .1 millimeter in size. This type is used directly on acid soils, which slowly attack the water-insoluble phosphate to make it available to plants. Next in simplicity is superphosphate, made by mixing ground phosphate rock with sulfuric acid to form a slurry that quickly hardens in a curing pile. After several weeks the hardened superphosphate is excavated and pulverized; often the powder is formed into granules. The pulverized or granulated material is marketed either alone, as a phosphate fertilizer containing about 18 percent water-soluble phosphorus pentoxide, or in conjunction with other fertilizer materials. The various processing steps convert insoluble tricalcium phosphate to water-soluble monocalcium phosphate and gypsum.

Gypsum, however, is of little use in soil except when deficiencies in calcium or sulfur exist or when salinity is excessive. It also has a diluting effect on the phosphorus pentoxide content. Therefore it was a substantial advance when methods were devised for producing monocalcium phosphate without gypsum. The technique is to dissolve phosphate rock in a mixture of sulfuric and phosphoric acid to form gypsum and additional phosphoric acid, which can be separated by filtration. Thereafter the gypsum is usually discarded; the phosphoric acid is concentrated and mixed with finely ground phosphate rock to form a slurry that soon hardens into a product known as triple superphosphate. Its content of water-soluble phosphorus pentoxide is about 48 percent. Moreover, the product is cheaper to transport and to apply per unit of phosphorus pentoxide than ordinary superphosphate.

Substantial tonnages of phosphate fertilizers are also made by treating

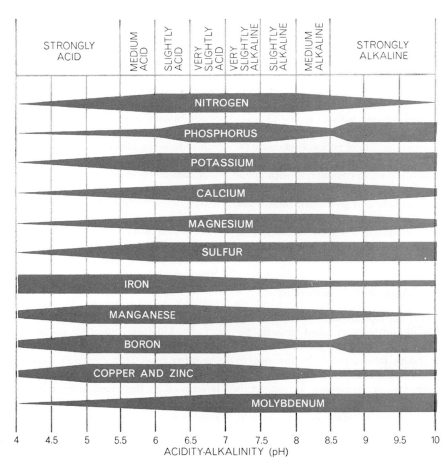

AVAILABILITY OF NUTRIENTS to plants is affected by the condition of the soil. The more soluble a nutrient is under a particular condition of soil acidity or alkalinity, the thicker is the horizontal band representing the nutrient. Solubility in turn is directly related to the availability of the nutrient in an ionic form that is assimilable by the plant.

phosphate rock with nitric acid and ammonia to yield a range of materials that contain nitrogen as well as phosphorus. Potash can be added to form high-analysis fertilizers with a content of primary nutrients as high as 60 percent, as for example in a 20-20-20 grade. Another popular fertilizer is diammonium phosphate, which is made by neutralizing phosphoric acid with ammonia to yield a material containing about 20 percent nitrogen and 50 percent water-soluble phosphorus pentoxide. Potash can be added to this product to make another high-analysis mixture containing all the primary nutrients.

Potassium exists in enormous quantities in the rocks and soils of the world. Often, however, it is in the form of insoluble minerals unsuitable for agriculture. Fortunately large deposits of soluble potassium chloride are available, mostly as sylvite and sylvinite or, in conjunction with magnesium, as carnallite and langbeinite. Such deposits are often mixed with sodium chloride in the form of halite, which is toxic to many crops and must be removed.

Extensive supplies of sylvite and carnallite were found first in Germany and later in France, the western U.S. and many other countries. Most of these deposits resulted from the evaporation of ancient seas during the Permian period (about 230 to 280 million years ago). In the Canadian province of Saskatchewan huge quantities of sylvite and carnallite were more recently found at depths of 3,000 to 4,000 feet, in the upper portion of a Devonian halite formation. Although these deposits are considerably deeper than U.S. and European potash sources, mining difficulties have now been overcome and the production of several million tons annually of Canadian potash will be of great benefit to world agriculture. Another Canadian development of growing importance is the large-scale production of potash by solution mining, which involves pumping water into the potash beds and bringing the resulting solution to the surface for evaporation and the recovery of potash in solid form.

After solid potash minerals are mined they are sometimes crushed and separated from their impurities by washing and froth-flotation, in which treatment with amine salts and air causes the sylvite particles to float away from the unwanted substances. In other cases potash is recovered by solution and crystallization. The relatively pure product is dried, treated with an amine anticaking agent and sold for agricultural purposes as muriate of potash containing 60 to 62 percent of potassium oxide. Most potash is used in conjunction with nitrogen and phosphorus compounds. Potassium sulfate and potassium nitrate are also used to a limited extent in agricultural situations where the chloride ions of potash would be harmful, as they are to tobacco.

Agronomic Considerations

It is appropriate now to consider the role of nutrients in plant growth, together with some other factors that must be taken into account in the use of fertilizers. As anyone experienced in agronomy or gardening knows, it is wasteful and sometimes even harmful to broadcast fertilizer indiscriminately.

The grower must know the condition of his soil and treat it accordingly. In most cases he must apply the bulk of the treatment before sowing, because as a rule most of the nutrient needed by a plant is taken up in the early stages of its growth. The correct nutrient balance is additionally important because a deficiency of any one plant food in the soil will reduce the effect of others, even if they are in oversupply.

A deficiency of nitrogen usually appears in plants as a yellowing of the leaves, accompanied by shriveling that proceeds upward from the lower leaves. The principal effects of nitrogen on plants include accelerated growth and increased yield of leaf, fruit and seed. Nitrogen also promotes the activity of soil bacteria. Nitrate nitrogen is quickly available to root systems, but it may therefore make the plants grow too rapidly. Moreover, nitrate is easily lost by leaching. Ammoniacal nitrogen, on the other hand, is immediately fixed in the soil by ion-exchange reactions and is released to the plants over a longer period than nitrate nitrogen. For these reasons it is sometimes the practice to inject free ammonia in anhydrous or aqueous form a few inches below the surface of a moist soil. With many crops optimum results are obtained by the proper combination of nitrate nitrogen and ammoniacal nitrogen in either solid or liquid form.

Phosphorus deficiency is often represented by purplish leaves and stems, slow growth and low yields. Phosphorus stimulates the germination of seedlings and encourages early root formation. Since these results are less evident than those induced by applications of nitrogen, many farmers, particularly in the Far East, use insufficient quantities of phosphate fertilizer.

Potassium deficiency can often be detected by a spotting or curling of lower leaves. Additional symptoms are weak stalks and stems, a condition that can cause heavy crop losses in strong winds and heavy rains. The application of potassium improves the yield of grain and seed, and it enhances the formation of starches, sugars and plant oils. It also contributes to the plant's vigor and its resistance to frost and disease.

As for deficiencies of secondary nutrients, a lack of magnesium may cause a general loss of color, weak stalks and white bands across the leaves in corn and certain other plants. A calcium deficiency may give rise to the premature death of young leaves and poor formation of seed. An inadequate supply of sulfur frequently leads to pale leaves, stunted growth and immature fruit.

Typical examples of micronutrient deficiency are heart rot in vegetables and fruits as a result of a shortage of boron and stunted growth of vegetables and citrus plants resulting from insufficient manganese and molybdenum. Micronutrient deficiencies may be hard to detect and even harder to rectify, because the balance between enough of a micronutrient and a toxic oversupply can be delicate.

An important consideration in the use of fertilizers is the acidity of the soil, which considerably influences the availability of many nutrients to the plant [see illustration on preceding page]. To complicate matters, nitrogen fertilizers such as ammonia, urea ammonium nitrate and other ammonia derivatives can themselves raise the acidity of soils, by means of complex ion-exchange reactions. In most cases the acidity of a soil can be controlled by adding appropriate amounts of lime, ground limestone or other forms of calcium carbonate.

Soil tilth, or structure, is also important. For example, the richer chernozem soils found in the middle of the

SUBSTANCE	APPROXIMATE POUNDS PER ACRE	SUPPLIED BY
NITROGEN	310	
PHOSPHORUS	120 (PHOSPHATE) 52 (PHOSPHORUS)	1,200 POUNDS OF 25-10-20 FERTILIZER
POTASSIUM	245 (POTASH) 205 (POTASSIUM)	
CALCIUM	58	APPROXIMATELY 150 POUNDS OF AGRICULTURAL LIMESTONE
MAGNESIUM	50	APPROXIMATELY 275 POUNDS OF EPSOM SALT, OR 550 POUNDS SULFATE OF POTASH-MAGNESIA
SULFUR	33	33 POUNDS OF SULFUR
IRON	3	15 POUNDS OF IRON SULFATE
MANGANESE	.45	APPROXIMATELY 1.3 POUNDS OF MANGANESE SULFATE
BORON	.10	APPROXIMATELY 1 POUND OF BORAX
ZINC	TRACE	SMALL AMOUNT OF ZINC SULFATE
COPPER	TRACE	SMALL AMOUNT OF COPPER SULFATE OR OXIDE
MOLYBDENUM	TRACE	VERY SMALL AMOUNT OF SODIUM OR AMMONIUM MOLYBDATE
OXYGEN	10,200	AIR
CARBON	7,800	AIR
WATER	3,225 TO 4,175 TONS	29 TO 36 INCHES OF RAIN

NUTRIENTS REQUIRED to produce 150 bushels of corn are indicated. Most plants take all their nutrients from the soil except carbon, oxygen and hydrogen, obtained from the air.

North American continent and in the Ukraine are in many cases well supplied with organic humus and lime salts and need only regular supplies of plant nutrients to replace those removed by agriculture and leaching. On the other hand, the podzol soils that cover the northeastern U.S., most of Britain and much of central Europe have been intensively leached by centuries of farming and exposure; they need not only liberal supplies of plant nutrients but also lime and organic humus. Desert soils may be rich in certain minerals and yet lacking in available nutrients and in the organic matter usually necessary to retain moisture and to provide good tilth. Such soils can be made productive, however, as has been amply demonstrated in Israel.

Prospective Developments

In spite of the many improvements made in chemical fertilizers during the past 50 years, several problems still confront the fertilizer industry. One major concern is achieving the controlled release of nutrients so that waste and also damage to young plants can be avoided. Methods now being tested include the use of slowly decomposing inorganic materials such as magnesium ammonium phosphate and synthetic organic compounds such as formamide and oxamide. Another technique being studied is the encapsulation of fertilizer particles with sulfur or plastic. Investigators are also exploring the possibilities of producing chemical fertilizers in which a plant nutrient would be "sequestered" in molecules of the chelate type. Chelation involves a tight molecular bonding that would protect the nutrient against rapid attack. In this way the desired plant food would be released slowly and in a prescribed manner by chemical reactions in the soil. An ultimate possibility is the production of "packaged" granules, each containing a seed and whatever substances are needed during the lifetime of the plant. They would be released in the proper amounts and sequence.

A new agricultural technique already in use on a small scale is "chemical plowing." Instead of turning stubble and cover crops into the ground mechanically, the farmer kills them by spraying them with the appropriate herbicides. Eventually the dead plant materials become sources of humus and plant nutrient. Any excess of herbicide is rendered harmless by the action of soil colloids. New seeds and fertilizer are drilled directly through the dead cover material, which also gives protection against erosion, frost and drought.

Efforts are also under way to reduce the cost of transporting fertilizers and their raw materials. The approach here is to try to produce them in highly concentrated liquid or solid form. They are then appropriately diluted or combined at the point of use.

Perhaps the most vital work is the education of farmers—particularly farmers in the developing countries—in modern agricultural methods, including the use of chemical fertilizers. In addition the developing nations must establish low-cost credit plans so that impoverished farmers can buy adequate supplies of fertilizer. Similarly, credit must be extended by the developed nations to the less developed ones on an even bigger scale than at present in order to help the less developed nations obtain the materials, equipment and expert advice they need to build their own chemical fertilizer plants. Until these steps are taken to spread modern agricultural technology, the developing nations will fall far short of the contribution they could make to the intensifying problem of producing enough food for the world's growing population.

11

INSECTS V. INSECTICIDES

ROBERT L. METCALF
October 1952

The increasing resistance of flies and mosquitoes to DDT is causing private and public concern. It is unlikely, however, that we will have to return to sticky paper and fly swatters

IT became a matter of common observation during the past summer that DDT no longer kills houseflies and mosquitoes with its original potency. There were complaints as long ago as 1948. Housewives, dairymen and hog farmers in many parts of the U. S. first blamed the pesticide manufacturers and pest-control contractors for skimping on their product or performance. Meanwhile, however, scientific investigation had shown that the explanation and cure of the situation must be sought in other quarters. By fundamental biochemical and genetic processes flies, mosquitoes and other pests have acquired a surprising immunity to DDT and many of the other synthetic organic pesticides that made their triumphant entry during World War II and its immediate aftermath. Today it is clear that these chemicals have won only a battle, not the war, against insects.

There was good reason at the time for believing that man at last had the weapons for ridding himself of hordes of insects and the plagues which they disseminate. DDT, lindane, toxaphene, chlordane and dieldrin had extraordinary effectiveness against a great variety of insects. They not only killed on fresh contact but provided long-lasting residual control when sprayed on the interiors of homes and buildings. Moreover, they could be produced cheaply and in unlimited quantities.

The first achievements which ushered in this new era of insect control seemed to be decisive. The mass application of DDT powder arrested an epidemic of louse-borne typhus in Naples in the initial period of Allied occupation in 1944. After the war the residual spraying of dwellings, barns and other shelters, and the larviciding of breeding places with DDT, virtually eliminated the anopheline mosquito vectors of malaria from Sardinia; on that island malaria is still a rarity. Conquest of this and other mosquito-transmitted diseases such as yellow fever, dengue, filariasis and encephalitis seemed certain. Wholesale spraying of towns and villages with DDT reduced the once ubiquitous housefly nearly to the vanishing point, and there was even talk of its extermination. In 1947 and 1948 we were hearing that "the flies in Iowa can now be counted on the fingers of one hand" or that "Idaho investigators were unable to find a single fly for their experiments." It was small wonder that scientists and laymen alike received with marked

TOXICITY of insecticides is tested at University of California by applying a microliter droplet containing a measured amount of poison to a housefly.

skepticism the first reports of the failure of DDT to kill flies.

IN 1947 R. Wiesmann of Switzerland reported that a strain of flies from Sweden, where poor control had been obtained with DDT the year before, was resistant to several hundred times the normal dosage of DDT. The same year the Italian entomologist G. Sacca described another strain of flies resistant to DDT. Such reports soon became common in the literature. Whereas from .03 to .1 micrograms of DDT was ordinarily lethal to the housefly, some of the resistant flies could survive the direct application of 100 micrograms. Such insects lived and reproduced normally in cages literally frosted with DDT.

Entomologists turned to other chemicals for fly control. At first lindane, chlordane and dieldrin all proved highly effective against these DDT-resistant flies, even when used in dosages of one-tenth to one-half of those originally required with DDT. Investigators were momentarily reassured. The flies might develop resistance to these newer chemicals after several years, but by then still newer insecticides would be available, or the flies would have lost their resistance to DDT. Then came discouragement. From southern California R. B. March and the author reported that lindane ceased to be effective after three monthly applications at a poultry ranch where there was a population of DDT-resistant flies. We found that this new strain of flies was also far more resistant to DDT than before and, furthermore, that the flies were resistant to dieldrin although they had never been exposed to this chemical. A similar experience was reported by H. F. Schoof and his coworkers of the U. S. Public Health Service. They had found dieldrin most promising for fly control in a city-wide campaign at Phoenix, Ariz. Residual applications in the fall of 1949 obtained effective fly control for 9 to 11 weeks. By the spring of 1950, however, the same applications showed only short-term results, and by midsummer dieldrin had no measurable effect on the fly population at all.

About this time it appeared that DDT-resistant races of mosquitoes were developing. This was first reported for the common house-mosquito in Italy by E. Mosna, who was able to continue satisfactory control by using chlordane and lindane. Florida salt-marsh mosquitoes and California pest mosquitoes also developed resistance, first to DDT and then to toxaphene and aldrin, which were subsequently used in control operations. At present larval mosquitoes from several areas of California show from 10 to 1,000 times their former resistance to these insecticides. As a result some mosquito-control agencies are now returning to the "old fashioned" oiling of the breeding waters.

From the Korean war area just recently has come word that a strain of human body-lice is showing resistance to DDT powder containing some 40 times the normally lethal concentration. Military sanitarians have been forced to employ pyrethrum and lindane powders for louse control.

The ability of important disease-carrying insects to develop immunity to toxicants that formerly threatened them with extinction is most disturbing to public health workers. Much of their thinking has been oriented around the successful use of insecticides. The same is true in the field of agriculture. The first results were so spectacular and apparently so decisive that it is clearly expedient to bend every effort to find out why the tables have been turned. Let us inquire closely then into what is known about insect resistance to insecticides and what can be done about it.

THAT insects may acquire resistance to insecticides is not a recent discovery. This was demonstrated early in this century by investigators who were coping with the several varieties of scale that bedevil fruit growers. In 1915 A. L. Melander of Washington State College showed that an acquired resistance accounted for the diminishing effectiveness of the lime-sulfur sprays that had been used for years to control San Jose scale on apples. A year later H. J. Quayle of the University of California at Riverside established the same explanation for the unsatisfactory results which were being obtained in limited areas with hydrogen-cyanide fumigation against California red scale infesting citrus. The use of the gas had been standard practice since 1886. Now such high concentrations of the poison are needed to secure a satisfactory kill, in some areas twice as much as before, that fumigation has been virtually abandoned because of the expense and to prevent damage to the trees.

More recently several other important cases of insect resistance have been brought to light. In the California citrus groves, again, thrips were effectively controlled for several years by a bait-spray of tartar emetic and sugar, devel-

INSECTICIDE	MOLECULE	NON-RESISTANT	DDT-RESISTANT 1948	LINDANE-RESISTANT 1949
DDT		0.03	11	>100
METHOXYCHLOR		0.07	0.96	1.4
PROLAN		0.09	0.15	0.1
LINDANE		0.01	0.08	0.25
DIELDRIN		0.03	0.05	0.86
PARATHION		0.015	0.02	0.023
PYRETHRINS		1.0	0.94	1.6

RELATIVE TOXICITY of various insecticides to houseflies is compared in terms of micrograms of toxicant per gram of solution. Resistance to DDT and lindane gives flies no immunity to parathion and natural pyrethrins.

oped in 1939. As early as 1941, however, the citrus thrips in a localized area were showing resistance; this resistance has now spread so widely that tartar emetic is of little use. DDT was widely employed in 1948, but by 1951 the thrips were giving signs of DDT-resistance. At present the most satisfactory thrip control is obtained with dieldrin. In South Africa the blue tick developed a marked resistance to sodium arsenite cattle-dips in 1938; arsenical solutions strong enough to injure the cattle did not give effective control. In 1946 lindane dips were found to be extraordinarily effective, a concentration of 50 parts per million being employed. Yet by 1948 1,000 parts per million would not give complete control of the tick.

An important feature in recent observation and study is the side resistance to other toxicants that insects develop as they acquire resistance to the one to which they are exposed. The specificity of resistance has obvious importance in the planning of practical control measures; it is also an important clue to an understanding of the process of resistance. In our work at the University of California at Riverside March and I have concentrated on this aspect of the housefly's versatile resistance to the new insecticides. In this work we applied one-microliter droplets of acetone containing measured amounts of the toxicants directly to the bodies of anesthetized flies. When we compared the relative amounts of insecticides required to kill 50 per cent of the fly population, we found that the strain of flies highly resistant to DDT showed

SYNERGISTS which activate DDT against resistant flies are similar in structure, harmless when used alone.

considerable resistance to a number of structural analogues of DDT, such as methoxychlor and others shown in the molecular diagrams on the opposite page. In general, the closer the structural resemblance to DDT, the higher the relative resistance. Among the members of this DDT group, methoxychlor departs furthest in structure. It proved, accordingly, to be the most effective against the DDT-resistant fly, and is still being used, though with diminishing returns, as a residual spray. We found that lindane, chlordane, heptachlor, aldrin and dieldrin, which belong to a somewhat different group of chlorinated hydrocarbon insecticides, were only slightly less effective against the original DDT-resistant strain and also gave satisfactory control of the same strain in the field. However, a high degree of resistance to these materials was quickly superimposed upon the DDT-resistance. It also appears that the members of this second group of chlorinated hydrocarbons have a common mode of action, since flies that develop resistance to one also become resistant to the others.

An interesting exception that helps to prove the rule is prolan. This toxicant may be considered an analogue of DDT in which the trichloromethyl group $(-CCl_3)$ has been replaced by a nitro-ethyl group $(-C(NO_2) CH_3)$. Initially we found it to be as effective against DDT-resistant strains as it was against non-resistant strains. The variation in the molecules suggests that this difference in action may be connected to the presence of the nitro group in place of the chloromethyl group which the other DDT analogues have in common. There is additional evidence, however, that the modes of action of these compounds are still in some way related. We found that after eight generations flies which are resistant to the DDT and lindane groups begin rapidly to show immunity to prolan; by the 12th generation immunity is practically complete. In contrast we find that neither of the strains of resistant flies is appreciably resistant to parathion, an entirely unrelated organic phosphorus toxicant, or to the natural pyrethrins, which come from chrysanthemum flowers.

A similar but less clear-cut picture has been developed with regard to the specificity of resistance in mosquitoes. None of the strains resistant to the various chlorinated hydrocarbons shows resistance to the organic phosphorus insecticides. The specificity of hydrogen-cyanide resistance of the California red scale has been studied by D. L. Lindgren of the University of California at Riverside, who has found that the resistant strain is also more resistant to ethylene oxide and methyl bromide. However, Lindgren could show no difference in susceptibility to oil or parathion sprays.

THUS FAR nothing in the study of insect resistance to chemicals suggests that it differs in any fundamental way from the same process in bacteria and protozoa, about which we know a great deal more. We can therefore expect to find that such resistance results either from the selection of the most resistant individuals present in a normal population, or from a continuous increase in resistance reflecting a specific interaction between the chemical and the chemistry of the organism. In many cases both processes may be involved. One way to measure the rate at which the resistance of an insect population increases is to expose each generation to a dosage of insecticide which permits the survival of only a few of the most resistant individuals to propagate the succeeding generation. Such experiments yield a characteristic curve, shown on the next page, which indicates that resistance develops in two steps. The first is a long drawn-out period of gradually increasing resistance which may extend over 20 to 30 generations and result in a 5- to 10-fold increase in resistance. In the second a very rapid increase within several generations results in virtual immunity to the toxicant. W. N. Bruce of the University of Illinois has made the interesting discovery that, by treating both larvae and adult flies and selecting the survivors for breeding, the development of resistance may be considerably accelerated. With this technique he has been able to develop a high degree of resistance to all of the chlorinated hydrocarbon insecticides. Against other toxicants such as the pyrethrins and the organic phosphorus compounds, he has found only a slight increase of resistance after exposing many generations. Thus in our laboratories the continuous selection with parathion of some 60 generations of adult flies yielded a mere 5-fold increase of resistance, and the exposure of both adults and larvae showed no greater increase. In contrast, the selection with lindane of 41 generations of flies resulted in a 10,000-fold increase above the initial resistance. We know that parathion is toxic to insects because it inactivates an essential nerve enzyme, cholinesterase. Perhaps it is not too much to hope that certain poisons may attack biochemical processes of such fundamental importance that the organism cannot develop true immunity.

FROM experience in the field and the laboratory we know that resistant insects are able to pass along this characteristic to their offspring. As yet the genetic picture of this accomplishment is far from clear, and it is confused by conflicting results secured by various investigators. The study is complicated because we have no practical way to measure the resistance of individual insects, and must work with populations rather than with individuals. The genetic factors are further obscured because much of the work has been carried out using field-collected strains. These are far from uniform in their characteristics, and various investigators have secured divergent results working with strains collected from different sources. It is apparent that, in selection for insecticide resistance, other characteristics have simultaneously been selected. Thus it has been reported that resistant houseflies may differ from susceptible strains in degree of pigmentation, in length of larval life cycle, in egg-laying ability, in resistance to heat and cold, in behavior, and even in the appearance of altered wing venation. Such variations often reflect differences in metabolism. It is not surprising that strains of normal and DDT-resistant flies should show differences in rate and volume of respiration and in the activity of their enzymes, such as cytochrome oxidase and cholinesterase. The significance of these variations in insecticide resistance is not yet understood; many of them may be entirely fortuitous.

R. C. Dickson of the University of California at Riverside has found that hydrogen-cyanide resistance in the red scale is inherited through a sex-linkage and depends on a single gene or a group of closely related genes. Matings of resistant and non-resistant scales produced female offspring of intermediate resistance and male offspring having their mothers' resistance. Working with a strain of DDT-resistant houseflies, C. Mary Harrison of the London School of Hygiene and Tropical Medicine has found that the resistance to knockdown by DDT was inherited in a simple Mendelian pattern. Non-resistance was dominant and the resistance appeared to be controlled by a single pair of genes. However, where mortality rather than knockdown was used as the criterion, most investigators have found that crosses of resistant and non-resistant flies have resulted in offspring of intermediate resistance and with a wide degree of heterogeneity. Resistance of this type is apparently governed by several pairs of genes.

FOR practical purposes it might be hoped that resistance to a given insecticide would decline after exposure to it ceases. Treatment with the chemical might then be resumed at suitable intervals. Investigation of this possibility, however, has not been encouraging. Laboratory studies with DDT-resistant houseflies have shown variable

results but indicate that in some strains the resistance may persist unchanged for 30 or more generations. Even after that period, although the average resistance of the population may decline, a number of individuals will be produced whose resistance is at the highest level. Similar results have been obtained with the laboratory-reared red scale, whose hydrogen-cyanide resistance has been maintained unchanged for 150 generations.

Results in the field have been no more encouraging. March and L. L. Lewallen have collected flies for several years from the same farms and dairies in southern California and found a progressive increase in resistance to DDT and lindane. Since almost no DDT has been used for fly control in this area for the past several years, it must be concluded that the build-up of resistance has resulted from the selection pressure of other insecticides. Similarly the hydrogen-cyanide resistance of the red scale has been maintained unchanged, and the resistant strain continues to increase in abundance although there has been little fumigation in the resistant areas for many years. The tartar emetic resistant citrus thrip has also persisted and spread despite the abandonment of this chemical in resistant areas.

Since the natural course of events does not improve the situation in our favor, we must inquire more thoroughly into the nature of resistance in the individual insect organism. Early efforts looked for simple physiological explanations. Thus it was suggested that resistant red scales were able to protect themselves against hydrogen cyanide by closing the spiracles, or breathing pores, for extended periods following exposure to low dosages of gas. However, it was subsequently found that the differential in resistance between the two strains of scale could be detected in short initial exposures before the closure of spiracles was observed. In the case of the DDT-resistant housefly it was first suggested that these flies had developed a thicker, more impervious cuticle which prevented the absorption of the chemical. However, direct injection of DDT into the body cavity of the fly shows no change in relative resistance, and several investigators have found that resistant flies absorb DDT through the cuticle just as rapidly as non-resistant flies.

IT WAS early apparent to investigators that we must attack the physiology of resistance in terms of its biochemistry and not merely its mechanics. One interesting lead came from the fact that DDT-resistant flies are often fairly susceptible to knockdown by the insecticide. The discovery of flies struggling

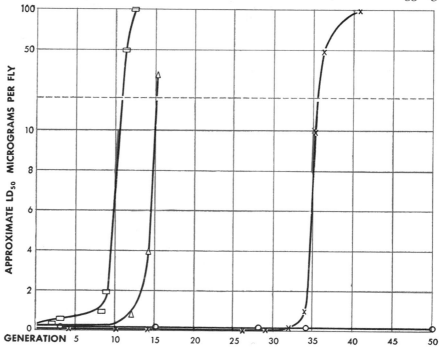

□ PROLAN FROM DDT-LINDANE RESISTANT LARVAE AND ADULTS TREATED
△ DDT FROM NON-RESISTANT LARVAE AND ADULTS TREATED
✕ LINDANE FROM NON-RESISTANT ADULTS TREATED
○ PARATHION FROM NON-RESISTANT ADULTS TREATED

INCREASING RESISTANCE in successive generations of flies exposed to various insecticides is plotted in terms of micrograms required per fly to kill half of the flies exposed (LD_{50} means lethal dose to 50 per cent of the flies).

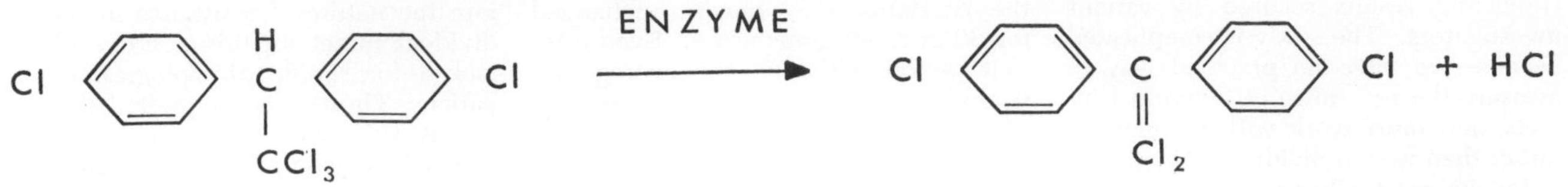

ENZYME

DETOXIFICATION OF DDT by enzyme reaction in resistant flies converts DDT molecule (*left*) to harmless DDE molecule. The reaction removes a hydrogen and a chlorine atom from the central group and yields hydrochloric acid (HCl) as a by-product. Toxicants less susceptible to reaction kill DDT-resistant flies.

on the floor of a residually treated barn would lead to the cursory opinion that they were non-resistant. When these flies were collected and caged for several hours, however, as many as 80 per cent would completely recover. This suggested the possibility that resistant flies were able to detoxify DDT into a non-lethal metabolite, a possibility which was confirmed by J. Sternberg and C. W. Kearns of the University of Illinois and by A. S. Perry and W. M. Hoskins of the University of California at Berkeley. These investigators applied measured amounts of DDT to both resistant and non-resistant flies and then, after macerating and extracting the body contents, determined the breakdown products of DDT by means of a spectrophotometer. They found that the resistant flies were able to detoxify the DDT at a high rate. This is accomplished by enzyme action which removes a single chlorine atom from the trichloromethyl group of the DDT molecule, converting it into the nontoxic ethylene derivative known as DDE, shown in the diagram above. The enzymes in resistant strains of flies carry through this dehydrochlorination process much more rapidly than those in non-resistant strains. Flies surviving the treatment had metabolized greater amounts of DDT than flies which had been killed. An interesting point which these workers demonstrated was that the survivors also still retained enough unaltered DDT in their bodies to kill non-resistant flies. There is no clear-cut explanation for this tolerance on the part of the flies. It may be that only a fraction of the total dosage of DDT is available at any time to the site where it is supposed to take effect, and that it is detoxified as fast as it arrives. Alternatively, the resistant flies may store the DDT in regions of the body where it can exert no harmful effects.

It was a logical step from the discovery that flies detoxify DDT by dehydrochlorination to the hope that the analogues of DDT which are most difficult to dehydrochlorinate would prove relatively more toxic to the resistant flies. This turns out to be the case. Methoxychlor, which is decomposed in alkali solutions only about 1/200 as fast as DDT, has a relative toxicity to the resistant flies of about 20 times that of DDT. In contrast, dibromo-DT, which

is decomposed about 1.5 times as fast as DDT, has a relative toxicity slightly less than DDT. Other toxic analogues of DDT behave in similar fashion. Prolan, the nitroanalogue which by definition is not subject to dehydrochlorination, fits perfectly into this scheme. It is almost equally toxic to DDT-resistant and to non-resistant flies. For some other reason, however, even prolan loses its potency after a few generations.

Another logical approach is to look for chemicals which will inhibit the process of detoxification. One such chemical is piperonyl cyclonene, commonly used as a synergist to amplify the toxicity of the natural pyrethrin insecticides. Hoskins and Perry found that it will also increase the toxicity of DDT to resistant flies, and were able to demonstrate that this chemical works by inhibiting the detoxification process. W. T. Summerford and his co-workers of the U. S. Public Health Service then made the equally interesting observation that DMC, a DDT analogue useful against mites, was even more effective than piperonyl cyclonene as a DDT synergist for resistant flies. March and his co-workers have tested hundreds of other DDT analogues and have found several that increase the effectiveness of DDT against resistant flies 100 or more times. These materials, three of which are shown on page 110, do not completely restore the potency of DDT against resistant flies. Interestingly enough the combination of DDT and a synergist has proved to be no more effective against non-resistant flies than DDT alone. A similar search is being made for synergists for lindane, and some progress is being made. Field tests have already demonstrated that synergists hold more than academic interest. Combinations of one part of synergist with five parts of DDT as a residual application have given from four to eight weeks of very satisfactory fly control in localities where very high DDT-resistance existed.

ADMITTEDLY fundamental investigation into the biochemistry and genetics of insect resistance to insecticides has not yet shown us much in the way of practical results. Such work is not calculated to yield immediate solutions. In the long run, however, it is certain that without increase in our

knowledge of these fundamental processes we would be merely fumbling in the dark.

Meanwhile we can expect to see high returns from purely empirical efforts along a number of lines. One largely neglected possibility is the use of two or more insecticides having entirely different modes of action, applied either in combination or in alternate treatments. Such applications are entirely feasible, and it seems logical that they would retard the development of resistance. Then there is evidence that insects are not able to develop resistance against certain insecticides. For example, oil sprays have been used for 30 years or more for the control of San Jose and California red scale without any indication of tolerance. As mentioned earlier in this article, laboratory studies have indicated that the housefly does not readily develop resistance to parathion. This material and related organic phosphorus insecticides are already showing much promise in the field control of resistant flies and mosquitoes. Other insecticides against which resistance is difficult to develop will doubtless be discovered.

Finally, it ought to be observed, time is on our side. The large majority of our important agricultural insect pests breed one to three generations a year. Since it apparently takes from 20 to 40 generations of intensive selection by an insecticide to develop a high level of resistance, such pests as the codling moth and the corn borer may require from 10 to 20 years to make our present chemicals obsolete. Thus, though the more universal nuisance of the fast-breeding housefly and mosquito might suggest otherwise, the synthetic organic insecticides still constitute a net gain in insect control. The normal rate of progress in development of new insecticides should continue to keep us ahead of the insects' amazing capacity to circumvent them. It can be predicted with confidence that we shall not have to resign ourselves again to sticky paper and fly swatters.

12

PESTICIDES AND THE REPRODUCTION OF BIRDS

DAVID B. PEAKALL

April 1970

*High concentrations of chlorinated hydrocarbon
residues accumulate in such flesh-eaters as hawks
and pelicans. Among the results are upsets in normal
breeding behavior and eggs too fragile to survive*

The birds of prey have had an uneasy coexistence with man. Apart from the training of certain hawks for falconry and the veneration of the eagle as a symbol of fortitude, the predatory birds have been preyed on by the human species. In many parts of the world farmers, hunters and bird-lovers have waged unceasing warfare on the rapacious birds as pests, and egg collectors have further threatened their survival by raiding their nests for the beautifully pigmented eggs. Nevertheless, over the centuries the birds of prey on the whole survived well. The peregrine falcon, for example, is known to have maintained a remarkably stable population; records of aeries that have been occupied more or less continuously by peregrines go back in some cases to the Middle Ages.

About two decades ago, however, the peregrines in Europe and in North America suddenly suffered a crash in population. The peregrine is now rapidly vanishing in settled areas of the world, and in some places, particularly the eastern U.S., it is already extinct [see illustration on page 115]. The abrupt population fall of the peregrine (known in the U.S. as the duck hawk) has been paralleled by sharp declines of the bald eagle, the osprey and Cooper's hawk in the U.S. and of the golden eagle and the kestrel, or sparrow hawk, in Europe. The osprey, or fish hawk, has nearly disappeared from its haunts in southern New England and on Long Island; along the Connecticut River, where 150 pairs nested in 1952, only five pairs nested in 1969.

The population declines of all these raptorial birds are traceable not to the killing of adults but to a drastic drop in reproduction. It has been found that the reproduction failures follow much the same pattern among the various species: delayed breeding or failure to lay eggs altogether, a remarkable thinning of the shells and much breakage of the eggs that are laid, eating of broken eggs by the parents, failure to produce more eggs after earlier clutches were lost, and high mortality of the embryos and among fledglings.

Examination of the geographic patterns suggests a cause for the birds' reproductive failure. The regions of population decline coincide with areas where persistent pesticides—the chlorinated hydrocarbons such as DDT and dieldrin—are widely applied. Attrition of the predatory birds has been most severe in the eastern U.S. and in western Europe, where these pesticides first came into heavy use two decades ago. Analysis confirmed the suspicions about the pesticides: the birds were found to contain high levels of the chlorinated hydrocarbons. In areas such as northern Canada, Alaska and Spain, where the use of these chemicals has been comparatively light, the peregrine populations have remained normal or nearly normal. Recent studies show, however, that even in the relatively isolated North American arctic region the peregrines now have fairly high levels of chlorinated hydrocarbons and their populations apparently are beginning to decline.

The birds of prey are particularly vulnerable to the effects of a persistent pesticide such as DDT because they are the top of a food chain. As George M. Woodwell of the Brookhaven National Laboratory has shown, DDT accumulates to an increasingly high concentration in passing up a chain from predator to predator, and at the top of the chain it may be concentrated a thousandfold or more over the content in the original source [see "Toxic Substances and Ecological Cycles," by George M. Woodwell; SCIENTIFIC AMERICAN, Offprint 1066]. The predatory birds, as carnivores, feed on birds that have fed in turn on insects and plants. Hence the birds of prey accumulate a higher dose of the persistent pesticides and are more likely to suffer the toxic effects than other birds.

The idea that the predatory birds' decline is due to an internal toxic effect, rather than to a change in their behavior or their habitat, has been verified by many experiments. One of the most interesting was a field test made by Paul Spitzer, now at Cornell University, working in cooperation with the Patuxent Wildlife Research Center in Maryland. He transferred eggs from nests of the failing osprey population in New England to nests of a successful population in the Chesapeake Bay area and placed the Chesapeake eggs in the New England nests. The Chesapeake eggs hatched as successfully in the New England nests as they would have at home with their own parents, whereas the New England eggs transferred to Chesapeake nests produced as few viable young as would have been expected if they had been incubated in their original nests in New England. The experiment thus indicated that the fate of the eggs was determined by an intrinsic factor in the egg itself.

The first clue to what was happening to the predatory birds' reproduction system came in the early 1960's when Derek Ratcliffe of the British Nature Conservancy, puzzled by the extraordinary number of broken eggs he found in peregrine nests, examined the shells of peregrine eggs that had been collected over a period of many years. He found that the eggs collected since the late 1940's show a sharp drop in thickness of the shell, averaging 19 percent. Similar findings were subsequently made

CRUSHED EGG in the nest of a brown pelican off the California coast had such a thin shell that the weight of the nesting parent's body destroyed it. The concentration of DDE in the eggs of this 300-pair colony reached 2,500 parts per million; no eggs hatched.

on peregrine eggs in North America and on the eggs of other species of predatory birds whose populations were decreasing. It became apparent that something must be wrong with the birds' calcium metabolism and that the effects of the suspected pesticides would bear looking into.

Experiments were started in several laboratories. At the Patuxent Wildlife Research Center, Richard D. Porter and Stanley N. Wiemeyer, working with kestrels, found that a mixture of DDT and dieldrin in doses measured in a few parts per million brought about a significant decrease in the shell thickness of the birds' eggs. Robert G. Heath of the Patuxent center tested the effects of

DDE, the principal metabolic product of DDT, on mallard ducks. DDE is now a ubiquitous feature of the earth's environment; it is estimated that there are a billion pounds of the substance in the world ecosystem, and traces of it have been found in animals everywhere, from polar bears in the Arctic to seals in the Antarctic. Heath found that DDE caused the failure of mallard eggs in two ways: by increasing the fragility of the eggs, leading to increased breakage soon after laying, and by the death of the embryos in intact eggs toward the end of the period of incubation. James H. Enderson of Colorado College and his associate Daniel D. Berger, studying the eggs of prairie falcons in the Southwest

desert, established that the amount of thinning of the shells and the mortality rate for the embryos were related to the quantity of DDE in the egg. Enderson and Berger also found that when they fed starlings loaded with dieldrin to falcons, the falcons' eggs showed similar thinning.

The ultimate in thinness of birds' eggshells was discovered recently in colonies of the brown pelican off the California coast. The DDE content in the eggs of this wild population (as measured by Robert Risebrough of the University of California at Berkeley) ranged as high as 2,500 parts per million, and the eggshells were so thin that the eggs could not be picked up without denting the

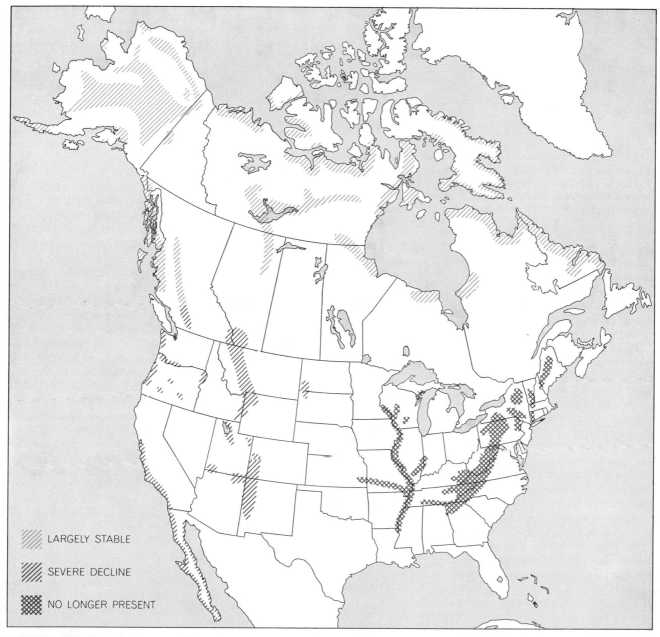

LARGELY STABLE

SEVERE DECLINE

NO LONGER PRESENT

NESTING AREAS of the peregrine falcon, or duck hawk, in the Northern Hemisphere of the New World are shown on this map.   Shades of color show the extent of interference with normal reproduction resulting from ingestion of pesticides by the birds.

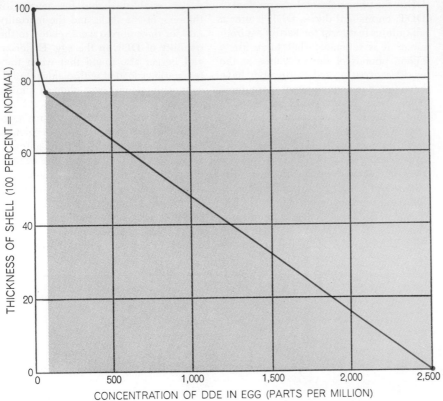

CONCENTRATION OF DDE IN EGG (PARTS PER MILLION)

SEVERE EFFECT of the concentration of relatively small amounts of the persistent chlorinated hydrocarbon pesticides is evident in this graph. When the parent's concentration is enough to add as few as 25 parts per million of pesticide to the egg, the shell becomes 15 percent thinner than normal. Soon after the shells become more than 20 percent thinner than normal (*area of light color*) eggs are usually not found in nests because of breakage.

suit of this clue led to the finding that chlordane induced rat liver cells to synthesize enzymes that speeded up the metabolism of hexobarbital. The enzymes brought about hydroxylation of the barbiturate, thereby making it more soluble in water and hastening its excretion. Further experiments showed that these enzymes could hydroxylate a wide variety of substances, including the sex hormones: estrogen, testosterone and progesterone.

Because the investigators were interested primarily in drug research and their reports were published mainly in pharmacological journals, these discoveries did not come to the attention of workers studying the effects of pesticides on wildlife until several years later. I myself came on the published findings only incidentally in the course of preparing lectures for medical students. The fact that chlordane could change the balance of sex hormones in animals immediately suggested a possible explanation of the mechanism whereby the chlorinated pesticides inhibit reproduction in birds. It was capable of explaining their reproductive failure in general and the alteration of the calcium balance in the egg in particular.

My colleagues and I at Cornell University launched on a program of experiments designed to explore the interesting questions suggested by this new aspect of the problem. To explain them I must briefly outline the complex chain of physiological events that characterizes breeding by birds. The cycle is initiated by a seasonal or climatic stimulus: the lengthening of daylight in spring in the northern Temperate Zone or rainfall in the arid and tropical regions. These signals cause an increase in the production of hormones in the nerve cells of the medial eminence of the bird's brain. The bloodstream carries these hormones to the anterior pituitary gland, which in turn dispatches to the gonads (the testes or ovaries) hormones that stimulate these organs to produce the sex hormones. The sex hormones not only generate physical changes in the reproductive organs and evoke breeding behavior but also promote the storage of a supply of calcium for the eggs.

Let us look first into the question of how a pesticide may affect the calcium supply. We carried out our experiments on the rather small Asian pigeon known as the ringdove, so that I shall describe the situation in this bird. The female forms the shell of the egg in the uterus within a period of 20 hours, and she needs 240 milligrams of calcium to pro-

shells [*see illustration on page 114*]. In a colony on the Anacapa Islands off the coast it was found that the 300 pairs of nesting pelicans had not produced a single viable egg. Their nests, visited shortly after the eggs were laid, contained many broken eggs.

Field studies and laboratory experiments suggest that the thinning of eggshells does not increase in direct proportion to the DDE dose. In fact, small doses can produce dramatic effects. A content of only 75 parts per million in the egg reduces the shell thickness by more than 20 percent; beyond that, as the dose increases the decrease in shell thickness is more gradual [*see illustration above*]. In the case of the brown pelican very heavy doses may thin the shell to a mere film.

Studies of white pelicans and cormorants have implicated the polychlorinated biphenyls (PCB's), now widely used as plasticizers, as another threat to birds of prey. These compounds cause thinning of the eggshells, although not as effectively as DDT and its metabolites do. Preliminary laboratory studies show that PCB's are particularly effective, however, in delaying the onset of breed-

ing. The PCB's are given off when plastic materials are burned, and they are widely distributed over the earth. They resemble DDT in molecular structure and produce similar physiological actions in animals.

Much interest has focused on the question of how the chlorinated pesticides produce their destructive effects in the predatory birds—a question that is of no small concern to man, who also is the top of a food chain. Oddly enough, the beginning of light on this question came about through an accidental discovery involving an animal totally unrelated to the birds: the laboratory rat. Larry G. Hart and James R. Fouts of the University of Iowa College of Medicine were investigating the effects of food deprivation on the metabolism of drugs in rats. The drug they were using was hexobarbital, and in one experiment they were startled to find that the rats' sleeping time after receiving a standard dose of the barbiturate was much shorter than it had been in previous tests. Reexamining the conditions of the experiment, they found that the only unusual factor was that the cages had been sprayed with chlordane to control bedbugs. Pur-

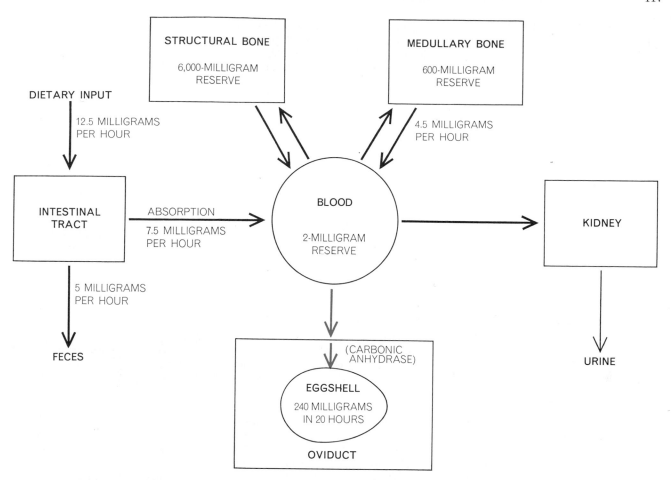

CALCIUM FOR EGGSHELL, which is formed around each egg in the last 20 hours before laying, is drawn in part from the bird's food supply and in part from calcium reserves in the bird's bones. The key to shell formation, however, is the enzyme carbonic an- hydrase, which makes the supply of calcium carried in the ringdove's bloodstream available to the bird's oviduct at a rate of 12 milligrams per hour. When laying ringdoves are injected with DDE, the action of the enzyme is severely inhibited, causing thin shells.

duce a shell of normal thickness. Since the calcium content of the circulating blood, even at the time of ovulation, is only two milligrams (barely a 10-minute supply), the bird must draw on other sources to meet the demand. About 60 percent of the demand is supplied by the bird's food intake; the rest is provided by a store of calcium in the marrow of the bones [see "How an Eggshell Is Made," by T. G. Taylor; SCIENTIFIC AMERICAN Offprint 1171]. This calcium reserve is laid down in the bone cavities early in the breeding cycle, and the amount of the deposit is controlled by the levels of estrogen in the blood and tissues. Obviously, therefore, a deficiency of estrogen will reduce the bird's calcium reserve. It seemed unlikely, however, that the reduction of this reserve alone could account for the drastic shell-thinning observed in eggs loaded with pesticides. If the *supply* of calcium were the sole problem, the birds could augment the supply by drawing on the calcium embodied in the skeleton; furthermore, birds on a very low calcium

diet have been found to cease egg-laying rather than laying eggs with abnormally thin shells. Was it possible, then, that the thinness of the eggshells was due less to the deficiency in supply than to a failure in delivery of calcium to the shell?

In our experiments we bred pairs of ringdoves in cages and delayed feeding the birds a pesticide until after they had completed at least one successful breeding cycle, thereby demonstrating their natural capability. For the experiment we separated the members of each pair, isolated them in individual cages where they had an eight-hour day instead of their normal 16-hour day and fed them a standard dose of DDT in their food. After three weeks we gave each bird an oral dose of radioactive calcium and returned the birds to cages with their original partners for pairing under long-day conditions. A number of days later we examined the birds, some before they laid eggs, others immediately after they finished laying their clutch. In both cases the birds showed a consid-

erable rise of enzyme activity in the liver. A substantially lower level of estrogen was found in the bloodstream of the birds that had not yet laid eggs. After the eggs had been laid, low estrogen levels were found in both experimental and control birds; this was to be expected because the level of estrogen falls at the time of egg-laying. We found that less labeled calcium was stored in the bone marrow of the experimental birds than in the marrow of control birds that had not been fed the pesticide.

Eggs laid by the pesticide-treated birds were notably thin-shelled, as was to be expected. We proceeded to experiments designed to determine whether this was due simply to the shortage of stored calcium or to something that prevented calcium from reaching the shell. In order to resolve this question we resorted to the tactic of injecting pesticides into females within a period of hours before they laid their eggs. In that short interval there would not be time for any significant change in the supply of calcium by way of an alteration of the

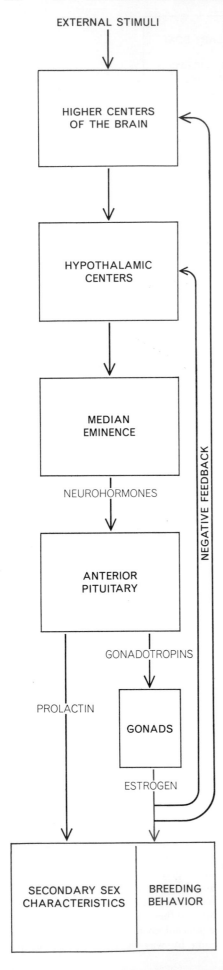

EXTERNAL STIMULI

HIGHER CENTERS
OF THE BRAIN

HYPOTHALAMIC
CENTERS

MEDIAN
EMINENCE

NEUROHORMONES

NEGATIVE FEEDBACK

ANTERIOR
PITUITARY

GONADOTROPINS

PROLACTIN

GONADS

ESTROGEN

SECONDARY SEX
CHARACTERISTICS   BREEDING
BEHAVIOR

BREEDING SUCCESS in birds involves the five sequential responses to external stimuli shown in the illustration at left. Breeding failures, due to late breeding or an inability to lay more eggs after earlier clutches are destroyed, result from the action of pesticides on the fifth response. They stimulate the activity of enzymes in the breeding bird's liver; the enzymes cut the amount of estrogen in the system below the level that is needed for normal sexual behavior.

estrogen levels through the activity of liver enzymes; consequently if the pesticide produced an effect, it would be not on the stored supply but on the delivery of calcium to the eggshell, which as we have noted is laid down within 20 hours of the laying of the egg. And with regard to delivery it was known that an enzyme, carbonic anhydrase, plays an important role in making calcium available to the eggshell in the oviduct. One could therefore look for a possible effect on the activity of this enzyme.

We tried two chlorinated hydrocarbons: dieldrin and DDE. Dieldrin, when injected into a ringdove shortly before it laid its egg, did not produce any significant thinning of the eggshell or inhibit the activity of carbonic anhydrase in the oviduct. DDE, on the other hand, severely depressed the activity of the enzyme and brought about a marked decrease in the thickness of the eggshell.

Our experiments with ringdoves also showed that the chlorinated hydrocarbons cause a significant delay in breeding by birds. Females that were fed pesticides did not lay eggs until 21.5 days (on the average) and sometimes as long as 25 days after pairing, whereas the normal interval, as indicated by the control birds, is 16.5 days on the average. The delay evidently was caused by the depression of the estrogen level resulting from the induction of liver enzymes by the pesticide. It turned out that dieldrin and the polychlorinated biphenyls were more powerful inducers of these enzymes than DDT was.

Delayed breeding is another factor in the predatory birds' population decline. Most birds do their breeding in the season when food is most plentiful, thus giving their young an optimal chance for survival. An artificial delay in their breeding consequently reduces the chances for reproductive success, and it is most serious for large birds, with their long egg-incubation period and the slower growth of the fledglings to maturity. It was found that the now extinct peregrine colonies along the Hudson River, the declining cormorant

rookery at Lake DuBay in Wisconsin and the failing pelican colonies in California were all notably late in breeding.

From this point of view it appears that dieldrin and the PCB's are greater threats to the predatory birds than DDT. Certain field and laboratory studies tend to bear out that deduction. Derek Ratcliffe and J. D. Lockie, in long-term observation of the nests of golden eagles in Scotland, found that although abnormal eggshell breakage began in 1952, about the time that DDT was introduced, marked decline in the breeding success of these birds did not begin until 1960, after the introduction of dieldrin. In laboratory experiments on the bobwhite quail James B. DeWitt and John L. George of the U.S. Fish and Wildlife Service found that one part per million of dieldrin was effective in reducing the success in hatching and survival of chicks, whereas it took 200 parts per million of DDT to produce the same effect. Robert Heath found in his studies of mallard ducks, however, that DDE severely impaired reproductive success at doses as low as 10 parts per million. Thus there appears to be a considerable difference in the effect of DDT and its metabolites on different species of birds.

We come to the following conclusions concerning the physiological mechanisms responsible for the various harmful effects on bird breeding that are brought about by the persistent insecticides. Abnormally late breeding and the failure of birds to lay eggs after their early clutches have been lost can be explained in terms of the induction of liver enzymes that lower the estrogen levels in the birds. The failure, or apparent failure, of birds to lay any eggs at all may be due either to depression of the estrogen level or to the circumstance that the eggs were broken and eaten by the parents shortly after they were laid, so that observers found no eggs in the nest on visiting them. The reduction in clutch size may also be accounted for by early breakage and eating of some of the eggs, as this has been noted mainly in cases where the nests were not checked frequently. The thinning of eggshells and breakage of the eggs evidently is due largely to the inhibition of carbonic anhydrase by DDT and its metabolites. We are left with some phenomena that are still unexplained. Why does a low dose of pesticide produce relatively more thinning of the eggshell than larger doses do? What is the mechanism that kills embryos in the shell? These questions need further investigation.

The effect of the pesticides in disturbing the calcium balance of birds probably is not of direct concern to man, because birds are a special case in their high calcium requirement at breeding time. It seems, however, that we should be concerned about the pesticides' effects on the hormone balance and on other physiological systems. The induction by pesticides of liver enzymes that lower the estrogen levels has been found in a wide variety of vertebrates, including a primate, the squirrel monkey. There is little doubt that this effect applies to man as well. Moreover, the chlorinated hydrocarbons are known to alter the glucose metabolism and inhibit an enzyme (adenosine triphosphatase, or ATPase) that plays a vital role in the energy economy of the human body.

The recent finding by investigators at the National Cancer Institute that a dose of 46 milligrams of DDT per kilogram of body weight can produce a fourfold increase in tumors of the liver, lungs and lymphoid organs of animals indicates that DDT should be banned for that reason alone. Human cancer victims have been found to have two to two and a half times more DDT in their fat than occurs in the normal population. Investigators in the U.S.S.R. recently reported that DDD, another metabolite of DDT, reduces the islets of Langerhans, the site of insulin synthesis.

The peregrine population crash has prompted two international conferences of concerned investigators, in 1965 and again in 1969. It is encouraging to note that in Britain, where severe restrictions were imposed in 1964 on the use of chlorinated hydrocarbon pesticides, the peregrine population has increased in the past two years. The Canadian government recently announced licensing restrictions that are expected to reduce the use of these pesticides by 90 percent, and many states in the U.S. are also instituting or considering such restrictions. Environmental problems do not respect political boundaries, and in the long run it will do little good if restrictions on the use of these hazardous toxins are applied only to certain regions or parts of the globe.

The long-term effects of the chlorinated hydrocarbons in the environment on human beings are admittedly much more difficult to detect or assess than the spectacular effects that have been seen in the predatory birds. Still, the story told by the birds is alarming enough. It seems obvious that agents capable of causing profound metabolic changes in such small doses should not be broadcast through the ecosystem on a billion-pound scale.

13

THIRD-GENERATION PESTICIDES

CARROLL M. WILLIAMS
July 1967

The first generation is exemplified by arsenate of lead; the second, by DDT. Now insect hormones promise to provide insecticides that are not only more specific but also proof against the evolution of resistance

Man's efforts to control harmful insects with pesticides have encountered two intractable difficulties. The first is that the pesticides developed up to now have been too broad in their effect. They have been toxic not only to the pests at which they were aimed but also to other insects. Moreover, by persisting in the environment—and sometimes even increasing in concentration as they are passed along the food chain—they have presented a hazard to other organisms, including man. The second difficulty is that insects have shown a remarkable ability to develop resistance to pesticides.

Plainly the ideal approach would be to find agents that are highly specific in their effect, attacking only insects that are regarded as pests, and that remain effective because the insects cannot acquire resistance to them. Recent findings indicate that the possibility of achieving success along these lines is much more likely than it seemed a few years ago. The central idea embodied in these findings is that a harmful species of insect can be attacked with its own hormones.

Insects, according to the latest estimates, comprise about three million species—far more than all other animal and plant species combined. The number of individual insects alive at any one time is thought to be about a billion billion (10^{18}). Of this vast multitude 99.9 percent are from the human point of view either innocuous or downright helpful. A few are indispensable; one need think only of the role of bees in pollination.

The troublemakers are the other .1 percent, amounting to about 3,000 species. They are the agricultural pests and the vectors of human and animal disease. Those that transmit human disease are the most troublesome; they have joined with the bacteria, viruses and protozoa in what has sometimes seemed like a grand conspiracy to exterminate man, or at least to keep him in a state of perpetual ill health.

The fact that the human species is still here is an abiding mystery. Presumably the answer lies in changes in the genetic makeup of man. The example of sickle-cell anemia is instructive. The presence of sickle-shaped red blood cells in a person's blood can give rise to a serious form of anemia, but it also confers resistance to malaria. The sickle-cell trait (which does not necessarily lead to sickle-cell anemia) is appreciably more common in Negroes than in members of other populations. Investigations have suggested that the sickle cell is a genetic mutation that occurred long ago in malarial regions of Africa. Apparently attrition by malaria-carrying mosquitoes provoked countermeasures deep within the genes of primitive men.

The evolution of a genetic defense, however, takes many generations and entails many deaths. It was only in comparatively recent times that man found an alternative answer by learning to combat the insects with chemistry. He did so by inventing what can be called the first-generation pesticides: kerosene to coat the ponds, arsenate of lead to poison the pests that chew, nicotine and rotenone for the pests that suck.

Only 25 years ago did man devise the far more potent weapon that was the first of the second-generation pesticides. The weapon was dichlorodiphenyltrichloroethane, or DDT. It descended on the noxious insects like an avenging angel. On contact with it mosquitoes, flies, beetles—almost all the insects—were stricken with what might be called the "DDT's." They went into a tailspin, buzzed around upside down for an hour or so and then dropped dead.

The age-old battle with the insects appeared to have been won. We had the stuff to do them in—or so we thought. A few wise men warned that we were living in a fool's paradise and that the insects would soon become resistant to DDT, just as the bacteria had managed to develop a resistance to the challenge of sulfanilamide. That is just what happened. Within a few years the mosquitoes, lice, houseflies and other noxious insects were taking DDT in their stride. Soon they were metabolizing it, then they became addicted to it and were therefore in a position to try harder.

Fortunately the breach was plugged by the chemical industry, which had come to realize that killing insects was —in more ways than one—a formula for

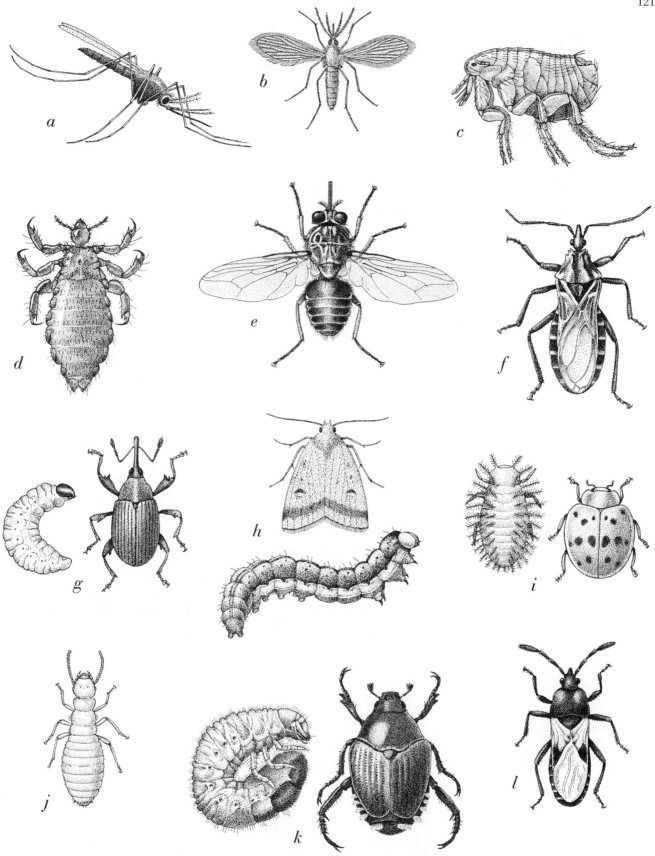

INSECT PESTS that might be controlled by third-generation pesticides include some 3,000 species, of which 12 important examples are shown here. Six (a–f) transmit diseases to human beings; the other six are agricultural pests. The disease-carriers, together with the major disease each transmits, are (a) the *Anopheles* mosquito, malaria; (b) the sand fly, leishmaniasis; (c) the rat flea, plague; (d) the body louse, typhus; (e) the tsetse fly, sleeping sickness, and (f) the kissing bug, Chagas' disease. The agricultural pests, four of which are depicted in both larval and adult form, are (g) the boll weevil; (h) the corn earworm; (i) the Mexican bean beetle; (j) the termite; (k) the Japanese beetle, and (l) the chinch bug. The species in the illustration are not drawn to the same scale.

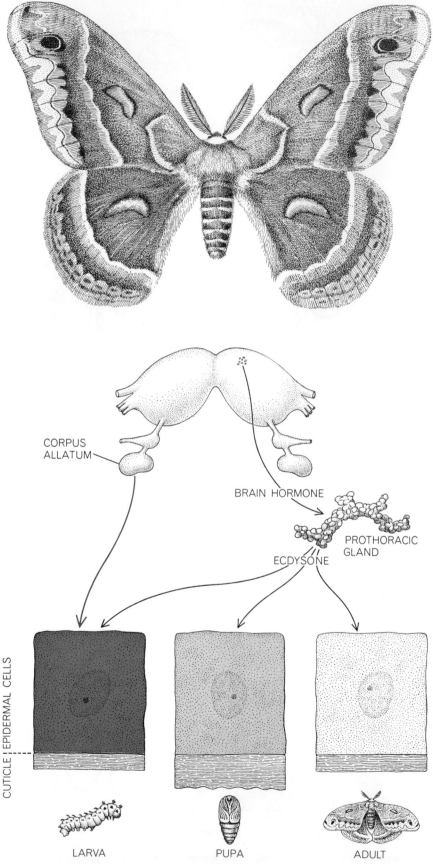

CORPUS
ALLATUM

BRAIN HORMONE

PROTHORACIC
GLAND

ECDYSONE

CUTICLE | EPIDERMAL CELLS

LARVA   PUPA   ADULT

HORMONAL ACTIVITY in a Cecropia moth is outlined. Juvenile hormone (*color*) comes
from the corpora allata, two small glands in the head; a second substance, brain hormone,
stimulates the prothoracic glands to secrete ecdysone, which initiates the molts through
which a larva passes. Juvenile hormone controls the larval forms and at later stages must be
in low concentration or absent; if applied then, it deranges insect's normal development.
The illustration is partly based on one by Howard A. Schneiderman and Lawrence I. Gilbert.

getting along in the world. Organic
chemists began a race with the insects.
In most cases it was not a very long race,
because the insects soon evolved an insensitivity to whatever the chemists had
produced. The chemists, redoubling
their efforts, synthesized a steady stream
of second-generation pesticides. By 1966
the sales of such pesticides had risen to
a level of $500 million a year in the U.S.
alone.

Coincident with the steady rise in the
output of pesticides has come a growing
realization that their blunderbuss toxicity can be dangerous. The problem has
attracted widespread public attention
since the late Rachel Carson fervently
described in *The Silent Spring* some actual and potential consequences of this
toxicity. Although the attention thus
aroused has resulted in a few attempts
to exercise care in the application of
pesticides, the problem cannot really be
solved with the substances now in use.

The rapid evolution of resistance to
pesticides is perhaps more critical. For
example, the world's most serious disease in terms of the number of people
afflicted continues to be malaria, which
is transmitted by the *Anopheles* mosquito—an insect that has become completely resistant to DDT. (Meanwhile the
protozoon that actually causes the disease is itself evolving strains resistant to
antimalaria drugs.)

A second instance has been presented
recently in Vietnam by an outbreak of
plague, the dreaded disease that is conveyed from rat to man by fleas. In this
case the fleas have become resistant to
pesticides. Other resistant insects that
are agricultural pests continue to take
a heavy toll of the world's dwindling
food supply from the moment the seed is
planted until long after the crop is harvested. Here again we are confronted by
an emergency situation that the old technology can scarcely handle.

The new approach that promises a
way out of these difficulties has
emerged during the past decade from
basic studies of insect physiology. The
prime candidate for developing third-
generation pesticides is the juvenile hormone that all insects secrete at certain
stages in their lives. It is one of the three
internal secretions used by insects to
regulate growth and metamorphosis
from larva to pupa to adult. In the living
insect the juvenile hormone is synthesized by the corpora allata, two tiny
glands in the head. The corpora allata
are also responsible for regulating the
flow of the hormone into the blood.

At certain stages the hormone must be

secreted; at certain other stages it must be absent or the insect will develop abnormally [see illustration on opposite page]. For example, an immature larva has an absolute requirement for juvenile hormone if it is to progress through the usual larval stages. Then, in order for a mature larva to metamorphose into a sexually mature adult, the flow of hormone must stop. Still later, after the adult is fully formed, juvenile hormone must again be secreted.

The role of juvenile hormone in larval development has been established for several years. Recent studies at Harvard University by Lynn M. Riddiford and the Czechoslovakian biologist Karel Sláma have resulted in a surprising additional finding. It is that juvenile hormone must be absent from insect eggs for the eggs to undergo normal embryonic development.

The periods when the hormone must be absent are the Achilles' heel of insects. If the eggs or the insects come into contact with the hormone at these times, the hormone readily enters them and provokes a lethal derangement of further development. The result is that the eggs fail to hatch or the immature insects die without reproducing.

Juvenile hormone is an insect invention that, according to present knowledge, has no effect on other forms of life. Therefore the promise is that third-generation pesticides can zero in on insects to the exclusion of other plants and animals. (Even for the insects juvenile hormone is not a toxic material in the usual sense of the word. Instead of killing, it derails the normal mechanisms of development and causes the insects to kill themselves.) A further advantage is self-evident: insects will not find it easy to evolve a resistance or an insensitivity to their own hormone without automatically committing suicide.

CHEMICAL STRUCTURES of the Cecropia juvenile hormone (left), isolated in 1967 by Herbert Röller and his colleagues at the University of Wisconsin, and of a synthetic analogue (right) made in 1965 by W. S. Bowers and others in the U.S. Dept. of Agriculture show close similarity. Carbon atoms, joined to one or two hydrogen atoms, occupy each angle in the backbone of the molecules; letters show the structure at terminals and branches.

JUVENILE HORMONE ACTIVITY has been found in various substances not secreted by insects. One (left) is a material synthesized by M. Romanuk and his associates in Czechoslovakia. The other (right), isolated and identified by Bowers and his colleagues, is the "paper factor" found in the balsam fir. The paper factor has a strong juvenile hormone effect on only one family of insects, exemplified by the European bug Pyrrhocoris apterus.

The potentialities of juvenile hormone as an insecticide were recognized 12 years ago in experiments performed on the first active preparation of the hormone: a golden oil extracted with ether from male Cecropia moths. Strange to say, the male Cecropia and the male of its close relative the Cynthia moth remain to this day the only insects from which one can extract the hormone. Therefore tens of thousands of the moths have been required for the experimental work with juvenile hormone; the need has been met by a small but thriving industry that rears the silkworms.

No one expected Cecropia moths to supply the tons of hormone that would be required for use as an insecticide.

Obviously the hormone would have to be synthesized. That could not be done, however, until the hormone had been isolated from the golden oil and identified.

Within the past few months the difficult goals of isolating and identifying the hormone have at last been attained by a team of workers headed by Herbert Röller of the University of Wisconsin. The juvenile hormone has the empirical formula $C_{18}H_{36}O_2$, corresponding to a molecular weight of 284. It proves to be the methyl ester of the epoxide of a previously unknown fatty-acid derivative [see upper illustration on this page]. The apparent simplicity of the molecule is deceptive. It has two double bonds and an

oxirane ring (the small triangle at lower left in the molecular diagram), and it can exist in 16 different molecular configurations. Only one of these can be the authentic hormone. With two ethyl groups ($CH_2 \cdot CH_3$) attached to carbons No. 7 and 11, the synthesis of the hormone from any known terpenoid is impossible.

The pure hormone is extraordinarily active. Tests the Wisconsin investigators have carried out with mealworms suggest that one gram of the hormone would result in the death of about a billion of these insects.

A few years before Röller and his colleagues worked out the structure of the authentic hormone, investigators at several laboratories had synthesized a number of substances with impressive juvenile hormone activity. The most potent of the materials appears to be a crude mixture that John H. Law, now at the University of Chicago, prepared by a simple one-step process in which hydrogen chloride gas was bubbled through an alcoholic solution of farnesenic acid. Without any purification this mixture was 1,000 times more active than crude Cecropia oil and fully effective in killing all kinds of insects.

One of the six active components of Law's mixture has recently been identified and synthesized by a group of workers headed by M. Romaňuk of the Czechoslovak Academy of Sciences. Romaňuk and his associates estimate that from 10 to 100 grams of the material would clear all the insects from 2½ acres. Law's original mixture is of course even more potent, and so there is much interest in its other five components.

Another interesting development that preceded the isolation and identification of true juvenile hormone involved a team of investigators under W. S. Bowers of the U.S. Department of Agriculture's laboratory at Beltsville, Md. Bowers and his colleagues prepared an analogue of juvenile hormone that, as can be seen in the accompanying illustration [*top of preceding page*], differed by only two carbon atoms from the authentic Cecropia hormone (whose structure was then, of course, unknown). In terms of the dosage required it appears that the Beltsville compound is about 2 percent as active as Law's mixture and about .02 percent as active as the pure Cecropia hormone.

All the materials I have mentioned are selective in the sense of killing only insects. They leave unsolved, however, the problem of discriminating between the .1 percent of insects that qualify as pests and the 99.9 percent that are helpful or innocuous. Therefore any reckless use of the materials on a large scale could constitute an ecological disaster of the first rank.

The real need is for third-generation pesticides that are tailor-made to attack only certain predetermined pests. Can such pesticides be devised? Recent work that Sláma and I have carried out at Harvard suggests that this objective is by no means unattainable. The possibility arose rather fortuitously after Sláma arrived from Czechoslovakia, bringing with him some specimens of the European bug *Pyrrhocoris apterus*— a species that had been reared in his laboratory in Prague for 10 years.

To our considerable mystification the bugs invariably died without reaching sexual maturity when we attempted to rear them at Harvard. Instead of metamorphosing into normal adults they continued to grow as larvae or molted into adult-like forms retaining many larval characteristics. It was evident that the bugs had access to some unknown source of juvenile hormone.

Eventually we traced the source to the paper toweling that had been placed in the rearing jars. Then we discovered that almost any paper of American origin—including the paper on which *Scientific American* is printed—had the same effect. Paper of European or Japanese manufacture had no effect on the bugs. On further investigation we found that the juvenile hormone activity originated in the balsam fir, which is the principal source of pulp for paper in Canada and the northern U.S. The tree synthesizes what we named the "paper factor," and this substance accompanies the pulp all the way to the printed page.

Thanks again to Bowers and his associates at Beltsville, the active material of the paper factor has been isolated and characterized [*see lower illustration on preceding page*]. It proves to be the methyl ester of a certain unsaturated fatty-acid derivative. The factor's kinship with the other juvenile hormone analogues is evident from the illustrations.

Here, then, is an extractable juvenile hormone analogue with selective action against only one kind of insect. As it happens, the family Pyrrhocoridae includes some of the most destructive pests of the cotton plant. Why the balsam fir should have evolved a substance against only one family of insects is unexplained. The most intriguing possibility is that the paper factor is a biochemical memento of the juvenile hormone of a former natural enemy of the tree—a pyrrhocorid predator that, for obvious reasons, is either extinct or has learned to avoid the balsam fir.

In any event, the fact that the tree synthesizes the substance argues strongly that the juvenile hormone of other species of insects can be mimicked, and perhaps has been by trees or plants on which the insects preyed. Evidently during the 250 million years of insect evolution the detailed chemistry of juvenile hormone has evolved and diversified. The process would of necessity have gone hand in hand with a retuning of the hormonal receptor mechanisms in the cells and tissues of the insect, so that the use as pesticides of any analogues that are discovered seems certain to be effective.

The evergreen trees are an ancient lot. They were here before the insects; they are pollinated by the wind and thus, unlike many other plants, do not depend on the insects for anything. The paper factor is only one of thousands of terpenoid materials these trees synthesize for no apparent reason. What about the rest?

It seems altogether likely that many of these materials will also turn out to be analogues of the juvenile hormones of specific insect pests. Obviously this is the place to look for a whole battery of third-generation pesticides. Then man may be able to emulate the evergreen trees in their incredibly sophisticated self-defense against the insects.

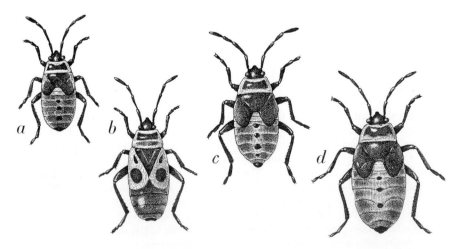

EFFECT OF PAPER FACTOR on *Pyrrhocoris apterus* is depicted. A larva of the fifth and normally final stage (*a*) turns into a winged adult (*b*). Contact with the paper factor causes the insect to turn into a sixth-stage larva (*c*) and sometimes into a giant seventh-stage larva (*d*). The abnormal larvae usually cannot shed their skin and die before reaching maturity.

14

PHEROMONES

EDWARD O. WILSON
May 1963

A pheromone is a substance secreted by an animal that influences the behavior of other animals of the same species. Recent studies indicate that such chemical communication is surprisingly common

It is conceivable that somewhere on other worlds civilizations exist that communicate entirely by the exchange of chemical substances that are smelled or tasted. Unlikely as this may seem, the theoretical possibility cannot be ruled out. It is not difficult to design, on paper at least, a chemical communication system that can transmit a large amount of information with rather good efficiency. The notion of such a communication system is of course strange because our outlook is shaped so strongly by our own peculiar auditory and visual conventions. This limitation of outlook is found even among students of animal behavior; they have favored species whose communication methods are similar to our own and therefore more accessible to analysis. It is becoming increasingly clear, however, that chemical systems provide the dominant means of communication in many animal species, perhaps even in most. In the past several years animal behaviorists and organic chemists, working together, have made a start at deciphering some of these systems and have discovered a number of surprising new biological phenomena.

In earlier literature on the subject, chemicals used in communication were usually referred to as "ectohormones." Since 1959 the less awkward and etymologically more accurate term "pheromones" has been widely adopted. It is used to describe substances exchanged among members of the same animal species. Unlike true hormones, which are secreted internally to regulate the organism's own physiology, or internal environment, pheromones are secreted externally and help to regulate the organism's external environment by influencing other animals. The mode of influence can take either of two general forms. If the pheromone produces a more or less immediate and reversible change in the behavior of the recipient, it is said to have a "releaser" effect. In this case the chemical substance seems to act directly on the recipient's central nervous system. If the principal function of the pheromone is to trigger a chain of physiological events in the recipient, it has what we have recently labeled a "primer" effect. The physiological changes, in turn, equip the organism with a new behavioral repertory, the components of which are thenceforth evoked by appropriate stimuli. In termites, for example, the reproductive and soldier castes prevent other termites from developing into their own castes by secreting substances that are ingested and act through the *corpus allatum,* an endocrine gland controlling differentiation [see "The Termite and the Cell," by Martin Lüscher; SCIENTIFIC AMERICAN, May, 1953].

These indirect primer pheromones do not always act by physiological inhibition. They can have the opposite effect. Adult males of the migratory locust *Schistocerca gregaria* secrete a volatile substance from their skin surface that accelerates the growth of young locusts. When the nymphs detect this substance with their antennae, their hind legs, some of their mouth parts and the antennae themselves vibrate. The secretion, in conjunction with tactile and visual signals, plays an important role in the formation of migratory locust swarms.

A striking feature of some primer pheromones is that they cause important physiological change without an immediate accompanying behavioral response, at least none that can be said to be peculiar to the pheromone. Beginning in 1955 with the work of S. van der Lee and L. M. Boot in the Netherlands, mammalian endocrinologists have discovered several unexpected effects on the female mouse that are produced by odors of other members of the same species. These changes are not marked by any immediate distinctive behavioral patterns. In the "Lee-Boot effect" females placed in groups of four show an increase in the percentage of pseudopregnancies. A completely normal reproductive pattern can be restored by removing the olfactory bulbs of the mice or by housing the mice separately. When more and more female mice are forced to live together, their oestrous cycles become highly irregular and in most of the mice the cycle stops completely for long periods. Recently W. K. Whitten of the Australian National University has discovered that the odor of a male mouse can initiate and synchronize the oestrous cycles of female mice. The male odor also reduces the frequency of reproductive abnormalities arising when female mice are forced to live under crowded conditions.

A still more surprising primer effect has been found by Helen Bruce of the National Institute for Medical Research in London. She observed that the odor of a strange male mouse will block the pregnancy of a newly impregnated female mouse. The odor of the original stud male, of course, leaves pregnancy undisturbed. The mouse reproductive pheromones have not yet been identified chemically, and their mode of action is only partly understood. There is evidence that the odor of the strange male suppresses the secretion of the hormone prolactin, with the result that the *corpus luteum* (a ductless ovarian gland) fails to develop and normal oestrus is restored. The pheromones are probably part of the complex set of control mechanisms that regulate the population density of animals [see "Population Density and Social Pathology," by John

B. Calhoun; SCIENTIFIC AMERICAN Offprint 506].

Pheromones that produce a simple releaser effect—a single specific response mediated directly by the central nervous system—are widespread in the animal kingdom and serve a great many functions. Sex attractants constitute a large and important category. The chemical structures of six attractants are shown on page 9. Although two of the six—the mammalian scents muskone and civetone—have been known for some 40 years and are generally assumed to serve a sexual function, their exact role has never been rigorously established by experiments with living animals. In fact, mammals seem to employ musklike compounds, alone or in combination with other substances, to serve several functions: to mark home ranges, to assist in territorial defense and to identify the sexes.

The nature and role of the four insect sex attractants are much better understood. The identification of each represents a technical feat of considerable magnitude. To obtain 12 milligrams of esters of bombykol, the sex attractant of the female silkworm moth, Adolf F. J. Butenandt and his associates at the Max Planck Institute of Biochemistry in Munich had to extract material from 250,000 moths. Martin Jacobson, Morton Beroza and William Jones of the U.S. Department of Agriculture processed 500,000 female gypsy moths to get 20 milligrams of the gypsy-moth attractant gyplure. Each moth yielded only about .01 microgram (millionth of a gram) of gyplure, or less than a millionth of its body weight. Bombykol and gyplure were obtained by killing the insects and subjecting crude extracts of material to chromatography, the separation technique in which compounds move at different rates through a column packed with a suitable adsorbent substance. Another technique has been more recently developed by Robert T. Yamamoto of the U.S. Department of Agriculture, in collaboration with Jacobson and Beroza, to harvest the equally elusive sex attractant of the American cockroach. Virgin females were housed in metal cans and air was continuously drawn through the cans and passed through chilled containers to condense any vaporized materials. In this manner the equivalent of 10,000 females were "milked" over a nine-month period to yield 12.2 milligrams of what was considered to be the pure attractant.

The power of the insect attractants is almost unbelievable. If some 10,000

molecules of the most active form of bombykol are allowed to diffuse from a source one centimeter from the antennae of a male silkworm moth, a characteristic sexual response is obtained in most cases. If volatility and diffusion rate are taken into account, it can be estimated that the threshold concentration is no more than a few hundred molecules per cubic centimeter, and the actual number required to stimulate the male is probably even smaller. From this one can calculate that .01 microgram of gyplure, the minimum average content of a single female moth, would be theoretically adequate, if distributed with maximum efficiency, to excite more than a billion male moths.

In nature the female uses her powerful pheromone to advertise her presence over a large area with a minimum expenditure of energy. With the aid of published data from field experiments and newly contrived mathematical models of the diffusion process, William H. Bossert, one of my associates in the Biological Laboratories at Harvard University, and I have deduced the shape and size of the ellipsoidal space within which male moths can be attracted under natural conditions [see bottom illustration on page 128]. When a moderate wind is blowing, the active space has a long axis of thousands of meters and a transverse axis parallel to the ground of more than 200 meters at the widest point. The 19th-century French naturalist Jean Henri Fabre, speculating on sex attraction in insects, could not bring himself to believe that the female moth could communicate over such great distances by odor alone, since "one might as well expect to tint a lake with a drop of carmine." We now know that Fabre's conclusion was wrong but that his analogy was exact: to the male moth's powerful chemoreceptors the lake is indeed tinted.

One must now ask how the male moth, smelling the faintly tinted air, knows which way to fly to find the source of the tinting. He cannot simply fly in the direction of increasing scent; it can be shown mathematically that the attractant is distributed almost uniformly after it has drifted more than a few meters from the female. Recent experiments by Ilse Schwinck of the University of Munich have revealed what is probably the alternative procedure used. When male moths are activated by the pheromone, they simply fly upwind and thus inevitably move toward the female. If by accident they pass out of the active zone, they either abandon the search or fly about at random until they pick up the scent again. Eventually, as they approach the female, there is a slight in-

INVISIBLE ODOR TRAILS guide fire ant workers to a source of food: a drop of sugar solution. The trails consist of a pheromone laid down by workers returning to their nest after finding a source of food. Sometimes the chemical message is reinforced by the touching of antennae if a returning worker meets a wandering fellow along the way. This is hap-

crease in the concentration of the chemical attractant and this can serve as a guide for the remaining distance.

If one is looking for the most highly developed chemical communication systems in nature, it is reasonable to study the behavior of the social insects, particularly the social wasps, bees, termites and ants, all of which communicate mostly in the dark interiors of their nests and are known to have advanced chemoreceptive powers. In recent years experimental techniques have been developed to separate and identify the pheromones of these insects, and rapid progress has been made in deciphering the hitherto intractable codes, particularly those of the ants. The most successful procedure has been to dissect out single glandular reservoirs and see what effect their contents have on the behavior of the worker caste, which is the most numerous and presumably the most in need of continuing guidance. Other pheromones, not present in distinct reservoirs, are identified in chromatographic fractions of crude extracts.

Ants of all castes are constructed with an exceptionally well-developed exocrine glandular system. Many of the most prominent of these glands, whose function has long been a mystery to entomologists, have now been identified as

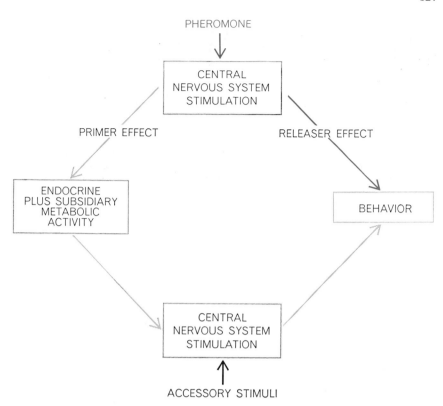

PHEROMONES INFLUENCE BEHAVIOR directly or indirectly, as shown in this schematic diagram. If a pheromone stimulates the recipient's central nervous system into producing an immediate change in behavior, it is said to have a "releaser" effect. If it alters a set of long-term physiological conditions so that the recipient's behavior can subsequently be influenced by specific accessory stimuli, the pheromone is said to have a "primer" effect.

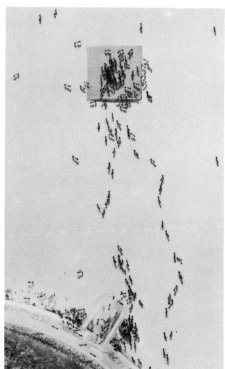

pening in the photograph at the far left. A few foraging workers have just found the sugar drop and a returning trail-layer is communicating the news to another ant. In the next two pictures the trail has been completed and workers stream from the nest in increasing numbers. In the fourth picture unrewarded workers return to the nest without laying trails and outward-bound traffic wanes. In the last picture most of the trails have evaporated completely and only a few stragglers remain at the site, eating the last bits of food.

ANTENNAE OF GYPSY MOTHS differ radically in structure according to their function. In the male (*left*) they are broad and finely divided to detect minute quantities of sex attractant released by the female (*right*). The antennae of the female are much less developed.

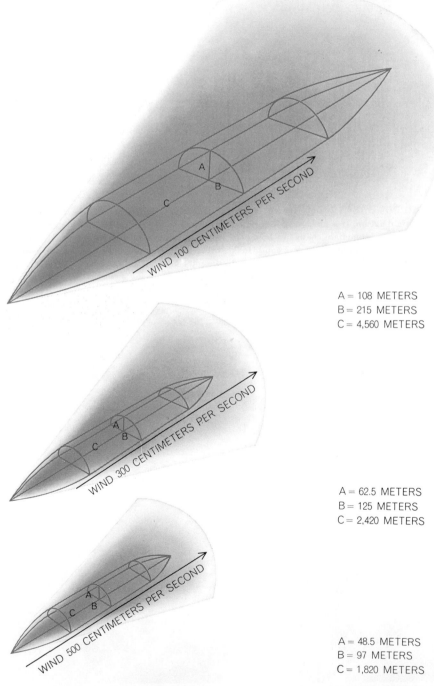

A = 108 METERS
B = 215 METERS
C = 4,560 METERS

A = 62.5 METERS
B = 125 METERS
C = 2,420 METERS

A = 48.5 METERS
B = 97 METERS
C = 1,820 METERS

ACTIVE SPACE of gyplure, the gypsy moth sex attractant, is the space within which this pheromone is sufficiently dense to attract males to a single, continuously emitting female. The actual dimensions, deduced from linear measurements and general gas-diffusion models, are given at right. Height (*A*) and width (*B*) are exaggerated in the drawing. As wind shifts from moderate to strong, increased turbulence contracts the active space.

the source of pheromones [*see illustration on page 130*]. The analysis of the gland-pheromone complex has led to the beginnings of a new and deeper understanding of how ant societies are organized.

Consider the chemical trail. According to the traditional view, trail secretions served as only a limited guide for worker ants and had to be augmented by other kinds of signals exchanged inside the nest. Now it is known that the trail substance is extraordinarily versatile. In the fire ant (*Solenopsis saevissima*), for instance, it functions both to activate and to guide foraging workers in search of food and new nest sites. It also contributes as one of the alarm signals emitted by workers in distress. The trail of the fire ant consists of a substance secreted in minute amounts by Dufour's gland; the substance leaves the ant's body by way of the extruded sting, which is touched intermittently to the ground much like a moving pen dispensing ink. The trail pheromone, which has not yet been chemically identified, acts primarily to attract the fire ant workers. Upon encountering the attractant the workers move automatically up the gradient to the source of emission. When the substance is drawn out in a line, the workers run along the direction of the line away from the nest. This simple response brings them to the food source or new nest site from which the trail is laid. In our laboratory we have extracted the pheromone from the Dufour's glands of freshly killed workers and have used it to create artificial trails. Groups of workers will follow these trails away from the nest and along arbitrary routes (including circles leading back to the nest) for considerable periods of time. When the pheromone is presented to whole colonies in massive doses, a large portion of the colony, including the queen, can be drawn out in a close simulation of the emigration process.

The trail substance is rather volatile, and a natural trail laid by one worker diffuses to below the threshold concentration within two minutes. Consequently outward-bound workers are able to follow it only for the distance they can travel in this time, which is about 40 centimeters. Although this strictly limits the distance over which the ants can communicate, it provides at least two important compensatory advantages. The more obvious advantage is that old, useless trails do not linger to confuse the hunting workers. In addition, the intensity of the trail laid by many workers provides a sensitive index of the amount of food at a given site and the rate of its depletion. As workers move to and from the food finds (consisting mostly of dead insects and sugar sources) they continuously add their own secretions to the trail produced by the original discoverers of the food. Only if an ant is rewarded by food does it lay a trail on its trip back to the nest; therefore the more food encountered at the end of the trail, the more workers that can be rewarded and the heavier the trail. The heavier the trail, the more workers that are drawn from the nest and arrive at the end of the trail. As the food is consumed, the number of workers laying trail substance drops, and the old trail fades by evaporation and diffusion, gradually constricting the outward flow of workers.

The fire ant odor trail shows other evidences of being efficiently designed. The active space within which the pheromone is dense enough to be perceived by workers remains narrow and nearly constant in shape over most of the length of the trail. It has been further deduced from diffusion models that the maximum gradient must be situated near the outer surface of the active space. Thus workers are informed of the space boundary in a highly efficient way. Together these features ensure that the following workers keep in close formation with a minimum chance of losing the trail.

The fire ant trail is one of the few animal communication systems whose information content can be measured with fair precision. Unlike many communicating animals, the ants have a distinct goal in space—the food find or nest site—the direction and distance of which must both be communicated. It is possible by a simple technique to measure how close trail-followers come to the trail end, and, by making use of a standard equation from information theory, one can translate the accuracy of their response into the "bits" of information received. A similar procedure can be applied (as first suggested by the British biologist J. B. S. Haldane) to the "waggle dance" of the honeybee, a radically different form of communication system from the ant trail [see "Dialects in the Language of the Bees," by Karl von Frisch; SCIENTIFIC AMERICAN Offprint 130]. Surprisingly, it turns out that the two systems, although of wholly different evolutionary origin, transmit about the same amount of information with reference to distance (two bits) and direction (four bits in the honeybee, and four or possibly five in the ant). Four bits of information will direct an ant or a bee into one of 16 equally probable sectors of a circle and two bits will identify one of four equally probable dis-

FIRE ANT WORKER lays an odor trail by exuding a pheromone along its extended sting. The sting is touched to the ground periodically, breaking the trail into a series of streaks.

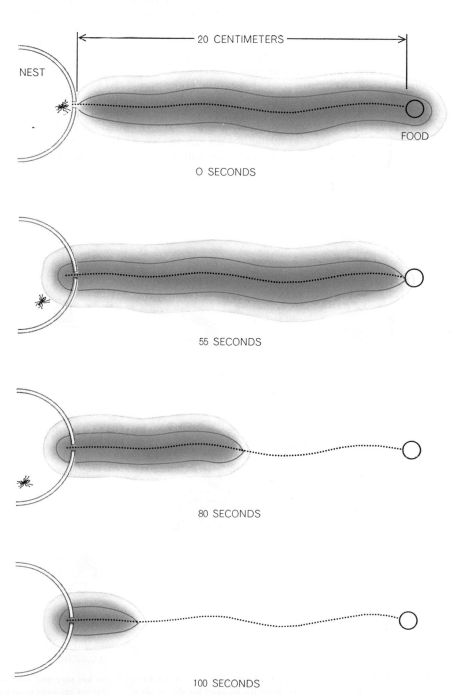

20 CENTIMETERS

NEST

FOOD

O SECONDS

55 SECONDS

80 SECONDS

100 SECONDS

ACTIVE SPACE OF ANT TRAIL, within which the pheromone is dense enough to be perceived by other workers, is narrow and nearly constant in shape with the maximum gradient situated near its outer surface. The rapidity with which the trail evaporates is indicated.

tances. It is conceivable that these information values represent the maximum that can be achieved with the insect brain and sensory apparatus.

Not all kinds of ants lay chemical trails. Among those that do, however, the pheromones are highly species-specific in their action. In experiments in which artificial trails extracted from one species were directed to living colonies of other species, the results have almost always been negative, even among related species. It is as if each species had its own private language. As a result there is little or no confusion when the trails of two or more species cross.

Another important class of ant pheromone is composed of alarm substances. A simple backyard experiment will show that if a worker ant is disturbed by a clean instrument, it will, for a short time, excite other workers with whom it comes in contact. Until recently most students of ant behavior thought that

the alarm was spread by touch, that one worker simply jostled another in its excitement or drummed on its neighbor with its antennae in some peculiar way. Now it is known that disturbed workers discharge chemicals, stored in special glandular reservoirs, that can produce all the characteristic alarm responses solely by themselves. The chemical structure of four alarm substances is shown on page 134. Nothing could illustrate more clearly the wide differences between the human perceptual world and that of chemically communicating animals. To the human nose the alarm substances are mild or even pleasant, but to the ant they represent an urgent tocsin that can propel a colony into violent and instant action.

As in the case of the trail substances, the employment of the alarm substances appears to be ideally designed for the purpose it serves. When the contents of the mandibular glands of a worker of the harvesting ant (*Pogonomyrmex badius*)

are discharged into still air, the volatile material forms a rapidly expanding sphere, which attains a radius of about six centimeters in 13 seconds. Then it contracts until the signal fades out completely some 35 seconds after the moment of discharge. The outer shell of the active space contains a low concentration of pheromone, which is actually attractive to harvester workers. This serves to draw them toward the point of disturbance. The central region of the active space, however, contains a concentration high enough to evoke the characteristic frenzy of alarm. The "alarm sphere" expands to a radius of about three centimeters in eight seconds and, as might be expected, fades out more quickly than the "attraction sphere."

The advantage to the ants of an alarm signal that is both local and short-lived becomes obvious when a *Pogonomyrmex* colony is observed under natural conditions. The ant nest is subject to almost innumerable minor disturbances. If the

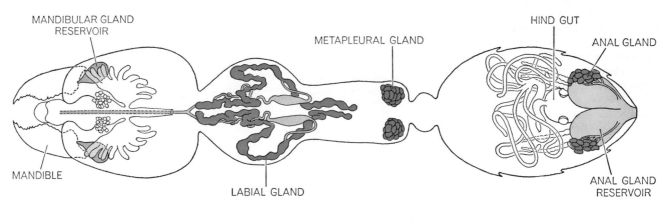

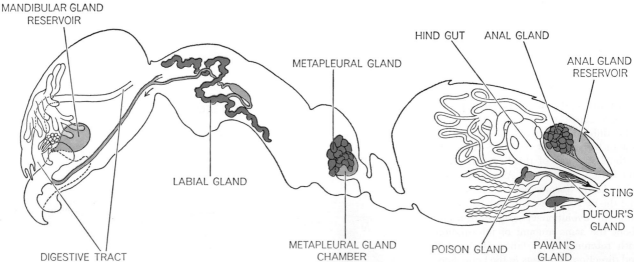

EXOCRINE GLANDULAR SYSTEM of a worker ant (*shown here in top and side cutaway views*) is specially adapted for the production of chemical communication substances. Some pheromones are stored in reservoirs and released in bursts only when needed; others are secreted continuously. Depending on the species, trail substances are produced by Dufour's gland, Pavan's gland or the poison glands; alarm substances are produced by the anal and mandibular glands. The glandular sources of other pheromones are unknown.

alarm spheres generated by individual ant workers were much wider and more durable, the colony would be kept in ceaseless and futile turmoil. As it is, local disturbances such as intrusions by foreign insects are dealt with quickly and efficiently by small groups of workers, and the excitement soon dies away.

The trail and alarm substances are only part of the ants' chemical vocabulary. There is evidence for the existence of other secretions that induce gathering and settling of workers, acts of grooming, food exchange, and other operations fundamental to the care of the queen and immature ants. Even dead ants produce a pheromone of sorts. An ant that has just died will be groomed by other workers as if it were still alive. Its complete immobility and crumpled posture by themselves cause no new response. But in a day or two chemical decomposition products accumulate and stimulate the workers to bear the corpse to the refuse pile outside the nest. Only a few decomposition products trigger this funereal response; they include certain long-chain fatty acids and their esters. When other objects, including living workers, are experimentally daubed with these substances, they are dutifully carried to the refuse pile. After being dumped on the refuse the "living dead" scramble to their feet and promptly return to the nest, only to be carried out again. The hapless creatures are thrown back on the refuse pile time and again until most of the scent of death has been worn off their bodies by the ritual.

Our observation of ant colonies over long periods has led us to believe that as few as 10 pheromones, transmitted singly or in simple combinations, might suffice for the total organization of ant society. The task of separating and characterizing these substances, as well as judging the roles of other kinds of stimuli such as sound, is a job largely for the future.

Even in animal species where other kinds of communication devices are prominently developed, deeper investigation usually reveals the existence of pheromonal communication as well. I have mentioned the auxiliary roles of primer pheromones in the lives of mice and migratory locusts. A more striking example is the communication system of the honeybee. The insect is celebrated for its employment of the "round" and "waggle" dances (augmented, perhaps, by auditory signals) to designate the location of food and new nest sites. It is not so widely known that chemical signals

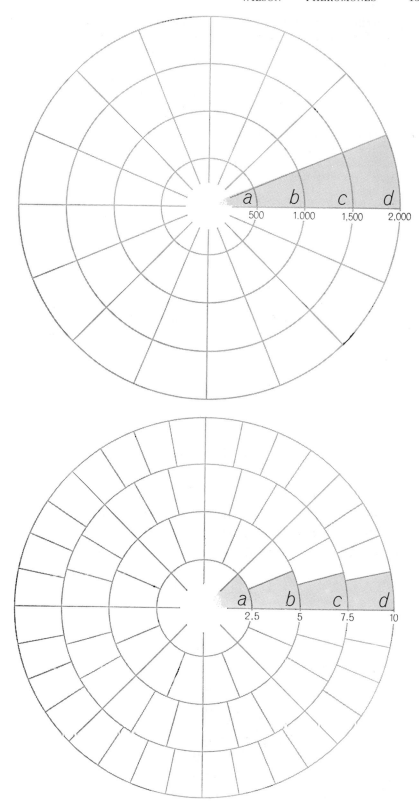

FORAGING INFORMATION conveyed by two different insect communication systems can be represented on two similar "compass" diagrams. The honeybee "waggle dance" (*top*) transmits about four bits of information with respect to direction, enabling a honeybee worker to pinpoint a target within one of 16 equally probable angular sectors. The number of "bits" in this case remains independent of distance, given in meters. The pheromone system used by trail-laying fire ants (*bottom*) is superior in that the amount of directional information increases with distance, given in centimeters. At distances *c* and *d*, the probable sector in which the target lies is smaller for ants than for bees. (For ants, directional information actually increases gradually and not by jumps.) Both insects transmit two bits of distance information, specifying one of four equally probable distance ranges.

play equally important roles in other aspects of honeybee life. The mother queen regulates the reproductive cycle of the colony by secreting from her mandibular glands a substance recently identified as 9-ketodecanoic acid. When this pheromone is ingested by the worker bees, it inhibits development of their ovaries and also their ability to manufacture the royal cells in which new queens are reared. The same pheromone serves as a sex attractant in the queen's nuptial flights.

Under certain conditions, including the discovery of new food sources, worker bees release geraniol, a pleasant-smelling alcohol, from the abdominal Nassanoff glands. As the geraniol diffuses through the air it attracts other workers and so supplements information contained in the waggle dance. When a worker stings an intruder, it discharges, in addition to the venom, tiny amounts of a secretion from clusters of unicellular glands located next to the basal plates of the sting. This secretion is responsible for the tendency, well known to beekeepers, of angry swarms of workers to sting at the same spot. One component, which acts as a simple attractant, has been identified as isoamyl acetate, a compound that has a banana-like odor. It is possible that the stinging response is evoked by at least one unidentified alarm substance secreted along with the attractant.

Knowledge of pheromones has advanced to the point where one can make some tentative generalizations about their chemistry. In the first place, there appear to be good reasons why sex attractants should be compounds that contain between 10 and 17 carbon atoms and that have molecular weights between about 180 and 300—the range actually observed in attractants so far identified. (For comparison, the weight of a single carbon atom is 12.) Only compounds of roughly this size or greater can meet the two known requirements of a sex attractant: narrow specificity, so that only members of one species will respond to it, and high potency. Compounds that contain fewer than five or so carbon atoms and that have a molecular weight of less than about 100 cannot be assembled in enough different ways to provide a distinctive molecule for all the insects that want to advertise their presence.

It also seems to be a rule, at least with insects, that attraction potency increases with molecular weight. In one series of esters tested on flies, for instance, a doubling of molecular weight resulted in as much as a thousandfold increase in efficiency. On the other hand, the molecule cannot be too large and complex or it will be prohibitively difficult for the insect to synthesize. An equally important limitation on size is

SIX SEX PHEROMONES include the identified sex attractants of four insect species as well as two mammalian musks generally believed to be sex attractants. The high molecular weight of most sex pheromones accounts for their narrow specificity and high potency.

the fact that volatility—and, as a result, diffusibility—declines with increasing molecular weight.

One can also predict from first principles that the molecular weight of alarm substances will tend to be less than those of the sex attractants. Among the ants there is little specificity; each species responds strongly to the alarm substances of other species. Furthermore, an alarm substance, which is used primarily within the confines of the nest, does not need the stimulative potency of a sex attractant, which must carry its message for long distances. For these reasons small molecules will suffice for alarm purposes. Of seven alarm substances known in the social insects, six have 10 or fewer carbon atoms and one (dendro-lasin) has 15. It will be interesting to see if future discoveries bear out these early generalizations.

Do human pheromones exist? Primer pheromones might be difficult to detect, since they can affect the endocrine system without producing overt specific behavioral responses. About all that can be said at present is that striking sexual differences have been observed in the ability of humans to smell certain

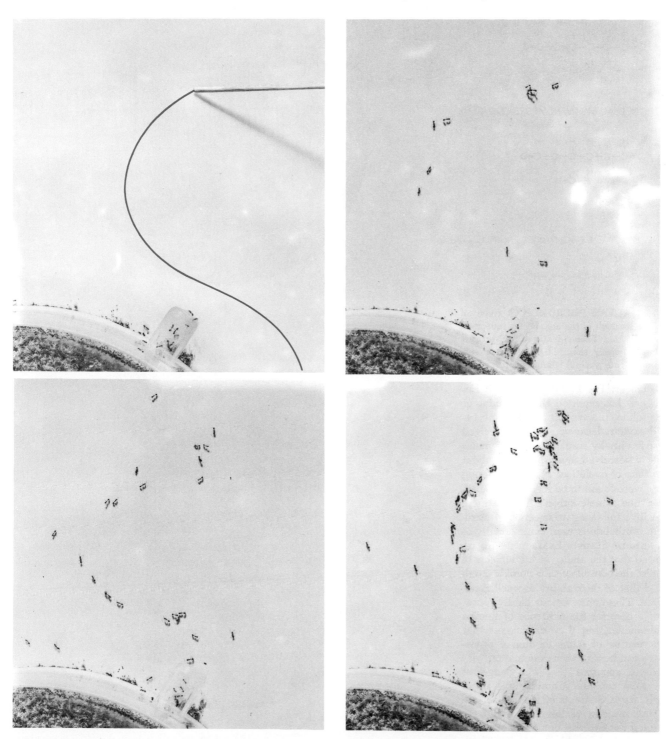

ARTIFICIAL TRAIL can be laid down by drawing a line (*colored curve in frame at top left*) with a stick that has been treated with the contents of a single Dufour's gland. In the remaining three frames, workers are attracted from the nest, follow the artificial route in close formation and mill about in confusion at its arbitrary terminus. Such a trail is not renewed by the unrewarded workers.

DENDROLASIN (*LASIUS FULIGINOSUS*)

CITRAL (*ATTA SEXDENS*)

CITRONELLAL (*ACANTHOMYOPS CLAVIGER*)

2-HEPTANONE (*IRIDOMYRMEX PRUINOSUS*)

FOUR ALARM PHEROMONES, given off by the workers of the ant species indicated, have so far been identified. Disturbing stimuli trigger the release of these substances from various glandular reservoirs.

substances. The French biologist J. Le-Magnen has reported that the odor of Exaltolide, the synthetic lactone of 14-hydroxytetradecanoic acid, is perceived clearly only by sexually mature females and is perceived most sharply at about the time of ovulation. Males and young girls were found to be relatively insensitive, but a male subject became more sensitive following an injection of estrogen. Exaltolide is used commercially as a perfume fixative. LeMagnen also reported that the ability of his subjects to detect the odor of certain steroids paralleled that of their ability to smell Exaltolide. These observations hardly represent a case for the existence of human pheromones, but they do suggest that the relation of odors to human physiology can bear further examination.

It is apparent that knowledge of chemical communication is still at an early stage. Students of the subject are in the position of linguists who have learned the meaning of a few words of a nearly indecipherable language. There is almost certainly a large chemical vocabulary still to be discovered. Conceiv-

ably some pheromone "languages" will be found to have a syntax. It may be found, in other words, that pheromones can be combined in mixtures to form new meanings for the animals employing them. One would also like to know if some animals can modulate the intensity or pulse frequency of pheromone emission to create new messages. The solution of these and other interesting problems will require new techniques in analytical organic chemistry combined with ever more perceptive studies of animal behavior.

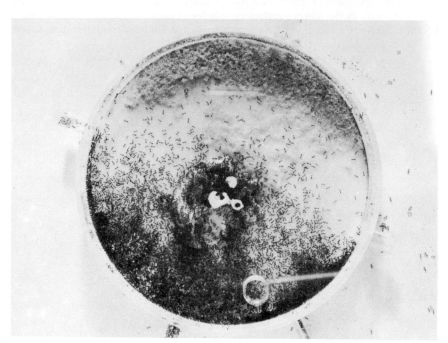

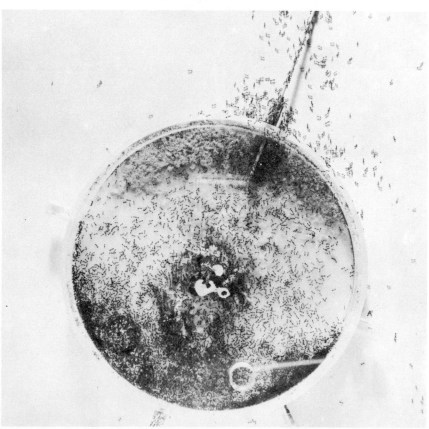

MASSIVE DOSE of trail pheromone causes the migration of a large portion of a fire ant colony from one side of a nest to another. The pheromone is administered on a stick that has been dipped in a solution extracted from the Dufour's glands of freshly killed workers.

III

ENERGY IN OUR SOCIETY

III

ENERGY IN OUR SOCIETY

INTRODUCTION

Just as the chemical reactions that sustain the biosphere require a constant input of energy, so do the operations that constitute the functioning of an industrial society. Industrial systems, however, require energy flows that far exceed those needed to maintain the conversion rates achieved by any components of the biosystem, and so have to rely on stored supplies of fuel. Earl Cook's article, "The Flow of Energy in an Industrial Society," in outlining the sources of energy exploited by the United States economy and tracing the patterns of utilization shows very clearly how dependent every aspect of our lives is on those continuing flows. Reading it, one can understand why energy use today is presenting the human race with one of the most perplexing dilemmas it has ever faced. Maintenance of our society demands that we keep on consuming our resources of energy-producing materials at a rate that is at least undiminished. Every such material now at our disposal, however, exists in finite supply, and our annual consumption of most of them is not an insignificant fraction of the total. In addition to the inexorable depletion of resources, our staggering rate of energy use creates a direct impact on us and the environment. Nearly every article in this section touches somehow on the pollution burden generated while fuels are converted into useful work. Various aspects of that problem are elaborated in Section IV; this group of articles concentrates on the prospects for continuing to supply the energy required for survival of a technologically advanced society, without poisoning it in the meantime.

M. King Hubbert has prepared an excellent summary of current estimates of the magnitude of world energy supplies, including all the sources that we can now or might in the future draw on. "The Energy Resources of the Earth" also assesses the potential each fuel has for meeting our increasing energy requirements, and provides a sobering illustration of the consequences of exponential growth.

Two articles reprinted here deal specifically with the nature of part of our fossil fuel supply. Today's developed societies draw heavily on the planet's reserves of fossil fuels, especially natural gas and petroleum. These materials, convenient to handle and relatively clean-burning, have been favored over the traditional fuel, coal, for at least twenty years. There is not as much oil and natural gas, however, and the return to extensive utilization of coal must soon begin. "Coal," by Lawrence P. Lessing, is an old article, written in 1955, but it remains surprisingly pertinent. The chemical nature of the resource is of course unchanged, though somewhat more is known about it now.* The reactions involved in its processing are still the same, and comparison with "Clean Power From Dirty Fuels" (written in 1972) reveals many similarities between engineering schemes for utilizing coal being thought about now and ones that were considered then.

The remaining source of fossil hydrocarbons lies in the huge, still virtually untapped, deposits described in "Tar Sands and Oil Shales" by Noel de Nevers. As other fluid fuels become increasingly short in supply, the relatively expensive processing required to free these unconventional forms is becoming more economically competitive, and the drive to bring them into the market is accelerating.

Meanwhile an active search is being carried on for ways both to

---

*The *Encyclopedia of Chemical Technology*, Volume 5, contains a concise up-to-date section on coal.

increase the efficiency of fossil fuel use and to decrease the attendant pollutant production. The use of electrical energy is essentially pollution-free and usually quite efficient. The generation of electricity, however, is plagued by outpourings of combustion by-products when done in fossil-fuel-fired plants. It is also limited in efficiency by the laws of thermodynamics. Arthur M. Squires, in "Clean Power from Dirty Fuels," discusses those problems and some chemical ways in which thermal generation of electricty might be improved. Quite another approach is taken in "Fuel Cells," in which Leonard G. Austin describes how electric current can be produced directly by electrochemical reactions. Although his basic premises are as valid as ever, practical application has come a long way since the article was published in 1959, and fuel cells are undergoing intensive development. Not only have they powered many space capsules, but their ability to provide adequate power for residential use and small business operation is being tested. In such applications, the fuel cells consume methane, which is steam reformed in an internal reactor to provide hydrogen for reaction with oxygen in the air.

Although at present fossil fuels provide for an overwhelming majority of our energy needs, more and more use is being made of nuclear reactions as a source of heat for generating electricity—we have already begun the transition to the next generation of energy production methods. Under current practice, however, our supplies of fissile materials are just as limited as those of hydrocarbon fuels. To circumvent that limitation, intensive research is being carried out to develop nuclear power reactors that will synthesize additional supplies of the fuel that they use. These efforts are described in "Fast Breeder Reactors," by Glenn T. Seaborg and Justin L. Bloom. Still farther in the future is the possibility of using the energy released in nuclear fusion reactions. "The Prospects of Fusion Power," by William C. Gough and Bernard J. Eastlund, summarizes our progress toward realizing that goal and outlines some of the difficulties that will have to be overcome on the way.

As the shift from fossil to nuclear fuels continues, society will necessarily become more and more geared to using energy in the form of electricity, in spite of the disadvantages posed by costly transmission and inconvenience in some applications. The need for portable fuel will continue, however, and it is commonly assumed that hydrogen will fill that role when supplies of the traditional fossil fuels begin to fail. "The Hydrogen Economy," by Derek P. Gregory, describes what the role of hydrogen might be as an adjunct to nuclear-produced electricity some years from now.

Successful and timely application of techniques like those described in this section should provide mankind with the energy required to insure survival of our civilization. Just maintaining the energy flow is not sufficient, however; our continued existence will also depend on conquering the pollution problems that energy use entails.

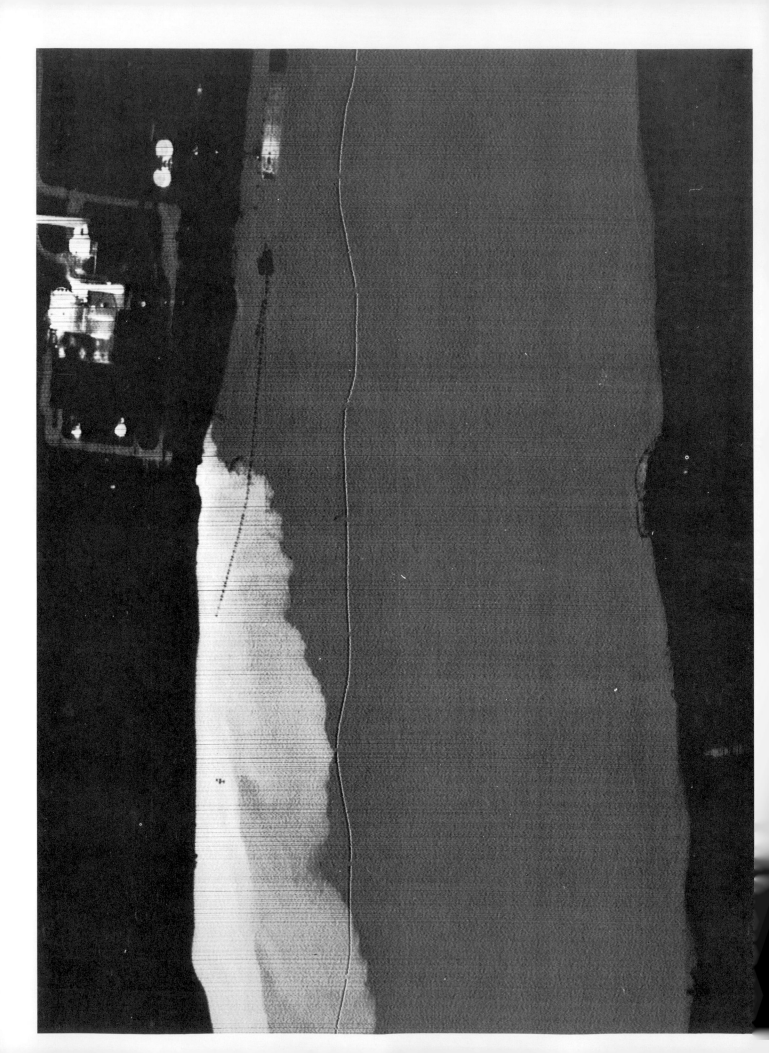

15

THE FLOW OF ENERGY IN AN INDUSTRIAL SOCIETY

EARL COOK

September 1971

The U.S., with 6 percent of the world's population, uses 35 percent of the world's energy. In the long run the limiting factor in high levels of energy consumption will be the disposal of the waste heat

This article will describe the flow of energy through an industrial society: the U.S. Industrial societies are based on the use of power: the rate at which useful work is done. Power depends on energy, which is the ability to do work. A power-rich society consumes—more accurately, degrades—energy in large amounts. The success of an industrial society, the growth of its economy, the quality of the life of its people and its impact on other societies and on the total environment are determined in large part by the quantities and the kinds of energy resources it exploits and by the efficiency of its systems for converting potential energy into work and heat.

Whether by hunting, by farming or by burning fuel, man introduces himself into the natural energy cycle, converting energy from less desired forms to more desired ones: from grass to beef, from wood to heat, from coal to electricity. What characterizes the industrial societies is their enormous consumption of energy and the fact that this consumption is primarily at the expense of "capital" rather than of "income," that is, at the expense of solar energy stored in coal, oil and natural gas rather than of solar radiation, water, wind and muscle power. The advanced industrial societies, the U.S. in particular, are further characterized by their increasing dependence on electricity, a trend that has direct effects on gross energy consumption and indirect effects on environmental quality.

The familiar exponential curve of increasing energy consumption can be considered in terms of various stages of human development [*see illustration on next page*]. As long as man's energy consumption depended on the food he could eat, the rate of consumption was some 2,000 kilocalories per day; the domestication of fire may have raised it to 4,000 kilocalories. In a primitive agricultural society with some domestic animals the rate rose to perhaps 12,000 kilocalories; more advanced farming societies may have doubled that consumption. At the height of the low-technology industrial revolution, say between 1850 and 1870, per capita daily consumption reached 70,000 kilocalories in England, Germany and the U.S. The succeeding high-technology revolution was brought about by the central electric-power station and the automobile, which enable the average person to apply power in his home and on the road. Beginning shortly before 1900, per capita energy consumption in the U.S. rose at an increasing rate to the 1970 figure: about 230,000 kilocalories per day, or about 65×10^{15} British thermal units (B.t.u.) per year for the country as a whole. Today the industrial regions, with 30 percent of the world's people, consume 80 percent of the world's energy. The U.S., with 6 percent of the people, consumes 35 percent of the energy.

In the early stages of its development in western Europe industrial society based its power technology on income sources of energy, but the explosive growth of the past century and a half has been fed by the fossil fuels, which are not renewable on any time scale meaningful to man. Modern industrial society is totally dependent on high rates of consumption of natural gas, petroleum and coal. These nonrenewable fossil-fuel resources currently provide 96 percent of the gross energy input into the U.S. economy [*see top illustration on page 141*]. Nuclear power, which in 1970 accounted for only .3 percent of the total energy input, is also (with present reactor technology) based on a capital source of energy: uranium 235. The energy of falling water, converted to hydropower, is the only income source of energy that now makes any significant contribution to the U.S. economy, and its proportional role seems to be declining from a peak reached in 1950.

Since 1945 coal's share of the U.S. energy input has declined sharply, while both natural gas and petroleum have increased their share. The shift is reflected in import figures. Net imports of petroleum and petroleum products doubled between 1960 and 1970 and now constitute almost 30 percent of gross consumption. In 1960 there were no imports of natural gas; last year natural-gas imports (by pipeline from Canada and as liquefied gas carried in cryogenic tankers) accounted for almost 4 percent of gross consumption and were increasing.

The reasons for the shift to oil and gas are not hard to find. The conversion of railroads to diesel engines represented a large substitution of petroleum for coal. The rapid growth, beginning during World War II, of the national

HEAT DISCHARGE from a power plant on the Connecticut River at Middletown, Conn., is shown in this infrared scanning radiograph. The power plant is at upper left, its structures outlined by their heat radiation. The luminous cloud running along the left bank of the river is warm water discharged from the cooling system of the plant. The vertical oblong object at top left center is an oil tanker. The luminous spot astern is the infrared glow of its engine room. The dark streak between the tanker and the warm-water region is a breakwater. The irregular line running down the middle of the picture is an artifact of the infrared scanning system. The picture was made by HRB-Singer, Inc., for U.S. Geological Survey.

network of high-pressure gas-transmission lines greatly extended the availability of natural gas. The explosion of the U.S. automobile population, which grew twice as fast as the human population in the decade 1960–1970, and the expansion of the nation's fleet of jet aircraft account for much of the increase in petroleum consumption. In recent years the demand for cleaner air has led to the substitution of natural gas or low-sulfur residual fuel oil for high-sulfur coal in many central power plants.

An examination of energy inputs by sector of the U.S. economy rather than by source reveals that much of the recent increase has been going into household, commercial and transportation applications rather than industrial ones [*see bottom illustration on opposite page*]. What is most striking is the growth of the electricity sector. In 1970 almost 10 percent of the country's useful work was done by electricity. That is not the whole story. When the flow of energy from resources to end uses is

charted for 1970 [*see illustration on pages 142 and 143*], it is seen that producing that much electricity accounted for 26 percent of the gross consumption of energy, because of inefficiencies in generation and transmission. If electricity's portion of end-use consumption rises to about 25 percent by the year 2000, as is expected, then its generation will account for between 43 and 53 percent of the country's gross energy consumption. At that point an amount of energy equal to about half of the useful work done in the U.S. will be in the form of waste heat from power stations!

All energy conversions are more or less inefficient, of course, as the flow diagram makes clear. In the case of electricity there are losses at the power plant, in transmission and at the point of application of power; in the case of fuels consumed in end uses the loss comes at the point of use. The 1970 U.S. gross consumption of 64.6×10^{15} B.t.u. of energy (or 16.3×10^{15} kilocalories, or 19×10^{12} kilowatt-hours) ends up as

32.8×10^{15} B.t.u. of useful work and 31.8×10^{15} B.t.u. of waste heat, amounting to an overall efficiency of about 51 percent.

The flow diagram shows the pathways of the energy that drives machines, provides heat for manufacturing processes and heats, cools and lights the country. It does not represent the total energy budget because it includes neither food nor vegetable fiber, both of which bring solar energy into the economy through photosynthesis. Nor does it include environmental space heating by solar radiation, which makes life on the earth possible and would be by far the largest component of a total energy budget for any area and any society.

The minute fraction of the solar flux that is trapped and stored in plants provides each American with some 10,000 kilocalories per day of gross food production and about the same amount in the form of nonfood vegetable fiber. The fiber currently contributes little to the energy supply. The food, however, fu-

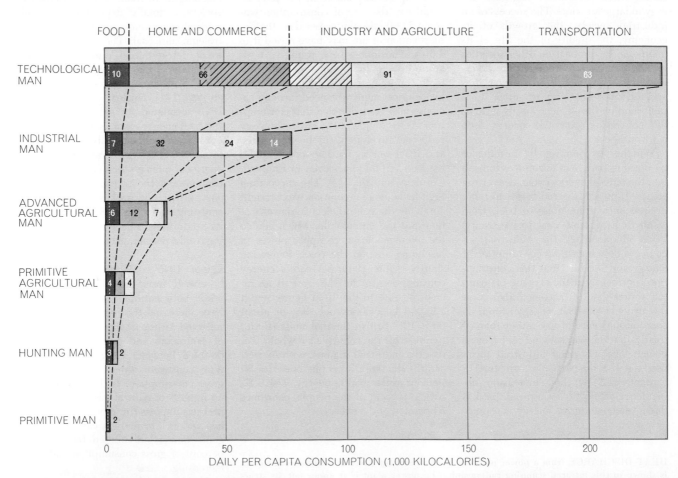

DAILY CONSUMPTION of energy per capita was calculated by the author for six stages in human development (and with an accuracy that decreases with antiquity). Primitive man (East Africa about 1,000,000 years ago) without the use of fire had only the energy of the food he ate. Hunting man (Europe about 100,000 years ago) had more food and also burned wood for heat and cooking. Primitive agricultural man (Fertile Crescent in 5000 B.C.) was growing crops and had gained animal energy. Advanced agricultural man (northwestern Europe in A.D. 1400) had some coal for heating, some water power and wind power and animal transport. Industrial man (in England in 1875) had the steam engine. In 1970 technological man (in the U.S.) consumed 230,000 kilocalories per day, much of it in form of electricity (*hatched area*). Food is divided into plant foods (*far left*) and animal foods (or foods fed to animals).

els man. Gross food-plant consumption might therefore be considered another component of gross energy consumption; it would add about 3×10^{15} B.t.u. to the input side of the energy-flow scheme. Of the 10,000 kilocalories per capita per day of gross production, handling and processing waste 15 percent. Of the remaining 8,500 kilocalories, some 6,300 go to feed animals that produce about 900 kilocalories of meat and 2,200 go into the human diet as plant materials, for a final food supply of about 3,100 kilocalories per person. Thus from field to table the efficiency of the food-energy system is 31 percent, close to the efficiency of a central power station. The similarity is not fortuitous; in both systems there is a large and unavoidable loss in the conversion of energy from a less desired form to a more desired one.

Let us consider recent changes in U.S. energy flow in more detail by seeing how the rates of increase in various sectors compare. Not only has energy consumption for electric-power generation been growing faster than the other sectors but also its growth rate has been increasing: from 7 percent per year in 1961–1965 to 8.6 percent per year in 1965–1969 to 9.25 percent last year [see top illustration on page 144]. The energy consumed in industry and commerce and in homes has increased at a fairly steady rate for a decade, but the energy demand of transportation has risen more sharply since 1966. All in all, energy consumption has been increasing lately at a rate of 5 percent per year, or four times faster than the increase in the U.S. population. Meanwhile the growth of the gross national product has tended to fall off, paralleling the rise in energy sectors other than fast-growing transportation and electricity. The result is a change in the ratio of total energy consumption to G.N.P. [see bottom illustration, page 144]. The ratio had been in a long general decline since 1920 (with brief reversals) but since 1967 it has risen more steeply each year. In 1970 the U.S. consumed more energy for each dollar of goods and services than at any time since 1951.

Electricity accounts for much of this decrease in economic efficiency, for several reasons. For one thing, we are substituting electricity, with a thermal efficiency of perhaps 32 percent, for many direct fuel uses with efficiencies ranging from 60 to 90 percent. Moreover, the fastest-growing segment of end-use consumption has been electric air conditioning. From 1967 to 1970 consumption for

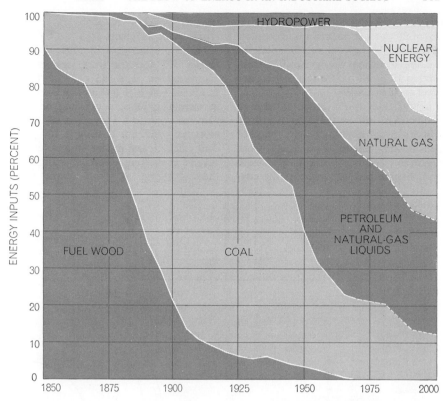

FOSSIL FUELS now account for nearly all the energy input into the U.S. economy. Coal's contribution has decreased since World War II; that of natural gas has increased most in that period. Nuclear energy should contribute a substantial percent within the next 20 years.

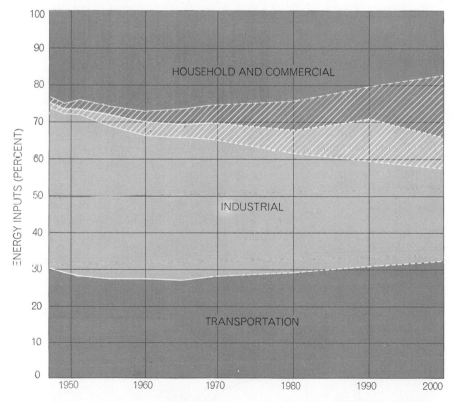

USEFUL WORK is distributed among the various end-use sectors of the U.S. economy as shown. The trend has been for industry's share to decrease, with household and commercial uses (including air conditioning) and transportation growing. Electricity accounts for an ever larger share of the work (hatched area). U.S. Bureau of Mines figures in this chart include nonenergy uses of fossil fuels, which constitute about 7 percent of total energy inputs.

air conditioning grew at the remarkable rate of 20 percent per year; it accounted for almost 16 percent of the total increase in electric-power generation from 1969 to 1970, with little or no multiplier effect on the G.N.P.

Let us take a look at this matter of efficiency in still another way: in terms of useful work done as a percentage of gross energy input. The "useful-work equivalent," or overall technical efficiency, is seen to be the product of the con-

version efficiency (if there is an intermediate conversion step) and the application efficiency of the machine or device that does the work [*see bottom illustration on page 145*]. Clearly there is a wide range of technical efficiencies in energy systems, depending on the conversion devices. It is often said that electrical resistance heating is 100 percent efficient, and indeed it is in terms, say, of converting electrical energy to thermal energy at the domestic hot-

water heater. In terms of the energy content of the natural gas or coal that fired the boiler that made the steam that drove the turbine that turned the generator that produced the electricity that heated the wires that warmed the water, however, it is not so efficient.

The technical efficiency of the total U.S. energy system, from potential energy at points of initial conversion to work at points of application, is about 50 percent. The economic efficiency of

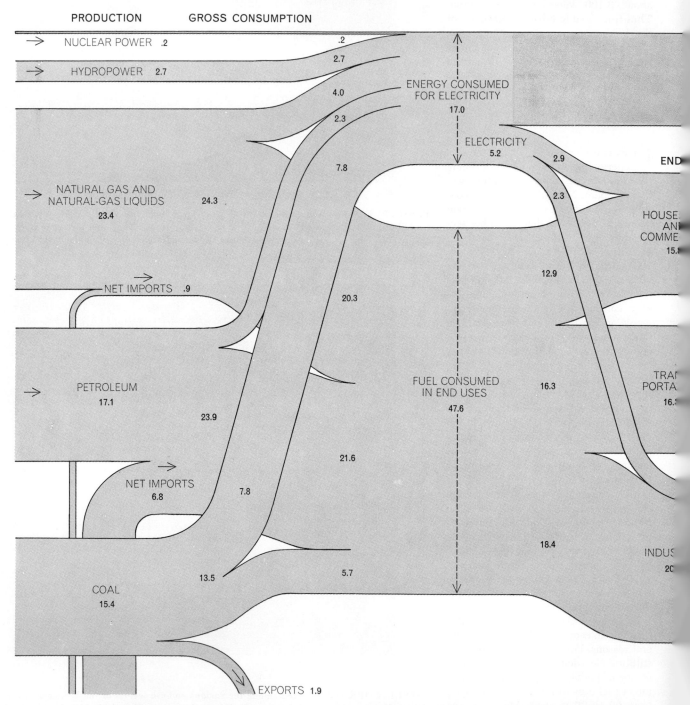

FLOW OF ENERGY through the U.S. system in 1970 is traced from production of energy commodities (*left*) to the ultimate conversion of energy into work for various industrial end products and waste heat (*right*). Total consumption of energy in 1970 was 64.6×10^{15} British thermal units. (Adding nonenergy uses of fossil fuels, primarily for petrochemicals, would raise the total to 68.8×10^{15} B.t.u.) The overall efficiency of the system was about 51 percent. Some of the fossil-fuel energy is consumed directly and

the system is considerably less. That is because work is expended in extracting, refining and transporting fuels, in the construction and operation of conversion facilities, power equipment and electricity-distribution networks, and in handling waste products and protecting the environment.

An industrial society requires not only a large supply of energy but also a high use of energy per capita, and the society's economy and standard of living are shaped by interrelations among resources, population, the efficiency of conversion processes and the particular applications of power. The effect of these interrelations is illustrated by a comparison of per capita energy consumption and per capita output for a number of countries [*see illustration on page 146*]. As one might expect, there is a strong general correlation between the two measures, but it is far from be-ing a one-to-one correlation. Some countries (the U.S.S.R. and the Republic of South Africa, for example) have a high energy consumption with respect to G.N.P.; other countries (such as Sweden and New Zealand) have a high output with relatively less energy consumption. Such differences reflect contrasting combinations of energy-intensive heavy industry and light consumer-oriented and service industries (characteristic of different stages of economic development) as well as differences in the efficiency of energy use. For example, countries that still rely on coal for a large part of their energy requirement have higher energy inputs per unit of production than those that use mainly petroleum and natural gas.

A look at trends from the U.S. past is also instructive. Between 1800 and 1880 total energy consumption in the U.S. lagged behind the population increase, which means that per capita energy consumption actually declined somewhat. On the other hand, the American standard of living increased during this period because the energy supply in 1880 (largely in the form of coal) was being used much more efficiently than the energy supply in 1800 (largely in the form of wood). From 1900 to 1920 there was a tremendous surge in the use of energy by Americans but not a parallel increase in the standard of living. The ratio of energy consumption to G.N.P. increased 50 percent during these two decades because electric power, inherently less efficient, began being substituted for the direct use of fuels; because the automobile, at best 25 percent efficient, proliferated (from 8,000 in 1900 to 8,132,000 in 1920), and because mining and manufacturing, which are energy-intensive, grew at very high rates during this period.

Then there began a long period during which increases in the efficiency of energy conversion and utilization fulfilled about two-thirds of the total increase in demand, so that the ratio of energy consumption to G.N.P. fell to about 60 percent of its 1920 peak although per capita energy consumption continued to increase. During this period (1920–1965) the efficiency of electric-power generation and transmission almost trebled, mining and manufacturing grew at much lower rates and the services sector of the economy, which is not energy-intensive, increased in importance.

"Power corrupts" was written of man's control over other men but it applies also to his control of energy re-

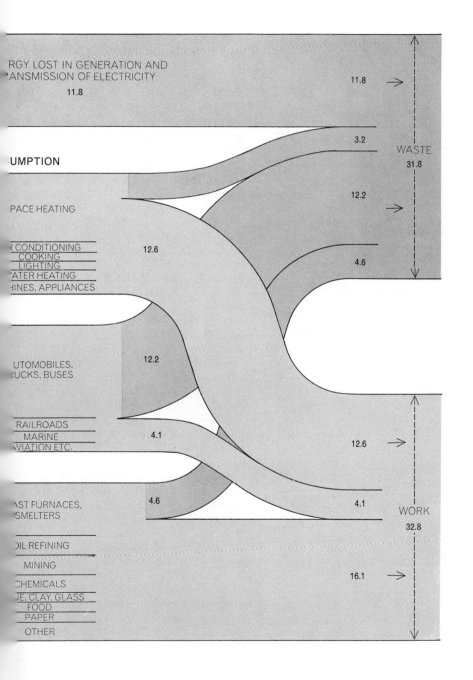

RGY LOST IN GENERATION AND
ANSMISSION OF ELECTRICITY
11.8

UMPTION

PACE HEATING

CONDITIONING
COOKING
LIGHTING
ATER HEATING
HINES, APPLIANCES

UTOMOBILES,
UCKS, BUSES

RAILROADS
MARINE
VIATION ETC.

ST FURNACES,
SMELTERS

IL REFINING

MINING

CHEMICALS
E, CLAY, GLASS
FOOD
PAPER
OTHER

11.8

3.2

WASTE
31.8

12.2

12.6

4.6

12.2

4.1

12.6

4.6

4.1

WORK
32.8

16.1

some is converted to generate electricity. The efficiency of electrical generation and transmission is taken to be about 31 percent, based on the ratio of utility electricity purchased in 1970 to the gross energy input for generation in that year. Efficiency of direct fuel use in transportation is taken as 25 percent, of fuel use in other applications as 75 percent.

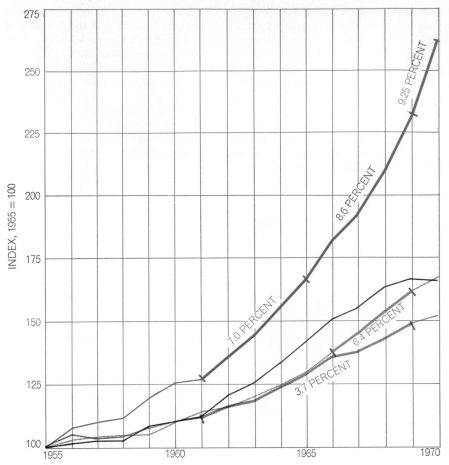

INCREASE IN CONSUMPTION of energy for electricity generation (*dark color*), transportation (*light color*) and other applications (*gray*) and of the gross national product (*black*) are compared. Annual growth rates for certain periods are shown beside heavy segments of curves. Consumption of electricity has a high growth rate and is increasing.

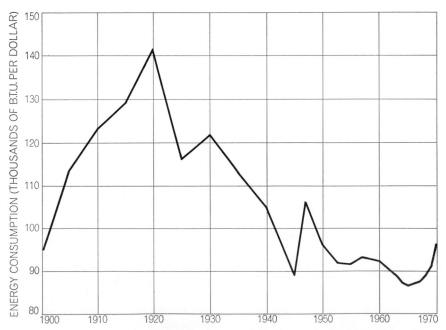

RATIO OF ENERGY CONSUMPTION to gross national product has varied over the years. It tends to be low when the G.N.P. is large and energy is being used efficiently, as was the case during World War II. The ratio has been rising steadily since 1965. Reasons include the increase in the use of air conditioning and the lack of advance in generating efficiency.

sources. The more power an industrial society disposes of, the more it wants. The more power we use, the more we shape our cities and mold our economic and social institutions to be dependent on the application of power and the consumption of energy. We could not now make any major move toward a lower per capita energy consumption without severe economic dislocation, and certainly the struggle of people in less developed regions toward somewhat similar energy-consumption levels cannot be thwarted without prolonging mass human suffering. Yet there is going to have to be some leveling off in the energy demands of industrial societies. Countries such as the U.S. have already come up against constraints dictated by the availability of resources and by damage to the environment. Another article in this issue considers the question of resource availability [see the article "The Energy Resources of the Earth," by M. King Hubbert, beginning on page 149]. Here I shall simply point out some of the decisions the U.S. faces in coping with diminishing supplies, and specifically with our increasing reliance on foreign sources of petroleum and petroleum products. In the short run the advantages of reasonable self-sufficiency must be weighed against the economic and environmental costs of developing oil reserves in Alaska and off the coast of California and the Gulf states. Later on such self-sufficiency may be attainable only through the production of oil from oil shale and from coal. In the long run the danger of dependence on dwindling fossil fuels—whatever they may be —must be balanced against the research and development costs of a major effort to shape a new energy system that is neither dependent on limited resources nor hard on the environment.

The environmental constraint may be more insistent than the constraint of resource availability. The present flow of energy through U.S. society leaves waste rock and acid water at coal mines; spilled oil from offshore wells and tankers; waste gases and particles from power plants, furnaces and automobiles; radioactive wastes of various kinds from nuclear-fuel processing plants and reactors. All along the line waste heat is developed, particularly at the power plants.

Yet for at least the next 50 years we shall be making use of dirty fuels: coal and petroleum. We can improve coal-combustion technology, we can build power plants at the mine mouth (so that the air of Appalachia is polluted instead of the air of New York City), we can make clean oil and gas from coal and oil

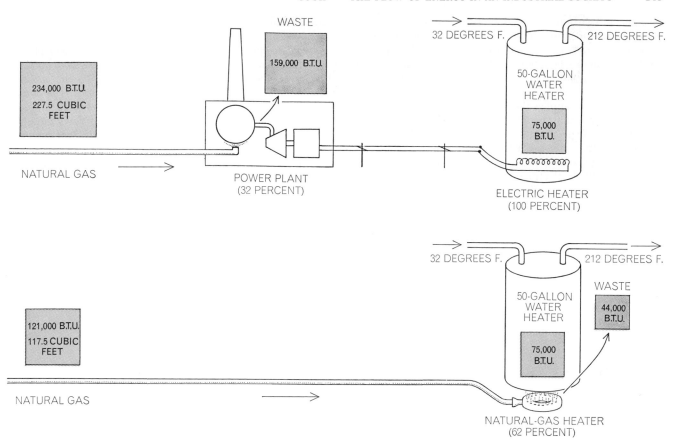

EFFICIENCIES OF HEATING WATER with natural gas indirectly by generating electricity for use in resistance heating (*top*) and directly (*bottom*) are contrasted. In each case the end result is enough heat to warm 50 gallons of water from 32 degrees Fahrenheit to 212 degrees. Electrical method requires substantially more gas even though efficiency at electric heater is nearly 100 percent.

from shale, and sow grass on the mountains of waste. As nuclear power plants proliferate we can put them underground, or far from the cities they serve if we are willing to pay the cost in transmission losses. With adequate foresight, caution and research we may even be able to handle the radioactive-waste problem without "undue" risk.

There are, however, definite limits to such improvements. The automobile engine and its present fuel simply cannot be cleaned up sufficiently to make it an acceptable urban citizen. It seems clear that the internal-combustion engine will be banned from the central city by the year 2000; it should probably be banned right now. Because our cities are shaped for automobiles, not for mass transit, we shall have to develop battery-powered or flywheel-powered cars and taxis for inner-city transport. The 1970 census for the first time showed more metropolitan citizens living in suburbs than in the central city; it also showed a record high in automobiles per capita, with the greatest concentration in the suburbs. It seems reasonable to visualize the suburban two-car garage of the future with one car a recharger for "downtown" and

	PRIMARY ENERGY INPUT (UNITS)	SECONDARY ENERGY OUTPUT (UNITS)	APPLICATION EFFICIENCY (PERCENT)	TECHNICAL EFFICIENCY (PERCENT)
AUTOMOBILE				
INTERNAL-COMBUSTION ENGINE	100		25	25
FLYWHEEL DRIVE CHARGED BY ELECTRICITY	100	32	100	32
SPACE HEATING				
BY DIRECT FUEL USE	100		75	75
BY ELECTRICAL RESISTANCE	100	32	100	32
SMELTING OF STEEL				
WITH COKE	100	94	94	70
WITH ELECTRICITY	100	32	32	32

TECHNICAL EFFICIENCY is the product of conversion efficiency at an intermediate step (if there is one) and application efficiency at the device that does the work. Losses due to friction and heat are ignored in the flywheel-drive automobile data. Coke retains only about 66 percent of the energy of coal, but the energy recovered from the by-products raises the energy conservation to 94 percent.

the other, still gasoline-powered, for suburban and cross-country driving.

Of course, some of the improvement in urban air quality bought by excluding the internal-combustion engine must be paid for by increased pollution from the power plant that supplies the electricity for the nightly recharging of the downtown vehicles. It need not, however, be paid for by an increased draft on the primary energy source; this is one substitution in which electricity need not decrease the technical efficiency of the system. The introduction of heat pumps for space heating and cooling would be

another. In fact, the overall efficiency should be somewhat improved and the environmental impact, given adequate attention to the siting, design and operation of the substituting power plant, should be greatly alleviated.

If technology can extend resource availability and keep environmental deterioration within acceptable limits in most respects, the specific environmental problem of waste heat may become the overriding one of the energy system by the turn of the century.

The cooling water required by power

plants already constitutes 10 percent of the total U.S. streamflow. The figure will increase sharply as more nuclear plants start up, since present designs of nuclear plants require 50 percent more cooling water than fossil-fueled plants of equal size do. The water is heated 15 degrees Fahrenheit or more as it flows through the plant. For ecological reasons such an increase in water released to a river, lake or ocean bay is unacceptable, at least for large quantities of effluent, and most large plants are now being built with cooling ponds or towers from which much of the heat of the water is dissi-

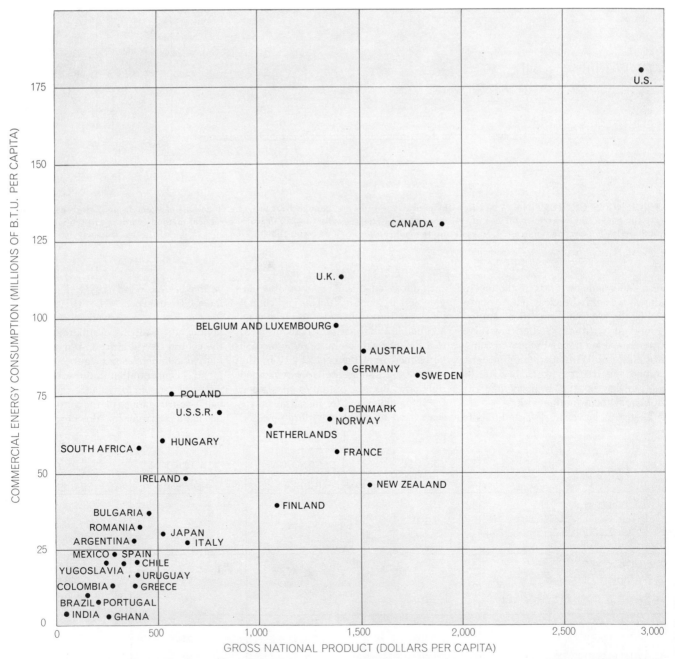

ROUGH CORRELATION between per capita consumption of energy and gross national product is seen when the two are plotted together; in general, high per capita energy consumption is a prerequisite for high output of goods and services. If the position plotted for the U.S. is considered to establish an arbitrary "line," some countries fall above or below that line. This appears to be related to a country's economic level, its emphasis on heavy industry or on services and its efficiency in converting energy into work.

pated to the atmosphere before the water is discharged or recycled through the plant. Although the atmosphere is a more capacious sink for waste heat than any body of water, even this disposal mechanism obviously has its environmental limits.

Many suggestions have been made for putting the waste heat from power plants to work: for irrigation or aquaculture, to provide ice-free shipping lanes or for space heating. (The waste heat from power generation today would be more than enough to heat every home in the U.S.!) Unfortunately the quantities of water involved, the relatively low temperature of the coolant water and the distances between power plants and areas of potential use are serious deterrents to the utilization of waste heat. Plants can be designed, however, for both power production and space heating. Such a plant has been in operation in Berlin for a number of years and has proved to be more efficient than a combination of separate systems for power production and space heating. The Berlin plant is not simply a conserver of waste heat but an exercise in fuel economy; its power capacity was reduced in order to raise the temperature of the heated water above that of normal cooling water.

With present and foreseeable technology there is not much hope of decreasing the amount of heat rejected to streams or the atmosphere (or both) from central steam-generating power plants. Two systems of producing power without steam generation offer some long-range hope of alleviating the waste-heat problem. One is the fuel cell; the other is the fusion reactor combined with a system for converting the energy released directly into electricity [see "The Conversion of Energy," by Claude M. Summers; SCIENTIFIC AMERICAN, September, 1971]. In the fuel cell the energy contained in hydrocarbons or hydrogen is released by a controlled oxidation process that produces electricity directly with an efficiency of about 60 percent. A practical fusion reactor with a direct-conversion system is not likely to appear in this century.

Major changes in power technology will be required to reduce pollution and manage wastes, to improve the efficiency of the system and to remove the resource-availability constraint. Making

the changes will call for hard political decisions. Energy needs will have to be weighed against environmental and social costs; a decision to set a pollution standard or to ban the internal-combustion engine or to finance nuclear-power development can have major economic and political effects. Democratic societies are not noted for their ability to take the long view in making decisions. Yet indefinite growth in energy consumption, as in human population, is simply not possible.

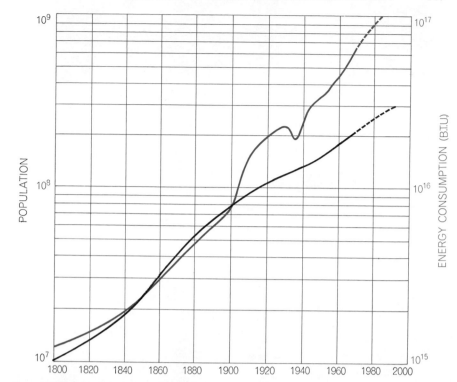

U.S. ENERGY-CONSUMPTION GROWTH (*curve in color*) has outpaced the growth in population (*black*) since 1900, except during the energy cutback of the depression years.

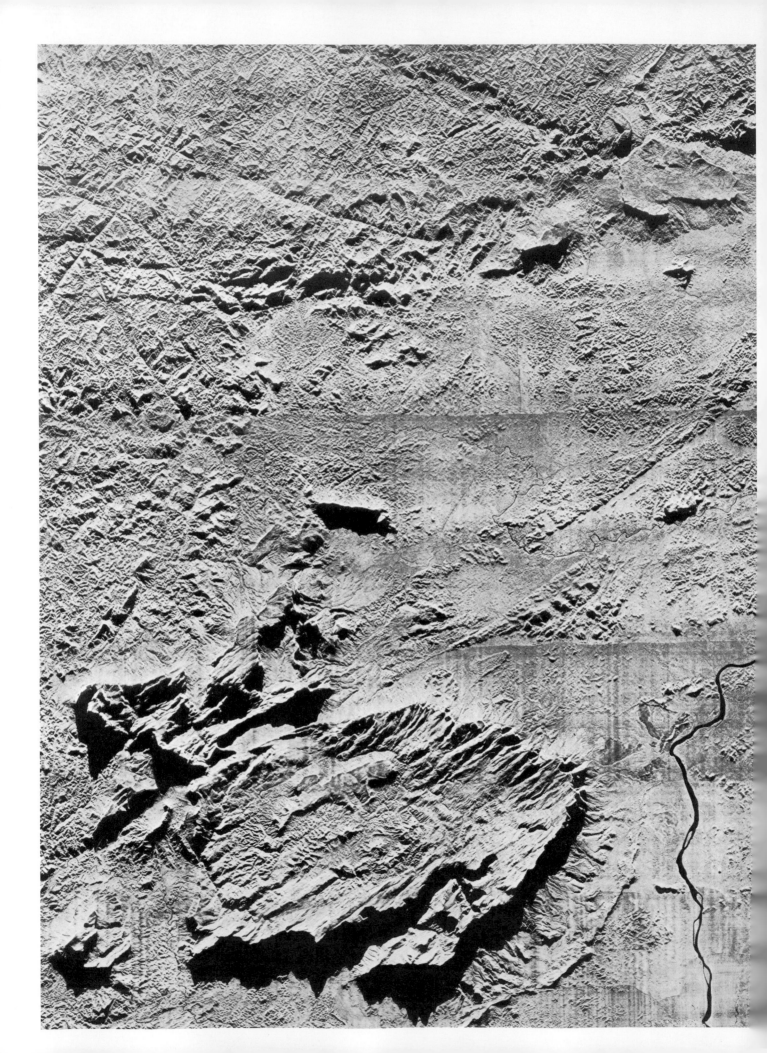

16

THE ENERGY RESOURCES OF THE EARTH

M. KING HUBBERT
September 1971

*They are solar energy (current and stored), the tides,
the earth's heat, fission fuels and possibly fusion fuels.
From the standpoint of human history the epoch of the
fossil fuels will be quite brief*

Energy flows constantly into and out of the earth's surface environment. As a result the material constituents of the earth's surface are in a state of continuous or intermittent circulation. The source of the energy is preponderantly solar radiation, supplemented by small amounts of heat from the earth's interior and of tidal energy from the gravitational system of the earth, the moon and the sun. The materials of the earth's surface consist of the 92 naturally occurring chemical elements, all but a few of which behave in accordance with the principles of the conservation of matter and of nontransmutability as formulated in classical chemistry. A few of the elements or their isotopes, with abundances of only a few parts per million, are an exception to these principles in being radioactive. The exception is crucial in that it is the key to an additional large source of energy.

A small part of the matter at the earth's surface is embodied in living organisms: plants and animals. The leaves of the plants capture a small fraction of the incident solar radiation and store it chemically by the mechanism of photosynthesis. This store becomes the energy supply essential for the existence of the plant and animal kingdoms. Biologically stored energy is released by oxidation at a rate approximately equal to the rate of storage. Over millions of years, however, a minute fraction of the vegetable and animal matter is buried under conditions of incomplete oxidation and decay, thereby giving rise to the fossil fuels that provide most of the energy for industrialized societies.

It is difficult for people living now, who have become accustomed to the steady exponential growth in the consumption of energy from the fossil fuels, to realize how transitory the fossil-fuel epoch will eventually prove to be when it is viewed over a longer span of human history. The situation can better be seen in the perspective of some 10,000 years, half before the present and half afterward. On such a scale the complete cycle of the exploitation of the world's fossil fuels will be seen to encompass perhaps 1,300 years, with the principal segment of the cycle (defined as the period during which all but the first 10 percent and the last 10 percent of the fuels are extracted and burned) covering only about 300 years.

What, then, will provide industrial energy in the future on a scale at least as large as the present one? The answer lies in man's growing ability to exploit other sources of energy, chiefly nuclear at present but perhaps eventually the much larger source of solar energy. With this ability the energy resources now at hand are sufficient to sustain an industrial operation of the present magnitude for another millennium or longer. Moreover, with such resources of energy the limits to the growth of industrial activity are no longer set by a scarcity of energy but rather by the space and material limitations of a finite earth together with the principles of ecology. According to these principles both biological and industrial activities tend to increase exponentially with time, but the resources of the entire earth are not sufficient to sustain such an increase of any single component for more than a few tens of successive doublings.

Let us consider in greater detail the flow of energy through the earth's surface environment [*see illustration on next two pages*]. The inward flow of energy has three main sources: (1) the intercepted solar radiation; (2) thermal energy, which is conveyed to the surface of the earth from the warmer interior by the conduction of heat and by convection in hot springs and volcanoes, and (3) tidal energy, derived from the combined kinetic and potential energy of the earth-moon-sun system. It is possible in various ways to estimate approximately how large the input is from each source.

In the case of solar radiation the influx is expressed in terms of the solar constant, which is defined as the mean rate of flow of solar energy across a unit of area that is perpendicular to the radiation and outside the earth's atmosphere at the mean distance of the earth from the sun. Measurements made on the earth and in spacecraft give a mean value for the solar constant of 1.395 kilowatts per square meter, with a variation of about 2 percent. The total solar radiation intercepted by the earth's diametric plane of 1.275×10^{14} square meters is therefore 1.73×10^{17} watts.

The influx of heat by conduction from the earth's interior has been determined from measurements of the geothermal gradient (the increase of temperature with depth) and the thermal conductivity of the rocks involved. From thousands of such measurements, both on land and on the ocean beds, the average rate of flow of heat from the interior of the earth has been found to be about .063 watt per square meter. For the earth's surface area of 510×10^{12} square meters the total heat flow amounts to some 32×10^{12} watts. The rate of heat convection by hot springs and volcanoes is estimated to be only about 1 percent of the rate of conduction, or about $.3 \times 10^{12}$ watts.

The energy from tidal sources has been estimated at 3×10^{12} watts. When all three sources of energy are expressed in the common unit of 10^{12} watts, the total power influx into the earth's surface environment is found to be 173,035 $\times 10^{12}$ watts. Solar radiation accounts for 99.98 percent of it. Another way of stating the sun's contribution to the energy budget of the earth is to note that at $173,000 \times 10^{12}$ watts it amounts to 5,000 times the energy input from all other sources combined.

About 30 percent of the incident solar energy ($52,000 \times 10^{12}$ watts) is directly reflected and scattered back into space as short-wavelength radiation. Another 47 percent ($81,000 \times 10^{12}$ watts) is absorbed by the atmosphere, the land surface and the oceans and converted directly into heat at the ambient surface temperature. Another 23 percent (40,-000×10^{12} watts) is consumed in the evaporation, convection, precipitation and surface runoff of water in the hydrologic cycle. A small fraction, about 370×10^{12} watts, drives the atmospheric and oceanic convections and circulations and the ocean waves and is eventually dissipated into heat by friction. Finally, an even smaller fraction—about

40×10^{12} watts—is captured by the chlorophyll of plant leaves, where it becomes the essential energy supply of the photosynthetic process and eventually of the plant and animal kingdoms.

Photosynthesis fixes carbon in the leaf and stores solar energy in the form of carbohydrate. It also liberates oxygen and, with the decay or consumption of the leaf, dissipates energy. At any given time, averaged over a year or more, the balance between these processes is almost perfect. A minute fraction of the organic matter produced, however, is deposited in peat bogs or other oxygen-deficient environments under conditions that prevent complete decay and loss of energy.

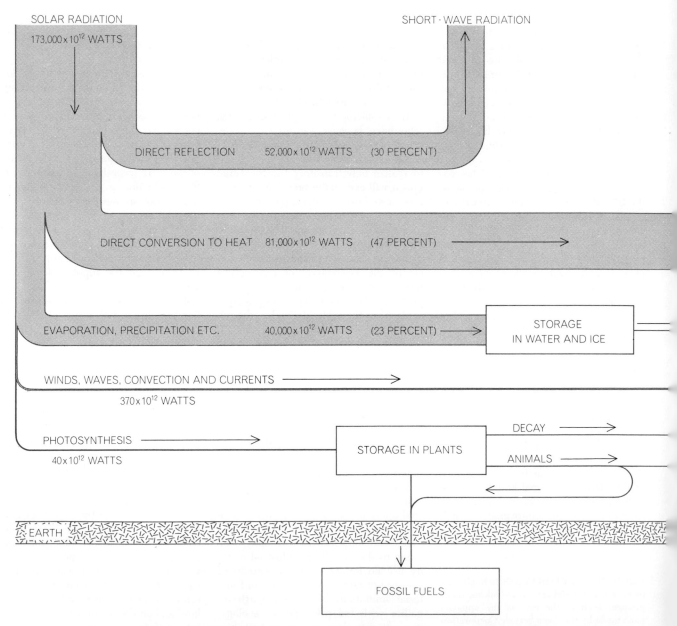

FLOW OF ENERGY to and from the earth is depicted by means of bands and lines that suggest by their width the contribution of each item to the earth's energy budget. The principal inputs are solar radiation, tidal energy and the energy from nuclear, thermal and gravitational sources. More than 99 percent of the input is solar radiation. The apportionment of incoming solar radiation is

Little of the organic material produced before the Cambrian period, which began about 600 million years ago, has been preserved. During the past 600 million years, however, some of the organic materials that did not immediately decay have been buried under a great thickness of sedimentary sands, muds and limes. These are the fossil fuels: coal, oil shale, petroleum and natural gas, which are rich in energy stored up chemically from the sunshine of the past 600 million years. The process is still continuing, but probably at about the same rate as in the past; the accumulation during the next million years will probably be a six-hundredth of the amount built up thus far.

Industrialization has of course withdrawn the deposits in this energy bank with increasing rapidity. In the case of coal, for example, the world's consumption during the past 110 years has been about 19 times greater than it was during the preceding seven centuries. The increasing magnitude of the rate of withdrawal can also be seen in the fact that the amount of coal produced and consumed since 1940 is approximately equal to the total consumption up to that time. The cumulative production from 1860 through 1970 was about 133 billion metric tons. The amount produced before 1860 was about seven million metric tons.

Petroleum and related products were not extracted in significant amounts before 1880. Since then production has increased at a nearly constant exponential rate. During the 80-year period from 1890 through 1970 the average rate of increase has been 6.94 percent per year, with a doubling period of 10 years. The cumulative production until the end of 1969 amounted to 227 billion (227×10^9) barrels, or 9.5 trillion U.S. gallons. Once again the period that encompasses most of the production is notably brief. The 102 years from 1857 to 1959 were required to produce the first half of the cumulative production; only the 10-year period from 1959 to 1969 was required for the second half.

Examining the relative energy contributions of coal and crude oil by comparing the heats of combustion of the respective fuels (in units of 10^{12} kilowatt-hours), one finds that until after 1900 the contribution from oil was barely significant compared with the contribution from coal. Since 1900 the contribution from oil has risen much faster than that from coal. By 1968 oil represented about 60 percent of the total. If the energy from natural gas and natural-gas liquids had been included, the contribution from petroleum would have been about 70 percent. In the U.S. alone 73 percent of the total energy produced from fossil fuels in 1968 was from petroleum and 27 percent from coal.

Broadly speaking, it can be said that the world's consumption of energy for industrial purposes is now doubling approximately once per decade. When confronted with a rate of growth of such magnitude, one can hardly fail to wonder how long it can be kept up. In the case of the fossil fuels a reasonably definite answer can be obtained. Their human exploitation consists of their being withdrawn from an essentially fixed initial supply. During their use as sources of energy they are destroyed. The complete cycle of exploitation of a fossil fuel must therefore have the following characteristics. Beginning at zero, the rate of production tends initially to increase exponentially. Then, as difficulties of discovery and extraction increase, the production rate slows in its growth, passes one maximum or more and, as the resource is progressively depleted, declines eventually to zero.

If known past and prospective future rates of production are combined with a reasonable estimate of the amount of a fuel initially present, one can calculate the probable length of time that the fuel can be exploited. In the case of coal reasonably good estimates of the

LONG-WAVE RADIATION

TIDAL ENERGY

TIDES, TIDAL CURRENTS, ETC.
3×10^{12} WATTS

CONVECTION IN VOLCANOES AND HOT SPRINGS
3×10^{12} WATTS

CONDUCTION IN ROCKS
32×10^{12} WATTS

TERRESTRIAL ENERGY

NUCLEAR, THERMAL AND GRAVITATIONAL ENERGY

indicated by the horizontal bands beginning with "Direct reflection" and reading downward. The smallest portion goes to photosynthesis. Dead plants and animals buried in the earth give rise to fossil fuels, containing stored solar energy from millions of years past.

amount present in given regions can be made on the basis of geological mapping and a few widely spaced drill holes, inasmuch as coal is found in stratified beds or seams that are continuous over extensive areas. Such studies have been made in all the coal-bearing areas of the world.

The most recent compilation of the present information on the world's initial coal resources was made by Paul Averitt of the U.S. Geological Survey. His figures [*see illustration below*] represent minable coal, which is defined as 50 percent of the coal actually present. Included is coal in beds as thin as 14 inches (36 centimeters) and extending to depths of 4,000 feet (1.2 kilometers) or, in a few cases, 6,000 feet (1.8 kilometers).

Taking Averitt's estimate of an initial supply of 7.6 trillion metric tons and assuming that the present production rate of three billion metric tons per year does not double more than three times, one can expect that the peak in the rate of production will be reached sometime between 2100 and 2150. Disregarding the long time required to produce the first 10 percent and the last 10 percent, the length of time required to produce the middle 80 percent will be roughly

the 300-year period from 2000 to 2300.

Estimating the amount of oil and gas that will ultimately be discovered and produced in a given area is considerably more hazardous than estimating for coal. The reason is that these fluids occur in restricted volumes of space and limited areas in sedimentary basins at all depths from a few hundred meters to more than eight kilometers. Nonetheless, the estimates for a given region improve as exploration and production proceed. In addition it is possible to make rough estimates for relatively undeveloped areas on the basis of geological comparisons between them and well-developed regions.

The most highly developed oil-producing region in the world is the coterminous area of the U.S.: the 48 states exclusive of Alaska and Hawaii. This area has until now led the world in petroleum development, and the U.S. is still the leading producer. For this region a large mass of data has been accumulated and a number of different methods of analysis have been developed that give fairly consistent estimates of the degree of advancement of petroleum exploration and of the amounts of oil and gas that may eventually be produced.

One such method is based on the principle that only a finite number of oil or gas fields existed initially in a given region. As exploration proceeds the shallowest and most evident fields are usually discovered first and the deeper and more obscure ones later. With each discovery the number of undiscovered fields decreases by one. The undiscovered fields are also likely to be deeper, more widely spaced and better concealed. Hence the amount of exploratory activity required to discover a fixed quantity of oil or gas steadily increases or, conversely, the average amount of oil or gas discovered for a fixed amount of exploratory activity steadily decreases.

Most new fields are discovered by what the industry calls "new-field wild-cat wells," meaning wells drilled in new territory that is not in the immediate vicinity of known fields. In the U.S. statistics have been kept annually since 1945 on the number of new-field wildcat wells required to make one significant discovery of oil or gas ("significant" being defined as one million barrels of oil or an equivalent amount of gas). The discoveries for a given year are evaluated only after six years of subsequent development. In 1945 it required 26

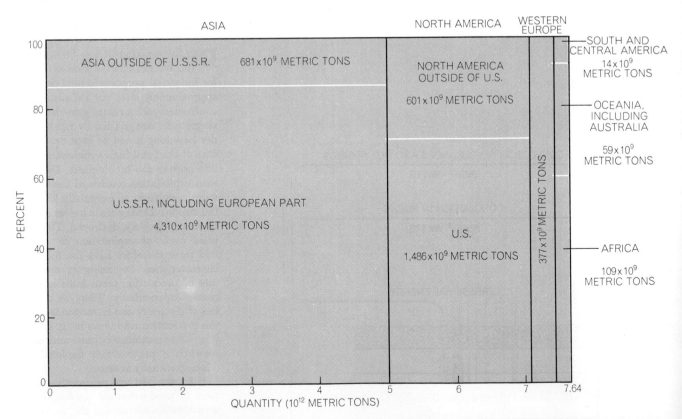

COAL RESOURCES of the world are indicated on the basis of data compiled by Paul Averitt of the U.S. Geological Survey. The figures represent the total initial resources of minable coal, which is defined as 50 percent of the coal actually present. The horizontal scale gives the total supply. Each vertical block shows the apportionment of the supply in a continent. From the first block, for example, one can ascertain that Asia has some 5×10^{12} metric tons of minable coal, of which about 86 percent is in the U.S.S.R.

new-field wildcat wells to make a significant discovery; by 1963 the number had increased to 65.

Another way of illuminating the problem is to consider the amount of oil discovered per foot of exploratory drilling. From 1860 to 1920, when oil was fairly easy to find, the ratio was 194 barrels per foot. From 1920 to 1928 the ratio declined to 167 barrels per foot. Between 1928 and 1938, partly because of the discovery of the large East Texas oil field and partly because of new exploratory techniques, the ratio rose to its maximum of 276 barrels per foot. Since then it has fallen sharply to a nearly constant rate of about 35 barrels per foot. Yet the period of this decline coincided with the time of the most intensive research and development in petroleum exploration and production in the history of the industry.

The cumulative discoveries in the 48 states up to 1965 amounted to 136 billion barrels. From this record of drilling and discovery it can be estimated that the ultimate total discoveries in the coterminous U.S. and the adjacent continental shelves will be about 165 billion barrels. The discoveries up to 1965 therefore represent about 82 percent of the prospective ultimate total. Making

due allowance for the range of uncertainty in estimates of future discovery, it still appears that at least 75 percent of the ultimate amount of oil to be produced in this area will be obtained from fields that had already been discovered by 1965.

For natural gas in the 48 states the present rate of discovery, averaged over a decade, is about 6,500 cubic feet per barrel of oil. Assuming the same ratio for the estimated ultimate amount of 165 billion barrels of crude oil, the ultimate amount of natural gas would be about 1,075 trillion cubic feet. Combining the estimates for oil and gas with the trends of production makes it possible to estimate how long these energy resources will last. In the case of oil the period of peak production appears to be the present. The time span required to produce the middle 80 percent of the ultimate cumulative production is approximately the 65-year period from 1934 to 1999—less than the span of a human lifetime. For natural gas the peak of production will probably be reached between 1975 and 1980.

The discoveries of petroleum in Alaska modify the picture somewhat. In particular the field at Prudhoe Bay appears likely by present estimates to contain

about 10 billion barrels, making it twice as large as the East Texas field, which was the largest in the U.S. previously. Only a rough estimate can be made of the eventual discoveries of petroleum in Alaska. Such a speculative estimate would be from 30 to 50 billion barrels. One must bear in mind, however, that 30 billion barrels is less than a 10-year supply for the U.S. at the present rate of consumption. Hence it appears likely that the principal effect of the oil from Alaska will be to retard the rate of decline of total U.S. production rather than to postpone the date of its peak.

Estimates of ultimate world production of oil range from 1,350 billion barrels to 2,100 billion barrels. For the higher figure the peak in the rate of world production would be reached about the year 2000. The period of consumption of the middle 80 percent will probably be some 58 to 64 years, depending on whether the lower or the higher estimate is used [see bottom illustration on page 157].

A substantial but still finite amount of oil can be extracted from tar sands and oil shales, where production has barely begun. The largest tar-sand deposits are in northern Alberta; they have total recoverable reserves of about 300 billion

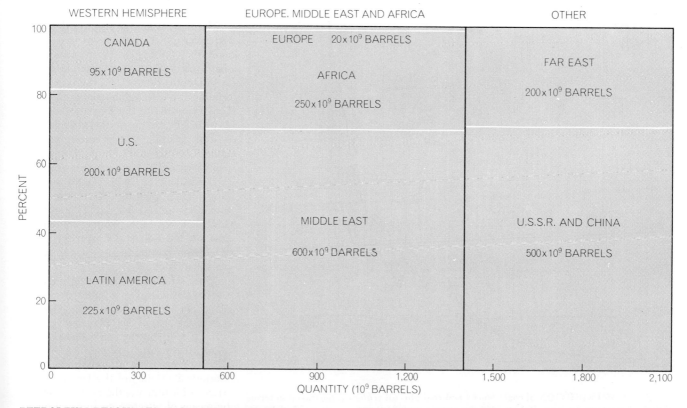

PETROLEUM RESOURCES of the world are depicted in an arrangement that can be read in the same way as the diagram of coal supplies on the opposite page. The figures for petroleum are derived from estimates made in 1967 by W. P. Ryman of the Standard Oil Company of New Jersey. They represent ultimate crude-oil production, including oil from offshore areas, and consist of oil already produced, proved and probable reserves, and future discoveries. Estimates as low as 1,350 × 10⁹ barrels have also been made.

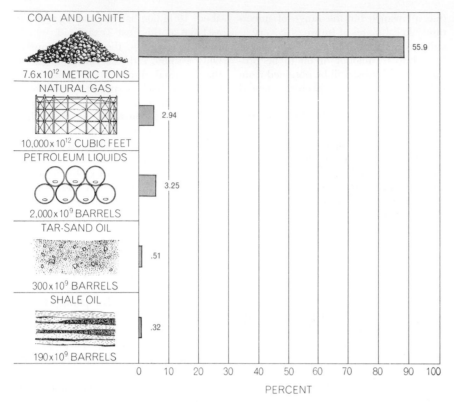

ENERGY CONTENT of the world's initial supply of recoverable fossil fuels is given in units of 10^{15} thermal kilowatt-hours (*color*). Coal and lignite, for example, contain 55.9 × 10^{15} thermal kilowatt-hours of energy and represent 88.8 percent of the recoverable energy.

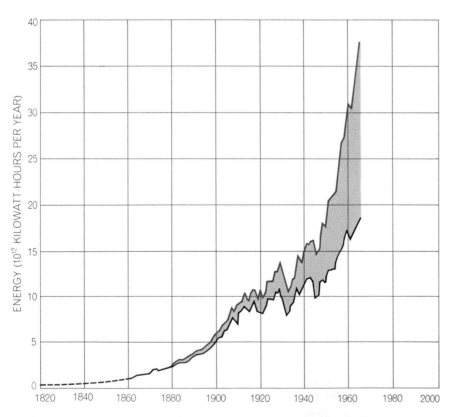

ENERGY CONTRIBUTION of coal (*black*) and coal plus oil (*color*) is portrayed in terms of their heat of combustion. Before 1900 the energy contribution from oil was barely significant. Since then the contribution from oil (*shaded area*) has risen much more rapidly than that from coal. By 1968 oil represented about 60 percent of the total. If the energy from natural gas were included, petroleum would account for about 70 percent of the total.

barrels. A world summary of oil shales by Donald C. Duncan and Vernon E. Swanson of the U.S. Geological Survey indicated a total of about 3,100 billion barrels in shales containing from 10 to 100 gallons per ton, of which 190 billion barrels were considered to be recoverable under 1965 conditions.

Since the fossil fuels will inevitably be exhausted, probably within a few centuries, the question arises of what other sources of energy can be tapped to supply the power requirements of a moderately industrialized world after the fossil fuels are gone. Five forms of energy appear to be possibilities: solar energy used directly, solar energy used indirectly, tidal energy, geothermal energy and nuclear energy.

Until now the direct use of solar power has been on a small scale for such purposes as heating water and generating electricity for spacecraft by means of photovoltaic cells. Much more substantial installations will be needed if solar power is to replace the fossil fuels on an industrial scale. The need would be for solar power plants in units of, say, 1,000 megawatts. Moreover, because solar radiation is intermittent at a fixed location on the earth, provision must also be made for large-scale storage of energy in order to smooth out the daily variation.

The most favorable sites for developing solar power are desert areas not more than 35 degrees north or south of the Equator. Such areas are to be found in the southwestern U.S., the region extending from the Sahara across the Arabian Peninsula to the Persian Gulf, the Atacama Desert in northern Chile and central Australia. These areas receive some 3,000 to 4,000 hours of sunshine per year, and the amount of solar energy incident on a horizontal surface ranges from 300 to 650 calories per square centimeter per day. (Three hundred calories, the winter minimum, amounts when averaged over 24 hours to a mean power density of 145 watts per square meter.)

Three schemes for collecting and converting this energy in a 1,000-megawatt plant can be considered. The first involves the use of flat plates of photovoltaic cells having an efficiency of about 10 percent. A second possibility is a recent proposal by Aden B. Meinel and Marjorie P. Meinel of the University of Arizona for utilizing the hothouse effect by means of selective coatings on pipes carrying a molten mixture of sodium and potassium raised by solar energy to a temperature of 540 degrees Celsius. By

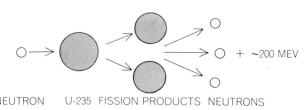

FISSION AND FUSION REACTIONS hold the promise of serving as sources of energy when fossil fuels are depleted. Present nuclear-power plants burn uranium 235 as a fuel. Breeder reactors now under development will be able to use surplus neutrons from the fission of uranium 235 (*left*) to create other nuclear fuels: plutonium 239 and uranium 233. Two promising fusion reactions, deuterium-deuterium and deuterium-tritium, are at right. The energy released by the various reactions is shown in million electron volts.

means of a heat exchanger this heat is stored at a constant temperature in an insulated chamber filled with a mixture of sodium and potassium chlorides that has enough heat capacity for at least one day's collection. Heat extracted from this chamber operates a conventional steam-electric power plant. The computed efficiency for this proposal is said to be about 30 percent.

A third system has been proposed by Alvin F. Hildebrandt and Gregory M. Haas of the University of Houston. It entails reflecting the radiation reaching a square-mile area into a solar furnace and boiler at the top of a 1,500-foot tower. Heat from the boiler at a temperature of 2,000 degrees Kelvin would be converted into electric power by a magnetohydrodynamic conversion. An energy-storage system based on the hydrolysis of water is also proposed. An overall efficiency of about 20 percent is estimated.

Over the range of efficiencies from 10 to 30 percent the amount of thermal power that would have to be collected for a 1,000-megawatt plant would range from 10,000 to 3,300 thermal megawatts. Accordingly the collecting areas for the three schemes would be 70, 35 and 23 square kilometers respectively. With the least of the three efficiencies the area required for an electric-power capacity of 350,000 megawatts—the approximate capacity of the U.S. in 1970— would be 24,500 square kilometers, which is somewhat less than a tenth of the area of Arizona.

The physical knowledge and technological resources needed to use solar energy on such a scale are now available. The technological difficulties of doing so, however, should not be minimized.

Using solar power indirectly means relying on the wind, which appears impractical on a large scale, or on the streamflow part of the hydrologic cycle. At first glance the use of streamflow appears promising, because the world's total water-power capacity in suitable sites is about three trillion watts, which

approximates the present use of energy in industry. Only 8.5 percent of the water power is developed at present, however, and the three regions with the greatest potential—Africa, South America and Southeast Asia—are the least developed industrially. Economic problems therefore stand in the way of extensive development of additional water power.

Tidal power is obtained from the filling and emptying of a bay or an estuary that can be closed by a dam. The enclosed basin is allowed to fill and empty only during brief periods at high and low tides in order to develop as much power as possible. A number of promising sites exist; their potential capacities range from two megawatts to 20,000 megawatts each. The total potential tidal power, however, amounts to about 64 billion watts, which is only 2 percent of the world's potential water power. Only one full-scale tidal-electric plant has been built; it is on the Rance estuary on the Channel Island coast of France. Its capacity at start-up in 1966 was 240 megawatts; an ultimate capacity of 320 megawatts is planned.

Geothermal power is obtained by extracting heat that is temporarily stored in the earth by such sources as volcanoes and the hot water filling the sands of deep sedimentary basins. Only volcanic sources are significantly exploited at present. A geothermal-power operation has been under way in the Larderello area of Italy since 1904 and now has a capacity of 370 megawatts. The two other main areas of geothermal-power production are The Geysers in northern California and Wairakei in New Zealand. Production at The Geysers began in 1960 with a 12.5-megawatt unit. By 1969 the capacity had reached 82 megawatts, and plans are to reach a total installed capacity of 400 megawatts by 1973. The Wairakei plant began operation in 1958 and now has a capacity of 290 megawatts, which is believed to be about the maximum for the site.

Donald E. White of the U.S. Geological Survey has estimated that the stored thermal energy in the world's major geothermal areas amounts to about 4×10^{20} joules. With a 25 percent conversion factor the production of electrical energy would be about 10^{20} joules, or three million megawatt-years. If this energy, which is depletable, were withdrawn over a period of 50 years, the average annual power production would be 60,000 megawatts, which is comparable to the potential tidal power.

Nuclear power must be considered under the two headings of fission and fusion. Fission involves the splitting of nuclei of heavy elements such as uranium. Fusion involves the combining of light nuclei such as deuterium. Uranium 235, which is a rare isotope (each 100,000 atoms of natural uranium include six atoms of uranium 234, 711 atoms of uranium 235 and 99,283 atoms of uranium 238), is the only atomic species capable of fissioning under relatively mild environmental conditions. If nuclear energy depended entirely on uranium 235, the nuclear-fuel epoch would be brief. By breeding, however, wherein by absorbing neutrons in a nuclear reactor uranium 238 is transformed into fissionable plutonium 239 or thorium 232 becomes fissionable uranium 233, it is possible to create more nuclear fuel than is consumed. With breeding the entire supply of natural uranium and thorium would thus become available as fuel for fission reactors.

Most of the reactors now operating or planned in the rapidly growing nuclear-power industry in the U.S. depend essentially on uranium 235. The U.S. Atomic Energy Commission has estimated that the uranium requirement to meet the projected growth rate from 1970 to 1980 is 206,000 short tons of uranium oxide (U_3O_8). A report recently issued by the European Nuclear Energy Agency and the International Atomic Energy Agency projects requirements of 430,000 short tons of uranium

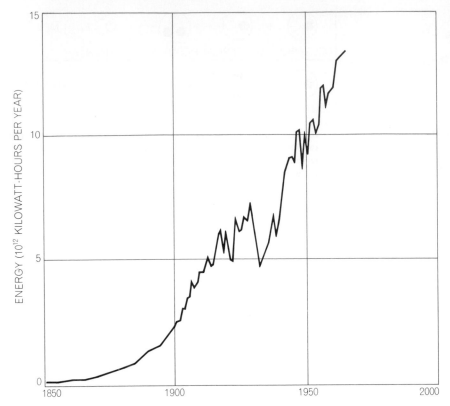

U.S. PRODUCTION OF ENERGY from coal, from petroleum and related sources, from water power and from nuclear reactors is charted for 120 years. The petroleum increment includes natural gas and associated liquids. The dip at center reflects impact of Depression.

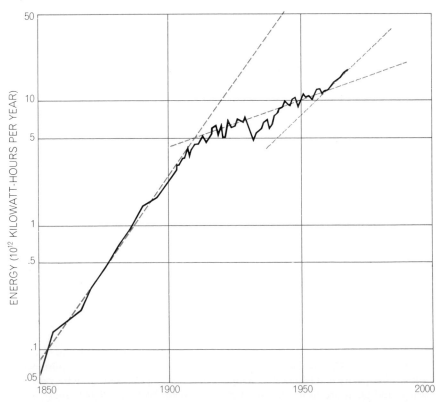

RATE OF GROWTH of U.S. energy production is shown by plotting on a semilogarithmic scale the data represented in the illustration at the top of the page. Broken lines show that the rise had three distinct periods. In the first the growth rate was 6.91 percent per year and the doubling period was 10 years; in the second the rate was 1.77 percent and the doubling period was 39 years; in the third the rate was 4.25 percent with doubling in 16.3 years.

oxide for the non-Communist nations during the same period.

Against these requirements the AEC estimates that the quantity of uranium oxide producible at $8 per pound from present reserves in the U.S. is 243,000 tons, and the world reserves at $10 per pound or less are estimated in the other report at 840,000 tons. The same report estimates that to meet future requirements additional reserves of more than a million short tons will have to be discovered and developed by 1985.

Although new discoveries of uranium will doubtless continue to be made (a large one was recently reported in northeastern Australia), all present evidence indicates that without a transition to breeder reactors an acute shortage of low-cost ores is likely to develop before the end of the century. Hence an intensive effort to develop large-scale breeder reactors for power production is in progress. If it succeeds, the situation with regard to fuel supply will be drastically altered.

This prospect results from the fact that with the breeder reactor the amount of energy obtainable from one gram of uranium 238 amounts to 8.1×10^{10} joules of heat. That is equal to the heat of combustion of 2.7 metric tons of coal or 13.7 barrels (1.9 metric tons) of crude oil. Disregarding the rather limited supplies of high-grade uranium ore that are available, let us consider the much more abundant low-grade ores. One example will indicate the possibilities.

The Chattanooga black shale (of Devonian age) crops out along the western edge of the Appalachian Mountains in eastern Tennessee and underlies at minable depths most of Tennessee, Kentucky, Ohio, Indiana and Illinois. In its outcrop area in eastern Tennessee this shale contains a layer about five meters thick that has a uranium content of about 60 grams per metric ton. That amount of uranium is equivalent to about 162 metric tons of bituminous coal or 822 barrels of crude oil. With the density of the rock some 2.5 metric tons per cubic meter, a vertical column of rock five meters long and one square meter in cross section would contain 12.5 tons of rock and 750 grams of uranium. The energy content of the shale per square meter of surface area would therefore be equivalent to about 2,000 tons of coal or 10,000 barrels of oil. Allowing for a 50 percent loss in mining and extracting the uranium, we are still left with the equivalent of 1,000 tons of coal or 5,000 barrels of oil per square meter.

Taking Averitt's estimate of 1.5 trillion metric tons for the initial minable coal in the U.S. and a round figure of 250 billion barrels for the petroleum liquids, we find that the nuclear energy in an area of about 1,500 square kilometers of Chattanooga shale would equal the energy in the initial minable coal; 50 square kilometers would hold the energy equivalent of the petroleum liquids. Adding natural gas and oil shales, an area of roughly 2,000 square kilometers of Chattanooga shale would be equivalent to the initial supply of all the fossil fuels in the U.S. The area is about 2 percent of the area of Tennessee

and a very small fraction of the total area underlain by the shale. Many other low-grade deposits of comparable magnitude exist. Hence by means of the breeder reactor the energy potentially available from the fissioning of uranium and thorium is at least a few orders of magnitude greater than that from all the fossil fuels combined.

David J. Rose of the AEC, reviewing recently the prospects for controlled fusion, found the deuterium-tritium reaction to be the most promising. Deuterium is abundant (one atom to each

6,700 atoms of hydrogen), and the energy cost of separating it would be almost negligible compared with the amount of energy released by fusion. Tritium, on the other hand, exists only in tiny amounts in nature. Larger amounts must be made from lithium 6 and lithium 7 by nuclear bombardment. The limiting isotope is lithium 6, which has an abundance of only 7.4 percent of natural lithium.

Considering the amount of hydrogen in the oceans, deuterium can be regarded as superabundant. It can also be extracted easily. Lithium is much less

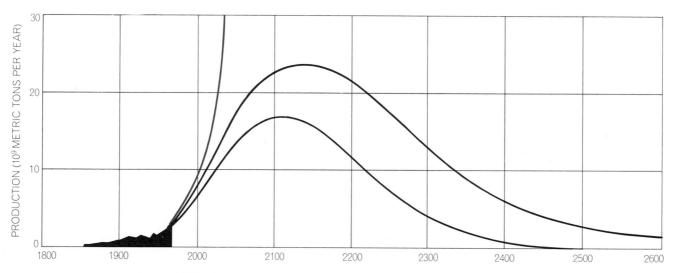

CYCLE OF WORLD COAL PRODUCTION is plotted on the basis of estimated supplies and rates of production. The top curve reflects Averitt's estimate of 7.6×10^{12} metric tons as the initial supply of minable coal; the bottom curve reflects an estimate of 4.3×10^{12} metric tons. The curve that rises to the top of the graph shows the trend if production continued to rise at the present rate of 3.56 percent per year. The amount of coal mined and burned in the century beginning in 1870 is shown by the black area at left.

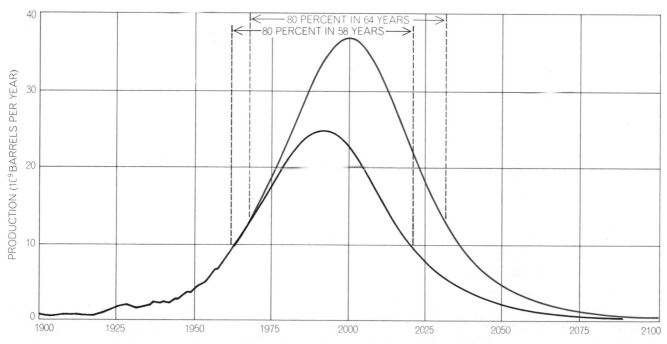

CYCLE OF WORLD OIL PRODUCTION is plotted on the basis of two estimates of the amount of oil that will ultimately be produced. The colored curve reflects Ryman's estimate of $2,100 \times 10^9$ barrels and the black curve represents an estimate of $1,350 \times 10^9$ barrels.

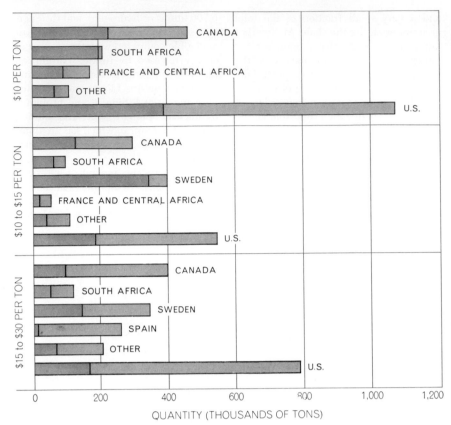

WORLD RESERVES OF URANIUM, which would be the source of nuclear power derived from atomic fission, are given in tons of uranium oxide (U_3O_8). The colored part of each bar represents reasonably assured supplies and the gray part estimated additional supplies.

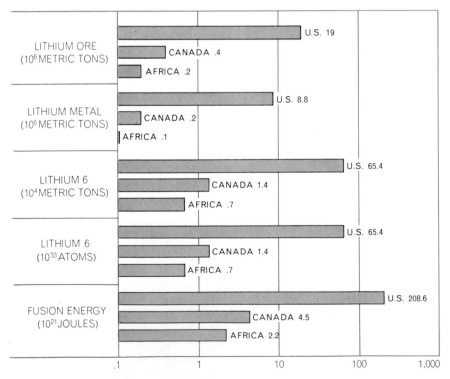

WORLD RESERVES OF LITHIUM, which would be the limiting factor in the deuterium-tritium fusion reaction, are stated in terms of lithium 6 because it is the least abundant isotope. Even with this limitation the energy obtainable from fusion through the deuterium-tritium reaction would almost equal the energy content of the world's fossil-fuel supply.

abundant. It is produced from the geologically rare igneous rocks known as pegmatites and from the salts of saline lakes. The measured, indicated and inferred lithium resources in the U.S., Canada and Africa total 9.1 million tons of elemental lithium, of which the content of lithium 6 would be 7.42 atom percent, or 67,500 metric tons. From this amount of lithium 6 the fusion energy obtainable at 3.19×10^{-12} joule per atom would be 215×10^{21} joules, which is approximately equal to the energy content of the world's fossil fuels.

As long as fusion power is dependent on the deuterium-tritium reaction, which at present appears to be somewhat the easier because it proceeds at a lower temperature, the energy obtainable from this source appears to be of about the same order of magnitude as that from fossil fuels. If fusion can be accomplished with the deuterium-deuterium reaction, the picture will be markedly changed. By this reaction the energy released per deuterium atom consumed is 7.94×10^{-13} joule. One cubic meter of water contains about 10^{25} atoms of deuterium having a mass of 34.4 grams and a potential fusion energy of 7.94×10^{12} joules. This is equivalent to the heat of combustion of 300 metric tons of coal or 1,500 barrels of crude oil. Since a cubic kilometer contains 10^9 cubic meters, the fuel equivalents of one cubic kilometer of seawater are 300 billion tons of coal or 1,500 billion barrels of crude oil. The total volume of the oceans is about 1.5 billion cubic kilometers. If enough deuterium were withdrawn to reduce the initial concentration by 1 percent, the energy released by fusion would amount to about 500,000 times the energy of the world's initial supply of fossil fuels!

Unlimited resources of energy, however, do not imply an unlimited number of power plants. It is as true of power plants or automobiles as it is of biological populations that the earth cannot sustain any physical growth for more than a few tens of successive doublings. Because of this impossibility the exponential rates of industrial and population growth that have prevailed during the past century and a half must soon cease. Although the forthcoming period of stability poses no insuperable physical or biological difficulties, it can hardly fail to force a major revision of those aspects of our current social and economic thinking that stem from the assumption that the growth rates that have characterized this temporary period can somehow be made permanent.

17

COAL

LAWRENCE P. LESSING
July 1955

It is by far the largest store of fossil organic substance, but its utilization lags. Scientific and technological advances suggest that it might best be used not as a fuel but as a chemical raw material

Coal, which powered most of the industrial revolution, is in trouble over most of the world. The U. S. learned in 1953 that for the first time in its history coal had dropped from a leading place as a source of energy to a position behind oil. Coal supplied only 34 per cent of the nation's total B.T.U.'s (British thermal units) as against 39.4 per cent for crude oil, 22.5 per cent for natural gas and 4.1 per cent for water power. This historic fall occurred during a year in which U. S. industrial activity was at an all-time high. This fact, plus coal's continuing depression, leads some to believe that coal will never again regain its pre-eminence. Coal, indeed, is one of the foremost technological problems of our time.

Coal is still a big and vital element in the operations of an industrial society. Some 400 million tons of coal, the amount annually being consumed in this country, is not a trifling quantity. It accounts, through conversion in steam power plants, for over three fourths of all U. S. electrical energy. It is indispensable, in the form of some 100 million tons of metallurgical coke, for the production of iron and steel. In addition, coal tar residues from these coking plants provide the essential raw materials for nearly a third of the whole U. S. organic chemical industry. In a dozen lesser ways coal is woven through all the basic strata of industry.

The major fact about coal, however, is that all the coal so far mined is as nothing to the store remaining underground. Some seven trillion tons runs the estimate; the quantity is so vast that no one really knows. This is by far the biggest reserve of all known organic mineral substances, which are alike in that all are preponderantly the hydrocarbon remains of former terrestrial life. In carbon content alone, mineable coal reserves represent 12×10^{12} tons of carbon against $.04 \times 10^{12}$ tons in the world's estimated reserves of oil. Since carbon is the backbone of all life and commerce—carbon compounds in one form or another making up some 90 per cent by value and 95 per cent by weight of all the products of human labor—this represents a great treasure. Long after the last reserves of oil and natural gas are exhausted, at a date not too far distant by most estimates, there will be coal. Even at the most pessimistic estimates, with proper conservation and development of other sources of energy there should be coal for some centuries to come.

Naturally such a massive deposit must be widely though unevenly distributed in the earth's crust. The tropics are fairly thin in coal. Africa has relatively little. Most major deposits seem to lie in the temperate zones but some are found even in polar regions. The U.S.S.R. is mining coal in the Arctic, and there are known deposits in the Antarctic. Europe's deep and dwindling mines stretch in a wide arc from Great Britain through the Lowlands, France and the Ruhr to the Ukraine. Asia has about a third of the world's coal, most of it in China, where the first coal was mined about 1100 B.C., but where, because of a lag in industrialization, most of it remains to be exploited. The Hopi Indians mined coal in Arizona two centuries before Columbus, and North and South America together have over half of the world's reserves. The U. S. alone has nearly half of all the bituminous and subbituminous, the most valuable of all coals.

Yet the U. S. coal industry today looks like anything but the administrator of so vast a treasure. For over 40 years it has been prosperous only during wars. Though the U. S. economy has expanded enormously, and the coal industry with it, coal in that half-century has had its biggest, most lucrative markets cut from under it by more convenient or more efficient fuels—oil and natural gas for heating, Diesel fuels for ships and railroads. Now it is faced with the emergence of atomic power and of methods for tapping directly the immense free stores of energy in the sun. Both of these developments are moving more rapidly than is generally realized. Both developments, as the century runs out, may well begin to cut into all of coal's remaining markets except steel.

An intricate shifting of the technological base of industry is in progress. For coal the handwriting has been on the wall for many years. Part of the trouble is that the coal industry came out of the 19th century with grimy, backward tenets of mining deep in its system. It was slow to modernize and mechanize, going steadily down in efficiency and up in costs. It persisted in regarding coal primarily as a fuel, in spite of the nudge given it by the early development of chemicals from coal tar wastes. While the petroleum industry began early to do heavy research on its basic raw material, the coal industry apparently knew little of its product beyond the fact that it burned. What is known about coal today—and most precise knowledge of it is recent—is almost wholly derived from the inquisitive chemical industry and from independent investigators.

Fossil Origins

If a lump of coal is closely examined, a regular cellular pattern may often be seen, denoting its plant origin. Coal was laid down, according to the geological evidence, by the debris of giant tree ferns and other rank vegetation flourishing in the hot delta bottoms of the late Paleozoic Era, some 300 million years ago. With the slow rhythm of subsidence, sedimentation and uplift, the great swamps were intermittently inundated by the sea, which flattened and covered the vegetation with thick layers of alluvial mud. The weight of these sediments, translated into rock and often folded, exerted those pressures and temperatures basic to all chemical processes, which slowly converted the vegetable matter to coal.

The enormous variety of its vegetable origins and the variables of geological processes make coal one of the world's most complex substances. Not only do coals vary widely in various regions, but also they vary in the same coal seam. Bounded by layers of shale, a coal seam shows an uneven but definite progression from a soft, peaty bottom, suggesting leaf mold, through a bright middle section, with recognizable fragments of decayed plants, bark and spores, to a hard upper layer of massive woody remains.

Four broad classes of coal are recognized: peat, a spongy, water-soaked substance of low heat (or carbon) content, generally regarded as the first stage of coal formation; lignite, a less watery form of peat also known as brown coal, which marks the transition to true coal; bituminous, still soft, but with a larger concentration of carbon and a smaller proportion of oxygen; and anthracite or hard coal, in which both oxygen and hydrogen have been nearly squeezed out, leaving almost solid carbon. (Anthracite is so preponderantly carbon that it is relatively inactive chemically and used only as a fuel.) This ranking of coal has no strict chronological order, for different coals were produced by various conditions at widely different times. Between the main classes are many gradations of type, which makes coal classification extremely difficult.

The constituents of coal have been known for a long time. They are mainly carbon, hydrogen, oxygen, nitrogen, sulfur and mineral ash. What long remained a mystery, however, were the processes by which the weight of carbon was progressively concentrated from 60 per cent in peat to 96 per cent in anthracite, while oxygen dropped from 6 to 3 per cent and hydrogen was maintained almost constant (except in anthracite) around a mean value of 5.5 per cent. This is the exact opposite of what happens in the normal decomposition of vegetable material above ground, where most of a dead plant's carbon is lost to the air via carbon dioxide, leaving behind humus.

In peat bogs, however, which may still be studied in many parts of the world, a different transformation takes place. Here plant material drops to the swamp floor and is soon covered by water, muck and other debris, which exclude air. Under these conditions only partial decomposition takes place. Most of the plant carbon is retained, while most of its oxygen is lost as water. If this is coal's first stage, then a number of chemical routes can be mapped by which woody material with a composition by weight of 50 per cent carbon, 6 per cent hydrogen, 42.75 per cent oxygen and 1.25 per cent ash loses 27 per cent of its volume as water and 16.5 per cent as carbon dioxide. This leaves a material containing 80.5 per cent carbon, 5.3 per cent hydrogen, 12 per cent oxygen and 2.2 per cent ash, which roughly corresponds to bituminous coal.

The whole process of coal formation has been partly confirmed by laboratory experiment. The late German fuel chemist Ernst Berl, working at Carnegie Institute of Technology in the late 1930s, succeeded in making synthetic coal. Compressing woody vegetable matter under pressures and temperatures analogous to those exerted by rocks in the Carbonaceous Period, he produced pellets of a material closely akin to brown coal. Since then coals ranging up through bituminous have been artificially made. Of course this process has no immediate practical significance; more energy is required to make the coal than could be regained by burning it. The process may nonetheless be useful as a means of converting vegetable matter directly into needed hydrocarbons.

Though most of coal consists of hydrocarbons, the lesser constituents, mainly in the ash, are astonishing in number. Recently a close analysis of West Virginia bituminous coals showed that the ash, produced in that state alone at the rate of 10 million tons a year, contains 36 different chemical elements. In concentrations over 1 per cent are sodium, potassium, calcium, aluminum, silicon, iron and titanium. In concentrations down to .01 per cent are 26 metals, including lithium, rubidium, chromium, cobalt, copper, gallium, germanium, lanthanum, nickel, tungsten and zirconium. Many of these are in amounts perhaps someday sufficient for economic recovery. Germanium, the new electronic metal, is the most immediately interesting. It is found largely concentrated in the bottom three inches of a coal seam. If this bottom is carefully separated by gravitational means, it yields coal with an ash containing 3 per cent germanium dioxide, worth $57.50 per ton as an ore and $900 per ton as finished product.

Altogether, including other elements only in trace amounts or occasional occurrence, coal is a compendium of no less than 72 items in the periodic table of elements. Thus coal, through the biochemical activity of plants that flourished in a different world eons ago, is not merely a store of carbon energy but a compact bundle of the most varied chemicals.

The Chemical Pattern

The analysis of coal into its elemental constituents, however, tells us almost nothing about it as a chemical substance. For this we must know the chemical structure of coal, the way in which its atoms are linked to form molecules. Until recently the exact chemical structure of coal, due to its great complexity and variability, was almost totally unknown.

Some clues to structure were contained in the discovery and development of coal tar chemistry just a century ago. In 1856 Sir William Perkin hit upon his historic synthesis of the textile dye mauve from a constituent (aniline) in the tarry messes that accumulated in the flues of gasworks and coke ovens. These tars yielded by fractional distillation a whole series of basic hydrocarbon compounds ranging from benzene, the lightest fraction, to anthracene, the heaviest. From anthracene a few years later two German chemists named Graebe and Liebermann synthesized the red dye alizarin, which theretofore had been obtained only from plant roots. Thus it was apparent that in its vegetable origins coal contained basic organic compounds which could be reconjugated into apparently natural substances by chemical means.

The Continuous Miner, made by the Joy

As the derivatives from coal were studied, their molecular structure fell into two wonderfully geometric divisions. From coal tars came a series of compounds, called aromatic because of their pungent odor, whose structure took the form of ring-shaped molecules. The basic unit was benzene (C_6H_6), consisting of a central hexagon of six carbon atoms with six hydrogen atoms attached. On this ring all the rest of the compounds in the series were built by the simple addition of one carbon atom at a time; *e.g.*, C_7H_8 for toluene, the next compound in the series, C_8H_{10} for xylene and so on. As the number of carbon atoms in the molecule reached 10, two benzene rings were fused, sharing two of the carbon atoms between them to form naphthalene. Three fused rings formed anthracene ($C_{14}H_{10}$), which is the largest molecule in the series.

From coal gas and the like came another series of carbon compounds, called aliphatic because of their fatty or oily character, whose distinguishing form was a linear or chainlike molecule. The basic unit in this series is methane, the simplest of hydrocarbons, consisting of a single carbon atom surrounded by four attached hydrogen atoms. From methane the series builds up, again by the increment of one carbon atom per molecule, into progressively longer and longer carbon chains—C_2H_6 (ethane), C_4H_{10} (butane), $C_{10}H_{22}$ (decane) and so on. These chain compounds and others were later found to be more abundant in natural gas and petroleum, fossil hydrocarbons whose origin differs from that of coal. Whereas coal derives entirely from land plants, petroleum is now generally believed to have originated in the decomposing remains of marine organisms.

These two broad structural forms—ring and chain, aromatic and aliphatic—proved to be the cornerstones of synthetic organic chemistry. With the simpler coal derivatives as starting points, molecules could be built up, linked and rearranged by various reactions adding or substituting other elements—such as oxygen, nitrogen, sulphur or chlorine—to form a vast variety of compounds with different properties. Though their number does not begin to approach the enormous variety and complexity of carbon compounds in nature, no fewer than 500,000 compounds have been synthesized. These compounds embrace the whole range of dyes, drugs, perfumes, plastics, fibers, rubbers, explosives, adhesives, detergents, solvents, insecticides and other products that constitute the modern organic chemical industry.

Throughout this development the exact molecular structure of coal remained a mystery. It was thought that the basic benzene and methane series secured from coal tars and coal gas were not representative of the composition of coal itself but were secondary products formed in the partial combustion or destructive distillation of coal by the heat of gas or coke ovens. Exactly how coal was put together was a matter difficult to ascertain. To begin with, coal was a highly variable mixture of complex plant substances—cellulose, lignin, resins, proteins and the like—converted by the coal-forming process into other complex substances, all bound together in the solid state. Any attempt to break the complex coal molecules apart by the usual combustion methods so drastically altered them that the resulting fragments could not be confidently identified as bearing any relation to the structure of the starting materials.

Was there a characteristic structure of coal molecules that might be found to hold, with variations, through the various ranks of coal? Was this structure small or large, taking as a scale the macromolecules of plastics and many natural products? Finally, was coal made up of only one, a few, or many different kinds of molecules? These basic questions had to wait upon refined means of disentangling the molecules from coal, and upon research interest.

Coal research has suffered some of the vicissitudes of its industry. In the early epoch of organic chemistry there was no urgent need for precise knowledge of coal structure. Research centered on derivatives. Then came the spectacular development of petroleum chemistry. Oil and natural gas offered a much more easily manipulated range of hydrocarbons than coal, and the industrial emphasis shifted to the aliphatics, which came to account for nearly two thirds of all organic-chemical production. Research emphasis likewise shifted, as illustrated last year by the publication of the definitive treatise *Hydrocarbons from Petroleum*, representing 25 years of investigation and over $500,000 in grants by the U. S. oil industry. No such treatise exists on coal.

The need for basic research on coal is steadily rising as the organic chemical industry faces the next big step in its development. In the last 10 years a mounting attack on the problems of coal, both here and abroad, has begun to clarify the chemical structure of coal. Techniques for coming to grips with this structure are multiplying. They include X-ray diffraction and infrared spectroscopy; solvent techniques for extracting the molecular constituents of coal; mild

Manufacturing Company, chews into a seam and delivers coal to the rear

oxidation reactions at temperatures from 200 to 500 degrees Fahrenheit to separate the constituents, and thermal decomposition in a vacuum at temperatures not exceeding 975 degrees F. Recently Irving Wender of the U. S. Bureau of Mines, which has an active coal research program, developed a new reduction method employing a metal (lithium) and an amine in a cleavage reaction in order to break out the skeletal molecules of coal.

The sum of the investigations by these techniques is that the basic molecular skeleton of coal is now generally agreed to be a ring-shaped structure whose nucleus is the six-carbon ring of the aromatic compounds. H. C. Howard of Carnegie Institute of Technology's Coal Research Laboratory, one of the pioneer agencies in this field, has shown that all the products of bituminous coal, obtained by whatever techniques and in all the temperature ranges, are predominantly of the benzene-ring type. There was no correlation between the temperatures used and the amounts of such products secured. Thus the evidence is strong that the ring structures are not formed in the heat of the experimental reactions, but are present in the basic molecule of coal itself.

The total molecule is by no means as simple as benzene. Recent investigations have begun to piece together some of its details. The broadest attack on the problem was launched in 1951 by the British Coal Utilization Research Association, which set out to analyze by all available means a series of vitrinites—bright black coal particles—from typical coals. These studies have yielded a working model of the basic molecule of coal.

The skeletal ring of carbon atoms is surrounded by hydrogen atoms and a variety of atomic groups which may replace them. These are hydroxyl units (OH), short aliphatic side chains, ether linkages and other appendages. The carbon ring may also be fused with other rings into discrete nuclear groups or clusters. These cluster units, divisible only by chemical means, probably form the basic molecules of coal. They range in size from about four fused rings in the low-ranking coals to five to 10 in bituminous and up to 30 in anthracite. The clusters are homogeneous for a given coal; i.e., all belong to the same skeletal species, though varying in size, shape and properties according to the different peripheral groups attached to them. The clusters form rather flat molecules which, infrared studies indicate, are stacked so

tightly that there may be some chemical bonds between the layers.

It is estimated that about 75 per cent of the carbon in coal is in the form of such condensed-ring clusters. The clusters are extended in a fairly orderly pattern, which has been visualized as a kind of chicken-wire structure formed by the hexagonal rings of the cluster nuclei. The pattern is broken, however, by smaller components such as phenol and quinone groups. In addition there are stray atoms of nitrogen, sulfur and metallic compounds within the rings or between clusters. Thus coal remains a complex mixture, behaving in a manner unlike that of any other class of chemical materials, with its molecular structure still not fully understood.

The effort to understand the structure of coal is worthwhile because the more that is known about how coal is put together, the more precisely can it be taken apart to yield desirable chemical substances. These studies will lead to the improvement of established chemical processes manipulating coal, as well as to the possible development of new ones. Already one improved process has made from coal over 200 basic chemicals, some entirely new or in quantities never before achieved. It is this fact that leads chemists and conservationists to conclude that the most wasteful thing to do with coal is merely to burn it.

Five Chemical Routes

There are five basic processes for extracting the hidden wealth of materials from coal. Most of them were originally devised in Germany, where, because petroleum was lacking, research centered on this alternative source of basic hydrocarbons. The processes range from methods for reducing the chemical parts of coal to the simplest gaseous molecules, then rebuilding them into desired products, to methods for extracting chemicals directly from coal in mixed groups, then separating them. In these coal processes, as well as in the more advanced ones of the oil industry, the old distinction between aromatic and aliphatic sources is largely lost. Oil products may now be made nearly as readily from coal as from petroleum; petroleum is already supplying by special processes such aromatics as benzene and toluene. Organic chemistry thus moves toward blending the two fields into one continuous spectrum of hydrocarbons.

This is the range of coal processes, in increasing order of complexity or specialization:

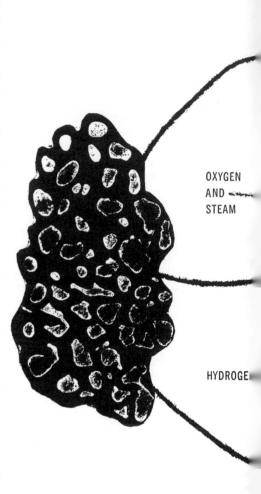

AIR

OXYGEN AND STEAM

HYDROGE[N]

THREE BASIC PROCESSES for extracting vari[ous] substances from coal are illustrated in this diagra[m]

Carbonization. This is the oldest process, embodied in the by-product coke oven, where charges of bituminous are roasted in a controlled atmosphere at 2,000 degrees F. to burn out part of the carbon, drive off volatile substances and give three main product groups: metallurgical coke, coal tar and gases. The principal gas is carbon monoxide (with some methane and nitrogen), used as fuel gas and to make ammonia, methanol and other chemicals. This carbonization process, whose leading chemical exponents in this country are the Allied

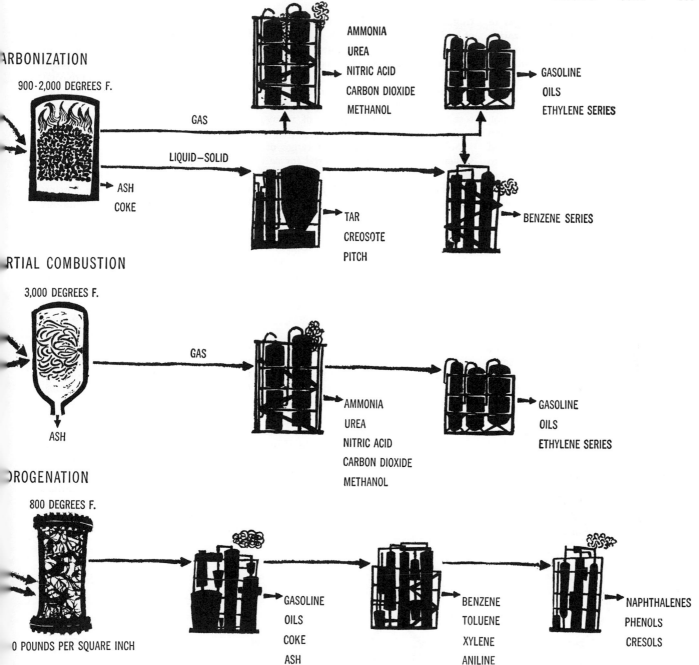

CARBONIZATION

900-2,000 DEGREES F.

GAS

AMMONIA
UREA
NITRIC ACID
CARBON DIOXIDE
METHANOL

GASOLINE
OILS
ETHYLENE SERIES

LIQUID—SOLID

ASH
COKE

TAR
CREOSOTE
PITCH

BENZENE SERIES

PARTIAL COMBUSTION

3,000 DEGREES F.

GAS

ASH

AMMONIA
UREA
NITRIC ACID
CARBON DIOXIDE
METHANOL

GASOLINE
OILS
ETHYLENE SERIES

HYDROGENATION

800 DEGREES F.

GASOLINE
OILS
COKE
ASH

BENZENE
TOLUENE
XYLENE
ANILINE

NAPHTHALENES
PHENOLS
CRESOLS

0 POUNDS PER SQUARE INCH

carbonization coal is heated primarily to produce metallurgical coke. Partial combustion is based on the old "water gas" reaction. In a modern version of the reaction oxygen is fed into the reactor. In hydrogenation coal is heated to 800 degrees F. at high pressure.

Chemical & Dye Corp. and the Koppers Co., Inc., is still capable of further development. Only about a third of the tar finds its way into useful chemicals, and nearly half of the coal tar chemicals remain commercially undeveloped. Low- or medium-temperature carbonization processes, with closer control of conditions and catalytic assistance, could now yield more of the desired or unexplored chemicals without lowering the quality of the coke.

Partial combustion. This also is based on an old process, the "water gas" reac-tion invented in the U. S. in 1873 but hardly changed until recently. A jet of steam is blown over an incandescent bed of coal or coke in a closed chamber to produce large quantities of mixed carbon monoxide and hydrogen, the base of domestic fuel gas. Since the steam cools the coals, which then have to be brought back to 3,000 degrees F. by a blast of air, the process is inefficiently intermittent and has lost ground steadily to higher-B.T.U. natural gas. However, this sim-ple, important process is the first cheap method for adding reactive hydrogen (from water) to coal to get a synthesis gas that can be manipulated into many chemicals. Germany made two advances in the process. In one it is made continu-ous by substituting pure oxygen for air and feeding the oxygen and steam stead-ily to the reaction. In the other the pres-sure is raised to 300 pounds per square inch to cause part of the carbon to com-bine with the hydrogen to form methane, thus enriching the gas. The efficiency of these developments is now being in-creased to produce synthesis gas of great chemical promise.

Fischer-Tropsch. If synthesis gas or the residual gas from coal carbonization are now fed across a cobalt or iron catalyst at low heat and pressure, a magical stream of products results: petroleum-like products ranging from gasoline to lubricating oil, plus a range of alcohols beginning with methanol and ethanol, the synthetic forms respectively of wood and grain alcohol. Heat, pressure and catalyst force the small gas molecules to link up into the short-chain molecules of the aliphatic series. By varying the controls, the preponderance of oil or chemical products may be altered. This process is the creation, dated 1933, of the German chemists Franz Fischer and the late Hans Tropsch. U. S. investigators, using fluid-bed catalysts and other techniques, are speeding up and refining the process.

Hydrogenation. This process, another product of German ingenuity developed by Friedrich Bergius in 1910, is the first chemical process to work directly from coal. A blast of pure hydrogen (separated from synthesis gas) is shot into a paste of pulverized coal and catalyst at medium temperature (850 degrees F.) and high pressure (3,000 pounds per square inch) literally to explode the coal molecules and attach hydrogen to their dismembered chains and rings. This, with its massive additions of hydrogen, produces the widest range of coal chemicals: gasoline, Diesel and heavy oils, benzene, phenols and the range of coal tar chemicals, aniline and a swath of nitrogen compounds, plus a small amount of hydrocarbon gas and high-grade coke. Depending on the catalyst and controls, oil products or other chemicals may be the major yield. Germany used this process for the bulk of its wartime gasoline, with Fischer-Tropsch supplying the rest. It is a process capable of many variations and refinements.

Solvent extraction. A great range of solvents has now been explored for selectively dissolving specific chemicals out of coal paste. None has yet reached commercial status. Solvents are expensive to handle, being merely temporary vehicles in the process. But with the increasing efficiency of solvent recovery, chemical extraction by this means may prove attractive. It is low in heat cost and is the least destructive method of obtaining certain compounds.

Toward Carbochemicals

Economic factors have retarded the application of most of these processes in the U. S. So long as oil and chemical products can be secured easily and directly from petroleum and natural gas, there is no strong incentive to take the more difficult, roundabout routes from coal. Nearly all involve net deficits in energy. Moreover, coal as a solid presents inescapable problems of physical handling and ash disposal, even in chemical processes. The multiplicity of chemicals from coal, which have unequal value, present other problems. One large hydrogenation plant, it has been estimated, could pour out enough phenol to supply nearly the whole U. S. requirement at present. But with increasing limitations in the price and availability of natural gas, and with the time approaching when petroleum will not be so freely plentiful, the chemical industry is making a start toward carbochemicals. Two experimental plants are under way. Both are located in the dark, narrow valley of the Kanawha River, flowing between West Virginia's bituminous mountains.

The first and broadest of these experiments is a great, black coal hydrogenation plant at Institute, W. Va., built at a cost of $11 million by the Carbide & Carbon Chemicals Co., a division of Union Carbide & Carbon Corp. This plant, as distinct from the German practice, is designed primarily to produce chemicals. It embodies a number of engineering advances, chiefly in the direction of economy. It cuts the reaction time from about 45 minutes to three minutes. A 300-ton-a-day stream of coal paste is almost continuously piped to two big steel reactors, heavy as naval guns, from which debouch a stream of sludgy mixed products. Through several separation stages, heavy oils, coke and ash are drawn off to one side and hydrocarbon gases to the other. This frees the main stream of liquid chemicals: volatile aromatics beginning with naphthalene in

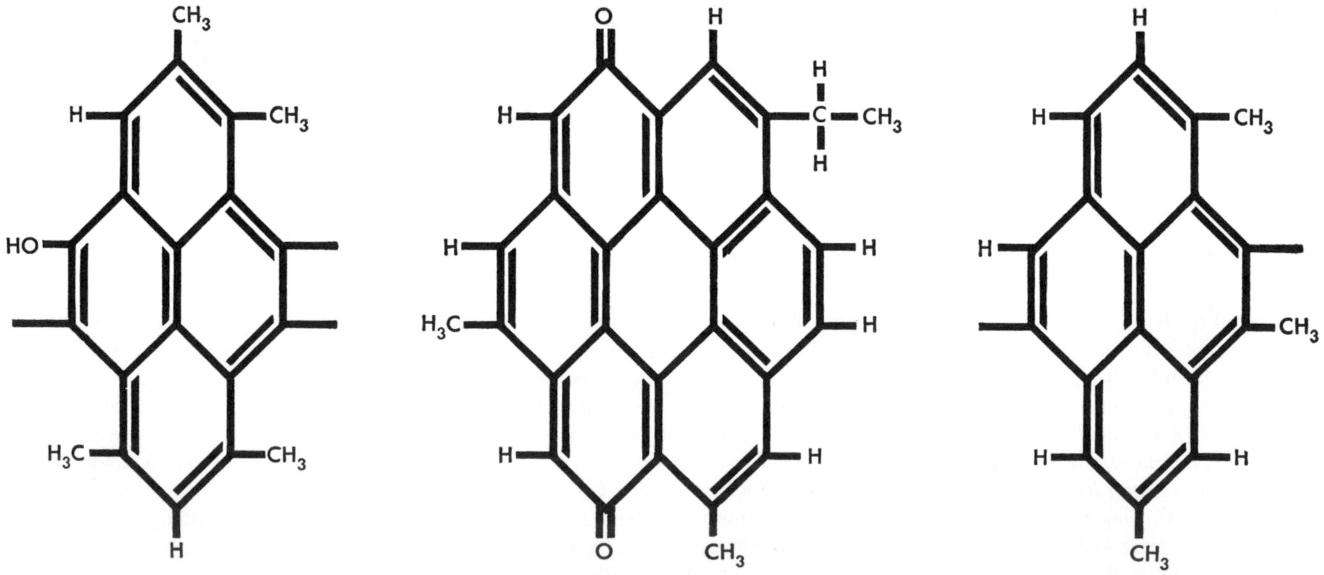

BASIC MOLECULAR STRUCTURE of coal is also characterized by the benzene ring. In coal, however, the rings are assembled in larger structures called cluster units. Depicted here are three typical cluster units. The long bonds that do not end in atoms or atomic groups are presumed to be connected to carbon or oxygen atoms in other cluster units. The flat cluster units are tightly stacked in layers that may be joined by chemical bonds. This gives coal a basic structure rather like contiguous layers of chicken wire.

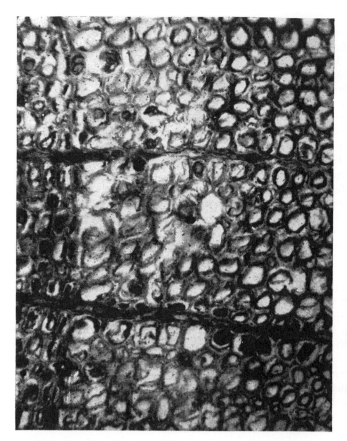

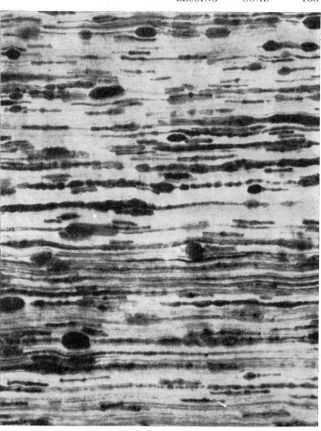

PEAT is revealed in the first of this series of micrographs made by the U. S. Bureau of Mines. It has the cellular structure of wood.

LIGNITE from North Dakota is streaked with dark bands of resinous material. All sections on this page are enlarged 200 diameters.

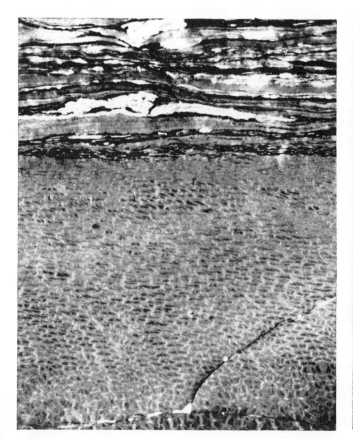

BITUMINOUS COAL from Kentucky has bands of coarse material (*bottom*) and fine (*top*). The coarse band has a woody structure.

CANNEL COAL is typical of unbanded coals. It has the pattern of finely divided plant remains such as resins, humus and spores.

one group, phenolics in another, nitrogen compounds in a third. Early last year the Institute plant shipped its first tank car of high-boiling phenols. Since then it has supplied six other bulk chemicals and a dozen more in experimental amounts.

Altogether Carbide spent over 15 years, interrupted by the war, studying coal in all its chemical aspects. Natural gas, around which it had built its big chemical plant at Charleston, W. Va., had a foreseeable end in that area. Coal hydrogenation, as worked out by its chemists, had three attractions. It supplied certain chemicals such as benzene, the demand for which was outrunning the supply of the coke ovens. It provided still other chemicals which were either wholly destroyed or reduced to barely recoverable amounts in the heat of the coke ovens. Thus it gave promise of developing a base for new products and markets which would permit a gradual shift from natural gas to coal. Some of the products already evolving are new types of phenolic plastics, new pharmaceuticals based on quinoline (nicotinic acid) and on picoline (tuberculosis drugs), and new rocket fuels.

Carbide continues its basic coal studies. It has investigated the underground gasification or burning of coal *in situ* in the earth, concluding that the process is still too uncontrolled and variable in the production of synthesis gas to allow a continuous chemical operation. It is studying other gasification schemes and variations on the Fischer-Tropsch. As natural gas dwindles, some such process will have to supply the large volumes of gas and intermediates needed for synthesis. It is not neglecting the more mundane

aspects of coal mining and transportation, which account for over 80 per cent of coal's high costs. Its engineers have developed an experimental robot miner, controlled entirely from above ground, which chews its way 1,000 feet into a coal seam, throwing out a stream of fine coal by means of a flexible conveyor carried behind. It mines up to two tons a minute, with only three outside operators. Carbide's engineers are also looking into the long-distance transportation of pulverized coal by pipeline.

In the next round of expansion, as Carbide scales up to a full-sized 1,000- to 5,000-ton-a-day hydrogenation plant, its engineers envision a continuous-flow operation characteristic of the chemical industry. A never-ending stream of coal from the mines will be processed and sent by pipeline to the chemical plant, which 24 hours a day will split the black mass into nearly 100 sparkling streams of chemicals.

The second big experiment in carbochemicals is taking place a few miles up river at Belle, W. Va. Here for a quarter of a century E. I. du Pont de Nemours & Co. has had one of its basic plants. It has operated on the old water gas reaction, synthesizing from the gas such chemicals as ammonia, urea and other nitrogen products (for fertilizers, plastics, explosives), methanol and other carbon compounds (for solvents, antifreeze, intermediates). The first raw materials for nylon (adipic acid and hexamethylene diamine) were derived here. But any process unchanged for 25 years is suspect to chemists, and the water gas process is not only discontinuous but dirty, throwing out clouds of partially burned

coke particles as the steam and air alternately play over the glowing coke. Thus nine years ago du Pont's Polychemicals Department began working on a cleaner, more efficient means of coal gasification. Later the help of the Babcock & Wilcox Company was enlisted in the design of a plant.

The new plant at Belle, which went into operation early this year, is an advanced model of the partial combustion of coal. A jet of oxygen and superheated steam meets a stream of pulverized coal being shot through a burner nozzle into a great tubular furnace. Synthesis gas is tapped off continuously at the top of the furnace, ash and molten slag is continuously washed out at the bottom. From this gas du Pont will make the same products as before except methanol, which is now more cheaply derived from petrochemical sources. The new unit will supply about a third of Belle's gas requirements. If it works out well—and operation thus far indicates that it has marked economic and control advantages over water gas—the process will supplant all of du Pont's remaining water gas units.

Du Pont long weighed the alternative of switching completely away from coal to natural gas, which is now the most economic source of both methanol and synthetic ammonia. But all its newer plants are based on petrochemistry, and du Pont decided that it had better experiment on coal as insurance against the future. Coal partial combustion may some day be called upon to supply the synthesis gas which, with the addition of Fischer-Tropsch-type processes, can be turned into both organic chemicals and liquid

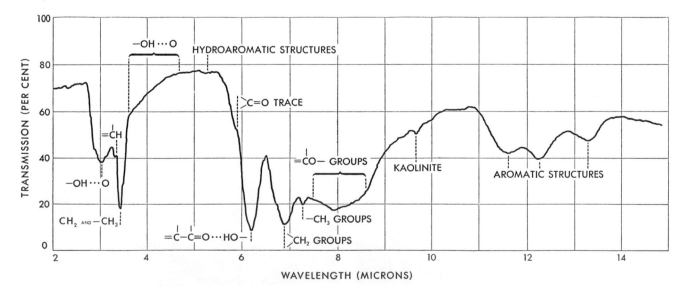

INFRARED SPECTRUM made by R. A. Friedel of the U. S. Bureau of Mines indicates molecules, atomic groups and chemical linkages in a bituminous coal from Pittsburgh, Pa. Kaolinite is an inorganic constituent. The aromatic structures are single and multiple.

fuels. The search for a cheap source of synthesis gas is in flux in many laboratories. The oil industry has done much work on coal gasification as insurance against a decline in petroleum supplies. One of the latest schemes is a process for the underground gasification of Western lignites and other subbituminous coals.

For a time during one of the recurrent oil-shortage scares it was thought that the oil industry would lead the way toward making chemicals from coal. But shifting reserve estimates, regional resource patterns and the even more shifty economics of hydrocarbon sources make prediction dangerous. The development of carbochemicals will probably follow the pattern of the so-called petrochemicals. The chemical industry will explore and establish markets. The oil industry will come in later, as it is pressed to make synthetic fuels, to supply by-product raw materials in huge quantity for the future carbochemical industry.

The Future of Coal

The absence of the coal industry proper in all these endeavors is most noticeable. There has been a certain amount of awakening recently. A few coal companies, such as the Pittsburgh Consolidation Coal Co., have undertaken research programs looking toward coal as a basic carbonaceous chemical material. A few producers have come to support the cooperative Bituminous Coal Research, Inc., which last year raised near Columbus, Ohio, the first general laboratory in the industry in an attempt to catch up on the research so signally lacking in coal for a century. But re-

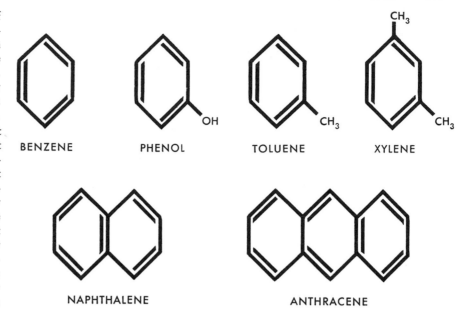

AROMATIC COMPOUNDS derived from coal tar are characterized by the ring-shaped molecule of benzene (C_6H_6). This structure is commonly represented by the hexagon at the upper left. At each corner of the hexagon is a carbon atom from which projects a hydrogen atom. The double lines in each of the hexagons indicate double chemical bonds.

search funds are meager compared with those in the oil and chemical industries.

What funds are available might better be turned to the fundamental study of coal chemistry, for it is within the molecule of coal that the future of the industry lies. Such fundamental studies would promote the balanced development of coal both as a source of new chemicals and of new energy, for these aspects of coal cannot be separated if this great natural resource is to be most efficiently utilized.

No one can exactly predict the pattern of the new coal age. Eugene Ayres, an authority on energy sources, thinks that low-temperature carbonization will come to the fore on the basis of its low cost and high thermal yields to supply char for electric steam generating stations at the same time that it supplies gas and tar for chemicals and synthetic liquid fuels. Others foresee great coal gasification plants close to the mines simultaneously turning out gases for chemical and liquid-fuel synthesis and by-product gas for heating. Still others see a range of coal-hydrogenation plants to supply a large group of organic chemicals, plus some oil products, some synthesis gas and vital amounts of metallurgical coke for the steel industry.

18

TAR SANDS AND OIL SHALES

NOEL DE NEVERS

February 1966

The world's largest potential liquid-hydrocarbon reserves are not recoverable by ordinary oil-producing methods. The pace of their exploitation depends on technical, economic and political forces

The advance of technology exerts a powerful force on the course of events in an industrial society, but it rarely operates alone. More often its effects are interwoven with economics and public policy. This interplay is nowhere more apparent than in the broad field of energy production; here the introduction and exploitation of a new technology is profoundly affected, on the one hand, by governmental considerations of national security, foreign exchange, taxes and conservation and, on the other, by the economic pressures of competition from alternate sources and methods, transport and marketing. The interrelation of these varied factors is illustrated clearly by the current situation involving two unconventional sources of petroleum: the Athabasca tar sands of northern Alberta in Canada and the Green River oil shales of Colorado, Utah and Wyoming.

The two deposits differ in their chemistry, physical state and history. Both, however, contain hydrocarbons that can be converted economically into petroleum products. Both occur at or near the surface, within reach of mining or shallow drilling. And both deposits are of staggering size, even compared with the world's total liquid-fuel reserves [*see upper illustration on page 175*]. Each contains potential petroleum products worth at least hundreds of billions of dollars and possibly trillions of dollars. The big questions, in both cases, are just when, how and by whom these

vast reserves are to be exploited. In Canada the issue of the tar sands has recently passed through a controversial phase and is now quiescent; in the U.S. the issue of the oil shales is approaching a time of decision and perhaps of political controversy.

The Athabasca deposit is a bed of bituminous sand—in effect a mixture of sand and paving asphalt—that covers thousands of square miles in northern Alberta. The black sand contains between 12 and 17 percent of oil by weight and the deposits are as much as 200 feet thick. The geologic history of the tar sands is still a debated subject; the general opinion is that the oil flowed into the sands after having been formed by the decomposition of marine organisms in deeper strata. Unlike deep oil formations, which are warm and under high pressure, the tar cannot be produced from wells; at its prevailing temperature of about 36 degrees Fahrenheit it is too stiff to flow.

Two basic approaches have been considered for recovering this asphaltic crude oil. One involves mining the sand and somehow washing the tar from it. The other calls for heating or otherwise treating the tar in place to decrease its viscosity and enable it to flow to wells.

An effective process of the first type was developed some years ago, largely by the Research Council of Alberta, a provincial government agency [see "The Athabaska Tar Sands," by Karl

A. Clark; SCIENTIFIC AMERICAN, May, 1949]. The mined tar sand is agitated in hot water through which air is bubbled. The bubbles carry oil globules to the surface in a watery froth and the sand grains settle out, along with most of the accompanying clay and silt. The water-laden asphaltic crude-oil froth is dried and then heated in a coking unit that produces coke for in-plant fuel and a cracked distillate that can be refined into a "synthetic" crude oil; this is further refined into the usual petroleum products [*see illustration on page 173*].

The fundamentals of this hot-water system have been known for some time; the barriers to commercial exploitation were the high cost of mining and production, the remoteness of the deposits from potential markets and the expense and difficulty of refining the tar to get a satisfactory yield of end products. Developments in large-scale mining machinery have lowered the cost of getting the sand into processing plants. The opening of conventional oil fields in southern Alberta has established markets and pipelines that tar sand oil can share. New developments in the catalytic treatment of heavy oils with hydrogen have made the refining process more efficient.

All these factors have combined to make the hot-water method appear economically feasible, and Great Canadian Oil Sands, Ltd., is now building a $191-million plant near Fort McMurray in Alberta to produce oil by this method.

The company expects to mine 100,000 tons of tar sand per day, scooping it from a surface vein 50 to 175 feet thick, and process it to recover 45,000 barrels per day of synthetic oil (a barrel is 42 U.S. gallons). The plant is expected to be in commercial production this year.

Ultimately, however, some way must be found to recover the tar directly from the ground as a fluid, because much of the deposit is too deep for economical open-pit mining. A number of schemes have been proposed and some have been tried on a small scale in the field. The most advanced of them is the process developed by the Shell Oil Company, which would treat the tar sands as a shallow oil field suitable for secondary recovery [see "The Secondary Recovery of Petroleum," by Noel de Nevers; SCIENTIFIC AMERICAN, July, 1965]. As it exists in nature the tar is so viscous that ordinary "fluid drive" methods utilizing water or gas to push it will not move it through its sand matrix to the producing wells. The Shell Company would create horizontal fractures at the bottom of the tar sand formation by pumping in water at a pressure high enough to lift the earth the way a hydraulic jack lifts a heavy load. It has demonstrated that in this way it can produce a horizontal fracture across the bottom of the tar sand deposit. Then it would heat the sand by injecting steam into the horizontal fracture. Finally, when the sand and tar were warm enough, it would inject al-ternate slugs of steam and of hot water containing sodium hydroxide into one set of wells and remove the solution from another set. It has found that the hot sodium hydroxide solution will pick up the previously warmed tar, forming an oil-water emulsion that flows much more easily than the tar itself. This emulsion would be brought to the surface from the collection wells; the tar would be separated and sent to a processing plant and the water and sodium hydroxide would be reused.

This process is similar to the steam-injection secondary-recovery processes now being developed by various oil companies for use in oil fields with very viscous crudes. The main difference is the horizontal fracture that would be required to start the flow. The tar sands

TAR SAND is exposed along a bank of the Athabasca River north of Fort McMurray in Alberta. The stratum of tar sand, about half-way up the bank, is light in color as a result of weathering. The tar is thought to have flowed upward into sand from older formations.

OIL SHALE deposits of the U.S. are concentrated in high plateaus of the Rocky Mountain region. The shale is a finely laminated sedimentary rock, visible in this scene as two pale, steep-faced strata just below the darker layer at the very peak of the mountain.

are shallow compared with typical oil fields, but much closer well spacing would be needed for the fracturing operation; fortunately shallow wells are not expensive to drill, so that the large number of wells per acre in a tar sand project would not cost much more than the smaller number of deeper wells in an oil field.

After making a successful small-scale field test of this process, the Shell Company proposed making a larger-scale test provided that the Alberta government would guarantee the company the right to go into commercial production if the trial was successful. This proposal was rejected by the Oil and Gas Conservation Board of Alberta—not on its technical merits but because, like the surface-mining process, it would only be economically feasible if commercial production were carried out on a very large scale. The Shell Company wanted permission to produce 96,000 barrels per day of synthetic crude oil, and the Oil and Gas Board decided that so much extra production would upset the oil market.

A second proposal for recovering the deeply buried tar sands was made by a group of oil companies. They wanted to detonate a small atomic bomb at the bottom of the deposit to fracture and heat a large volume of tar sand. Much of the sand would become warm enough for the oil to become quite fluid, suitable for being pumped out of the heated area with conventional oil-well pumps. This proposal has also been shelved, at least temporarily.

The Athabasca tar sands dwarf all similar deposits. There are, however, significant tar sand deposits in Utah, California and elsewhere, variously called "bituminous sandstones," "asphalt rock," "oil sands" or "sand asphaltum." There is no sharp line of demarcation between a conventional oil field with a very viscous crude oil that will flow into wells and a tar sand with a slightly more viscous crude oil that will not flow into wells in commercial quantities. In the U.S. this has led to legal controversies over oil leases, which normally allow the extraction of oil and gas but not of other minerals. Does an oil lease cover tar sands from which the oil can be recovered only by mining or by some secondary-recovery technique? The Department of the Interior, which supervises oil leases on Government properties, says no. Then how much oil has to be recoverable by simple flow into a well before a "tar sand" becomes a "viscous oil field" and thus comes under the

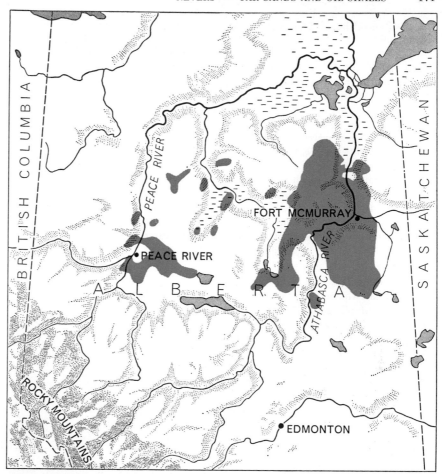

ATHABASCA TAR SANDS of northern Alberta (*color*) are in a generally undeveloped region of forest and muskeg. There are conventional oil fields in the Edmonton area.

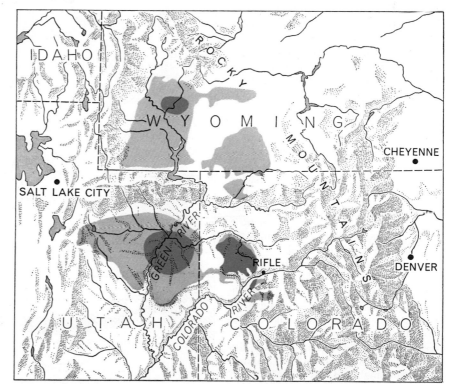

GREEN RIVER OIL SHALES vary in quality. The dark color denotes deposits 15 feet or more thick with at least 25 gallons of oil per ton. The lighter tint indicates less rich shale.

terms of an oil lease? This is still a debated point.

Oil shales are far more widely distributed in the world than tar sands; they occur in many countries and in sedimentary formations of many geologic ages [see "Oil from Shale," by H. M. Thorne; SCIENTIFIC AMERICAN, February, 1952]. In the U.S. there are oil shale beds in at least 30 states. The largest by far is the deposit in the Green River formation of Colorado, Utah and Wyoming. It is the greatest known concentration of hydrocarbons in the world, dwarfing the huge oil pools of the Middle East. The Green River shales—the equivalent of some two trillion barrels of oil, perhaps half of it recoverable by currently known methods—are in effect mountains of oil: some 16,500 square miles of uplifted, eroded sedimentary rock impregnated with organic matter. Actually the rock is not shale and the organic matter is not oil. The Green River rock is a marlstone, a fine-grained, compacted mixture of carbonates, clays and other minerals. The organic matter is not merely a very viscous petroleum like the Canadian tar but a rubbery solid called kerogen, essentially insoluble in petroleum solvents and intimately mixed with the mineral grains.

In terms of chemical structure the difference between petroleum and kerogen is primarily one of geometry. Typical petroleum molecules are linear chains with some rings and branches but little linking between chains. In kerogen, on the other hand, the chains are cross-linked to a significant extent. When kerogen is heated to between 850 and 900 degrees F., the links are broken and the solid undergoes a chemical transformation: a pyrolysis, or thermal cracking, that yields an oil (typically about 66 percent of the kerogen's weight), a fuel gas (9 percent) and a cokelike solid (25 percent). The oil is quite stiff and is high in sulfur and nitrogen content, but after treatment it is as good a feedstock for refining as most good grades of crude oil.

The Green River shales were formed during the Eocene epoch some 50 million years ago by the deposition of silt and organic matter—mostly algae—in large, shallow freshwater lakes. (Petroleum was formed by organic material deposited in oceans.) Today the former lake bottoms have been uplifted to become high plateaus carved by erosion into steep-walled mesas and deep canyons. The richest shale beds are those of the Piceance Creek Basin of

northwestern Colorado. Here the layers of shale containing at least 25 gallons of oil per ton range from a few feet to some 2,000 feet in thickness; they are exposed in outcrops on hillsides and canyon walls, but the richest deposits of all are buried under some 1,000 feet of overburden.

Liquid fuels were produced from oil shales as long ago as 1838 in France, and over the years shale has been processed on an industrial scale in several parts of the world, notably Scotland, Estonia, Australia and Manchuria. There was a small shale-oil industry in the U.S. in 1860, but that ended as

liquid petroleum became plentiful. Experimental shale projects have been conducted intermittently, however, first by the Bureau of Mines of the Department of the Interior and, with increasing intensity in the past few years, by a number of oil companies.

Like the oil in tar sands, shale oil can be recovered either by mining the shale and then processing it or by heating the kerogen in place. So far attempts to produce shale oil have been largely limited to the first approach: the stone is mined, crushed and then heated in a closed retort to extract the organic material. Experimental work by the Bureau of Mines has established

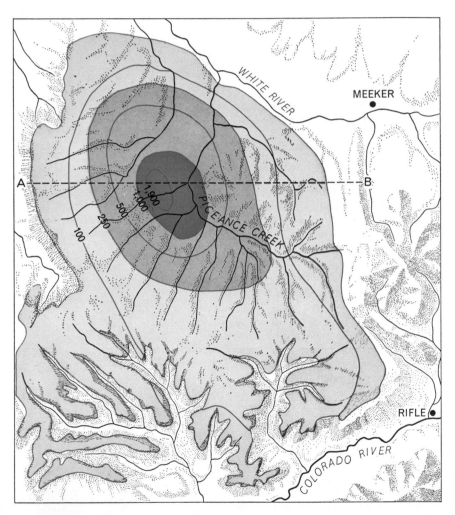

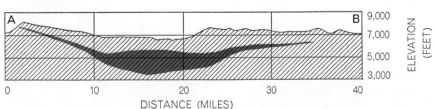

PICEANCE CREEK BASIN of Colorado has the thickest deposits of rich shale. The contour lines indicate the thickness (*in feet*) of deposits averaging at least 25 gallons of oil per ton. The section (*bottom*) is based on cores taken along the line *AB* on the map.

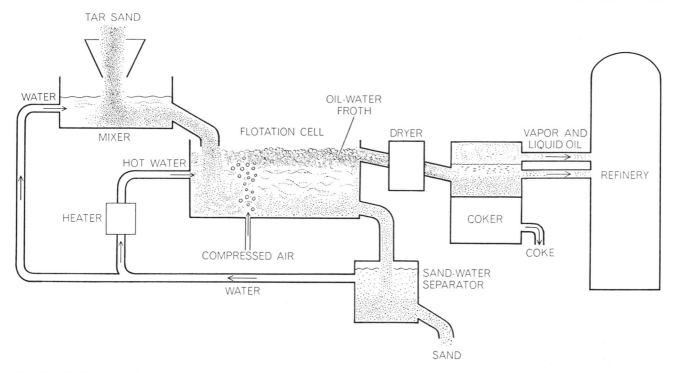

HOT-WATER PROCESS for washing oil out of tar sand is shown in simplified form. The tar-impregnated sand is mixed with hot water; the slurry flows into an open, water-filled flotation tank in which oil globules attach to air bubbles and rise to form a froth as sand sinks to the bottom. After further separation of water and fine sand in a dryer, the asphaltic crude oil is heated in a coker to produce coke (for use as in-plant fuel) and petroleum fractions to be refined into a "synthetic" crude oil, with sulfur as a by-product.

that, at least in the rich Colorado beds, the shale can be taken from the ground economically by the "room-and-pillar" technique used in some coal and salt mines. A shaft is driven horizontally into an outcrop and the shale is blasted and scooped out of great 120-foot-square "rooms," with 60-foot-square "pillars" left unmined in the center to support the roof. It appears that current mining technology is adequate for getting oil shale out of the ground.

The heating of the shale in retorts, on the other hand, presents many complex problems. A practical retorting process must receive and heat many thousands of tons of raw shale a day and get rid of the "stripped" shale. It must recover all the vaporized oil and condense it to liquid form. It must recover a large part of the heat from the treated shale in order to keep fuel costs down. And it must do all this with a minimum of coolant, since the shale country is extremely poor in water.

The early shale retorts were closed vessels like coke ovens, heated by combustion gases from a furnace outside the retort. The high cost of the fuel they require makes them impractical today. To be economical the retort must obtain its heat by burning the solid coke residue left on the stripped shale or part of the gas produced when the kerogen

is heated. This is accomplished in different ways in the three basic retort designs that seem to hold promise for commercial production.

One thermally self-sufficient method is the gas-combustion process developed by the Bureau of Mines at a pilot plant near Rifle, Colo. The crushed shale is introduced at the top of a large retort that resembles a vertical lime kiln or blast furnace. Air and recycled gas from the retorting process move up through the shale, the gas is ignited to provide heat and more heat is derived from the burning of the coke residue. The heat decomposes the kerogen in the shale above the combustion zone [see drawing at left in top illustration on next page]. The oil leaves the retort as a mist carried off by the effluent gas. After the oil has been separated part of the gas is recirculated into the retort, where it cools the stripped shale and is itself heated to near combustion temperature. In 1964 the former Bureau of Mines facility was leased to the Colorado School of Mines Research Foundation, which is developing the gas-combustion process further in a joint research effort with a number of oil companies. The group expects to be ready for commercial production within five years.

One difficulty in the gas-combustion process is that the product oil has to move upward out of the retort. The effluent gases must be kept quite warm in order to prevent the oil from condensing and dripping down into the combustion zone, and the oil droplets must be condensed and separated outside the retort. To avoid these requirements the Union Oil Company of California developed a retort in which the shale moves up and hot gas (from the burning of the spent shale's coke residue) is pulled downward by blowers. The gas gives up its heat to decompose the kerogen and is cooled enough by the incoming cold shale so that the oil condenses, drips to the bottom of the retort and is collected as a liquid. This provides better efficiency in heat recovery than the gas-combustion retort and saves the cost of an external oil condenser. The critical design element in this process is a "rock pump" that loads the shale and rams it up through the retort [see drawing at right in top illustration on next page]. The Union Oil Company operated a pilot plant in Colorado, testing room-and-pillar mining methods as well as retorting and refining processes, between 1955 and 1958.

Both internally fired retort processes decompose the shale in the presence of air. This reduces yield because it allows

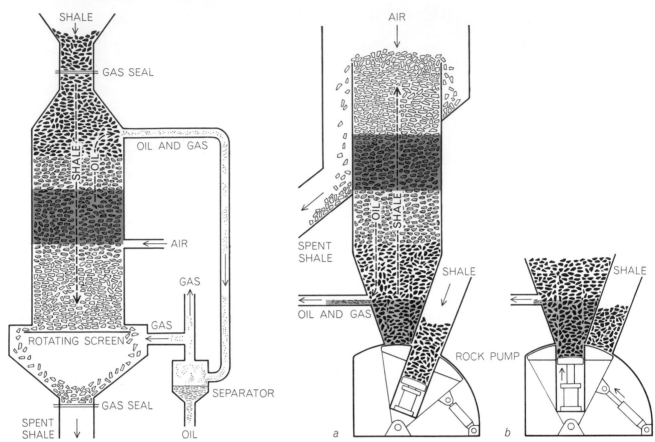

TWO SHALE RETORTS in which heat is generated internally are shown schematically. In the Bureau of Mines process (*left*) the crushed shale moves downward. It is decomposed above the combustion zone (*color*) by heat from the burning of both recycled shale gas and coke on the spent shale. In the Union Oil Company process (*right*) the shale moves upward. Hot gas from the burning coke is drawn down to decompose the shale. A "rock pump" feeds the retort, taking a load of shale (*a*) and ramming it upward (*b*).

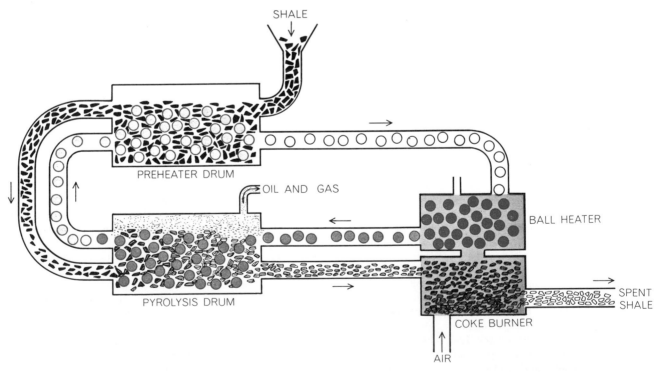

HOT-BALL PROCESS developed by the Oil Shale Corporation separates the combustion stage and the decomposition stage. In the version diagrammed countercurrents of shale and hot ceramic balls are tossed together in rotating drums, where the shale is pulverized and decomposed. The coke left on the shale is subsequently burned to bring the ceramic balls back to retorting heat.

some valuable constituents to be burned up, and it also leads to some undesirable chemical reactions between the product oil and the air. Combustion and decomposition are separated in a process developed by the Oil Shale Corporation. The crushed shale is fed into a rotating horizontal kiln along with ceramic balls that have been heated to 1,200 degrees F. As the kiln turns, the balls pulverize the shale and heat it, decomposing the kerogen and driving off the oil as a vapor [see bottom illustration on opposite page]. Both the retorted shale and the cooled ceramic balls leave the drum, whereupon the residual coke on the shale can be burned to provide the heat that brings the balls back to retorting temperature. The Oil Shale Corporation has joined the Standard Oil Company of Ohio and the Cleveland Cliffs Iron Company to form the Colony Development Company, which is operating a "semiworks" plant based on the ceramic-ball process and which expects to go into commercial shale-oil production in 1967.

In all the mining-retorting schemes the cost of mining the shale, crushing it and transporting it to the processing plant is three or four times higher than the cost of retorting. The preliminary steps could all be eliminated if some way could be found to retort the shale in the ground. The underground-combustion process widely used in the secondary recovery of petroleum seems to be the only feasible method. In this process several holes would be drilled into a shale formation and air would be pumped down alternate holes. The shale would be ignited at these injection holes and the combustion front would be directed toward recovery wells by the continued injection of compressed air. The hot combustion gases would decompose the kerogen adjacent to the burning zone, and the resulting shale oil and gas would be driven to the recovery wells. The high flow resistance of the fine shale matrix in which the kerogen is dispersed makes in-place retorting extremely difficult, and tests to date have been quite discouraging. Nevertheless, the Sinclair Oil Corporation and others are continuing research on such techniques.

There have been proposals that the high flow resistance of the shale could be overcome through the use of nuclear explosives. A nuclear device small enough to be lowered into a buried shale deposit could be detonated to fracture the rock in place and might break up the shale into fine enough pieces to allow efficient underground-combustion recovery. There are a number of questions that must be answered before a field test can even be considered, but if all the imponderables in a nuclear-explosion technique can be worked out, it might result in very economical production of shale oil.

An in-place retorting process, in addition to saving money, would relieve the producer of a serious waste-disposal problem. A plant producing 50,000 barrels of oil per day would have to dispose of 70,000 cubic yards of spent shale every day—a pile three feet wide, three feet high and 40 miles long! Bureau of Mines tests indicate that after a few years of weathering the waste shale could support vegetation, but the problem of where to put it all would still be formidable. In principle, at least, most of the spent shale could be put back into the holes from which it had been dug; this would raise costs, however, and the chances are it will not be done.

The decision to drill an oil well or start a manufacturing plant is largely a private one, made within the framework of existing laws and economic circumstances. The decision to undertake a tar sand or oil shale project is not so simple. Significant questions of public policy are involved and must be settled through political processes. Moreover, small-scale projects are apparently not economically sound; any tar sand or oil shale operation is going to have to be sizable, producing a minimum of some 50,000 barrels of oil per day. The fact that these unconventional reserves cannot come into the economy gradually—they either come with a bang or not at all—makes the political decisions all the more difficult.

The Athabasca tar sands belong outright to the Alberta provincial government, which owns all subsurface minerals in the province. For many years Alberta's conventional oil fields could have produced more oil than could be marketed at a satisfactory price; the

IMPORTANCE of Athabasca tar and Green River shale reserves is evident if they are compared with known petroleum reserves—the world total and the U.S. (*color*) and Canadian (*dark gray*) shares (*top bar*). All figures are in billions of barrels of oil. Shale-oil estimates vary widely; this one includes shale with more than 15 gallons of oil per ton. Potential reserves of oil made from coal are even greater, but that process is not yet competitive.

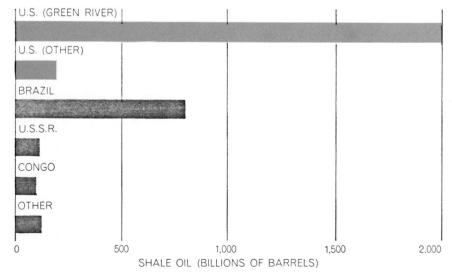

DISTRIBUTION of known potential shale-oil reserves is shown by this chart based on U.S. Geological Survey statistics. Here shales assaying 10 or more gallons per ton are included.

REFINERY is part of the shale research center built by the Bureau of Mines at Anvil Points, Colo., and now operated by a research group. Shale is mined from cliffs in background.

government restricted production and was not interested in bringing the tar sands into large-scale oil production. Now the markets have grown. Petroleum supply and demand are in approximate balance and the introduction of moderate amounts of oil from the tar sands will probably not upset the price of oil. Moreover, some experts believe that liquid petroleum production in Alberta is near its peak, since it is becoming increasingly harder to find new oil fields to replace those whose production is beginning to decline. This is a disputed view, but if it is correct, it behooves the government to bring the tar sands into production in order to keep for Alberta its share of the world oil market.

Under these conditions the Alberta government decided to grant licenses to allow production of oil from tar sands up to 5 percent of the total oil production of the province. It was this decision that led to the granting of a production license to Great Canadian Oil Sands and to the refusal of a license—on political and economic grounds—to the Shell Company.

For the Alberta government such issues are resolved into relatively simple yes-or-no decisions on licenses to operate. For the U.S. Government the shaping of policy on oil shale develop-

ment is more complicated. It requires decisions on the leasing of land, on oil-importation regulations and on taxes.

About 77 percent of the known, commercially exploitable Green River oil shale is on Federally owned land. The other 23 percent is owned by the states or is owned or leased by oil companies. Since 1930 the Federal oil shale lands have been closed to mineral leasing by executive order. The technological advances of the past few years have stimulated increased interest in these lands, and strong pressure is being put on the Department of the Interior to open them up to leasing. Much of the pressure comes from political leaders, chambers of commerce and business groups in the intermountain region. All are anxious to see the oil shales developed as quickly as possible to spur the region's economy. They reason that the oil companies will not spend the millions of dollars required to develop extraction processes unless they own the mineral rights to all the shale. A number of oil companies have pleaded the same case vigorously; others have asked that small leases be granted for research and development. Some independent petroleum engineers and economists support this position.

Others argue the opposite case. They believe that the necessary processes will be developed—are being developed, in fact—whether or not the land is leased, and that after commercial feasibility has been demonstrated the Government will be able to sell mineral leases at much higher prices. Conservationists maintain that the companies' desire to lease is based on their traditional desire to control land, not on any current economic necessity; they point out that some shale land is so rich that one standard 5,120-acre lease tract might contain oil equivalent to 40 percent of the known U.S. petroleum reserves. The conservationists therefore urge the Government to reconsider its standard leasing arrangements and in the meantime to refuse to lease until the oil companies begin to exploit the 23 percent of the shale land to which they now have access.

Shale-oil politics are complicated by the running controversy over oil imports. Petroleum from the Persian Gulf, South America, North Africa and Indonesia can be delivered to U.S. coastal cities at about half the price of domestic oil. The domestic producers of crude oil have succeeded in having laws passed limiting oil imports to about 13 percent of current U.S. consumption. These producers are likely to be as opposed to shale-oil development as they are to crude oil imports.

The attitudes of the diversified oil companies, which refine petroleum products in addition to producing crude, vary with their foreign holdings. The international "have" companies (those owning large reserves of low-cost foreign oil) are generally in no hurry to see shale oil on the market; they would rather bring in their foreign oil. (Many of them own substantial shale reserves and are holding them as a hedge against future developments.) The "have not" companies (diversified producer-refiners without large foreign holdings) are more anxious to develop new sources in the U.S. That is why the Union Oil Company, one of the largest have-nots, developed its shale-oil process. Such companies want assurances that imports will continue to be limited before they will make the capital investments needed to begin shale-oil production.

The question of oil imports affects shale-oil policy even more directly. Each U.S. petroleum refiner receives from a branch of the Department of the Interior an oil-import quota: a permit to bring in a certain quantity of foreign oil at its foreign price. Since this is about half the price of domestic crude, such a permit is equivalent to a cashier's check for half the domestic price of the oil. Import quotas are based on a complicated formula that includes the amount of domestic oil refined by the company. The Department of the Interior could decide that any shale oil refined by a company should or should not be included as domestic oil refined; if it is not, then a refiner who purchases shale oil rather than domestic petroleum will lose part of his import quota. In that case a shale-oil seller would have to offer his product at a price much lower than that of a comparable domestic petroleum to be competitive. Import-quota policy is therefore another administrative lever that could hasten or retard the production of shale oil.

Finally, there are the tax aspects. In order to encourage the search for petroleum, which is a risky business, Congress created the "depletion allowance" that permits a petroleum producer to exempt from Federal income taxation 27½ percent of the value of his crude-oil production. Shale-oil production is similarly encouraged, but according to Internal Revenue Service rulings only by an exemption of 15 percent of the value of the shale as mined and crushed. The Oil Shale Corporation, the Union Oil Company and other companies have argued that both the lower percentage and the fact that it is applied to the rock rather than to the retorted oil are discriminatory; the Union Oil Company has implied that it would have gone into commercial production with its technologically satisfactory process if shale oil had the same tax treatment as crude petroleum.

One branch of the Department of the Interior can encourage or not encourage shale-oil production by leasing or not leasing land; another branch of the department can affect the situation through import-quota policy; the Internal Revenue Service can play a large role by regulating the industry's taxes. Clearly some national goals and ground rules for the shale-oil industry are needed, and a broad Federal policy will have to be enunciated before long.

The tar sands are on their way to commercial development, with a large-scale commercial plant under construction. Shale oil is not far behind; it seems likely that there will be significant production by 1970. In both cases the magnitude of the deposits, their potential value, their possible disruptive effects on existing industries and the resulting political pressures will pose continuing problems for the governments involved.

LAYER OF SAND, about 175 feet thick, to be mined in Alberta is generally buried under less than 150 feet of clay and rock. Here sand is being loaded for a pilot plant; commercial mining will be done by two giant bucket-wheel excavators that can dig 100,000 tons per day.

SHALE DEPOSITS will probably be mined in Colorado by techniques developed by the Bureau of Mines. Large underground "rooms" are mined by drilling, blasting and loading with heavy equipment. The bolts and plates in the roof lessen the likelihood of cave-ins.

19

CLEAN POWER FROM DIRTY FUELS

ARTHUR M. SQUIRES
October 1972

*Considerations of both efficiency and pollution control
suggest that a major effort should be mounted to
generate electric power with turbines operated on
"power gas" produced from coal or oil*

Industrial nations face the need to curtail air pollution caused by the burning of dirty fossil fuels to generate electric power and at the same time a scarcity of clean fossil fuels. The main offender, sulfur, can in principle be removed from stack gases after combustion, but that is at best a difficult and expensive process; several once promising techniques intended to accomplish it have had to be abandoned and others are running into trouble. In this situation the question arises: Why not remove the pollutants before the power is generated, at an earlier stage in the combustion process?

A historic approach to clean energy doing just that is emerging as the best hope for dealing with pollutants from power generation with two dirty fossil fuels: coal and residual oil. The 19th-century industrialist sometimes needed to apply clean heat at high temperatures. He could not use the dirty products of coal combustion for such purposes as heat-treating metals or producing ceramics and fine glassware, and so he resorted to a two-step combustion process in order to obtain clean, intense heat. Instead of supplying air to a shallow bed of coal and burning the coal to convert its carbon and hydrogen into carbon dioxide and water vapor, he blew air and steam through a deep bed of coal to obtain a fuel gas composed mainly of carbon monoxide, hydrogen and nitrogen. Cooled and scrubbed with water to remove dust, the clean gas could be burned itself to provide the desired clean heat.

Even at the dawn of the age of electricity power engineers saw the possibilities inherent in two-stage combustion. Ludwig Mond, the great chemist and industrialist who dominated chemical technology in England until his death in 1909, made improvements in the production of what he called "power gas" and used it to fuel the reciprocating gas engines that generated electricity for his electrochemical works. At the time (1890) the reciprocating gas engine's 8 percent efficiency in converting fuel energy to electricity was not much less than that of the newly invented steam turbine, but the steam turbine made rapid progress. Mond worked on turbines that were driven directly by the hot gases of combustion rather than by steam, but he could not get very far because the metals then available could not withstand gas temperatures high enough to make the gas turbine competitive.

It was not until the 1930's that advances in metallurgy made the gas turbine feasible for some stationary power applications. The gas turbines powering military aircraft that appeared during World War II could handle gases at an inlet temperature of about 500 degrees Celsius. Spurred by the desire for improved aircraft performance, metallurgists have formulated a series of materials able to withstand ever higher temperatures. Moreover, techniques have been introduced for cooling turbine blades and other parts exposed to the high temperature of gases entering the turbine. As a result aircraft turbines of the latest design operate at 1,200 degrees C. during takeoff and at temperatures not much below that while cruising. Land-based machines can now be specified for steady operation at around 1,000 degrees or for operation at 1,100

GAS AND STEAM TURBINES, rated respectively at 25 and 85 megawatts, are combined in a high-efficiency power plant operated by West Texas Utilities at Lake Nasworthy in Texas. The turbines and generators were made by the Westinghouse Electric Corporation. The gas turbine is at the upper center, the steam turbine at the top right; the generators are to the left of each turbine. Hot gases exhausted from the gas turbine are piped to a boiler to produce the steam that powers the steam turbine.

degrees if they are run intermittently to meet peaks in the demand for power. The steady advance in gas-turbine inlet temperatures can be expected to continue, according to engineers of the United Aircraft Corporation, who believe a temperature of 1,300 degrees can be attained in land-based machines within 10 years and up to 1,700 degrees by about 1990.

Gas turbines are also growing steadily in size. Units generating 80 megawatts are now available in Europe and the U.S.; the U.S.S.R. has a 100-megawatt unit; the General Electric Company is studying plans for units of 100 megawatts and larger. United Aircraft has proposed designs for 250- and 300-megawatt machines at projected costs of about $25 per kilowatt. This figure compares with not less than $75 per kilowatt for a steam boiler and steam turbine. (These costs exclude electricity generators and other power-station facilities.) Even at today's sizes and inlet temperatures a complete gas-turbine power in-

stallation costs only about half as much as a steam plant. As sizes and temperatures increase the advantage of the gas turbine should improve.

In spite of the gas turbine's advantage in capital cost, utility men tend to regard gas-turbine power as expensive because it requires expensive clean fuel, and so they utilize gas turbines primarily to supply peak-load power. If a gas turbine is run only about 1,500 or 2,000 hours a year, its low efficiency in converting fuel energy to electricity—commonly about 25 percent—is not an important negative factor. Moreover, for operation at only about 2,000 hours a year it does not pay to provide a boiler to capture heat from the hot gases leaving the turbine. On the other hand, if such a boiler is provided [see illustration on next page], additional power can be recovered with a steam turbine. Even at present gas-turbine temperatures such a gas-steam system has an efficiency beyond the 39 percent attained by the best existing steam-power installations. At

the temperatures projected by United Aircraft, system efficiencies that are well over 50 percent appear to be within sight.

Such considerations create a strong economic incentive to find ways of providing gas turbines with a clean power gas made at high pressure from coal or residual oil (the relatively inexpensive dregs of the refining process). Technologies exist today, even though they are not ideal technologies because they were developed for other purposes, that can be exploited to build experimental installations immediately. It is important to realize that an economic incentive would exist for building such installations even in the absence of pollution advantages. In addition, however, gas-steam systems can provide electricity with absolutely no emission of dust. Moreover, they can be fitted at moderate cost with equipment to suppress emission of sulfur dioxide. Most fortunate of all, they will emit far smaller

amounts of nitrogen oxides than conventional stations (probably by about two orders of magnitude). Because of their higher efficiency they will discharge less waste heat into the environment.

A feasible but imperfect method for producing power gas from coal already exists: the "gravitating bed" gasifier manufactured by Lurgi Gesellschaft für Mineralöltechnik in West Germany. Lurgi has built more than 50 units to provide town gas (for domestic use) or synthesis gas (for making gasoline) and is now putting into operation in Germany a pioneering installation to supply power gas to a turbine that will generate electricity at a rate of 74 megawatts. A steam turbine for 98 megawatts will operate in conjunction with the gas turbine.

In the Lurgi gasifier as it is adapted for power-gas production coal gravitates downward against a rising flow of air and steam introduced through slots in a rotating grate [see illustration on opposite page]. Directly above the grate oxygen in the air is consumed in a shallow combustion zone that converts the last carbon in the descending solid into carbon dioxide. Ash, typically containing a few percent carbon, is discharged below the grate. Hot gases (carbon dioxide, steam and nitrogen) rise upward from the combustion zone through the carbon bed, giving up heat to sustain the endothermic (heat-absorbing) reactions of steam and carbon dioxide with carbon to yield hydrogen and carbon monoxide. When the rising gases have cooled to about 700 degrees C., these reactions effectively cease. Further exchange of heat with the incoming raw coal drives methane and tars from the coal and cools the gases to about 500 degrees. The gases are then quenched with water to reduce their temperature to about 160 degrees.

Sulfur compounds, primarily hydrogen sulfide, can be scrubbed from the crude power gas by any one of several alkaline liquors at a cost far below the cost of scrubbing sulfur dioxide from the stack gases of a conventional power station. There are several reasons for the lower cost. Chemical methods for absorbing hydrogen sulfide are freer of troublesome complications than chemical methods for absorbing sulfur dioxide. Hydrogen sulfide can be converted more readily to elemental sulfur, the byproduct a power station can most readily market (or stockpile in the absence of markets). The molecular quantity of power gas from the Lurgi system is only about 40 percent that of stack gas, and because the power gas is under 20 atmospheres of pressure its volume is only 1.7 percent of stack gas. Rough cost studies indicate that equipment to desulfurize Lurgi power gas can be expected to cost less than $20 per kilowatt, compared with costs of from $40 to $70 for systems now being built to capture sulfur dioxide. Moreover, it should be noted that some of the sulfur dioxide

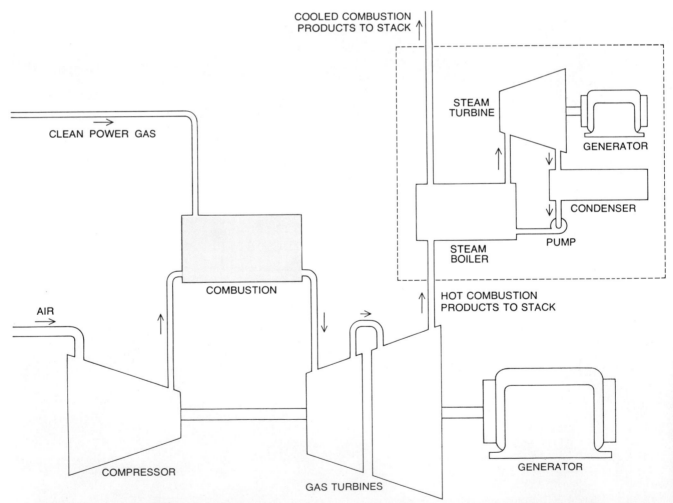

COMBINATION OF GAS AND STEAM TURBINES promises efficient power production with low pollution. Compressed air and a clean power gas are burned and the hot combustion products expand against the blades of the turbine. One section of the turbine drives the compressor, the other the generator. In most current gas-turbine installations spent combustion products are exhausted, with loss of efficiency. Availability of inexpensive power gas could make it economic to add a steam turbine, increasing the efficiency.

systems do not yield sulfur as a by-product and make it necessary to dispose of calcium sulfate. Others yield an undesirable by-product: sulfuric acid.

Although the Lurgi gasifier has been used on a wide range of coals, it may not be suitable for processing some strongly caking coals mined in the eastern U.S. The Commonwealth Edison Company of Chicago has engaged Lurgi to conduct engineering studies for producing clean power gas from Illinois coal, and one task will be to determine the suitability of moderately caking Illinois coal for use in the Lurgi gasifier. As an immediate stopgap answer to the pollution problem at an existing steam power station Lurgi proposes that clean power gas made at 20 atmospheres be let down in pressure through a turbine generating a relatively small amount of electricity and then be used to fire the station's steam boiler.

The Lurgi gasifier has some major faults. One is that the products of combustion of Lurgi power gas will contain a large amount of water vapor. The Lurgi system discharges ash in the form of a loose, nonagglomerated powder, and the temperature in the shallow combustion zone must be kept below a critical level (generally around 1,100 degrees) at which the ash will agglomerate and form clinkers. A large amount of steam must be supplied with the combustion air to keep down the combustion-zone temperature. Most of this steam is converted to hydrogen in the endothermic gasification zone above the combustion zone. Although hydrogen is a desirable constituent of town gas or synthesis gas, the original Lurgi objectives, it gives rise to an undesirable release of water vapor from the stack of a power station burning Lurgi power gas.

Still another input of water vapor arises from the necessity to quench the crude power gas from 500 degrees to 160. Rapid cooling of the gases is essential because they contain chemically active molecules that would polymerize to form heavy tars if the gases were allowed, for example, to pass slowly through a heat exchanger to raise steam. Such tars would deposit on the heat-exchange surfaces and crack to form coke, ruining the heat exchange and eventually plugging the passageway.

Another problem is that the Lurgi gasifier must be fed with coal from which particles smaller than an eighth of an inch have been removed. Therefore if the Lurgi system is to work on run-of-mine coal, a pelletizing or ag-

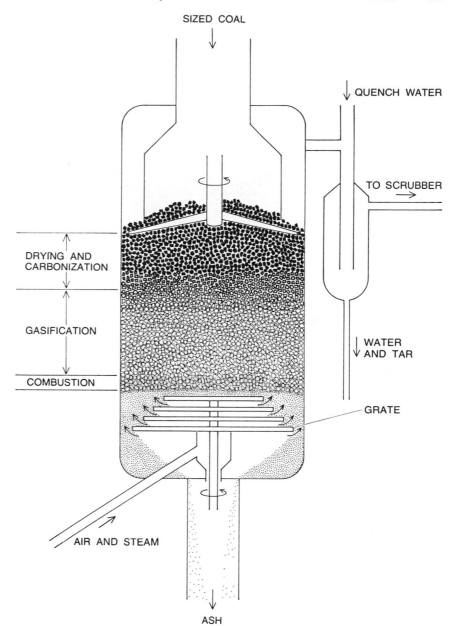

LURGI GASIFIER can make a power gas from coal at high pressure with which to drive a turbine. The coal is treated in a "gravitating bed." In such a bed the descending coal is first dried and turned into coke by the hot gases rising from lower in the bed; in the next layer down carbon is gasified to yield hydrogen and carbon monoxide; below that the remaining carbon is burned to provide the heat for the reactions above. The gas is cooled and then scrubbed to remove hydrogen sulfide. The ashes are expelled by a rotating grate.

glomerating step must be provided to deal with coal fines. Finally, the Lurgi system has a limited coal-processing capacity. The West German installation will require five gasification vessels 13 feet in outside diameter to provide power gas for an electric-generating capacity of 170 megawatts. Scaling up the Lurgi system to larger capacities may be difficult and uncertain.

At the City College of the City University of New York, Robert A. Graff, Robert Pfeffer and I have been looking into a fluidized-bed process that offers possibilities for improvement on the Lurgi system. The fluidized bed provides an attractive technique for bringing run-of-mine coal, merely crushed to sizes smaller than about three-quarters of an inch, into intimate contact with air and steam. In a fluidized bed rising gases buoy up granular material, setting it in motion. Large-scale movement of the solid conveys heat from exothermic zones (such as a combustion zone) to endothermic zones (such as a zone for reaction of steam and carbon dioxide

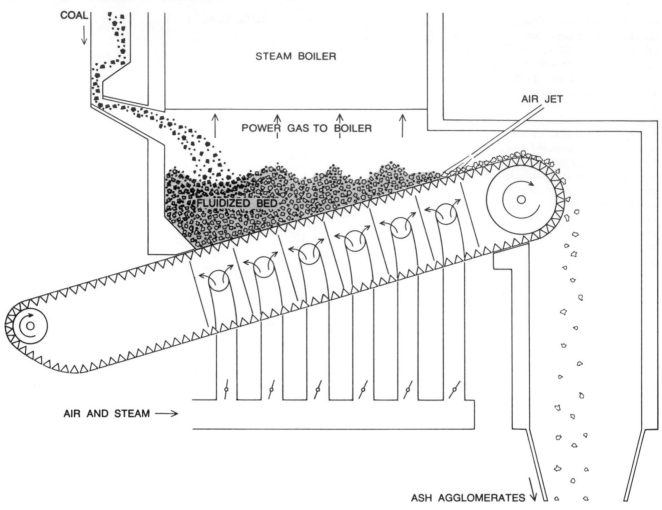

FLUIDIZED BED, in which fuel is buoyed by a rising current of air, is utilized in the Ignifluid boiler developed by Albert Godel and Babcock-Atlantique in France. Coal is gasified with air and a little steam in a fluidized bed. The resulting fuel gas, mainly carbon monoxide and nitrogen, moves into a boiler, where it is burned with additional air to make steam. Godel found that in a bed operated at about 1,100 degrees C. and with air at a velocity of 10 feet per second ashes agglomerate and fall to traveling grate.

with carbon), so that the temperature of a fluidized bed is uniform [see "Fluidization," by H. William Flood and Bernard S. Lee; SCIENTIFIC AMERICAN, July, 1968].

A single fluidized-bed reaction vessel could easily provide power gas for 1,000 megawatts. A fluidized-bed gasifier can deliver gases free of tars or chemically active molecules that would polymerize to tars. No sudden quenching of the gases is required. The amount of steam supplied directly to the gasifier can be a small fraction of the steam needed for the Lurgi system because combustion heat generated near the air inlet is carried away by the motion of the solids. Moreover, there is a good chance that a fluidized-bed gasifier can convert substantially all the steam supplied to it into hydrogen and carbon monoxide, thereby reducing the water-vapor content of the power gas to an absolute minimum. Reducing the water vapor in

power gas reduces the loss of latent heat and significantly increases the efficiency of the process [see illustration on page 185].

The hope that a fluidized-bed gasifier can thus make full use of its steam supply rests on data shown to me in 1958 by F. J. Dent, who was then director of the British Gas Council's research station at Solihull. Dent had fluidized coke with steam at atmospheric pressure in a small tube that was heated from the outside. He found that as he raised the temperature the utilization of the steam climbed to a value of 99.6 percent at 1,050 degrees C. The gas leaving the tube at that temperature was composed almost entirely of hydrogen and carbon monoxide. This was surprising, because only about 60 percent of the same amount of steam passed through a fixed bed of coke at 1,050 degrees was utilized. In the fixed-bed experiment the steam flowed downward through static

granules of coke; in the fluidized-bed experiment the steam flowed upward, buoying the same quantity of coke granules and setting them in rapid motion. Dent explained the difference between fixed-bed and fluidized-bed performance by hypothesizing that each granule of coke in the fluidized bed was reactivated whenever the mixing of the solid in the bed brought the granule near the bottom, into contact with fresh steam. In the fixed bed, on the other hand, the static coke granules near the gas exit, where hydrogen and carbon monoxide left the bed along with unreacted steam, were continuously in contact with hydrogen, which is known to reduce the reactivity of carbon.

I did not see much commercial significance in Dent's data in 1958, because I thought I knew from my earlier participation in a large experiment in gasifying anthracite fines that a fluidized-bed gasifier working at 1,050 degrees would be impracticable. The reason was that if the formation of ash agglomerates at this temperature was to be avoided, the solid in the bed would need to consist of a high percentage of carbon; I saw no way of removing ash from the bed without removing a large amount of carbon along with it. That would give rise either to a prohibitively large loss of carbon or to the necessity of gasifying or burning the carbon residue in an additional step.

What I did not know was that as early as 1955 Albert Godel of Paris had built an ingenious new boiler in which coal was gasified with air and a little steam in a fluidized bed resting on a traveling grate. Our experiment with anthracite had been at a fluidizing-gas velocity of about a foot per second, and any tendency for ash matter to agglomerate was quickly fatal. Godel, on the other hand, made the marvelous discovery that ash matter of substantially all coals forms agglomerates in a bed of coke at about 1,100 degrees that is fluidized by air at 10 feet per second. The ash agglomerates are roughly spherical and remain in motion until they grow to a size such that they sink to the traveling grate and are carried to an ashpit. Godel had therefore neatly solved the problem of separating ash from a bed of carbon, which had appeared to prevent the application of Dent's favorable levels of steam utilization.

If our projected experiments confirm the hope that a fluidized bed at high pressure and 1,100 degrees can afford comparable steam utilization (or even merely utilization that is high in comparison with other gasification techniques), then the development of hardware to exploit Godel's discovery at high pressure would emerge as a goal of prime importance. In one possible design ash agglomerates would be removed from a shaft blown with air [*see illustration at right*]. Godel has used such a shaft to reduce the carbon level in ashes from relatively nonreactive anthracites with a high ash content. The conical bottom of the fluidized bed, fitted with a large number of horizontal pipes for the introduction of air and steam, has been used successfully by the British National Coal Board in a low-temperature coal-carbonization process.

As fine particles are released by the consumption of carbon a "fast fluidized bed" would be established in the upper part of the coal-gasification vessel. This is our name for a new fluidization technique developed recently by a second Lurgi firm. A fine powder is fluidized at a velocity much higher than would previously have been considered attractive for such a solid. The trick is to provide a large cyclone and a pipe to recirculate powder to the bottom of the bed. Powder flows at a large throughput upward through the region occupied by the fast fluidized bed, which lacks the sharply defined upper surface of the usual fluidized bed.

At City College we are also conducting experiments to determine the feasibility of cleaning power gas from our

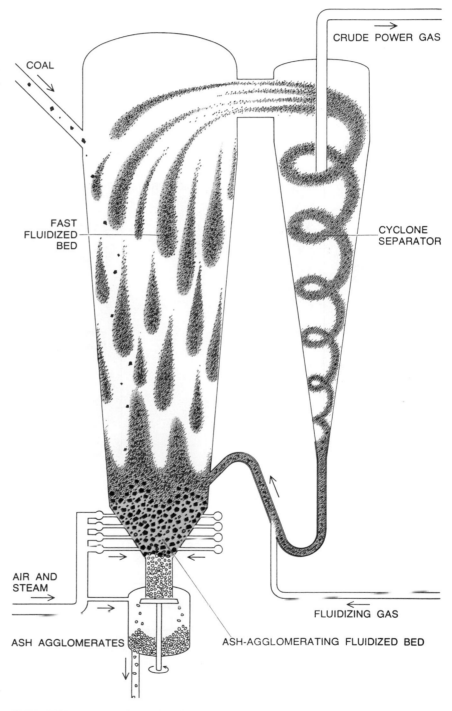

FLUIDIZED-BED-GASIFIER DESIGN being studied by the author and his colleagues would exploit Godel's ash-agglomeration principle at high pressure. The clinkers sink through a central shaft into which air is introduced to burn remaining traces of carbon. In the ash-agglomerating bed above the grate, air and steam admitted through numerous spoke-like tubes fluidize and gasify coal, releasing fine carbon particles that are further gasified in a fast fluidized bed above. Particles blown out of the vessel along with the gas are collected in the cyclone separator and are returned, boosted by fluidizing gas, to the gasification vessel.

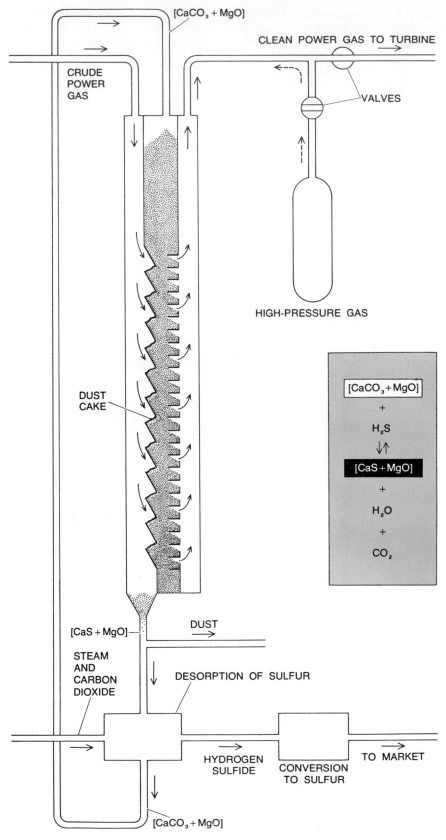

[CaCO₃ + MgO]

CRUDE
POWER
GAS

CLEAN POWER GAS TO TURBINE

VALVES

HIGH-PRESSURE GAS

DUST
CAKE

$$[CaCO_3+MgO]$$
$$+$$
$$H_2S$$
$$\downarrow\uparrow$$
$$[CaS+MgO]$$
$$+$$
$$H_2O$$
$$+$$
$$CO_2$$

[CaS + MgO]   DUST

STEAM
AND
CARBON
DIOXIDE

DESORPTION OF SULFUR

HYDROGEN   CONVERSION   TO MARKET
SULFIDE   TO SULFUR

[CaCO₃ + MgO]

SULFUR AND DUST can be simultaneously removed from power gas by granules of half-calcined dolomite in a "panel bed" formed by louvers. The crude gas enters at one side of the bed, leaving a deposit of dust at the entry surface; the sulfur in the gas is absorbed by the dolomite and the clean gas leaves the other side of the bed. Sulfur-laden dolomite and the dust deposits are dislodged from time to time by a blast of high-pressure gas (*broken arrows*). The sulfur is removed from the dolomite as hydrogen sulfide and the sulfur-free dolomite is recycled. The equation (*right*) shows the reversible sulfur-dolomite reaction.

proposed ash-agglomerating gasifier at high temperature. Our hope is to find a technique for removing dust and hydrogen sulfide simultaneously by passing the dirty power gas through a bed of a granular solid derived from dolomite rock.

Dolomite is the double carbonate of magnesium and calcium: $CaCO_3 \cdot MgCO_3$. If the rock is heated gently, the magnesium carbonate decomposes to release carbon dioxide and form magnesium oxide. The resulting solid is a half-calcined dolomite, $[CaCO_3 + MgO]$. (The brackets are used to indicate that it is not a true chemical species but an intermingling of microscopic crystallites of calcium carbonate and magnesium oxide.) The solid is porous, so that all the crystallites are accessible to a power gas being treated for the removal of sulfur. We have found that the calcium carbonate crystallites in this solid are extraordinarily reactive to hydrogen sulfide. Since calcium carbonate is almost completely unreactive to hydrogen sulfide, it would appear that it is the porosity of the half-calcined dolomite that accounts for its reactivity. The absorption reaction can readily be conducted in reverse to desorb sulfur (as hydrogen sulfide) from the solid, which can then be used again.

The panel-bed filter we have developed for the desulfurizing reaction can also be highly efficient in removing dust [see illustration at left]. If the filter is operated at relatively low velocities, a cake of the filtered dust forms on the surfaces of the bed of granular solid where the gas enters the filter. We have observed efficiencies beyond 99.9 percent for the filtration of an airborne suspension of 1.1-micron particles through a cake of dust resting on sand.

Other techniques for gasifying coal must also be considered. Texaco Incorporated has studied a slagging gasifier in which finely powdered coal is reacted in a chamber with air and steam at high pressure; the company is said to have conducted a large test around 1957, but the results have not been published. The crucial questions have to do with the life of the refractory material lining the chamber and the efficiency of the utilization of carbon. If these are good, the efficiency of the Texaco gasifier can be expected to fall somewhere between the efficiencies of the Lurgi system and the fluidized bed. The City College techniques for cleaning power gas at high temperatures would be applicable to Texaco gas.

The Bureau of Mines and several research organizations under contract to the Office of Coal Research are studying coal-gasification techniques with the objective of producing pipeline gas, which is substantially pure methane. That is a more difficult task, for which one wants a processing scheme that maximizes the quantity of methane made directly from coal. This usually requires several steps in which the coal is brought into contact with gases, the flows of solid and gases being countercurrent. Such schemes are not useful for making power gas, since the exact heating value of power gas does not matter very much; the main objectives, instead, are to achieve simplicity and the lowest possible cost.

Power gas can also be produced from residual oil. In Texaco's "partial oxidation" process oil is reacted with oxygen and steam at high pressure to furnish synthesis gas (hydrogen and carbon monoxide) for conversion to ammonia. The Shell Oil Company licenses a similar process. An experimental installation in which the oxygen is replaced by air could provide early experience in firing a gas turbine with power gas made from oil.

An experiment in which oil is gasified with air at atmospheric pressure in a fluidized bed of lime is in progress at British Esso's research laboratories at Abingdon in England [*see bottom illustration on next page*]. The design of the fluidized bed is unsuitable for operation at high pressure, but the Abingdon chemistry might be conducted in another design, perhaps one incorporating a fast fluidized bed. Esso uses a roasting process to drive sulfur dioxide from sulfided lime; this unfortunately exposes the solid to such a high temperature that reactivity suffers, limiting its usefulness to only a few reaction cycles.

The Abingdon experiment has shown, however, that the hydrocarbons produced by cracking residual oil over lime

ENERGY BALANCE (COAL INPUT=100)	CONVENTIONAL STEAM BOILER AND STEAM TURBINE WITHOUT RECOVERY OF SULFUR	GAS TURBINE AT 1,538 DEGREES CELSIUS FOLLOWED BY CONVENTIONAL STEAM BOILER AND TURBINE WITH SULFUR RECOVERY	
		LURGI GASIFIER (GAS CLEANED BY SCRUBBING)	ASH-AGGLOMERATING FLUIDIZED BED (GAS CLEANED AT HIGH TEMPERATURE)
ELECTRICITY SENT OUT	39.5	45.0	50.5
HEATING VALUE OF SULFUR	0	1.0	1.1
LOSS OF SENSIBLE HEAT IN STACK GASES	5.0	4.6	4.7
LOSS OF LATENT HEAT (WATER VAPOR)	3.8	14.0	4.5
LOSS OF HEAT AT CONDENSER AND OTHER HEAT EXCHANGERS	47.7	28.4	35.2
LOSS OF UNBURNED FUEL AND HEAT LEAKAGE	2.0	5.0	2.0
MECHANICAL LOSSES AND POWER FOR AUXILIARY EQUIPMENT	2.0	2.0	2.0
EFFICIENCY, ALLOWING CREDIT FOR HEATING VALUE OF SULFUR	39.5 PERCENT	45.5 PERCENT	51.1 PERCENT

ENERGY BALANCES calculated for three systems demonstrate the advantage of improved power-generating machinery even with the Lurgi system and the further advantage of a new fuel-conversion system. In the first case the gain is the result of heat conservation; in the second case it is the result of reducing the water vapor in stack gases. The calculation of the final efficiency (*bottom*) of the systems represented by the two right-hand columns allows a credit for the heating value of the sulfur that is recovered.

COMPOSITION (PERCENT BY VOLUME)	LURGI GRAVITATING-BED GASIFIER	ASH-AGGLOMERATING FLUIDIZED-BED GASIFIER
METHANE	4.4	.5
CARBON MONOXIDE	10.7	31.8
HYDROGEN	15.7	15.6
CARBON DIOXIDE	10.7	.5
WATER VAPOR	27.8	.5
NITROGEN	30.2	50.4
HYDROGEN SULFIDE	.5	.7
HEATING VALUE (B.t.u. PER CUBIC FOOT)	129	157

COMPOSITION AND HEATING VALUE of crude power gas from the Lurgi system and a proposed fluidized-bed system are compared. The large water-vapor content of the Lurgi-system gas means a loss of latent heat; the high carbon dioxide content requires a scrubbing liquor that absorbs hydrogen sulfide rapidly but not carbon dioxide. The choice of a liquor is complicated by the fact that some sulfur in the Lurgi gas is in the form of organic compounds. The fluidized-bed gas is easier to clean and has a higher heating value.

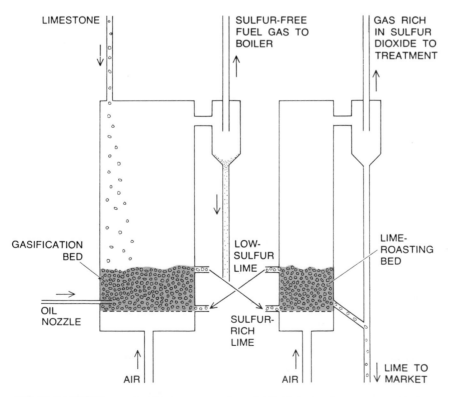

OIL IS GASIFIED at atmospheric pressure in a fluidized bed of lime in a system being developed by British Esso. Sulfur in the oil is removed as calcium sulfide and a clean hydrocarbon fuel gas is produced. The sulfided lime is roasted in a second vessel, releasing a sulfur-rich gas that is subjected to further treatment. The low-sulfur lime produced by the roasting can be recycled a few times but must be supplemented by makeup limestone.

are free of sulfur. This result is encouraging for the potential of a scheme we have studied at City College: a system for cracking residual oil in a coke-agglomerating fluidized bed that has a superposed, contiguous bed of fine dolomite. The process would yield pea-sized coke pellets of low sulfur content, a light aromatic oil and a rich fuel gas, and in addition a power gas that would fire the gas turbine that furnishes air to the process.

We call the system an "Oilplex," and we believe it illustrates an exciting opportunity for the power station of the future [see illustration on opposite page]. The station will include both gas-turbine and steam-turbine power cycles. It will be big. It will handle great quantities of fuel. Such a station will provide a new context for chemical processing in which the chemical engineer will play under a new set of rules. First, air at high pressure will be available "free" from the compressor of the gas turbine (or from an axial-flow compressor of the type used in a gas turbine). Second, the gas turbine will constitute a sink for excess gas at high pressure, including a lean fuel gas—even a gas so lean as normally to be considered incombustible. Third, the steam-raising system will constitute a sink for excess heat at temperature levels so high that its discharge would cause a serious loss of efficiency in normal chemical-engineering practice. The availability of such a heat sink will make feasible a broad range of new chemistries. These new rules constitute an invitation to invention, and I have no doubt that the power station of the future will become the scene for chemical operations leading to a greater efficiency in the utilization of fuel. In this context removing dust and sulfur from the power gas that is supplied to the gas turbine will seem to be a mere incidental.

The situation has an ironic aspect. The 19th-century industrial chemist habitually resorted to transformations brought about by the application of intense heat, but after 1900 research on high-temperature chemistry gave way to attempts to achieve chemical transformations with the aid of catalysts at ever lower temperatures. The new rules will revive interest in high-temperature chemistry. Perhaps we shall see a return to some historic technologies: the Brin process used for making oxygen between about 1885 and 1910, in which oxygen was absorbed from air by barium oxide; a pressurized version of the steam-iron process for hydrogen, as proposed by

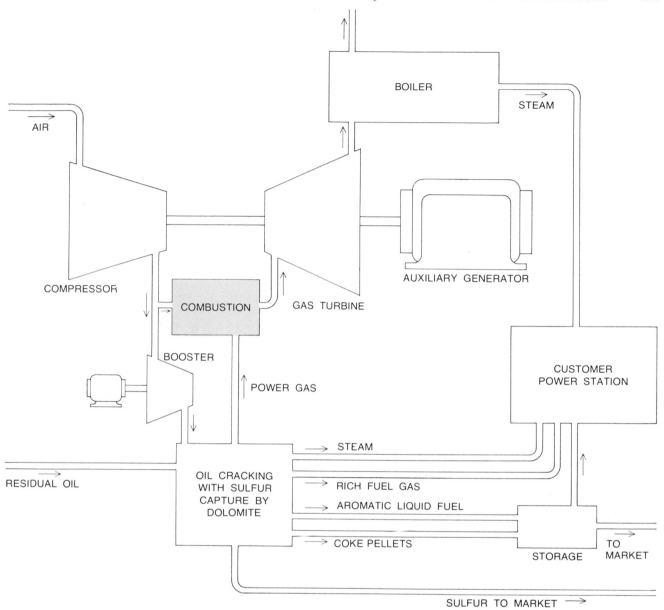

AIR

BOILER

STEAM

COMPRESSOR

COMBUSTION

GAS TURBINE

AUXILIARY GENERATOR

BOOSTER

POWER GAS

CUSTOMER
POWER STATION

RESIDUAL OIL

OIL CRACKING
WITH SULFUR
CAPTURE BY
DOLOMITE

STEAM

RICH FUEL GAS

AROMATIC LIQUID FUEL

COKE PELLETS

STORAGE

TO
MARKET

SULFUR TO MARKET →

"OILPLEX" SYSTEM would produce gas-turbine and steam-turbine power and marketable by-products and would be highly efficient. It is based on the cracking of residual oil at about 700 degrees C. and an elevated pressure and would yield three low-sulfur products: pea-size coke pellets, a light aromatic oil and a rich fuel gas. Each of these would be in a quantity equivalent to about a fifth of the oil's heating value. The remaining heating value would be available as high-level heat (steam) and as clean power gas for firing a gas turbine that compresses air for the system and provides auxiliary power. The rich fuel gas and steam would be used promptly to generate power; the light oil and coke could be stored for use during periods of peak demand or be shipped to market.

the Institute of Gas Technology; the shifting of carbon monoxide to hydrogen by the action of steam in the presence of lime, a popular development objective until about 1930.

The new rules will lead naturally to new kinds of fuel-processing complexes serving society's energy needs. Eventually it will become too wasteful to burn raw coal directly in order to generate electricity: the hydrogen chemically bound in raw coal will be too valuable simply to burn to steam and send up a stack as water vapor. In the end power generation must be based on a coke resi-

due from an operation in which fuels of higher value are "creamed off" the raw fossil fuel. For example, a "Coalplex" might produce pipeline gas, a light aromatic liquid such as gasoline, electricity and low-sulfur coke for metallurgy. There might even be roles in such a complex for Mond's dream of a fuel cell that would convert power gas to electricity at an efficiency beyond 50 percent and for a magnetohydrodynamic electric generator operating on power gas made from coke. Such advanced techniques for the production of electricity will have to compete, however, with gas tur-

bines that afford comparable efficiencies.

None of this will happen unless the engineering community devotes much more attention to fuel technology. Higher temperatures for gas turbines are on the way through the efforts of men now engaged in gas-turbine development and design, and will probably be achieved even without additional Government support. But good technologies for providing clean power gas will be developed only if a great deal more talent is recruited for the work, and probably only with a large input of Federal money.

20

FUEL CELLS

LEONARD G. AUSTIN
October 1959

Devices that convert chemical energy directly into electricity, thus circumventing the inefficiency of the heat engines used to drive electric generators, are now under intensive development

Civilization gets most of the energy it consumes from the energy of the chemical bonds in coal, petroleum and natural gas. But in the process of putting that chemical energy to work, it throws most of it away. The energy is first converted, by combustion of the fuel, into heat. The heat is then converted, by several kinds of heat engine, into mechanical energy, which may in turn be converted into electricity. These transformations yield less than half of the original energy as useful work. But the fault does not lie in the energy-converting machines. Though the most modern central power-stations manufacture electricity at an efficiency of only 35 to 40 per cent, the performance of boilers, turbines and generators has been improved over the years until it now approaches the maximum which can be expected from the heat-steam-electricity cycle. Internal-combustion engines have reached a corresponding peak of efficiency at 25 to 30 per cent, and high-temperature gas turbines are approaching their limit at 40 per cent. The ceiling on efficiency is partly imposed by the second law of thermodynamics, which dictates the downhill flow of energy throughout the cosmos. At the operating temperatures of heat engines—temperatures set by the strength of materials and the economics of heat transfer—this law decrees that more than half of the original chemical energy must be lost in irrevocably wasted heat. Further energy is lost to the friction that is encountered in any machine.

With conventional energy-converting technology approaching a dead end, power engineers are seeking ways to bypass the heat cycle and to convert the chemical energy of fuels directly into electricity. The notion is not a new one. In 1839 the English investigator Sir William Grove constructed a chemical battery in which the familiar water-forming reaction of hydrogen and oxygen generated an electric current. Fifty years later, also in England, the chemists Ludwig Mond and Carl Langer developed another version of this device which they called a fuel cell. But the dynamo was

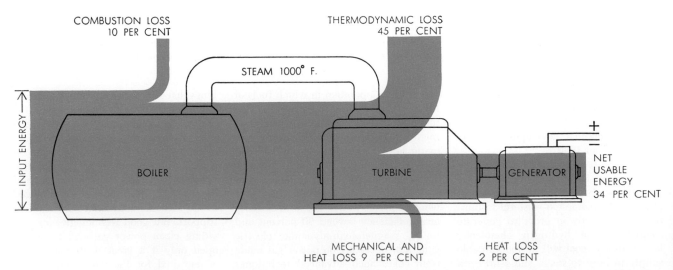

EFFICIENCY OF FUEL CELL is potentially greater than that of conventional generating equipment. Fuel cell (*top left*) converts 45 to 75 per cent of its input energy (*color*) into electricity compared to 34 per cent for typical steam turbogenerators (*bottom*).

then coming into its own, and although research continued spasmodically the difficulties encountered deterred any extensive effort to develop fuel cells. Since 1944, however, the fuel cell has come under active development again, and at least one is now in practical use.

The first voltaic pile and its modern descendant, the dry battery, are fuel cells in a sense: they convert chemical energy directly into electricity. But they use expensive "fuels" such as zinc, lead or mercury that are refined by the expenditure of considerable energy from fossil fuels or hydroelectric power. A true fuel cell uses the basic fuel directly, or almost directly. In theory the fuel cell may approach 100 per cent efficiency in converting the chemical energy of the fuel into electricity; actual efficiencies of 75 per cent—more than twice that of the average steam power-station—are quite feasible.

Fuel cells hold other attractions for contemporary engineering. An artificial satellite, for example, requires a small, light battery that can deliver a high electrical output. The fuel cell can meet these specifications from energy compactly stored in a liquid or gaseous fuel and in oxygen, as opposed to the cumbersome plates of an ordinary battery.

In public transportation the electric motor possesses a number of advantages over the gasoline or Diesel engine, including higher speed, more rapid acceleration, quietness and absence of noxious exhaust gases. However, the high capital cost of the electrical distribution system has caused a decline in electric transport during the past two decades. A few battery-powered delivery trucks still operate in some cities, but they suffer competitively from the low power-to-weight ratio of their lead batteries and from the long periods required for recharging. A fuel cell that could operate efficiently on gasoline or oil and could be "recharged" by the filling of its tank might reverse the present trend toward gasoline and Diesel locomotives, trucks and buses. Ultimately fuel cells might make the quiet, non-air-polluting electric automobile a reality.

The realization of these attractive possibilities will require a great deal of development work. To understand some of the difficulties to be surmounted, let us consider the fuel cell in which hydrogen and oxygen combine to produce an electric current and water.

As everyone knows, hydrogen and oxygen burn to produce water. They do so because separately they possess more

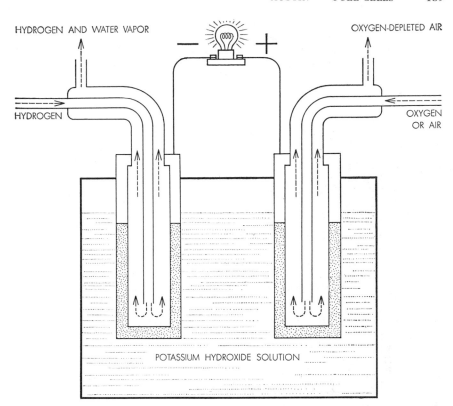

HYDROGEN-OXYGEN FUEL CELL, shown schematically, consists of two porous carbon electrodes (*dotted areas*) separated by an electrolyte such as potassium hydroxide. Hydrogen enters one side of the cell; oxygen, the other. Atoms of both gases diffuse into the electrodes, reacting to form water and to liberate electrons which flow through the circuit.

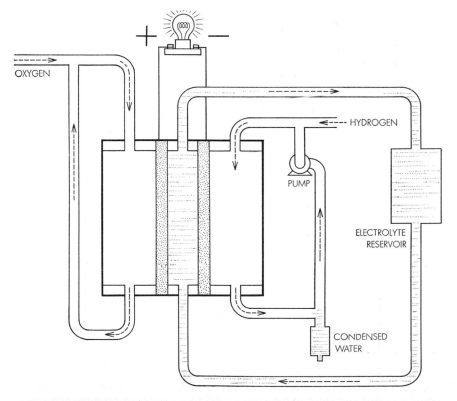

ANOTHER HYDROGEN-OXYGEN CELL was developed recently by Francis T. Bacon of the University of Cambridge. His cell consists basically of an electrolyte solution held between two thin electrodes of porous nickel. Gases under pressure diffuse through the electrodes and react with the electrolyte, which is held in tiny pores in the opposite surface.

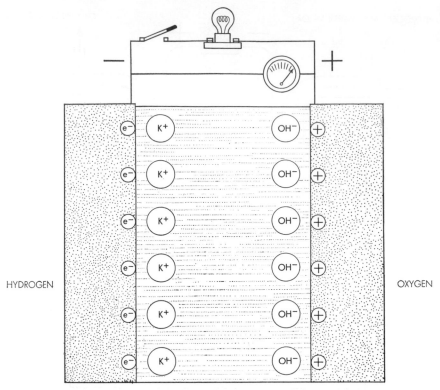

WHEN FUEL-CELL CIRCUIT IS OPEN, the hydrogen electrode accumulates a surface layer of negative charges that attracts positively charged potassium ions in the electrolyte solution. Similarly, the oxygen electrode attracts negative ions to balance its positive charge. These layers prevent further reaction between the gases and the electrolyte.

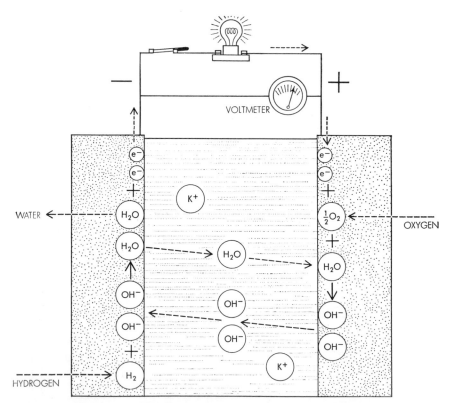

WHEN CIRCUIT IS CLOSED, the gases and electrolyte react to produce a flow of electrons. A catalyst embedded in the electrode dissociates hydrogen gas molecules into individual atoms, which combine with hydroxyl ions in the electrolyte to form water. The process yields electrons to the electrode. The electrons flow through the circuit to the positive electrode, where they combine with oxygen and water to form hydroxyl ions. The ions complete the circuit by migrating through the electrolyte to the negative electrode.

energy than water and therefore "prefer" to exist in combination. However, at ordinary temperatures and pressures, additional "activation" energy is needed to raise the molecules to the energy state at which the reaction will ignite; this energy barrier ordinarily prevents the reaction from proceeding at room temperature. Activation energy may be illustrated by the following analogy. In a large number of people there may be one man capable of clearing a seven-foot high-jump bar, several capable of clearing six feet, thousands who can jump five feet, hundreds of thousands who can jump four feet, and so on. Molecules are like that with respect to their individual energy content: only a small fraction of them have high energies at room temperature. If the energy barrier for a reaction is comparable to an eight-foot hurdle, no reaction occurs. Raising the temperature has the effect of increasing the "jumping ability" of the molecules until some can clear the activation-energy barrier. At about 500 degrees centigrade a hydrogen-oxygen mixture will combine explosively, and the chemical energy is converted to heat.

In a hydrogen-oxygen fuel cell essentially the same chemical reaction is made to take place, but the reaction is stepwise at a lower energy of activation for each step. This can be considered as analogous to requiring the molecule to jump several barriers only four feet high, instead of one barrier eight feet high. The reaction thus proceeds quite quickly at room temperature. The cell is also designed so that one of the essential steps in the reaction is the transfer of electrons, from the negative terminal of the cell to the positive terminal, by an electrical connection. The flow of electrons, which is of course an electric current, can be used to drive an electric motor, light a lamp or operate a radio. Instead of the chemical energy of the reaction being immediately converted to heat, a large part of it is carried by the electrons, which can give up the energy as useful electrical work.

The cell consists of two porous electrodes separated by an electrolyte, which in this case is a concentrated solution of sodium hydroxide or potassium hydroxide. On the negative side of the cell, hydrogen gas diffuses through the electrode; hydrogen molecules (H_2), assisted by a catalyst embedded in the electrode surface, are adsorbed on the surface in the form of hydrogen atoms (H). The atoms react with hydroxyl ions (OH^-) in the electrolyte to form water, in the process giving up electrons to the electrode; the water goes into the electrolyte. This reaction is also aided by the catalyst.

The flow of these electrons around the external circuit to the positive electrode constitutes the electric output of the cell and supports the oxygen half of the reaction. On the positive side of the cell oxygen (O_2) diffuses through the electrode and is adsorbed on the electrode surface. In a somewhat indirect reaction the adsorbed oxygen, plus the inflowing electrons, plus water in the electrolyte, form hydroxyl ions. Here again a catalyst helps the reaction to proceed. The hydroxyl ions complete the circle by migrating through the electrolyte to the hydrogen electrode [*see bottom illustration on opposite page*].

If the external circuit is open, the hydrogen electrode accumulates a surface layer of negative charges that attracts a layer of positively charged sodium or potassium ions in the electrolyte; an equivalent process at the oxygen electrode similarly balances its accumulated positive charge. These electrical "double layers" prevent further reaction between the gases and the electrolyte. The presence of the electrical layers provides the potential that forces the electrons through the external circuit when connection is made.

When the circuit is closed and the resistance across the external circuit between the electrodes is high, the reaction proceeds at a moderate rate, and a high percentage of the reaction energy is released as electricity, with only a little lost as heat. Part of the energy is expended at all times, however, in driving the chemical reactions over the barrier of the activation energies of the reactions inside the cell, and this energy appears as heat within the cell. The function of the catalysts in the electrodes is to lower the energy barriers, thus decreasing the amount of useful energy that is converted to heat. As resistance in the external circuit goes down, the current flow increases and a rising proportion of the energy is consumed in overcoming the energy barriers within the cell. With the increase in the reaction rate, heat losses go up rapidly. At zero resistance (short circuit) the reaction proceeds so rapidly that it becomes equivalent to combustion, producing only heat. Thus the reaction energy of the fuel cell resembles the energy of water behind a dam. By allowing the water to escape slowly through the blades of a turbine, we compel it to do useful work. If we open the floodgates, the water gushes out without performing any work.

In addition to the expenditure of energy needed to drive the reaction over activation-energy barriers, the fuel cell must consume some energy to force gas molecules through the electrodes to the reaction area, to transport hydroxyl ions from one electrode to the other and to overcome the electrical resistance of the electrodes themselves. These losses reduce the cell voltage below the theoretical ideal. A common working standard of voltage efficiency for fuel cells, however, is 75 per cent.

In practice, at the present stage of the art, other considerations loom larger than simple efficiency. For instance, a standard criterion is the power output per cubic foot of cell when the cell is converting 75 per cent of the thermodynamically available energy into electricity. Another important factor is the length of time a cell can operate before its performance falls off due to the deterioration of the electrode or the electrolyte.

In a typical hydrogen-oxygen cell the electrodes consist of porous carbon impregnated with catalysts: fine particles of platinum or palladium in the hydrogen electrode and cobalt oxide, platinum or silver in the oxygen electrode. To prevent flooding of the pores by the electrolyte, which would cut down the active surface, the electrodes are waterproofed with a layer of paraffin wax about one molecule thick. This thin film allows ions and individual water molecules to pass through to the internal surfaces of the electrode, but prevents the water from flooding the pores. To bring the electrodes closer together and thus speed ion transport, the electrodes are typically arranged as concentric tubes or adjacent plates. Cells of this type developed by Karl Kordesch of the National Carbon Company have won the distinction of being the first practical fuel cells; the U. S. Army uses them to power its "silent sentry" portable radar sets. Some have been in operation for more than a year with no appreciable decline in performance.

Low-temperature hydrogen-oxygen cells are limited in their applications, although they may find widespread special uses. Hydrogen is a costly fuel and the power-to-volume ratio of the cell (about one kilowatt-hour per cubic foot) makes it too bulky for use in vehicles.

An obvious way to improve the performance of hydrogen-oxygen cells is to

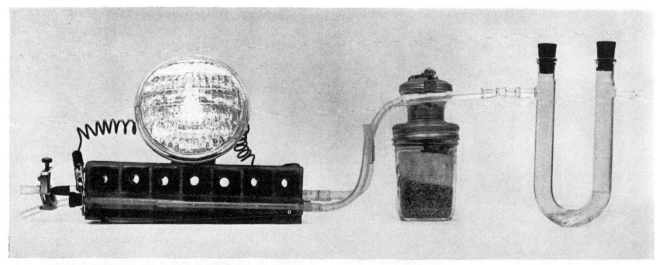

LABORATORY MODEL of a simple fuel cell reacts hydrogen with the oxygen in air. Hydrogen is generated in jar at right by dropping water onto calcium hydride. The gas then flows through carbon tubes in a block of Lucite, where it reacts with an electrolyte. The electrolyte in turn reacts with the oxygen that diffuses into other carbon tubes. The power output of the cell is three watts.

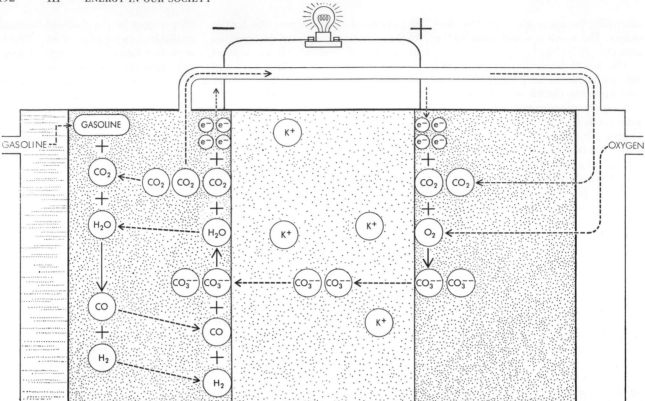

HIGH-TEMPERATURE FUEL CELL operates above 500 degrees centigrade and uses fuels such as gasoline or natural gas. The cell contains two electrodes tightly pressed against a "solid" electrolyte, which is usually a molten salt such as potassium carbonate. The fuel in the cell is usually broken down (by reaction with steam and carbon dioxide) to produce hydrogen and carbon monoxide. These gases then diffuse into the negative electrode, where they react with carbonate ions in the electrolyte, forming carbon dioxide and water and giving up electrons. The electrons flow through the circuit to the positive electrode, where they combine with oxygen and carbon dioxide to form carbonate ions. The carbonate ions complete the cell's electrical circuit by flowing back to the negative electrode.

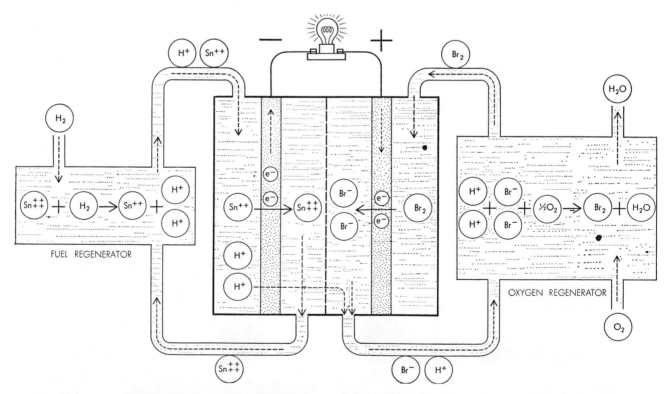

REDOX CELL is so named because in it the fuel and oxygen react with oxidizing and reducing agents in two so-called regenerators. The hydrogen reduces (adds electrons to) tin ions, which then give up electrons to the electrode. The electrons flow to the positive electrode. On the positive side of the cell, oxygen oxidizes (takes electrons from) bromide ions, converting them to bromine. In turn, the electrons flowing into the positive electrode reduce the bromine to bromide ions, which are then returned for regeneration.

operate them at higher pressures (which speed up gas transport through the electrodes) and higher temperatures (which speed up the electrochemical reactions). By appropriate design and insulation the waste heat liberated in the cell can be used to maintain the cell at the proper operating temperature.

The best-known cell of this type has been developed by Francis T. Bacon of the University of Cambridge. It operates at temperatures up to 250 degrees C. with gas pressures up to 800 pounds per square inch. The electrodes are of porous nickel about 1/16-inch thick and are usually in the form of disks or plates. A thin surface layer on the electrode, penetrated by very fine pores, constitutes the reaction area. The electrolyte, a concentrated solution of potassium hydroxide, can enter these pores, but pressure differences within the electrode prevent it from flooding the larger pores in the body of the electrode, through which gas percolates to the reaction area. The Bacon cell produces six times as much power per cubic foot as the low-temperature cell. With this relatively high output, the cell should have bright prospects as a standby source of auxiliary power in airplanes. It can deliver as much as 150 watts per pound, as against 10 watts for the lead-acid storage batteries currently in use.

To produce economical power on a large scale, fuel cells must "burn" cheap fuels such as natural gas, vaporized gasoline or the mixture of gases obtained from the gasification of coal. The extraction of energy from such fuels calls for operating temperatures above 500 degrees C. Since aqueous electrolytes would boil away at these temperatures, the electrolyte consists of some molten salt, usually a carbonate of sodium or potassium mixed with lithium carbonate to lower the melting point. In the most efficient of these cells, the electrolyte is held in a matrix of porous refractory material. The electrodes, made of a variety of metals or metallic oxides, are tightly pressed against the "solid" electrolyte.

In these cells the fuel does not necessarily combine directly with oxygen as hydrogen does in the hydrogen-oxygen cell. Usually the fuel is "cracked" to hydrogen and carbon monoxide by reaction with steam and carbon dioxide, which the fuel cell produces as by-products. This cracking may be conducted outside the cell, or inside the cell on the electrode surface. In the current-generating reaction the hydrogen and carbon mon-

oxide diffuse into the cell at the negative electrode, where they react with carbonate ions in the electrolyte, forming carbon dioxide and water and giving up electrons to the electrode. At the positive electrode, oxygen or air takes up the electrons flowing in from the external circuit and reacts with the carbon dioxide to produce the carbonate ions. The migration of carbonate ions through the electrolyte from the positive to the negative electrode completes the circuit [see top illustration on opposite page].

High-temperature fuel cells, intensively investigated only since World War II, still perform poorly. The best of them produce no more than half a kilowatt per cubic foot—half the yield of the low-temperature hydrogen-oxygen cell and a twelfth the yield of the Bacon cell. However, the progress already made in hydrogen-oxygen cells suggests that further research can improve the performance of high-temperature cells by a factor of 10 or more.

In the "redox" cell—named for reduction and oxidation—the fuel and oxygen do not react directly with each other. Rather, the fuel and oxygen are made to react with other substances in "regenerators" outside the cell to produce chemical intermediates, which in turn generate current in the cell. The over-all reaction is the same as that of combustion, however, because the intermediates are regenerated. A typical cell of this type, developed in England under the leadership of Sir Eric Rideal, utilizes tin salts and bromine as intermediates. The fuel reduces (i.e., adds electrons to) tin ions, which then give up the added electrons to the negative electrode and return to react with more fuel. The oxygen similarly oxidizes (i.e., takes electrons from) bromide ions, converting them to bromine, which then takes up electrons from the positive electrode and returns as bromide ions for regeneration [see bottom illustration on opposite page]. A similar cell, using titanium salts instead of tin, is under development by the General Electric Company.

In principle redox cells should be able to achieve high efficiencies. The intermediates can be chosen so that the electrode reactions are rapid and yield high currents with little energy loss. With suitable catalysts and operating conditions it may be possible to carry out the regeneration reactions at satisfactory efficiencies. However, the problems involved in the regenerators have not yet been solved. Moreover, the two electrolyte systems must be separated from

each other by an impermeable membrane to keep the bromine from mixing and reacting with the tin or titanium ions. All known membranes of this sort have a rather high electrical resistance. It has not yet been demonstrated that the redox cell represents any improvement over simpler types.

Engineers are working on a number of other reaction cycles and combinations of cycles. Each of them presents knotty technical difficulties. But the fundamental processes of electrochemistry are fairly well understood, probably because electrochemical experiments require no expensive apparatus and thus fit well into university budgets. The future development of the fuel cell is thus a question of applied rather than basic research.

Low-temperature and moderate-temperature hydrogen-oxygen cells should come into use during the next few years as low-weight, easily "charged" batteries. The development of strong, lightweight containers, perhaps made of plastic-impregnated glass fibers, would reduce the poundage if not the cubic footage needed to store the reaction gases. Where cost is not too important, the hydrogen could be stored as solid lithium hydride and the oxygen as solid calcium superoxide. Moderate-temperature cells may well be used to power submarines. Such vessels, like nuclear submarines, could cruise for extended periods without surfacing and would be far quieter in operation than nuclear vessels.

Hydrogen-oxygen cells may also furnish a means of capturing the power of the sun. Investigators at the Stanford Research Institute have developed a catalytic process for decomposing water into hydrogen and oxygen by sunlight. Used in conjunction with fuel cells, which would recombine the hydrogen and oxygen into water, a solar photolysis plant covering two square kilometers of desert could provide as much energy as a 100,-000-kilowatt power station in continuous operation. The over-all efficiency of such a plant, estimated at 25 per cent, would be two and a half times that of present solar batteries or solar boilers.

In auxiliary installations at nuclear power-stations, hydrogen-oxygen cells may help to bring down the cost of nuclear power. The high capital cost of nuclear-power plants requires that they be operated at near-peak capacity if they are to yield cheap electricity. Power generated during daily or seasonal periods of low demand might be used to electrolyze water into hydrogen and oxygen, which would then be made to yield the stored energy via fuel cells during peak-demand periods. The large volume of gas generated might be stored in "sausage skins" of plastic film buried underground to eliminate wind damage.

If the performance of high-temperature fuel cells can be substantially improved, large-scale electric power might be generated near sources of cheap natural gas. The power produced would of course be in the form of direct rather than alternating current, and though high-voltage direct current is somewhat easier to transmit than alternating current, fuel cells apparently cannot produce high voltages. Large numbers of cells must be connected in series, and above 700 volts there is electrical leakage through the insulation separating the terminals. Large-scale power from fuel cells should therefore find its first application in electrochemical processes such as the production of aluminum, which utilize large quantities of direct current at low voltage. Electrochemical industries may then congregate near natural-gas sources as they now cluster around hydroelectric installations.

The hydrogen-oxygen cell may make an unorthodox contribution of its own to the chemical industries. With slight modifications a low-temperature cell can employ instead of hydrogen a liquid fuel such as methyl (wood) alcohol, which it oxidizes to formic acid. The power output is very low, but the formic acid is almost free of impurities. Ethyl alcohol can similarly be oxidized to acetic acid, an important raw material in the manufacture of plastics and lacquers. Such processes, amounting to a sort of electrolysis in reverse, may prove useful in the manufacture of other industrial chemicals. Since the energy released would be extracted as electricity rather than heat, unwanted side reactions could be held to a minimum. It would be ironic if fuel cells should find their principal application in the production of chemicals rather than of power.

Attempts to construct cells that would operate directly on coke or coal, the cheapest of fuels, have been disappointing. Far more promising is the mixture of hydrogen, carbon monoxide and hydrocarbons that can be made from coal. With suitable equipment to remove tar and grit, it could be piped directly from the gasification plant and used hot. However, a really low-cost process for generating gas from coal has yet to be devised.

A high-output fuel cell operating on liquid fuel would find immediate application in trucks and locomotives. The technology of electric traction is well developed and is waiting for a compact power-unit utilizing a cheap fuel that can be easily stored and pumped. Designers will have to figure out a simple way to warm up the cell to operating temperature; the high-temperature cell is not a self-starter.

The possibilities of fuel cells are great, but not all these possibilities are going to be realized. Although much small-scale development work remains to be done, some cells have reached the stage where further progress will require large amounts of money and faith. No doubt some of this money will be wasted, and some of the faith will be misplaced. Fuel cell development is not a field for the faint-hearted.

EXPERIMENTAL FUEL CELL operates at room temperature and atmospheric pressure. It produces 20 watts of electrical power, enough to light three bicycle lamps. The two Lucite boxes composing the cell contain an electrolyte solution and nine porous carbon electrodes, four for hydrogen and five for oxygen. This cell, and the one on page 191, were photographed in the laboratories of the National Carbon Company in Parma, Ohio.

21

FAST BREEDER REACTORS

GLENN T. SEABORG AND JUSTIN L. BLOOM

November 1970

*Nuclear reactors that use fast neutrons to produce
more fuel than they consume are a promising approach
to producing electric power with a minimum of
strain on energy resources and the environment*

The need to generate enormous additional amounts of electric power while at the same time protecting the environment is taking form as one of the major social and technological problems that our society must resolve over the next few decades. The Federal Power Commission has estimated that during the next 30 years the American power industry will have to add some 1,600 million kilowatts of electric generating capacity to the present capacity of 300 million kilowatts. As for the environment, the extent of public concern over improving the quality of air, water and the landscape hardly needs elaboration, except for one point that is often overlooked: it will take large amounts of electrical energy to run the many kinds of purification plants that will be needed to clean up the air and water and to recycle wastes.

A related problem of equal magnitude is the rational utilization of the nation's finite reserves of coal, oil and gas. In the long term they will be far more precious as sources of organic molecules than as sources of heat. Moreover, any reduction in the consumption of organic fuels brings about a proportional reduction in air pollution from their combustion products.

Nuclear reactors of the breeder type hold great promise as the solution to these problems. Producing more nuclear fuel than they consume, they would make it feasible to utilize enormous quantities of low-grade uranium and thorium ores dispersed in the rocks of the earth as a source of low-cost energy for thousands of years. In addition, these reactors would operate without adding noxious combustion products to the air. It is in the light of these considerations that the U.S. Atomic Energy Commission, the nuclear industry and the electric utilities have mounted a large-scale effort to develop the technology whereby it will be possible to have a breeder reactor generating electric power on a commercial scale by 1984.

Nuclear breeding is achieved with the neutrons released by nuclear fission. The fissioning of each atom of a nuclear fuel, such as uranium 235, liberates an average of more than two fast (high-energy) neutrons. One of the neutrons must trigger another fission to maintain the nuclear chain reaction; some neutrons are nonproductively lost, and the remainder are available to breed new fissionable atoms, that is, to transform "fertile" isotopes of the heavy elements into fissionable isotopes. The fertile raw materials for breeder reactions are thorium 232, which is transmuted into uranium 233, and uranium 238, which is transmuted into plutonium 239 [*see illustrations on next page*].

We have mentioned that breeding occurs when more fissionable material is produced than is consumed. A quantitative measure of this condition is the doubling time: the time required to produce as much net additional fissionable material as was originally present in the reactor. At the end of the doubling time the reactor has produced enough fissionable material to refuel itself and to fuel another identical reactor. An efficient breeder reactor will have a doubling time in the range of from seven to 10 years.

Two different breeder systems are involved, depending on which raw material is being transmuted. The thermal breeder, employing slow neutrons, operates best on the thorium 232–uranium 233 cycle (usually called the thorium cycle). The fast breeder, employing more energetic neutrons, operates best on the uranium 238–plutonium 239 cycle (the uranium cycle). Nonproductive absorption of neutrons is less in fast reactors than it is in thermal reactors, resulting in a decrease in the doubling time.

The concept of the breeder reactor is almost as old as the idea of the nuclear chain reaction. In the early stages, soon after World War II, many types of breeder reactor were visualized. Some were thermal and some were fast. Another important differentiation involved the type of coolant employed to carry off the heat of fission and deliver it to a power-generating system. Among the coolants proposed were water and molten salts for thermal breeding and inert gas (such as helium), liquid metal (such as sodium) and steam for fast breeding.

In the U.S. and several other countries decisions were made rather early that a fast breeder reactor cooled with liquid metal was the most attractive concept to pursue. This concept is known to atomic energy workers as the LMFBR (liquid-metal-cooled fast breeder reactor). Since the greater part of breeder-reactor development is now proceeding on the basis of this concept, this article is mainly devoted to the liquid-metal fast breeder. A serious alternative effort is being pursued, chiefly by utility companies, to develop the technology of a gas-cooled fast breeder reactor using pressurized helium as the coolant. In the U.S. two thermal-breeder-reactor concepts operating on the thorium cycle are being developed: the light-water breeder reactor at the Bettis Atomic Power Laboratory and the molten-salt reactor at the Oak Ridge National Laboratory.

In the design of a liquid-metal-cooled fast breeder reactor several features are noteworthy. The core of a fast reactor can be quite small. For economic reasons the reactor must be operated at a much higher power density than ordinary fission reactors are. The active core volume is therefore only a few cubic meters and is roughly proportional to the power output. The power density is about .4 megawatt per liter.

To carry off the heat while maintaining the fuel at a reasonable temperature sodium must flow through the core at a rate of tens of thousands of cubic meters per hour. In order to provide channels

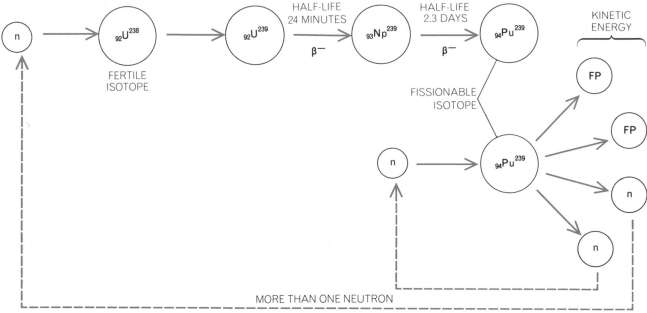

URANIUM CYCLE for breeding in a fast breeder reactor relies on fast, or highly energetic, neutrons. In the cycle an atom of fertile uranium 238 absorbs a neutron and emits a beta particle to become neptunium, which then undergoes beta decay to become fissionable plutonium 239. When an atom of plutonium 239 absorbs a neutron, it can fission, releasing energy, fission products (*FP*) and at least two neutrons. One of the neutrons is needed to continue the chain reaction, but the others are available to transform a fertile isotope into a fissionable one, thereby "breeding" fuel. Within a few years a breeder doubles its original fuel inventory.

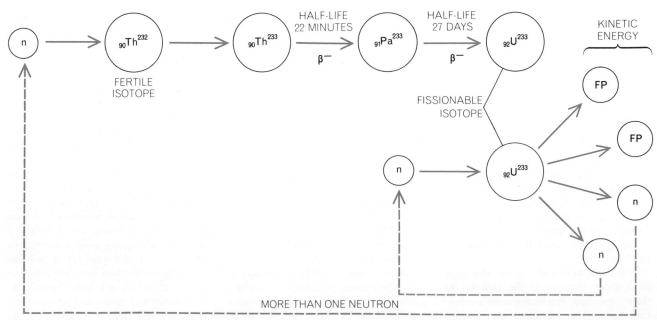

THORIUM CYCLE of breeding is similar to the uranium cycle except that it works best in a thermal breeder reactor, where it relies on thermal, or relatively slow, neutrons. Thorium 232 is the fertile isotope that becomes protactinium and then uranium 233.

for the flow of sodium the fuel is divided into thousands of slender vertical rods, which are usually referred to as pins. Each pin is sealed in stainless steel or another high-temperature alloy.

The fuel is preferably in a ceramic form such as oxide or carbide, since these ceramics are stable during long exposures to heat and radiation, have very high melting points and are relatively inert in liquid metal. The fissionable component of the fuel can be enriched uranium 235, plutonium 239 or a mixture of the two. Typically the fuel is diluted with uranium 238, so that part of the breeding takes place within the core. The uranium 238 also serves a safety function in the core, which we shall explain in more detail below. For maximum economy and performance the fuel must be able to accept neutron irradiation at many times the rate common in present nuclear reactors of commercial scale. Furthermore, the consumption of fuel between reprocessing steps is to be at least twice that of thermal reactors. The development of a fuel meeting these stringent criteria requires the testing of numerous fuel combinations in reactors and accounts for a major element of the breeder development program.

A second major feature of a fast breeder reactor is the "blanket" that surrounds the core. Much of the breeding takes place here, and so the blanket consists of uranium 238 in stainless-steel tubes. (It can be uranium that is depleted in the isotope uranium 235 as a result of enrichment procedures designed to make uranium 235 as a fuel for nuclear reactors; large stocks of such depleted uranium are now available.) Since there is a certain amount of fission in the blanket, it too must be cooled by the flow of sodium. The blanket also has an important nuclear function. Not all the neutrons entering the blanket are captured; a fairly large proportion are reflected back into the reactor core, enhancing the neutron economy there.

The sodium coolant has excellent heat-transfer characteristics. Moreover, it can be used at a fairly low pressure even though it emerges from the reactor at a temperature (above 500 degrees Celsius) that with water would give rise to high pressures. Indeed, the sodium pressure arises solely from the force required to maintain the high rate of flow through the maze of tubes in the core and the blanket. Compared with coolants such as water and gas, sodium requires low pumping power. It is not particularly corrosive to the reactor.

Sodium does, however, have certain

CORE AND BLANKET of a fast breeder reactor, Experimental Breeder Reactor II, are the heart of the breeding operation. The dark hexagonal area is the core, where fuel elements can be installed and removed by the gripper mechanism at right center. Rods clustered at left center are connected to control rods in the core. Around the core is the blanket, consisting of rods containing uranium 238, which is converted to plutonium during the breeding.

EXPERIMENTAL BREEDER REACTOR II is operated by the U.S. Atomic Energy Commission at the National Reactor Testing Station in Idaho. The primary components of the nuclear reactor are under the floor in a tank that is 26 feet high and 26 feet in diameter and contains liquid sodium, which is the coolant. Vertical assembly at right center contains mechanisms for operating the control rods and for handling fuel elements within the reactor. The reactor is used to test fuels and materials for breeder reactors of commercial scale.

198

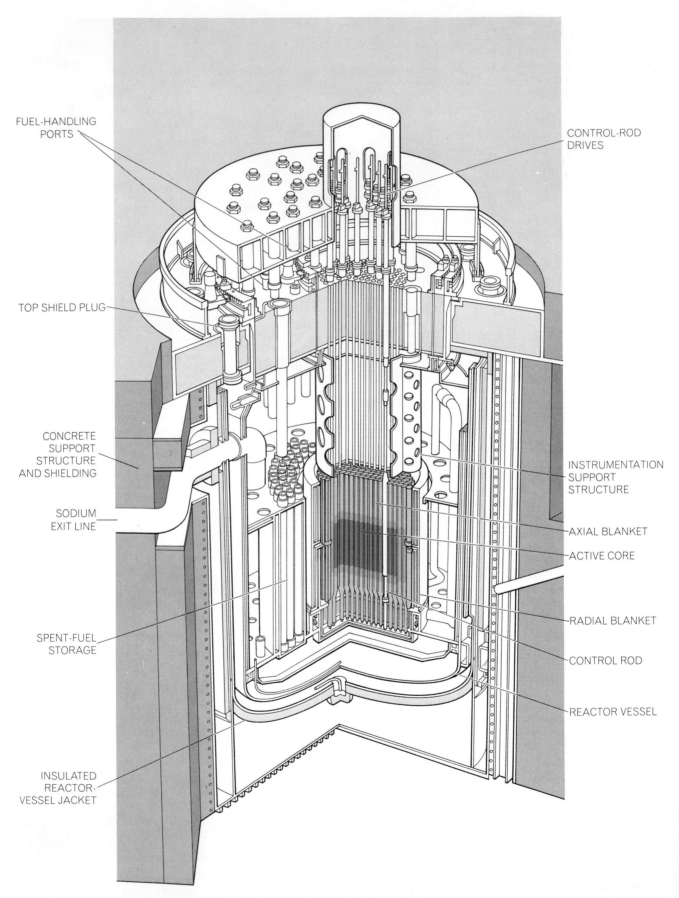

FUEL-HANDLING PORTS

CONTROL-ROD DRIVES

TOP SHIELD PLUG

CONCRETE SUPPORT STRUCTURE AND SHIELDING

SODIUM EXIT LINE

INSTRUMENTATION SUPPORT STRUCTURE

AXIAL BLANKET

ACTIVE CORE

SPENT-FUEL STORAGE

RADIAL BLANKET

CONTROL ROD

REACTOR VESSEL

INSULATED REACTOR-VESSEL JACKET

LIQUID-METAL REACTOR of the fast-breeder type is depicted on the basis of a design for a demonstration plant that would produce some 500 megawatts of electricity. A full-scale commercial plant, scheduled for operation by 1984, would be of 1,000-megawatt capacity. This design is of the loop type, meaning that the reactor proper, which is contained in a large tank of liquid sodium, is separated from the primary heat exchangers and the associated pumps by loops of piping through which sodium coolant flows.

disadvantages that markedly influence the design of a reactor. Since sodium is opaque, provision must be made for the maintenance and refueling of the reactor without benefit of visual observation. Sodium is of course highly reactive chemically, and it becomes intensely radioactive when it is exposed to neutrons, even though its "cross section," or neutron-absorption capacity, is relatively low. Hence the sodium must be kept out of contact with air or water, and radiation shielding must be used to protect workers who are near sodium that has been through the core and blanket of an operating reactor.

Interspersed through the core region are numerous rods with safety and control functions. They maintain the power output at the desired level and provide the means for starting and stopping the reactor. The rods are filled with neutron-absorbing material such as boron carbide or tantalum metal.

All materials have markedly lower neutron-absorption probabilities for fast neutrons than for thermal ones. The lower cross sections reduce the effectiveness of fast-reactor control rods of sizes comparable to those in thermal reactors. On the other hand, a large amount of excess fuel is present in the core of a thermal reactor to compensate for the fuel that will be consumed by fission and to counteract the poisoning effect of the fission products. (The fission products capture neutrons without yielding significant amounts of energy.) With extra fuel there must be extra controls. Fast breeder reactors require fewer control rods because their greater effectiveness in converting uranium 238 to fissionable plutonium 239 compensates for depletion of the initial fuel charge and because fast neutrons are not absorbed by fission products as much as thermal neutrons are.

During a fission reaction not all the neutrons are released at the precise instant that each nucleus disintegrates. A small proportion of the neutron population is created by the decay of fission products. One thus distinguishes delayed neutrons from the "prompt" neutrons emitted directly by the fissioning nuclei. It is the delayed neutrons that keep the chain reaction from escalating into an essentially instantaneous propagation of one generation of neutrons to the next.

The fraction of delayed neutrons depends appreciably on which nucleus is fissioning. Most thermal reactors are fueled with uranium 235, whereas the fast breeder will be fueled with plutoni-

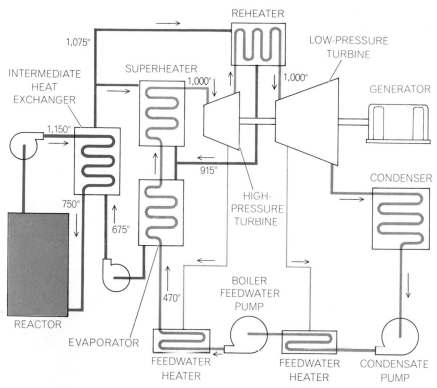

FLOW PLAN for a liquid-metal fast breeder reactor entails pumping the sodium coolant (*color*) through the reactor, where it becomes radioactive, and then to an intermediate heat exchanger, where heat is transferred to a separate stream of sodium (*dark gray*) that is not radioactive. The heat of that stream is put into a water and steam cycle (*light gray*) that is employed to generate electricity. The numerals give temperatures in degrees Fahrenheit.

um 239. The fraction of delayed neutrons produced by the fission of uranium 235 is about .0065 and by plutonium 239 fission about .003. The smaller fraction of delayed neutrons present in a fast reactor is not of major concern under normal operation. It does increase the sensitivity of the reactor to adjustments of the control rods and to other inputs that affect reactivity, such as temperature variations in the core.

Two different designs of containers for the core-and-blanket assembly and the primary heat-transfer system are under consideration: the pot type and the loop type [*see illustrations on next page*]. In the pot type a large tank filled with sodium encloses (1) the reactor vessel, (2) sodium pumps that take sodium from the pool and move it through the core and blanket and (3) intermediate heat exchangers that transfer heat from the radioactive sodium to another sodium stream. In the loop type only the reactor vessel is filled with sodium; the liquid metal is circulated by pumps through heat-exchange loops mounted outside the reactor container. The pot type has the advantage of a much greater heat capacity in the event of pump failure, but it also requires a much larger inventory of sodium.

In both the pot and the loop design the liquid-metal fast breeder reactor employs a complex heat-transfer arrangement to isolate the sodium that flows through the core from the steam-generating equipment. This is the role of the intermediate heat exchangers. They transfer heat from the radioactive sodium to nonradioactive sodium, which then flows through the steam generator. Subsidiary streams of sodium are required to superheat the steam and to reheat it from time to time as it works against the blades of the turbine.

Both the pot and the loop design require sealing of the part of the structure that is in direct contact with the radioactive core and blanket. In routine operation there would be no release of radioactive fission products to the environment. Because of the inherently low pressure of the sodium coolant, the reactor vessel and its associated piping need be designed to withstand only moderate operating stresses, in marked distinction to the pressure vessels and other primary-system components of a pressurized-water reactor, a boiling-water reactor or a gas-cooled reactor.

At present the pot design seems to be attracting the most interest. It is inherently a less complicated arrangement than the loop design. Nonetheless, it

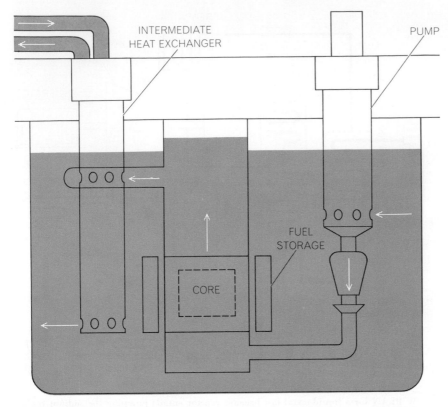

POT SYSTEM is one of two designs for containing the core-and-blanket assembly of the reactor and the primary heat-transfer system. The pot is a tank that is filled with sodium and also contains the reactor, pumps that take sodium from the pool and move it through the reactor, and intermediate heat exchanger where heat is transferred to nonradioactive sodium.

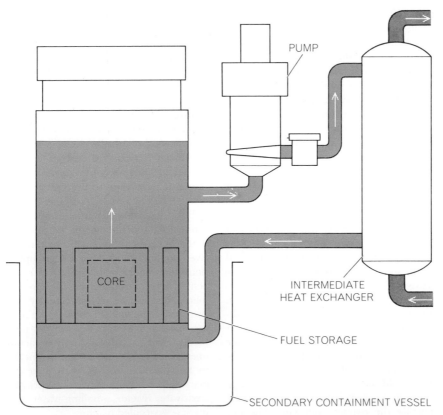

LOOP SYSTEM has most of its heat-exchange apparatus outside the reactor. Only the reactor vessel is filled with sodium, which is circulated by pumps through heat-exchange loops mounted outside the reactor vessel. In the present state of breeder-reactor technology both of the designs in the schematic illustrations on this page are being pursued.

does present certain problems, notably in gaining access to the reactor for maintenance.

The gas-cooled fast breeder reactor is receiving attention (comparatively modest so far) as a parallel and complementary concept to the liquid-metal fast breeder. Gas-cooled thermal reactors are already in operation, and a gas-cooled fast breeder would not represent a large step in terms of coolant technology. The design and testing of the fuel for a gas-cooled fast breeder have much in common with the work on fuel for the liquid-metal fast breeder.

The essential difference between the two fast breeders is that the gas-cooled one uses helium gas at a pressure of from 70 to 100 atmospheres rather than molten sodium to transport the heat from the reactor core to the steam generators. Since the gas does not become radioactive and cannot react chemically with the water in the steam generator, there is no need for an intermediate heat exchanger. The resulting simplification of the system is a helpful offset against the need to design for a higher coolant pressure with gas.

The use of helium as a coolant has other special advantages for a fast breeder reactor. Helium does not interact with the fast neutrons in the reactor core, resulting in both simplified control of the reactor and enhanced breeding of new fissionable fuel from fertile material. In addition helium is transparent and chemically inert, providing visibility during refueling and maintenance operations, a simpler engineering design and freedom from corrosion problems.

In a gas-cooled fast breeder the reactor core, helium circulators and steam generators are all contained in a prestressed-concrete reactor pressure vessel. These major components and their arrangement are almost the same as in a thermal gas-cooled reactor.

The development of a gas-cooled fast breeder reactor could result in substantial additional savings beyond those that would be achieved by liquid-metal fast breeders. Neutrons are moderated, or slowed, less in helium than they are in sodium. Hence the doubling time is short. It is also possible to foresee the development of a gas-cooled fast breeder with a direct power cycle wherein the gas coolant flows from the reactor directly to a gas turbine that drives the electrical generator. Such a cycle should help to reduce the capital cost of fast breeder reactors.

Three major reactors will carry the burden of the Atomic Energy Commission's program to develop a liquid-metal fast breeder reactor. Two of them are already in operation: the Experimental Breeder Reactor II (EBR-II) and the Zero Power Plutonium Reactor (ZPPR).

EBR-II is a fast-neutron test reactor operated by the Argonne National Laboratory at the commission's National Reactor Testing Station in Idaho. This reactor, which as of July 1 had a cumulative record of more than 35,000 megawatt-days of operation, is the focal point of the program of testing fuels and irradiating materials for the liquid-metal fast breeder reactor. At present almost 800 experimental fuel pins and more than 100 capsules containing hundreds of structural, control-rod and shield-material specimens are being irradiated in the reactor. EBR-II achieved its design power of 62.5 megawatts (thermal) last year.

The Argonne National Laboratory is also operating the ZPPR, which went into operation last year. (Zero power in this context means that the reactor does not generate a significant amount of heat.) It is the nation's largest zero-power fast reactor and the only one in the world that is big enough and has a large enough inventory of plutonium (at least 3,000 kilograms) to allow full-scale mock-ups of the plutonium fuel arrangements that will be used in the large commercial breeders envisioned for the 1980's and beyond. The reactor will provide important information on the behavior of neutrons in breeder-reactor cores.

The third reactor is now being designed on the basis of data obtained from EBR-II, the ZPPR and smaller facilities. Called the Fast Flux Test Facility, it will operate at a very high neutron flux (defined as the number of neutrons passing through a square centimeter of area per second) to produce the radiation effects on fuel and structural materials that will take place in a commercial breeder reactor. The reactor, which will cost about $100 million, will operate at a power level of 400 megawatts (thermal) with no conversion to electric power. It will be built at the Atomic Energy Commission's site in Richland, Wash.; construction should start next year and full power should be achieved by the middle of the decade.

Following the lessons learned in the development of thermal reactors, the commission has taken the first steps toward construction of one or more liquid-metal fast-breeder demonstration plants. The cost will be shared by the Government and industry. The first such plant, with a capacity of from 300 to 500 megawatts (electric), will accumulate valuable operating experience with both the reactor and the power-conversion equipment. Such a plant will not compete economically with existing nuclear or conventional plants because of its relatively small size and early stage of development. The full-scale liquid-metal fast breeder reactor of the 1980's will be rated at 1,000 megawatts (electric) or more.

Much consideration is being given to safety in the fast breeder development program. The waste products of fission are the elements in the middle of the periodic table that represent the split nuclei of the fuel atoms. Many isotopes

| | NAME | COUNTRY | POWER | | INITIAL OPERATION | TYPE (POT OR LOOP) |
			MEGAWATTS (THERMAL)	MEGAWATTS (ELECTRICAL)		
OPERATING	BR-5	U.S.S.R.	5	—	1959	LOOP
	DFR	U.K.	60	15	1959	LOOP
	EBR-II	U.S.	62.5	20	1964	POT
	FERMI	U.S.	200	66	1963	LOOP
	RAPSODIE	FRANCE	40	—	1967	LOOP
	SEFOR	U.S.	20	—	1969	LOOP
	BOR-60	U.S.S.R.	60	12	1970	LOOP
UNDER CONSTRUCTION	BN-350	U.S.S.R.	1,000	150	1971	LOOP
	PFR	U.K.	600	250	1972	POT
	PHENIX	FRANCE	600	250	1973	POT
	BN-600	U.S.S.R.	1,500	600	1973/75	POT
	FFTF	U.S.	400	—	1974	LOOP
PLANNED	KNK-II	WEST GERMANY	58	20	1972	LOOP
	JEFR	JAPAN	100	—	1973	LOOP
	PEC	ITALY	140	—	1975	MODIFIED POT
	SNR	WEST GERMANY	730	300	1975	LOOP
	DEMO #1	U.S.	750-1,250	300-500	1976	NOT DECIDED
	JPFR	JAPAN	750	300	1976	LOOP
DECOMMISSIONED	CLEMENTINE	U.S.	.025	—	1946	LOOP
	EBR-I	U.S.	1	.2	1951	LOOP
	BR-2	U.S.S.R.	.1	—	1956	LOOP
	LAMPRE-I	U.S.	1	—	1961	LOOP

LIQUID-METAL FAST REACTORS built or planned are summarized. Those that produce electricity have far less capacity than the 1,000-megawatt commercial fast-breeder plant that the development program of the U.S. seeks to have in operation by 1984.

of these elements are radioactive. The permanent control of the fission products has been recognized as essential from the early stages of reactor development and is routinely achieved in the fuel cycle. In addition to making fission products, fast breeders will also contain large amounts of plutonium, which in certain forms is radiologically toxic. The standard procedure in both thermal and fast reactors is to ensure the confinement of all potentially hazardous substances under all foreseeable conditions, including earthquakes.

Perhaps the most significant safety feature of commercial nuclear reactors is that they are self-regulating, that is, they are designed to compensate inherently for any incident that could lead to an un-

intentional increase of power output. In water reactors the compensation is usually achieved through the decrease in reactivity caused by the decrease in the density of water as its temperature increases. In a fast reactor the change in density of the coolant with temperature may lead in the opposite direction.

Compensation is provided in a fast breeder by the Doppler effect, which results from the increase in the rate at which neutrons are absorbed by uranium 238 as the temperature of the fuel in the core rises. Since a sudden power increase will necessarily be accompanied by increased fuel temperature, there will be increased neutron absorption and a consequent tendency toward reduction of power. A small sodium-cooled fast reac-

tor has been built in Arkansas with private funds to measure this effect under conditions analogous to those in a large power reactor. It is called the Southwest Experimental Fast Oxide Reactor.

The fact that a decrease in coolant density or coolant content can result in an increase in reactivity leads to other safety considerations for sodium-cooled fast reactors. For example, if one postulates a bubble of gas or another void whereby an area of the core might overheat without detection, some of the fuel pins in the area would be expected to fail. Further disturbances of flow might ensue. A continued sequence of such events would not necessarily result in an automatic shutdown of the reactor. Thus

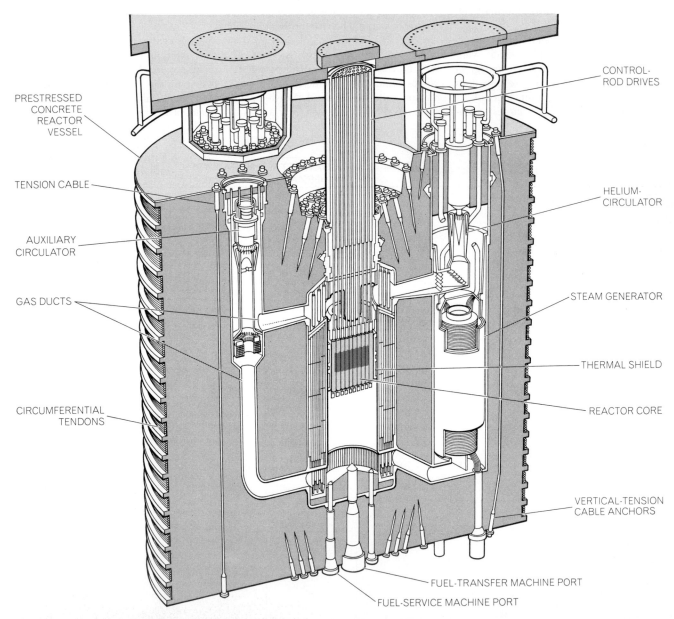

GAS-COOLED REACTOR is depicted in the form it might take for a demonstration breeder plant with a capacity of 300 megawatts of electric power. The chief difference between such a reactor and a **liquid-metal one is that the coolant here is helium at high pressure instead of liquid sodium at low pressure. Because of the pressure the reactor is contained within a prestressed-concrete vessel.**

it is necessary to preclude by design the propagation of fuel-pin failures.

This control can be achieved by a number of techniques. The addition of a moderator such as beryllium oxide increases the magnitude of the neutron Doppler effect. A change in the ratio of coolant to fuel can reduce the void effect. Other methods include distributing the fuel in such a way that the potential reactivity of coolant voids is decreased by increasing the number of neutrons leaking from the core.

In a gas-cooled fast reactor there is no problem with voids because a bubble cannot form in gas. There is, however, another condition to be guarded against: a sudden loss of coolant pressure resulting from an event such as rupture of the pressure vessel. This possibility is minimized by the use of the prestressed-concrete type of reactor vessel.

Having made sure that perturbations in normal operating conditions do not escalate, one looks to the possibility of other problems. One is a loss of cooling by mechanical blockage. Such accidents have occurred, but they will become less likely as engineering experience is gained. In this type of accident any significant release of fission products is precluded by providing several layers of structural containment for the entire reactor system.

Another possibility is an increase in power to the point where heat is being generated faster than it can be carried away by the coolant. Such an accident took place in EBR-I some years ago. Here again the answer has been found in improved design. Even beyond this, structural containment sufficient for any foreseeable accident will be provided.

Much consideration is also being given to environmental factors in the design of fast breeder reactors. Because fast breeder reactors will operate at far higher temperatures than are encountered in contemporary water reactors, they will have greater thermodynamic efficiency. Today's water reactors operate at an overall efficiency of about 32 percent, meaning that 32 percent of the thermal energy produced is converted to electrical energy. Modern fossil-fueled plants operate at about 39 percent efficiency. Hence water reactors add more waste heat to the environment per unit of electrical energy produced than fossil-fueled plants do. Fast breeder reactors will probably attain efficiencies equal to that of the most modern fossil-fueled plant, thereby reducing the nuclear waste-heat problem.

The release of radioactivity to the air from fast breeders will be near zero.

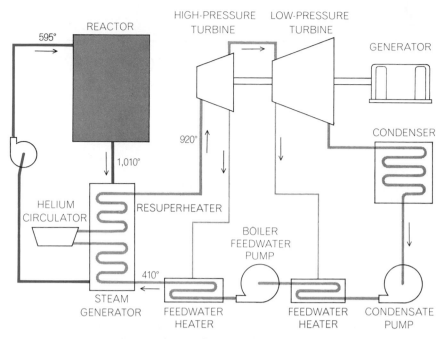

PATTERN OF FLOW in a gas-cooled fast breeder reactor entails the transfer of the heat from the helium coolant (*color*) to a water-steam cycle (*gray*). The system operates without an intermediate heat-exchange cycle. Numerals give temperatures in degrees Fahrenheit.

Even the small amounts of radioactive fission-product gases (primarily krypton 85 and tritium) now released under controlled conditions from water-moderated reactors will be eliminated because the necessity of hermetically sealing the core area will give an inherently effective method for collecting and disposing of the gases, which can then be rendered harmless. Moreover, since the coolant in a fast breeder is kept in a closed system, and since the water used to generate steam is never exposed to neutrons, there should be no formation of radioactivity in aqueous effluents from the plant.

The economic potential of fast breeder reactors lies mainly, but not entirely, in the fact that they would conserve resources of nuclear fuel. Over the next 50 years the use of breeders as planned can be expected to reduce by 1.2 million tons the amount of uranium that would be consumed without breeders. That is the energy equivalent of about three billion tons of coal.

The development of a breeder economy also appears to offer a direct dollar gain of large proportions. Studies have indicated that the cost of research and development of the liquid-metal fast breeder will be more than $2 billion through the year 2020 for the Atomic Energy Commission alone, with large industrial expenditures added. If the first commercial breeder is introduced by 1984 as planned, however, reductions in the cost of electrical energy thereafter

(to 2020) are estimated at $200 billion in 1970 dollars.

The present cost of producing electricity in the U.S. ranges from five to 10 mills per kilowatt-hour delivered to the transmission system, depending on the type, age and location of the plant. This range covers most plants, although there are a few outside of either extreme. The liquid-metal fast breeder reactor is predicted to produce power at a saving of from .5 to one mill per kilowatt-hour. Large breeder-reactor systems that eventually bring the cost of electricity down by as much as two mills per kilowatt-hour will make it possible to extract, use and reuse resources in ways that cannot be afforded today. It will be possible to tap substantial resources in the oceans and on land and to use land not now habitable or productive.

Indeed, we believe breeders will result in a transition to the massive use of nuclear energy in a new economic and technological framework. The transition may be slow, and it will require the introduction of a series of innovations in the technologies of industry, agriculture and transportation. The innovations will include large-scale, dual-purpose desalting plants; electromechanization of farms and of means of transportation; electrification of the metal and chemical industries, and more effective means for utilizing wastes. The key to these possibilities is abundant low-cost electrical energy, and the route to that is by way of the breeder reactor.

22

THE PROSPECTS OF FUSION POWER

WILLIAM C. GOUGH AND BERNARD J. EASTLUND
February 1971

*Recent advances in the performance of several
experimental plasma containers have brought the
fusion-power option very close to the "break even"
level of scientific feasibility*

The achievement of a practical fusion-power reactor would have a profound impact on almost every aspect of human society. In the past few years considerable progress has been made toward that goal. Perhaps the most revealing indication of the significance of this progress is the extent to which the emphasis in recent discussions and meetings involving workers in the field has tended to shift from the question of purely scientific feasibility to a consideration of the technological, economic and social aspects of the power-generation problem. The purpose of this article is to examine the probable effects of the recent advances on the immediate and long-term prospects of the fusion-power program, with particular reference to mankind's future energy needs.

The Role of Energy

The role of energy in determining the economic well-being of a society is often inadequately understood. In terms of *total* energy the main energy source for any society is the sun, which through the cycle of photosynthesis produces the food that is the basic fuel for sustaining the population of that society. The efficiency with which the sun's energy can be put to use, however, is determined by a feedback loop in which auxiliary energy sources form a critical link [*see illustration on page 206*]. The auxiliary energy (derived mainly from fossil fuels, water power and nuclear-fission fuels) "opens the gate" to the efficient use of the sun's energy by helping to produce fertilizers, pesticides, improved seeds, farm machinery and so on. The result is that the food yield (in terms of energy content) produced per unit area of land in a year goes up by orders of magnitude. This auxiliary energy input, when it is transformed into food energy, enables large populations to live in cities and develop new ways to multiply the efficiency of the feedback loop. If a society is to raise its standard of living by increasing the efficiency of its agricultural feedback loop, clearly it must expand its auxiliary energy sources.

The dilemma here is that the economically less developed countries of the world cannot *all* industrialize on the model of the more developed countries, for the simple reason that the latter countries, which contain only a small fraction of the world's population, currently maintain their high standard of living by consuming a disproportionately large share of the world's available supply of auxiliary energy. Just as there is a direct, almost linear, relation between a nation's use of auxiliary energy and its standard of living, so also there is a similar relation between energy consumption and the amount of raw material the nation uses and the amount of waste material it produces. Thus the more developed countries consume a correspondingly oversized share of the world's reserves of material resources and also account for most of the world's environmental pollution.

In order to achieve a more equitable and stable balance between the standards of living in the more developed countries and those in the less developed countries, only two alternatives exist. The more developed countries could reduce their consumption of auxiliary energy (thereby lowering their standard of living as well) or they could contribute to the development of new, vastly greater sources of auxiliary energy in order to help meet the rising demands for a better standard of living on the part of the rapidly growing populations of the less developed countries.

When one projects the world's long-term energy requirements against this background, another important factor must be taken into account. There are finite limits to the world's reserves of material resources and to the ability of the earth's ecological system to absorb pollutants safely. As a consequence future societies will be forced to develop "looped," or "circular," materials economies to replace their present, inherently wasteful "linear" materials economies [*see bottom illustration on page 215*]. In such a "stationary state" system, limits on the materials inventory, and hence on the total wealth of the society, would be set by nature. Within these limits, however, the standard of living of the population would be higher if the rate of flow of materials were lower. This maximizing of the life expectancy of the materials inventory could be accomplished in two ways: increasing the durability of individual commodities and developing the technological means to recycle the limited supply of material resources.

The conclusion appears radical. Future societies must *minimize* their physical flow of production and consumption. Since a society's gross national product for the most part measures the flow of physical things, it too would be reduced.

But all nations now try to *maximize* their gross national product, and hence their rate of flow of materials! The explanation of this paradox is that in the existing linear economies the inputs for increasing production must come from the environment, which leads to depletion, while an almost equal amount of materials in the form of waste must be returned to the environment, which leads to pollution. This primary cause of pollution is augmented by the pollution that is produced by the energy sources used to drive the system.

In order to make the transition to a stationary-state world economy, the wealthier nations will have to develop the technology—and the concomitant auxiliary energy sources—necessary to operate a closed materials economy. This capability could then be transferred to the poorer nations so that they could develop to the level of the wealthier nations without exhausting the world's supply of resources or destroying the environment. Thus some of the causes of international conflict would be removed, thereby reducing the danger of nuclear war.

Of course any effort to bring about a rapid change from linear economies to looped economies will encounter the massive economic, social and political forces that sustain the present system. The question of how to distribute the stock of wealth, including leisure, within a stationary-state economy will remain. In summary, the world's requirements for energy are intimately related to the issues of population expansion, economic development, materials depletion, pollution, war and the organization of human societies.

The Energy Options

What are the available energy options for the future? To begin with there are the known finite and irreplaceable energy sources: the fossil fuels and the better-grade, or easily fissionable, nuclear fuels such as uranium 235. Estimates of the life expectancy of these sources vary, but it is generally agreed that they are being used up at a rapid rate—a rate that will moreover be accelerated by increases in both population and living standards. In addition, environmental considerations could further restrict the use of these energy sources.

Certain other known energy sources, such as water power, tidal power, geothermal power and wind power are "infinite" in the sense of being continuously replenished. The total useful *amount* of energy available from these sources, however, is insufficient to meet the needs of the future.

Direct solar radiation, resulting from the fusion reactions that take place in the core of the sun, is an abundant as well as effectively "infinite" energy source. The immediate practical obstacle to the direct use of the sun's energy as an effective auxiliary energy source is the necessity of finding some way to economically concentrate the available low energy density of solar radiation. Controlled fusion is another potentially "infinite" energy source; its energy output arises from the reduction of the total mass of a nuclear system that accompanies the merger of two light

U.S. TOKAMAK, a toroidal plasma-confinement machine used in fusion research, was recently put into operation at the Plasma Physics Laboratory of Princeton University. Until about a year ago this machine, formerly known as the Model-C stellarator and now called the Model ST tokamak, had been the largest of the stellarator class of experimental plasma containers developed primarily at the Princeton laboratory. The decision to convert it to the closely related tokamak design followed the 1969 announcement by the Russian fusion-research group of some important new results obtained from their Model T-3 machine, the most advanced of the tokamak class of plasma containers developed mainly at the I. V. Kurchatov Institute of Atomic Energy near Moscow. In large part because of the cooperative nature of the world fusion-research program, this conversion was accomplished quickly and the Model ST has already produced results comparable to those obtained by the Russians. Several other tokamak-type machines are being built in this country.

RUSSIAN STELLARATOR is now the largest representative of this class of experimental plasma containers in the world. It is named the Uragan (or "hurricane") stellarator and is located at the Physico-Technical Institute of the Academy of Sciences of the Ukrainian S.S.R. at Kharkov. In both photographs on this page the large circular structures surrounding and almost completely obscuring the toroidal plasma chambers are the primary magnet coils. The main difference between the tokamak design and the stellarator design is that in a tokamak a secondary plasma-stabilizing magnetic field is generated by an electric current flowing axially through the plasma itself, whereas in a stellarator this secondary magnetic field is set up by external helical coils situated just inside the primary coils and hence not visible.

nuclei. The most likely fuel for a fusion-power energy source is deuterium, an abundant heavy isotope of hydrogen easily separated from seawater.

In addition to these two primary "infinite" energy sources, secondary "infinite" energy sources could be made by using neutrons to transmute less useful elements into other elements capable of being used effectively as fuels. Thus for fission systems the vast reserves of uranium 238 could be converted by neutron bombardment into easily fissionable plutonium 239; similarly, thorium 232 could be converted into uranium 233. For fusion systems lithium could be converted into tritium, another heavy isotope of hydrogen with a comparatively low resistance to entering a fusion reaction and a comparatively high energy output once it does.

The prime hope for extending the world's reserves of nuclear-fission fuels is the development of the neutron-rich fast breeder fission reactors [see the article "Fast Breeder Reactors," by Glenn T. Seaborg and Justin L. Bloom, beginning on page 195]. Another potential source of abundant, inexpensive neutrons is a fusion-fission hybrid system, an

alternative that will be discussed further below.

Fusion Energy

Nuclear fusion, the basic energy process of the stars, was first reproduced on the earth in 1932 in an experiment involving the collision of artificially accelerated deuterium nuclei. Although it was thereby shown that fusion energy could be released in this way, the use of particle accelerators to provide the nuclei with enough energy to overcome their Coulomb, or mutually repulsive, forces was never considered seriously as a practical method for power generation. The reason is that the large majority of the nuclei that collide in an accelerator scatter without reacting; thus it is impossible to produce more energy than was used to accelerate the nuclei in the first place.

The uncontrolled release of a massive amount of fusion energy was achieved in 1952 with the first thermonuclear test explosion. This test proved that fusion energy could be released on a large scale by raising the temperature of a high-density gas of charged particles (a plas-

ma) to about 50 million degrees Celsius, thereby increasing the probability that fusion reactions will take place within the gas.

Coincident with the development of the hydrogen bomb, the search for a more controlled means of releasing fusion energy was begun independently in the U.S., Britain and the U.S.S.R. Essentially this search involves looking for a practical way to maintain a comparatively low-density plasma at a temperature high enough so that the output of fusion energy derived from the plasma is greater than the input of some other kind of energy supplied to the plasma. Since no solid material can exist at the temperature range required for a useful energy output (on the order of 100 million degrees C.) the principal emphasis from the beginning has been on the use of magnetic fields to confine the plasma.

The variety of magnetic "bottles" designed for this purpose over the years can be arranged in several broad categories in order of increasing plasma density [see illustration on pages 210 and 211]. First there are the basic plasma devices. These are low-density, low-temperature systems used primarily to study the fundamental properties of plasmas. Their configuration can be either linear (open) or toroidal (closed). Linear basic-plasma devices include simple glow-discharge systems (similar in operation to ordinary fluorescent lamps) and the more sophisticated "Q-machines" ("Q" for "quiescent") found in many university plasma-physics laboratories. Toroidal representatives of this class include the "multipole" devices, developed primarily at Gulf Energy & Environmental Systems, Inc. (formerly Gulf General Atomic Inc.) and the University of Wisconsin, and the spherator, developed at the Plasma Physics Laboratory of Princeton University.

Next there are the medium-density plasma containers; these are defined as systems in which the outward pressure of the plasma is much less than the inward pressure of the magnetic field. A typical configuration in this density range is the linear magnetic bottle, which is usually "stoppered" at the ends by magnetic "mirrors": regions of somewhat greater magnetic-field strength that reflect escaping particles back into the bottle. In addition extra current-carrying structures are often used to improve the stability of the plasma. These structures were originally proposed on theoretical grounds in 1955 by Harold Grad of New York University. They were first used successfully in an experimental test in 1962 by the Russian physicist M. S. Ioffe.

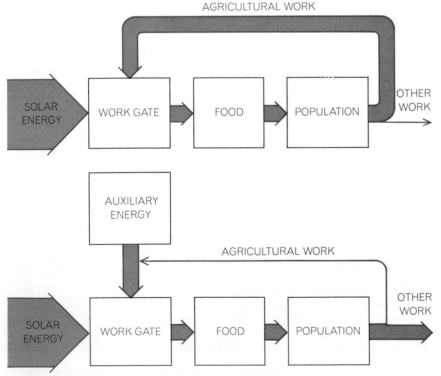

ROLE OF AUXILIARY ENERGY in determining the economic well-being of a society is illustrated by these two diagrams of agricultural feedback loops. In an economically less developed country (*top*) the bulk of the population must be devoted to the agricultural transformation of the sun's energy into food in order to support itself at a subsistence level. In an economically more developed industrial country (*bottom*) auxiliary energy sources "open the gate" to the more efficient utilization of the sun's energy making it possible for the entire population to maintain a higher standard of living and freeing many people to live in cities and develop new ways to multiply the efficiency of the feedback loop.

| | | LIFE EXPECTANCY OF KNOWN RESERVES (YEARS) | | LIFE EXPECTANCY OF POTENTIAL RESERVES (YEARS) | | LIFE EXPECTANCY OF TOTAL RESERVES (YEARS) | |
		AT .17Q	AT 2.8Q	AT .17Q	AT 2.8Q	AT .17Q	AT 2.8Q
FINITE ENERGY SOURCES	FOSSIL FUELS (COAL, OIL, GAS)	132	8	2,700	165	2,832	173
	MORE ACCESSIBLE FISSION FUELS (URANIUM AT $5 TO $30 PER POUND OF U_3O_8 BURNED AT 1.5 PERCENT EFFICIENCY)	66	4	66	4	132	8
	LESS ACCESSIBLE FISSION FUELS (URANIUM AT $30 TO $500 PER POUND OF U_3O_8 BURNED AT 1.5 PERCENT EFFICIENCY)	43,000	2,600	129,000	7,800	172,000	10,400
"INFINITE" NATURAL ENERGY SOURCES	WATER POWER, TIDAL POWER, GEOTHERMAL POWER, WIND POWER	INSUFFICIENT		INSUFFICIENT		INSUFFICIENT	
	SOLAR RADIATION	10 BILLION	10 BILLION			10 BILLION	10 BILLION
	FUSION FUELS (DEUTERIUM FROM OCEAN)	45 BILLION	2.7 BILLION			45 BILLION	2.7 BILLION
"INFINITE" ARTIFICIAL ENERGY SOURCES (ELEMENTS TRANSMUTED FROM OTHER ELEMENTS BY NEUTRON BOMBARDMENT)	FISSION FUELS (PLUTONIUM 239 FROM URANIUM 238; URANIUM 233 FROM THORIUM 232)	8.8 MILLION	536,000	21 MILLION	1.3 MILLION	30 MILLION	1.8 MILLION
	FUSION FUELS (TRITIUM FROM LITHIUM) a) ON LAND b) IN OCEAN	48,000 120 MILLION	2,900 7.3 MILLION	UNKNOWN	UNKNOWN	48,000+ 120 MILLION	2,900+ 7.3 MILLION

WORLD ENERGY RESERVES are listed in this table in terms of their life expectancy estimated on the basis of two extreme assumptions, which were chosen so as to bracket a reasonable range of values. First, the assumption was made that the world population would remain constant at its 1968 level of 3.5 billion persons and that the energy-consumption rate of this population would remain constant at the estimated 1968 rate of .17 Q (Q is a unit of heat measurement equal to 10^{18} BTU, or British Thermal Units). Second, the assumption was made that the world population would eventually reach seven billion and that this population would consume energy at a per capita rate of 400 million BTU per year (about 20 percent higher than the present U.S. rate), giving a total world energy-consumption rate of 2.8 Q per year. (A commonly projected world energy-consumption rate for the year 2000 is one Q.) Current fission-converter reactors use only between 1 and 2 percent of the uranium's potential energy content, since the component of the ore that is burned as fuel is primarily high-grade, or easily fissionable, uranium 235. The world fission-fuel reserves were derived by multiplying the U.S. reserves times the ratio of world land area to the U.S. land area (approximately 16.2 to one). For fusion-converter reactors lithium-utilization studies show that natural lithium, a mixture of lithium 6 and lithium 7, would be superior to pure lithium 6 in a tritium-breeding reactor "blanket" and would yield an energy output of about 86.4 million BTU per gram. The figure for known world lithium reserves is based on a study carried out in 1970 by James J. Norton of the U.S. Geological Survey. The potential reserves of lithium are unknown, since there has been no exploration program comparable to that undertaken for, say, uranium. Lithium, however, is between five and 15 times more abundant in the earth's crust than uranium. Finally, the life expectancy of the earth — and hence that of potentially useful solar radiation — is predicted to be at most 10 billion years.

The straight rods used by Ioffe in his experiment have come to be called Ioffe bars, but such stabilizing structures can assume various other shapes. For example, in one series of medium-density linear devices they resemble the seam of a baseball; accordingly these devices, developed at the Lawrence Radiation Laboratory of the University of California at Livermore, are named Baseball I and Baseball II.

Medium-density plasma containers with a toroidal geometry include the stellarators, originally developed at the Princeton Plasma Physics Laboratory, and the tokamaks, originally developed at the I. V. Kurchatov Institute of Atomic Energy near Moscow. The only essential difference between these two machines is that in a stellarator a secondary, plasma-stabilizing magnetic field is set up by external helical coils, whereas in a tokamak this field is generated by an electric current flowing through the plasma itself. The close similarity between these two designs was emphasized recently by the fact that the Princeton Model-C stellarator was rather quickly converted to a tokamak system following the recent announcement by the Russians of some important new results from their Tokamak-3 machine.

The astron concept, also developed at the Lawrence Radiation Laboratory, at Livermore, is another example of a medium-density plasma container. In overall geometry it shares some characteristics of both the linear and the toroidal designs.

Higher-density plasma containers, defined as those in which the plasma pressure is comparable to the magnetic-field pressure, have also been built in both the linear and the toroidal forms. In one such class of devices, called the "theta pinch" machines, the electric current is in the theta, or azimuthal, direction (around the axis) and the resulting magnetic field is in the zeta, or axial, direction (along the axis). The Scylla and Scyllac machines at the Los Alamos Scientific Laboratory are respectively examples of a linear theta-pinch design and a toroidal theta-pinch design.

As the density of the plasma is increased further, one reaches a technological limit imposed by the inability of the materials used in the magnet coils to withstand the pressure of the magnetic field. Consequently very-high-density plasma systems are often fast-pulsed and obtain their principal confining forces from "self-generated" magnetic fields (fields set up by electric currents in the plasma itself), from electrostatic fields or from inertial pressures. In this very-high-density category are the "zeta pinch" machines, devices in which the electric current is in the zeta direction and the resulting magnetic field is in the theta direction. An example of this type of configuration is the Columba device at Los Alamos.

Other very-high-density, fast-pulsed systems include the "strong focus" designs, in which a stream of plasma in a cylindrical, coaxial pipe is heated rapidly by shock waves as it is brought to a sharp focus by self-generated magnetic forces, and laser designs, in which a pellet of fuel is ionized instantaneously by a pulse from a high-power laser, producing an "inertially confined" plasma. Still another confinement scheme that has been investigated in this general density range includes an electrostatic device in which the plasma is confined by inertial forces generated by concentric spherical electrodes.

The Fusion-Power Balance

What are the fundamental requirements for a meaningful release of fusion energy in a reactor? First, the plasma must be hot enough for the production of fusion energy to exceed the energy loss due to bremsstrahlung radiation (radiation resulting from near-collisions between electrons and nuclei in the plasma). The temperature at which this transition occurs is called the ignition temperature. For a fuel cycle based on fusion reactions between deuterium and tritium nuclei the ignition temperature is about 40 million degrees C. Second, the plasma must be confined long enough to release a significant net output of energy. Third, the energy must be recovered in a useful form.

In the first years of the controlled-fusion research program one of the major goals was to achieve the ignition temperature in a fairly dense laboratory plasma. Steady progress was made toward this goal, culminating in 1963, when the ignition temperature (for a deuterium-tritium fuel mixture) was reached in one of the Scylla devices at Los Alamos. This test, which was performed in a pure deuterium plasma to avoid the generation of excessive neutron flux, resulted in the release of fusion energy: about a thousandth of a joule per pulse, or 370 watts of fusion power during the three-microsecond duration of the pulse. If the test had been performed using a deuterium-tritium mixture, it would have released approximately a half-joule of fusion energy per pulse, or 180,000 watts of fusion power.

Today a large number of different devices have either achieved the deuterium-tritium ignition temperature or are very close to it [see bottom illustration on opposite page]. The main difficulties encountered in reaching this goal were comparatively straightforward energy-loss processes involving impurity atoms that entered the plasma from the walls of the container. A large research effort in the areas of vacuum and surface technology was a major factor in surmounting the ignition-temperature barrier.

The problem of confining a plasma long enough to release a significant net amount of energy has proved to be even more difficult than the problem of achieving the ignition temperature. Extremely rapid energy-loss processes—known collectively as "anomalous diffusion" processes—appeared to prevent the attainment of adequate confinement times. Plasma instabilities were the primary cause of this rapid plasma leakage [see "The Leakage Problem in Fusion Reactors," by Francis F. Chen; SCIENTIFIC AMERICAN, July, 1967]. Within the past few years, however, several large containment devices have reduced these instabilities to such a low amplitude that other more subtle effects, such as convective plasma losses and magnetic-field imperfections, can be studied. As a result it has been shown that there is no basic law of physics (such as an instability-initiated anomalous plasma loss) that prevents plasma confinement for times long enough to release significant net fusion energy. In fact, "classical," or ideal, plasma confinement has been achieved in several machines; this is the best confinement possible and yields a plasma-loss rate much lower than that required for a fusion reactor.

The twin achievements of ignition temperature and adequate confinement time, it should be noted, have taken place in quite different machines, each

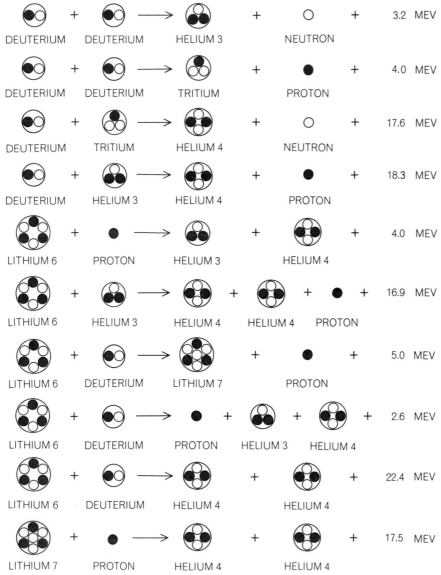

FUSION REACTIONS regarded as potentially useful in full-scale fusion reactors are represented in this partial list. The two possible deuterium-deuterium reactions occur with equal probability. The deuterium-tritium fuel cycle has been considered particularly attractive because this mixture has the lowest ignition temperature known (about 40 million degrees Celsius). Other fuel cycles, including many not shown in this list, have been attracting increased attention lately, since certain plasma-confinement schemes actually operate better at higher temperatures and offer the advantage of direct conversion to electricity. The energy released by each reaction is given at right in millions of electron volts (MeV).

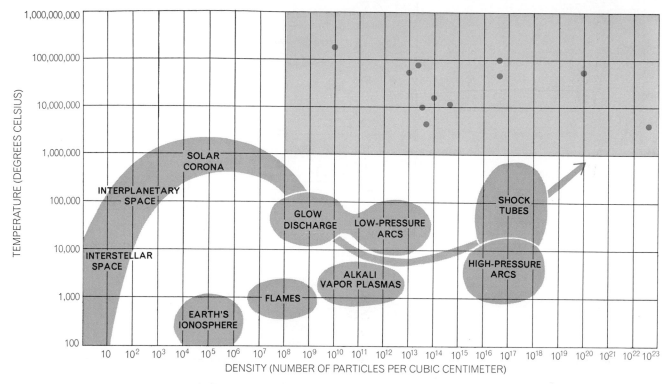

INDUSTRIALLY UNEXPLORED RANGE of plasma temperatures and densities has already been made available by the fusion-power research program. These experimental plasmas (*colored dots*), which range in temperature from 500,000 to a billion degrees C. and in density from 10^9 to 10^{22} ions per cubic centimeter, are compared here with various other industrial and natural plasmas.

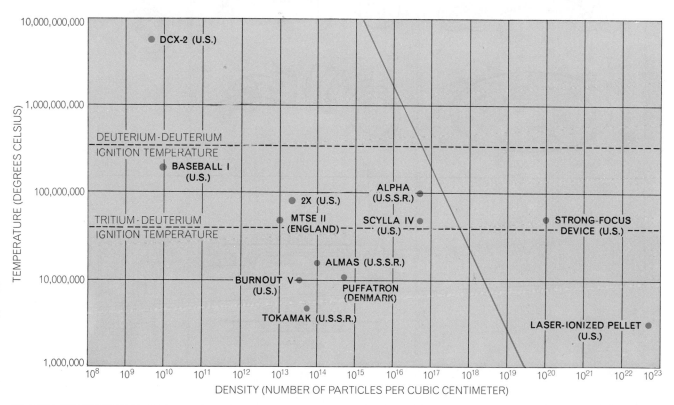

PLASMA EXPERIMENTS that have achieved temperatures near or above the fusion ignition temperatures of a deuterium-tritium fuel (*bottom horizontal line*) and a deuterium-deuterium fuel (*top horizontal line*) are identified by the name of the experimental device and the country in which the experiment took place in this enlargement of the upper right-hand section of the illustration at top. The diagonal colored line represents the limit beyond which the materials used to construct the magnet coils can no longer withstand the magnetic-field pressure required to confine the plasma (assumed to be 300,000 gauss in this case). Beyond this limit only fast-pulsed systems (in which the magnetic fields are generated by intense currents inside the plasma itself) or systems operating on entirely different principles (such as laser-produced, inertially confined plasmas) are possible. The record of six billion degrees C. was achieved with the aid of a high-energy ion-injection system associated with DCX-2 device at the Oak Ridge National Laboratory.

210

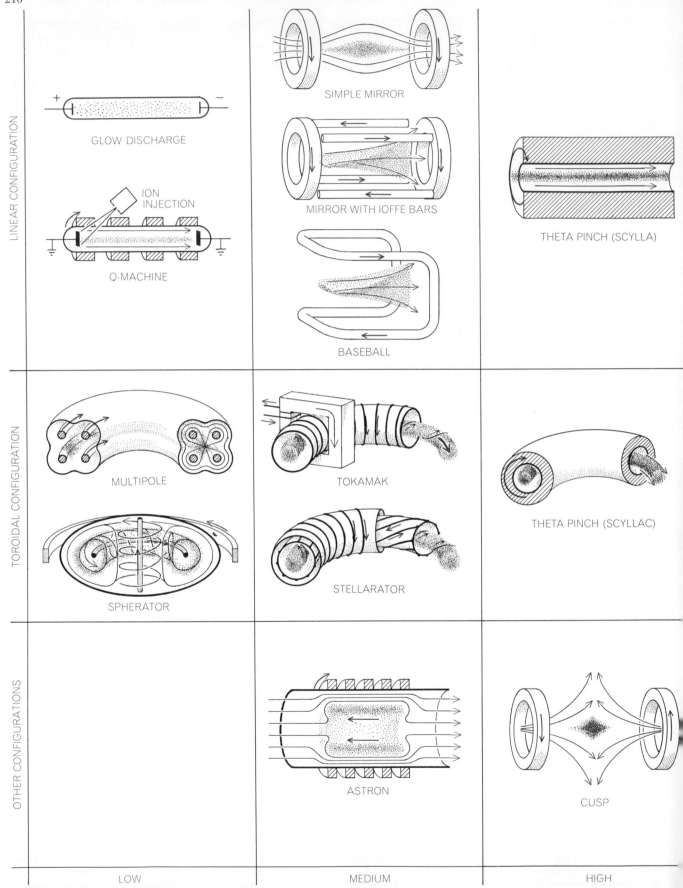

LINEAR CONFIGURATION

GLOW DISCHARGE

ION INJECTION

Q-MACHINE

SIMPLE MIRROR

MIRROR WITH IOFFE BARS

BASEBALL

THETA PINCH (SCYLLA)

TOROIDAL CONFIGURATION

MULTIPOLE

SPHERATOR

TOKAMAK

STELLARATOR

THETA PINCH (SCYLLAC)

OTHER CONFIGURATIONS

ASTRON

CUSP

LOW

MEDIUM

HIGH

PRINCIPAL SCHEMES devised in the past 18 years to confine plasmas for fusion research are arranged in the illustration on these two pages in order of increasing plasma density (*left to right*) and overall geometry (*top to bottom*). Only a few examples are depicted in each category. In every case the plasma is in color, the colored arrows signify the direction of the electric current and the black arrows denote the direction of the resultant magnetic field. Various structural details have been omitted for clarity. For each example shown there are a large number of variations either already in existence or in the conceptual stage. Furthermore, the

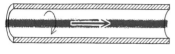

ZETA PINCH

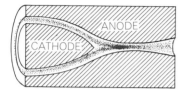

CATHODE ANODE

PLASMA FOCUS

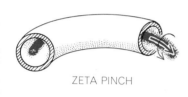

ZETA PINCH

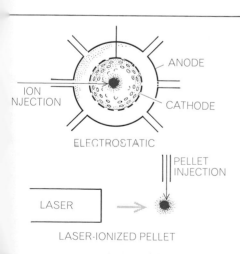

ION
NJECTION

ANODE

CATHODE

ELECTROSTATIC

PELLET
INJECTION

LASER

LASER-IONIZED PELLET

VERY HIGH

fact that an example is given in one category does not necessarily mean that that configuration is not applicable to some other category; there are, for instance, toroidal Q-machines and medium-density cusp designs.

specially designed to maximize the conditions for reaching one goal or the other. How does one compare the performances of these machines in order to gauge how near one is to the combined conditions needed to operate a practical fusion-power reactor? The basic criterion for determining the length of time a plasma must be confined at a given density and temperature to produce a "break even" point in the power balance was laid down in 1957 by the British physicist J. D. Lawson. Combining data on the physics of fusion reactions with some estimates of the efficiency of energy recovery from a hypothetical fusion reactor, Lawson derived a factor, which he called R, that denoted the ratio of energy output to energy input needed to compensate for all possible plasma losses. Lawson's criterion is still in general use as a convenient yardstick for measuring the extent to which losses must be controlled in order to make possible the construction of a fusion reactor. Although more recent calculations consider many other physical constraints in order to arrive at the break-even power balance, these criteria still give values very close to those derived by Lawson.

For a deuterium-tritium fuel mixture Lawson found that at temperatures higher than the ignition temperature the product of density and confinement time must be equal to 10^{14} seconds per cubic centimeter in order to achieve the break-even condition. This criterion defines a surface in three-dimensional space, the coordinates being the logarithmic values of density, temperature and confinement time [see illustration on page 213]. The goal of a break-even release of energy will have been achieved when the set of conditions for a given machine reaches this surface. It should be emphasized that the exact location and shape of the surface is a function of both the fuel cycle used and the recovery efficiency of the hypothetical reactor system. Fuels other than the deuterium-tritium mixture would increase the temperature needed to achieve a break-even power balance.

The extraordinary progress made recently by various groups in learning how to raise the combination of density, temperature and confinement time to a set of values approaching this break-even surface can be appreciated by referring to the illustration of the Lawson-criterion surface. The several plasma systems shown range in density from about 10^9 ions per cubic centimeter to 5×10^{22} ions per cubic centimeter. (Below a density of about 10^{11} ions per cubic centimeter the power density would be so low that it would require an impractically

large reactor.) The particular density range chosen for investigation in each case is a function of the scientific preferences of the investigators concerning the best route to fusion power and the available technology (magnets, power supplies, lasers and so forth). Thus there are various trajectories to the break-even surface being followed through the three-dimensional "parameter space" of the illustration. Closing the gap between where each trajectory is now and the break-even surface depends in some cases (for example the tokamak devices) on obtaining a better understanding of the physical principles required to develop reliable scaling rules, whereas in other cases (such as the linear theta-pinch devices) all that may be required is an economic solution to the engineering problem of building a large enough system.

Fusion-Reactor Designs

How would a full-scale fusion reactor operate? In the first place fusion reactors, like fission reactors, could be run on a variety of fuels. The nature of the fuel used in the core of a fusion reactor would, however, have a decisive effect on the method used to recover the fusion energy and the uses to which the recovered energy might be put. Most research on reactor technology has centered on the use of a deuterium-tritium mixture as a fuel. The reason is that the mixture has the lowest ignition temperature, and hence the lowest rate of energy loss by radiation, of any possible fusion fuel. Nonetheless, other combinations of light nuclei have been considered for many years as potential fusion fuels. Prominent among these are reactions involving a deuterium nucleus and a helium-3 nucleus and reactions involving a single proton (a hydrogen nucleus) and a lithium-6 nucleus. Because containment based on the magnetic-mirror concept actually operates better at higher temperatures, a number of other fuels have been attracting increased attention [see illustration on page 208].

Depending on the fuel used, a fusion reactor could release its energy in several ways. For example, neutrons, which are produced at various rates by different fusion reactions, can cross magnetic fields and penetrate matter quite easily. A reactor based on, say, a deuterium-tritium fuel cycle would release approximately 80 percent of its energy in the form of highly energetic neutrons. Such a reactor could be made to produce electricity by absorbing the neutron energy in a liquid-lithium shield, circulating the

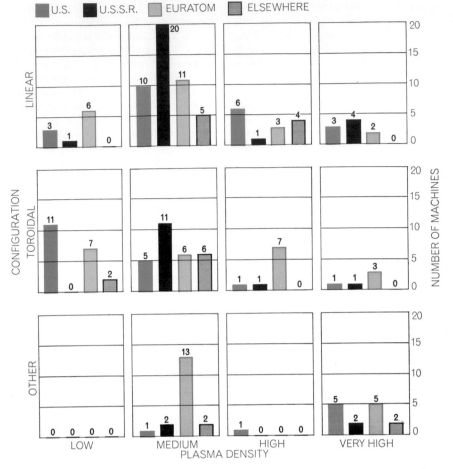

INVENTORY of the number of machines now operating throughout the world in each of the broad categories represented in the illustration on the preceding two pages is given in this table. The total number in each category is broken down into subtotals for the U.S., the U.S.S.R., the European Atomic Energy Community, or Euratom, countries (Belgium, France, Germany, Italy, Luxembourg and the Netherlands) and the rest of the world (principally Japan, Sweden and Australia). Britain, although not officially a member of Euratom, is included in the Euratom subtotal. The figures are drawn mainly from a recent survey compiled by Amasa S. Bishop and published by the International Atomic Energy Commission. The U.S. fusion-power program currently represents about a fifth of the world total.

liquid lithium to a heat exchanger and there heating water to produce steam and so drive a conventional steam-generator electric power plant [see top illustration on page 214].

This general approach could also lead to an attractive new technique for converting the world's reserves of uranium 238 and thorium 232 to suitable fuels for fission reactors—the fusion-fission hybrid system mentioned above. By employing the abundance of inexpensive, energetic neutrons produced by the deuterium-tritium fuel cycle to synthesize fissionable heavy nuclei, a fusion reactor could act as a new type of breeder reactor. This could have the effect of lowering the break-even surface defined by Lawson's criterion, bringing the fusion-breeding scheme actually closer to feasibility than the generation of electricity solely by fusion reactions. Cheap fuel might thus be made for existing fission

reactors in systems that could be inherently safe. A "neutron-rich" economy created by fusion reactors would have other potential benefits. For example, it has been suggested that large quantities of neutrons could be useful for "burning" various fission products, thereby alleviating the problem of disposing of radioactive wastes.

Fuel cycles that release most of their energy in the form of charged particles offer still other avenues for the recovery of fusion energy. For example, Richard F. Post of the Lawrence Radiation Laboratory at Livermore has proposed a direct energy-conversion scheme in which the energetic charged particles produced in a fusion-reactor core are slowed directly by an electrostatic field set up by an array of large electrically charged plates [see bottom illustration on page 214]. By a judicious arrangement of the voltages applied to the

plates such a system could theoretically be made to operate at a conversion efficiency of 90 percent.

J. Rand McNally, Jr., of the Oak Ridge National Laboratory has suggested that a long sequence of fusion reactions similar to those that power the stars could be reproduced in a fusion reactor. The data necessary to evaluate fuel cycles operating in this manner, however, do not exist at present.

The characteristics of a full-scale fusion reactor would depend not only on the fuel cycle but also on the particular plasma-confinement configuration and density range chosen. Thus it is probable that there eventually will exist a number of different forms of fusion reactor. For example, medium-density magnetic-mirror reactors and very-high-density laser-ignited reactors could be expected to operate at power levels as low as between five and 50 megawatts, which could make them potentially useful for fusion-propulsion schemes.

For central-station power generation the medium-density reactors would most likely operate on a deuterium-tritium fuel cycle in order to take advantage of the mixture's low ignition temperature. Because of the high neutron output associated with this fuel, a heat-cycle conversion system would be appropriate.

A reactor of this type would operate most efficiently with a power output in the billion-watt range. Before such a reactor can be built, it will be necessary to prove that the plasma will remain stable as present devices are scaled to reactor sizes and temperatures. Problems likely to be encountered in this effort involve the long-term equilibrium of the plasma, the interaction of the plasma with the walls of the container and the necessity of pumping large quantities of liquid lithium across the magnetic field.

Medium-density linear reactors would be better suited for fuel cycles that yield a major part of their energy output in the form of charged particles, since this approach would allow the direct recovery of the kinetic energy of these reaction products through schemes such as Post's. Such fuel cycles could be based on a deuterium-deuterium reaction, a deuterium-helium reaction or a proton-lithium reaction. A system operating on this principle could be made to produce direct-current electricity at a potential of about 400 kilovolts, which would be ideal for long-distance cryogenic (supercooled) power transmission.

Although the break-even conditions would be lowered in this case (because of the high energy-conversion efficiency), it still remains to be shown that existing experiments can be scaled to large sizes and higher temperatures. Some major technological obstacles that need to be overcome include the construction of large atomic-beam injectors and extremely strong magnetic mirrors.

For reactors operating on the basis of any of the higher-density schemes, such as the theta-pinch machines or the fast-pulsed systems, major technological hurdles include the development of efficient energy-storage and energy-transfer techniques and problems related to heating techniques such as lasers.

In addition to generating electric power and possibly serving in a propulsion system, fusion reactors are potentially useful for other applications. For example, fusion research has already made available plasmas that range in temperature from 500,000 to a billion degrees C. and in density from 10^9 to 10^{22} ions per cubic centimeter. Almost all industrial processes that use plasmas fall outside this range [*see top illustration on page 209*]. In order to suggest how this industrially unexplored range might be exploited, we recently put forward the concept of the "fusion torch." The gen-

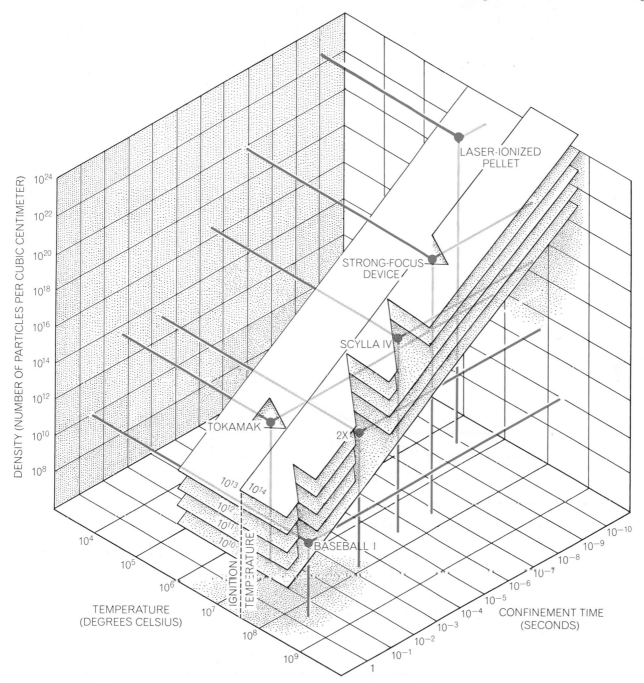

BASIC CRITERION for determining the length of time a plasma must be confined at a given density and temperature to achieve a "break even" point in the fusion-power balance is represented in this three-dimensional graph. The graph is based on a method of analysis devised in 1957 by the British physicist J. D. Lawson. For a deuterium-tritium fuel mixture in the temperature range from 40 million degrees C. to 500 million degrees C., Lawson found that the product of density and confinement time must be close to 10^{14} seconds per cubic centimeter to achieve the break-even condition (based on an assumed energy-conversion efficiency of 33 percent). This criterion corresponds to the top layer in the stack of planes in the illustration. The lower planes, which correspond to successively smaller values of density times confinement time, are included in order to give some idea of the positions of the best confirmed results from several experimental devices with respect to the combination of parameters needed to operate a full-scale fusion reactor.

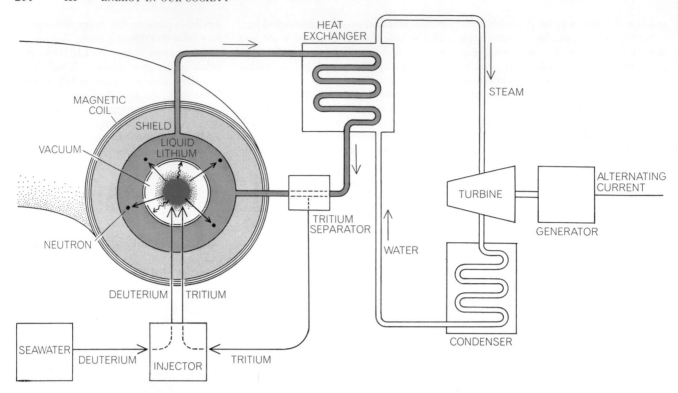

THERMAL ENERGY CONVERSION would be most effective in a fusion reactor based on a deuterium-tritium fuel cycle, since such a fuel would release approximately 80 percent of its energy in the form of highly energetic neutrons. The reactor could produce elec- tricity by absorbing the neutron energy in a liquid-lithium shield, circulating the liquid lithium to a heat exchanger and there heating water to produce steam and thus drive a conventional steam-gener- ator plant. The reactor core could be either linear or toroidal.

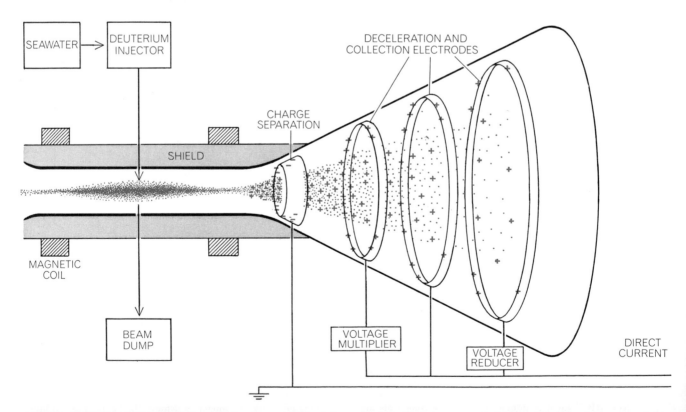

DIRECT ENERGY CONVERSION would be more suitable for fusion fuel cycles that release most of their energy in the form of charged particles. In this novel direct energy-conversion scheme, first proposed by Richard F. Post of the Lawrence Radiation Lab- oratory of the University of California at Livermore, the energetic charged particles (primarily electrons, protons and alpha particles) produced in the core of a linear fusion reactor would be released through diverging magnetic fields at the ends of the magnetic bot- tle, lowering the density of the plasma by a factor of as much as a million. A large electrically grounded collector plate would then be used to remove only the electrons. The positive reaction prod- ucts (at energies in the vicinity of 400 kilovolts) would finally be collected on a series of high-voltage electrodes, resulting in a direct transfer of the kinetic energy of the particles to an external circuit.

eral idea here is to use these ultrahigh-density plasmas, possibly directly from the exhaust of a fusion reactor, to vaporize, dissociate and ionize any solid or liquid material [*see top illustration at right*]. The potential uses of such a fusion-torch capability are intriguing. For one thing, an operational fusion torch in its ultimate form could be used to reduce all kinds of wastes to their constituent atoms for separation, thereby closing the materials loop and making technologically possible a stationary-state economy. On a shorter term the fusion torch offers the possibility of processing mineral ores or producing portable liquid fuels by means of a high-temperature plasma system.

The fusion-torch concept could also be useful in transforming the kinetic energy of a plasma into ultraviolet radiation or X rays by the injection of trace amounts of heavy atoms into the plasma. The large quantity of radiative energy generated in this way could then be used for various purposes, including bulk heating, the desalting of seawater, the production of hydrogen or new chemical-processing techniques. Because such new industrial processes would make use of energy in the form of plasmas rather than in the form of, say, chemical solvents, they would be far less likely to pollute the environment. Although the various fusion-torch possibilities are largely untested and many aspects may turn out to be impractical, the concept is intended to stimulate new ideas for the industrial use of the ultrahigh-temperature plasmas that have already been developed in the fusion program as well as those plasmas that would be produced in large quantities by future fusion reactors.

Environmental Considerations

The environmental advantages of fusion power can be broken down into two categories: those advantages that are inherent in all fusion systems and those that are dependent on particular fuel cycles and reactor designs. Among the inherent advantages, one of the most important is the fact that the use of fusion fuel requires no burning of the world's oxygen or hydrocarbon resources and hence releases no carbon dioxide or other combustion products to the atmosphere. This advantage is shared with nuclear-fission plants.

Another advantage of fusion power is that no radioactive wastes are produced as the result of the fuel cycles contemplated. The principal reaction products would be neutrons, nonradioactive heli-

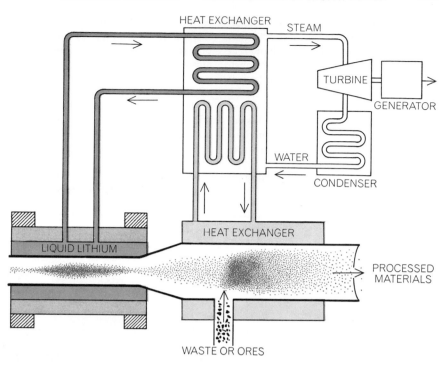

POTENTIAL NONPOWER USE of fusion energy is represented by the concept of the "fusion torch," which was put forward recently by the authors as a suggestion intended to stimulate new ideas for the industrial exploitation of the ultrahigh-temperature plasmas already made available by the fusion-research program as well as those that would be produced by fusion reactors. The general idea is to use some of the energy from these plasmas to vaporize, dissociate and ionize any solid or liquid material. In its ultimate form the fusion torch could be used to reduce any kind of waste to its constituent atoms for separation.

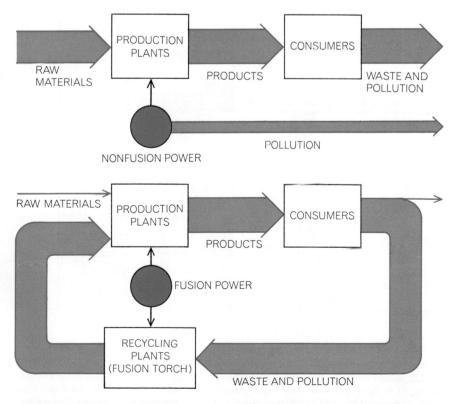

CLOSED MATERIALS ECONOMY could be achieved with the aid of the fusion-torch concept illustrated at the top of this page. In contrast to present systems, which are based on inherently wasteful linear materials economies (*top*), such a stationary-state system would be able to recycle the limited supply of material resources (*bottom*), thus alleviating most of the environmental pollution associated with present methods of energy utilization.

um and hydrogen nuclei, and radioactive tritium nuclei. It is true that tritium emits low-energy ionizing radiation in the form of beta particles (electrons), but since tritium is also a fusion fuel, it could be returned to the system to be burned. This situation is strongly contrasted with that in nuclear fission, which by its very nature must produce a multitude of highly radioactive waste elements.

Fusion reactors are also inherently incapable of a "runaway" accident. There is no "critical mass" required for fusion. In fact, the fusioning plasma is so tenuous (even in the "high density" machines) that there is never enough fuel present at any one time to support a nuclear excursion. This situation is also in contrast to nuclear-fission reactors, which must contain a critical mass of fissionable material and hence an extremely large amount of potential nuclear energy.

Among the system-dependent environmental advantages of fusion power must be counted the fact that the only radioactive fusion fuel considered so far is tritium. The amount of tritium present in a fusion reactor can range from near zero for a proton-lithium fuel cycle to a maximum for a deuterium-tritium cycle, where a "blanket" for the production of tritium must be included. Tritium, however, is one of the least toxic of the radioactive isotopes, whereas the fission fuel plutonium is one of the most toxic radioactive materials known.

The most serious radiological prob-

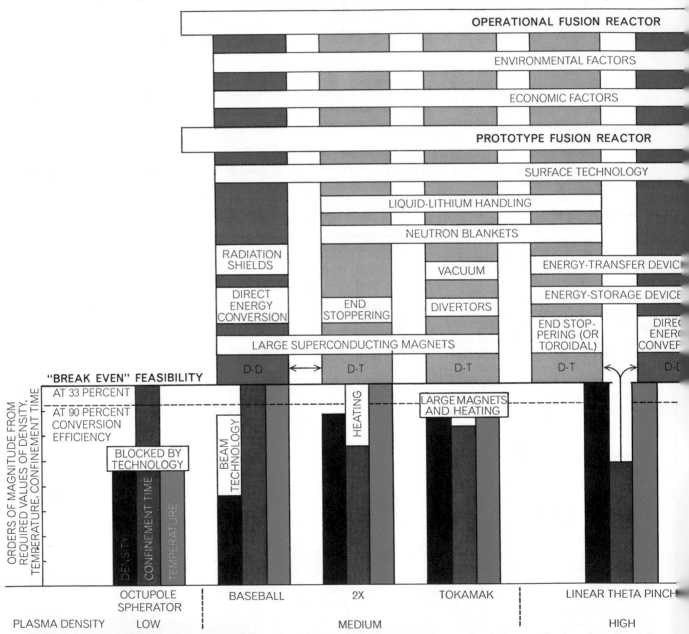

REMAINING PROBLEMS that must be solved before the goal of useful, economic fusion power can be achieved are depicted schematically in this illustration. The major experimental routes to the goal are ordered according to plasma density. Various experimental devices are represented by bars indicating the best combination of plasma density, confinement time and temperature achieved by each device; the logarithmic scale at lower left gauges how far each of these essential parameters is from the values needed to attain break-even feasibility. Technological problems that must be solved in each case are labeled. The achievement of a prototype reactor will be a function not only of plasma technology but also of the fuel cycle and the method of energy conversion chosen. Thus medium-density magnetic-mirror devices could be built to operate with either a deuterium-tritium (D-T) fuel mixture or a deuterium-deuterium (D-D) fuel mixture; the arrows signify these alternatives. If a D-D cycle is chosen, then direct energy conversion is possible, and once the converters are developed very few obstacles would remain to delay the construction of a prototype reactor. If, on the other hand, a D-T cycle is chosen, then conventional thermal energy conversion would be needed, and the listed technological

lems for fusion would exist in a reactor burning and producing tritium. A representative rate of tritium consumption for a 2,000-megawatt deuterium-tritium thermal plant would be about 260 grams per day. Tritium "holdup" in the blanket and other elements of the tritium loop would dictate the tritium inventory. Holdup is estimated to be about 1,000 grams in a 2,000-megawatt plant. If necessary, the doubling time of breeding tritium could be less than two months in order to meet the needs of an expanding

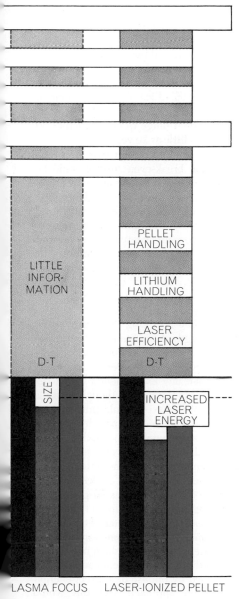

LITTLE INFORMATION

D-T

SIZE

PELLET HANDLING

LITHIUM HANDLING

LASER EFFICIENCY

D-T

INCREASED LASER ENERGY

PLASMA FOCUS   LASER-IONIZED PELLET

VERY HIGH

hurdles would have to be overcome. Other systems operating on the D-T cycle would have to climb past similar hurdles. The high-density linear theta-pinch device could take either the thermal-conversion path or the direct-conversion path (arrows). The final step, from a prototype to an operational reactor, would proceed through a region in which economic and environmental considerations can be expected to be paramount.

economy. The amount of tritium produced by the plant is controllable, however, and need not exceed the fuel requirements of the plant.

Careful design to prevent the leakage of tritium fuel from a deuterium-tritium reactor is mandatory. Engineering studies that take into account economic considerations indicate that the leakage rate can be reduced to .0001 percent per day. The conclusion is that even for an all-deuterium-tritium fusion economy the genetic dose rate from worldwide tritium distribution would be negligible.

In fact, for a given total power output the tritium inventory for an all-deuterium-tritium fusion economy (including both the inventory within the plant and that dispersed in the biosphere) would be between one and 100 times what it would be for an all-fission economy. It is true that tritium would be produced in a deuterium-tritium fusion reactor at a rate of from 1,000 to 100,000 times faster than in various types of fission reactor. Since tritium is burned as a fuel, however, it has an effective half-life of only about three days rather than the normal 12 years.

A technology-dependent but possibly serious limitation on deuterium-tritium fusion plants could be the release of tritium to the local environment. The level would be quite low but the long-term consequences from tritium emission to the environment in the vicinity of a deuterium-tritium reactor needs to be explored. In general the biological-hazard potential of the tritium fuel inventory in a deuterium-tritium reactor is lower by a factor of about a million than that of the volatile isotope iodine 131 contained in a fission reactor. Of course there is no expectation in either case that such a release would occur.

The radioactivity induced in the surrounding structures by a fusion reactor is dependent on both the fuel cycle and the engineering design of the plant. This radioactivity could range from zero for a fuel cycle that produces no neutrons up to very high values for a deuterium-tritium cycle if the engineering design is such that the type and amount of structural materials could become highly activated under neutron bombardment. Cooling for "after heat" will be required for systems that have intense induced radioactivity. Even if the cooling system should fail, however, there could be no nuclear excursion that would disperse the radioactivity outside the plant.

Other system-dependent environmental advantages of fusion power include safety in the event of sabotage or natural disaster, reduced potential for the diversion of weapons-grade materials and low

waste heat. In fact, the potential exists for fusion systems to essentially eliminate the problem of thermal pollution by going to charged-particle fuel cycles that result in direct energy conversion. Finally, there is the advantage of the materials-recycling potential of the fusion-torch concept.

The Timetable to Fusion Power

The construction and operation of a power-producing controlled-fusion reactor will be the end product of a chain of events that is already to a certain extent discernible. For controlled fusion, however, there can never be an instant equivalent to the one that demonstrated the "feasibility" of a fission-power reactor (the Stagg Field experiment of Enrico Fermi in 1942). To reach the plasma conditions required for a net release of fusion power it is necessary to first develop many new technologies. In this context the term "scientific feasibility" cannot be precisely defined. To some investigators it means simply the achievement of the basic plasma conditions necessary to reach the break-even surface in the illustration on page 213. To others it represents reaching the same surface—but with a system that can be enlarged to a full-scale, economic power plant. To a few it represents the attainment of a full understanding of all the phenomena involved.

Although these differing interpretations of what is needed to give confidence in our ability to construct a fusion reactor may be somewhat confusing, each interpretation nevertheless contains a modicum of truth. To depict the complexity of this drive toward the goal of fusion power we have prepared the highly schematic illustration on the opposite page. The goal is to achieve useful, economic fusion power. The major routes to the goal are ordered in the illustration according to plasma density. Various individual experiments have climbed past various obstacles to reach positions close to the break-even level. In fact, in some instances two of the three essential parameters (density, temperature and confinement time) have already been achieved. The ignition temperature has been achieved in a number of cases. The rest of the climb to the break-even level in some cases involves a better understanding of the physics of the plasma-confinement system, but in others it may involve only engineering problems. Indeed, the location of the break-even level is a function of the technology used. Direct energy conversion, for example, would lower this level.

The next portion of the climb, the

construction of a prototype reactor, will be a function of the route taken to scientific feasibility. For instance, if a deuterium-tritium mixture is the fuel, this would require the development of components such as lithium blankets, large superconducting magnets, radiation-resistant vacuum liners, fueling techniques and heat-transfer technology. If fuels that release most of their energy in the form of charged particles are considered, however, then in the case of mirror reactors direct-conversion equipment may be part of a device used to demonstrate break-even feasibility. The step from that device to a prototype reactor could then be very short because the conversion equipment would be already developed. Other devices would face similar problems of differing magnitude in prototype construction. The final step from a prototype to an operational reactor would proceed through a much more nebulous region in which economic and environmental considerations would influence the comparative desirability of different power plants.

At present the main factor limiting the rate of progress toward fusion power is financial. The annual operating and equipment expenditures for the U.S.

fusion program, when one uses the consumer price indexes to adjust these dollars for inflation, has remained fairly constant for the past eight years [*see illustration below*]. The total amount spent on the program since its inception is the cost equivalent of a single Apollo moon shot. The annual funding rate of about $30 million per year is the equivalent of 15 cents per person per year in the U.S.

The road to fusion power is a cumulative one in that successive advances can be built on earlier advances. At present the U.S. has a fairly broad program of investigations approaching the break-even surface for net energy release. It is essential that larger (and thus more expensive) devices be built if the goal of the break-even surface is to be reached. The surface should be broken through in a number of places so that the relative advantages of the possible routes beyond that surface to an eventual fusion-power reactor can be assessed.

Clearly the timetable to fusion power is difficult to predict. If the level of effort on fusion research remains constant or decreases slightly, the requirement for larger devices and advanced engineering will automatically cause a premature

narrowing of the density range under investigation. This increases the risk of reaching the goal in a given time scale. To put it another way, it extends one's estimate of the probable time scale. If the level of fusion research expands sufficiently to maintain a fairly broad program across the entire density range, the probability of success increases and the probable time scale decreases. If fusion power is pursued as a "national objective," expanded programs could be carried out across the entire density range accompanied by parallel strong programs of research on the remaining engineering and materials problems to determine as quickly as possible the best routes to practical fusion-power systems. Therefore, depending on one's underlying assumptions on the level of effort and the difficulties ahead, the time it would take to produce a large prototype reactor could range from as much as 50 years to as little as 10 years.

There is at least one case in which the fusion break-even surface could be reached without making any new scientific advances and without developing any new technologies. This "brute force" approach, which might not be the optimum route to an eventual power reactor, would involve simply extending the length of the existing theta-pinch linear devices. It has been estimated that to reach the break-even surface by this method such a system would have to be about 2,000 feet long—less than a fifth of the length of the Stanford Linear Accelerator. This one fusion device, however, would cost an order of magnitude more than any experimental fusion device built to date. Even though a simple scaling of this type would introduce no new problems in plasma physics, one could not exclude the possibility of unexpected difficulties arising solely from the extended length of the system.

The length of such a device could be shortened by as much as 90 percent by installing magnetic mirrors at the ends, by increasing the diameter of the plasma or by making the system toroidal, but these steps would introduce new physical conditions. The system could also be shortened by the use of a direct energy-conversion approach, but this would introduce an unproved technology. At present a significant portion of the fusion-power program is concentrating on developing the new physics and technology that would reduce the cost of such break-even experiments. This continuing effort is sustained by the growing conviction that the eventual attainment of a practical fusion-power reactor is not blocked by the laws of nature.

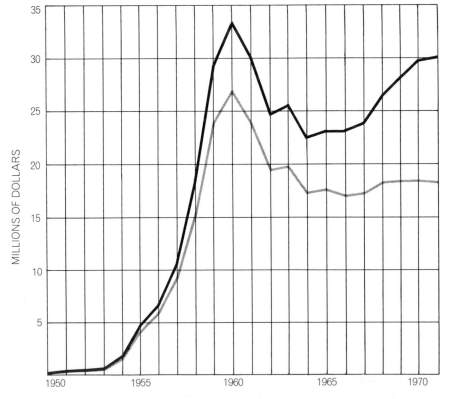

FINANCIAL SUPPORT is currently the main factor limiting the rate of progress toward the goal of fusion power. The solid curve shows the annual operating and equipment expenditures for the U.S. fusion program. The gray curve shows these expenditures adjusted for inflation. The adjustment shows that fusion research has been funded by the Atomic Energy Commission at an essentially constant rate for the past eight years. Smaller research programs have been funded by both private industry and other Government agencies.

23

THE HYDROGEN ECONOMY

DEREK P. GREGORY

January 1973

*A case is made for an energy regime in which all
energy sources would be used to produce hydrogen,
which could then be distributed as a nonpolluting
multipurpose fuel*

The basic dilemma represented by what has been termed the "world energy crisis" can be simply stated: At the very time that the world economy in general and the economies of the industrialized countries in particular are becoming increasingly dependent on the consumption of energy, there is a growing realization that the main sources of this energy—the earth's nonrenewable fossil-fuel reserves—will inevitably be exhausted, and that in any event the natural environment of the earth cannot readily assimilate the byproducts of fossil-fuel consumption at much higher rates than it does at present without suffering unacceptable levels of pollution.

What is not generally recognized is that the eventual solution of the energy problem depends not only on developing alternative sources of energy but also on devising new methods of energy conversion. There is, after all, plenty of "raw" energy around, but either it is not in a form convenient for immediate use or it is not in a location close enough to where it is needed. Most of the research-and-development effort in progress in the U.S. on the energy problem is devoted to finding ways to convert chemical energy (derived from fossil fuels), nuclear energy (derived from fission or fusion reactions) and solar energy (derived directly from the sun) into electrical energy.

At present nuclear-fission plants supply about 1.6 percent of the electricity consumed in the U.S. (Of the remainder, fossil-fuel plants supply about 82 percent and hydroelectric plants about 16 percent.) Assuming that the development of economically feasible "breeder" reactors will soon eliminate any short-term concern about the resource limitation of nuclear energy, then by the year 2000 nuclear plants may be supplying as much as half of the nation's electricity.

If this projection is correct, and if the "energy gap" of the future is to be filled with nuclear power made available to the consumer in the form of electricity, then the U.S. will have gone a long way toward becoming an "all-electric economy." This trend can be detected already: the demand for electricity is currently growing in the U.S. at a much higher rate than the overall energy demand [*see illustration on next page*]. It has been estimated that whereas the overall U.S. energy consumption will double by the year 2000, the demand for electricity will increase about eightfold, raising the electrical share of total energy consumption from about 10 percent to more than 40 percent.

The question naturally arises: How desirable is this trend toward a predominantly electrical economy? Specifically, are there any other forms of energy that can be delivered to the point of use more cheaply and less obtrusively than electrical energy can? Consider such major energy-consumption categories as transportation, space heating and heavy industrial processes, all of which are primarily supplied today with fossil-fuel energy, mainly for reasons of economy and portability. As the fossil fuels run out, they will become more expensive, making the direct use of nuclear electrical energy relatively more economical. In this situation a case can be made for utilizing the nuclear-energy sources indirectly to produce a synthetic secondary fuel that would be delivered more cheaply and would be easier to use than electricity in many large-scale applications. In this article I shall discuss the merits of what I consider to be the leading candidate for such a secondary fuel: hydrogen gas.

In many respects hydrogen is the ideal fuel. Although it is not a "natural" fuel, it can be readily synthesized from coal, oil or natural gas. More important, it can be produced simply by splitting molecules of water with an input of electrical energy derived from an energy source such as a nuclear reactor. Perhaps the greatest advantage of hydrogen fuel, however, at least from an environmental standpoint, is the fact that when hydrogen burns, its only combustion product is water! None of the traditional fossil-fuel pollutants—carbon monoxide (CO), carbon dioxide (CO_2), sulfur dioxide (SO_2), hydrocarbons, particulates, photochemical oxidants and so on—can be produced in a hydrogen flame, and the small amount of nitrogen oxide (NO) that is formed from the air entering the flame can be controlled. Moreover, assuming that the energy options are restricted to the use of effectively "unlimited" materials such as air and water, hydrogen is by far the most readily synthesized fuel.

In principle, then, one can envision an energy economy in which hydrogen is manufactured from water and electrical energy, is stored until it is needed, is transmitted to its point of use and there is burned as a fuel to produce electricity, heat or mechanical energy [*see illustration on opposite page*]. Such a hypothetical model is not without its problems and disadvantages, but on balance the benefits appear to be so great that I believe at the same time that we are moving toward an "electric economy" we should also be moving toward a "hydrogen economy."

Just as the food and beverage industry has found it uneconomical to collect and reuse empty containers, so the present energy industry cannot afford to collect and recycle used "energy containers": the by-products of the combustion necessary to produce the energy. The

drawback in both cases is that the "no deposit, no return" system throws the burden of recovery and recycling onto the environment. Apart from the obvious harmful effect on the earth's atmosphere, this kind of energy cycle suffers from the further disadvantage of having an extremely slow step of several million years' duration for the re-formation of fossil fuels from atmospheric carbon dioxide [*see illustration on page 222*]. That is the basic reason we are running out of fossil-fuel reserves. In the hydrogen cycle, in contrast, only water is deposited into the atmosphere, where it rapidly equilibrates with the abundant and mobile water supply on the earth's crust. At another location the water is reconverted to hydrogen. The system is characterized by negligible delay and does not disturb the environment, yet it relies on the environment to carry out the "return empty" function. Assuming the availability of an abundant supply of nuclear or solar energy, this system can be operated as rapidly as the demand requires without depleting any natural resources.

The idea of using hydrogen as a synthetic fuel is far from new. In 1933 Rudolf A. Erren, a German inventor working in England, suggested the large-scale manufacture of hydrogen from off-peak electricity. He had done extensive work on modifying internal-combustion engines to run on hydrogen, and the main object of his suggestion was to eliminate automobile-exhaust pollution and to relieve pressure on the importation of oil into Britain. (It is interesting to note that 40 years later the U.S. is concerned with the same two problems: automobile pollution and an increasing dependence on oil imports.)

Others have suggested using hydrogen as a fuel or as a means of storing energy. F. T. Bacon, a pioneer in the development of fuel cells in England since the 1930's, has always had as his ultimate objective the development of a hydrogen-energy storage system using reversible electrolyzer fuel cells. More recently the U.S. Atomic Energy Commission sponsored a series of studies during the 1960's of "nuplexes"—nuclear-agricultural-industrial complexes that derive all their energy from a single nuclear reactor. The AEC studies included the concept of water electrolysis to provide hydrogen as a precursor to the manufacture of fertilizers and chemicals. Within the past two years several articles have appeared in engineering and scientific journals proposing active studies of the production, transmission, storage and utilization of hydrogen in both combustion appliances and engines. Such studies are in progress at several universities and industrial research laboratories in the U.S. and abroad, including my own institution, the Institute of Gas Technology in Chicago, where our work is sponsored by the American Gas Association.

The difficulty of transporting hydrogen has historically prevented its use as a fuel. Clearly some better method than compressing it in steel cylinders has to be found. Storage and transportation as liquid hydrogen are already in use; metal hydrides and synthetic organic or inorganic hydrides have also been considered and have promise. There is no reason, however, why hydrogen should not be distributed in the same way that natural gas is distributed today: by underground pipelines that reach most industries and more than 80 percent of the homes in this country.

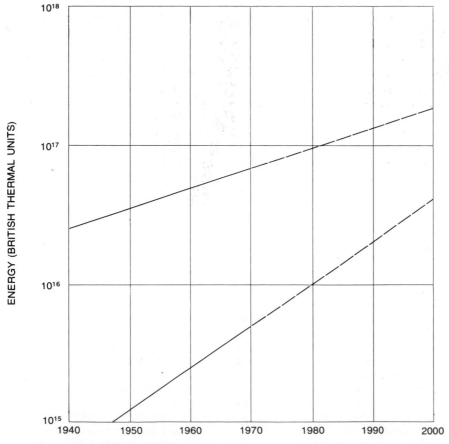

ACCELERATING TREND toward an "all electric" economy is evident in this graph, which shows that the demand for electricity (*bottom line*) is growing in the U.S. at a much higher rate than the overall energy demand (*top line*). Assuming that the trend continues, the U.S. is heading for a predominantly electrical economy sometime in the 21st century. The data are from the U.S. Department of Commerce and the Edison Electric Institute.

Before weighing the merits of the hydrogen-economy concept, it is instructive to consider the alternative: the all-electric economy. Suppose for a mo-

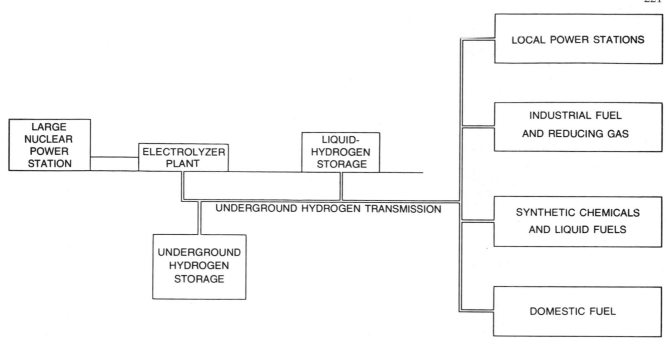

HYDROGEN ENERGY ECONOMY would operate with hydrogen as a synthetic secondary fuel produced from water in large nuclear or solar power stations (*left*). The hydrogen would be fed into a nationwide network of underground transmission lines (*center*), which would incorporate facilities for storing the energy, either in the form of hydrogen gas underground or in the form of liquid hydrogen aboveground. The hydrogen would then be distributed as it is needed to energy consumers for use either as a direct heating fuel, as a raw material for various chemical processes or as a source of energy for the local generation of electricity (*right*).

ment that one does not consider synthesizing a secondary chemical fuel; then one must face the prospect of generating and transmitting very large quantities of electricity. To meet the rising demand for electricity in the U.S. new generating stations are already being constructed in sizes larger than ever before. A few years ago a 500-megawatt power station was considered a giant. Today 1,000-megawatt stations are typical, and the electrical industry is contemplating 10,000-megawatt installations for the future.

In spite of the intensive efforts of their designers, the efficiency of steam-driven electric-power stations is still fairly low: about 40 percent for a modern fossil-fuel plant and 33 percent for a nuclear plant [see "The Conversion of Energy," by Claude M. Summers; Scientific American Offprint 668]. As a result the waste heat released from these large plants, or clusters of plants, is considerable. Accordingly they must be located near large bodies of water where ample cooling is available or in open country where cooling to the atmosphere will have no adverse local effects. Concern over the safety of nuclear reactors is also having a strong influence on the location of such plants. Because of these constraints the huge power stations of the future are likely to be built at distances of 50 miles or more from the load centers. Power stations located on offshore

platforms floating in the ocean are already planned for the U.S. East Coast.

Power must be moved from the generating stations to the load centers. High-voltage overhead cables are expensive, in terms of both equipment costs and the land they occupy, and they are vulnerable to storm damage. Moreover, the electrical industry is encountering considerable resistance to the continued stringing of overhead power-transmission lines in many areas. Underground cables for carrying bulk power cost at least nine times (and sometimes up to 20 times) as much as overhead lines and thus are far too expensive to be used over long distances. Underground transmission is used only where the expense is justified by other considerations, such as aesthetic appearance or very expensive right-of-way. Much work is being done to develop cryogenic superconducting cables, which would allow large currents to be carried underground at a reasonable cost. At present, however, the technology is still at an early stage of development.

Some form of electrical storage would be of great value to the electrical industry, because power stations work most efficiently when operated at constant output at their full rated load. Since consumer demand varies widely both seasonally and during the day, however, the generating rate must be adjusted continuously. The only practical way available today to store large quantities of electrical energy is the pumped-storage plant, a reversible hydroelectric station; unfortunately only a limited number of sites are geographically suitable for such systems.

Thus it appears that several of the problems faced by the electrical industry—the siting of power stations, the expense of underground transmission and the lack of storage—are being amplified by factors that lead to larger and more remote power stations. The hydrogen-economy concept could help to alleviate these problems.

Hydrogen can be transmitted and distributed by pipeline in much the same way that natural gas is handled today. The movement of fuel by pipeline is one of the cheapest methods of energy transmission; hydrogen pipelining would be no exception. A gas-delivery system is usually located underground and is therefore inconspicuous. It also occupies less land area than an electric-power line. Hydrogen can also be stored in huge quantities by the very same techniques used for natural gas today.

Let us take a look at the existing gas-transmission network in the U.S. In 1970 a total mileage of 252,000 miles of trunk pipeline was in operation, carrying a total of 22.4 trillion cubic feet of gas during the year [see *illustration on pages 224 and 225*]. Such a pipeline system is needed because natural-gas sources are located in certain parts of the country, whereas markets for the gas exist in other areas.

In the hydrogen economy hydrogen gas would be produced from large nuclear-energy (or solar-energy) plants located in places that provide optimum cooling and other environmental facilities. Even coal-fueled hydrogen generators, located close to the mine mouths, could be integrated into this power-generation network. A pipeline transmission system would grow up to link these locations to the cities in a way analogous to the growth of the natural-gas transmission system.

The technology for the construction and operation of natural-gas pipelines has been well developed and proved. A typical trunk pipeline, 600 to 1,000 miles long, consists of a welded steel pipe up to 48 inches in diameter that is buried underground with appropriate protection against mechanical failure and/or electrochemical corrosion. Gas is pumped along the line by gas-driven compressors spaced along the line typically at 100-mile intervals, using some of the gas in the line as their fuel. Typical line pressures are 600 to 800 pounds per square inch, but some systems operate at more

than 1,000 pounds per square inch. A typical 36-inch pipeline has a capacity of 37,500 billion British thermal units (B.t.u.) per hour, or in electrical equivalent units 11,000 megawatts, roughly 10 times as much as a single-circuit 500-kilovolt overhead transmission line.

Natural gas is not the only gas to be moved in bulk pipelines, although no other gas is moved on such a scale. Carbon dioxide, carbon monoxide, hydrogen and oxygen are all delivered in bulk by pipeline. So far industry has had no incentive to pipeline hydrogen in huge quantities over great distances, but where it now pipelines hydrogen over short distances it uses conventional natural-gas pipeline materials and pressures. There is no technical reason why hydrogen cannot be pipelined over any distance required.

Because of the lower heating value of hydrogen (325 B.t.u. per cubic foot compared with about 1,000 B.t.u. per cubic foot for natural gas) three times the volume of hydrogen must be moved in order to deliver the same energy. Hydrogen's density and viscosity are so much lower, however, that the same pipe can handle three times the flow rate of hydrogen, although a somewhat larger compressor energy is required. Thus where existing

pipelines happen to be suitably located, they could be converted to hydrogen with the same energy-carrying capacity.

In the hydrogen economy it will be possible to store vast quantities of hydrogen to even out the daily and seasonal variations in load. Natural gas is stored today in two ways: in underground gas fields and as a cryogenic liquid. At 337 locations in the U.S. natural gas is stored in underground porous-rock formations with a total capacity of 5,681 billion cubic feet. Whether hydrogen can be stored in underground porous rock can be finally ascertained only by future field trials. At present, however, 30 billion cubic feet of helium, a low-density gas with leakage characteristics similar to those of hydrogen, is stored quite satisfactorily in an underground reservoir near Amarillo, Tex.

Cryogenic storage of natural gas is a rapidly growing technique; at 76 locations in the U.S. "peak shaving" operations involving liquefied natural gas are in use or under construction. There is no technical reason why a similar peak-shaving technique cannot be employed with liquid hydrogen. Liquid hydrogen used to be considered a hazardous laboratory curiosity, but it is already being used as a convenient means of storing

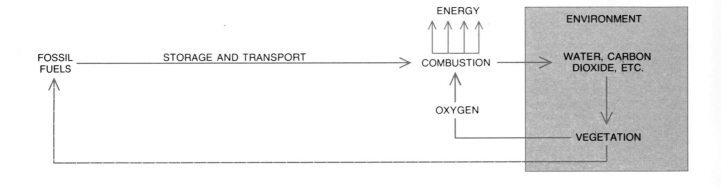

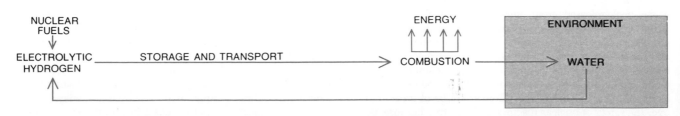

ENVIRONMENTAL EFFECTS of the present fossil-fuel energy cycle and the proposed hydrogen-fuel energy cycle are compared here. When fossil fuels are burned to release their stored energy (top), the environment is relied on to accommodate the combustion by-products. The re-formation of the fossil fuels from atmospheric carbon dioxide takes millions of years (broken line). On the other hand, when hydrogen is burned as a fuel (bottom), the only combustion product is water, which is easily assimilated by the environment. The fuel cycle is completed rapidly without depleting limited resources or accumulating harmful waste products.

and transporting hydrogen over long distances. Liquid hydrogen is regularly shipped around the U.S. in railroad tank cars and road trailers. The technology for the liquefaction and tankage of hydrogen has already been developed, mainly for the space industry. Indeed, the largest liquid-hydrogen storage tank is at the John F. Kennedy Space Center; it has a capacity of 900,000 gallons, equivalent to 37.7 billion B.t.u. or 11 million kilowatt-hours [*see illustration at right*]. Although the energy content of this tank is only about 4 percent of the energy content of a typical liquid-natural-gas peak-shaving plant, its energy capacity is 73 percent of the capacity of the world's largest pumped-storage hydroelectric plant, located at Ludington, Mich.

The cryogenic approach to energy storage has the advantage of being applicable in any location, no matter what the geography or geology, factors that limit both underground gas storage and pumped hydroelectric storage.

The simplest way to manufacture hydrogen using nuclear energy is by electrolysis, a process in which a direct electric current is passed through a conductive water solution, causing it to decompose directly into its elementary constituents: hydrogen and oxygen. Complete separation of the two gases is achieved, since they are evolved separately at the two electrodes. Salts or alkalis, which have to be added to the water to increase conductivity, are not consumed; thus the only input material required is pure water.

A number of large-scale electrolytic hydrogen plants are operated today in locations where hydrogen is needed (for example in the manufacture of ammonia and fertilizers) and where cheap electric power (usually hydroelectric power) is available. One of the largest commercial electrolyzer plants in the world is operated by Cominco, Ltd., in British Columbia [*see illustration on page 226*]. This plant consumes about 90 megawatts of power and produces about 36 tons of hydrogen per day for synthesis into ammonia. The by-product oxygen is used in metallurgical processes. Similar large plants are located in Norway and Egypt. Many smaller plants exist where hydrogen is produced from unattended equipment.

The theoretical power required to produce hydrogen from water is 79 kilowatt-hours per 1,000 cubic feet of hydrogen gas. In practice the large industrial plants are only about 60 percent

ENERGY STORAGE in the form of liquefied hydrogen is already a routine practice in the space industry. This vacuum-insulated cryogenic tank at the John F. Kennedy Space Center, for example, contains 900,000 gallons of liquid hydrogen for fueling the Apollo rockets. It is the largest facility of its kind in existence. In terms of energy its contents are equivalent to 37.7 British thermal units (B.t.u.) of heat or 11 million kilowatt-hours of electricity.

efficient; a typical power-consumption figure is 150 kilowatt-hours per 1,000 cubic feet of hydrogen. This power requirement represents a major part of the plant's operating cost. Thus there is a considerable incentive—indeed, a real need—to increase the operating efficiency of such plants if one is to consider using electrolytic hydrogen as a fuel.

The fuel cell, the subject of intensive research and development as part of the space program over the past 15 years, is really an electrolyzer cell operating in reverse. The simplest fuel cell to build and operate is one that operates on hydrogen and oxygen, yielding water and electric power as its products. Hydrogen-oxygen fuel cells were selected and developed for both the Gemini and the Apollo programs because of their high efficiency, which reduces the amount of fuel needed aboard the spacecraft to supply its electric power. Much effort has gone into developing fuel cells with high efficiencies. This same technology can be applied to increase the efficiency

of the reverse process: electrolysis. Electrolytic cells are operating in aerospace laboratories today with an efficiency of more than 85 percent.

Increasing the electrolyzer efficiency alone has relatively little merit as long as the present power-station efficiency in converting nuclear heat to electric power is only about 33 percent. This efficiency loss can, however, also be circumvented. For example, Cesare Marchetti at the Euratom laboratories in Italy has designed a chemical process for the thermal splitting of water to hydrogen and oxygen directly using the heat energy produced by a nuclear reactor. If water is to be split into its elements directly, it must be heated to very high temperatures—about 2,500 degrees Celsius—to achieve dissociation. Not only are such temperatures not available from nuclear reactors but also the gases cannot conveniently be separated from each other before they recombine. It is possible to conceive of a two-stage reaction in which a metal, say, reacts with steam at a reasonable temperature to produce hydrogen and a metal oxide. The hydrogen is easily separated from the metal oxide, which in turn could be decomposed to oxygen and the metal by the application of heat. Unfortunately there does not appear to be any suitable metal that undergoes such a series of reactions at temperatures low enough to be compatible with nuclear reactors, whose construction materials limit operating temperatures to about 1,000 degrees C.

Marchetti's concept, therefore, is a far more complex reaction sequence involving calcium bromide ($CaBr_2$), water (H_2O) and mercury (Hg), in which, except for the hydrogen and oxygen, all the reactants are recycled. Each of the reactions proceeds at temperatures below 730 degrees C., which can be achieved in a nuclear reactor. Although the process appears to be feasible, development work is still required to try to bring the overall efficiency up and the cost down to practical limits.

The quantities of hydrogen that the hydrogen economy would require are immense. For example, if we were to produce today an amount of hydrogen equivalent to the total production of natural gas in the U.S., we would have to provide during one year the same fuel value as 22.5 trillion cubic feet of gas, or 22.5 quadrillion (10^{15}) B.t.u. of energy. This corresponds to about 70 trillion cubic feet of hydrogen, which, if we could produce it at a steady rate all year round from nuclear electrolytic plants, would require an electrical input of more than a million megawatts. The present total electrical generating capacity in the U.S. is 360,000 megawatts, so that we are envisioning a fourfold increase in generating capacity, which would require the construction of more than 1,000 new 1,000-megawatt power stations. That is in addition to the rapidly increasing demand for electric power for other uses. During the past five years, in contrast, the electrical generating capacity in the U.S. has grown by "only" 105,000 megawatts.

Such a formidable task of increasing capacity, however, does not follow solely from our turning to a hydrogen economy. As our huge consumption of fossil fuels declines in future years, we must provide at least an equivalent alternative energy source. Such numbers give a taste of the energy revolution that must take place within the next half-century.

At present the cheapest bulk hydrogen is made from natural gas. Clearly since hydrogen from such a source cannot be cheaper than the starting materi-

al, it cannot therefore be expected to replace natural gas as a fuel. Electrolytic hydrogen is even more expensive, unless very cheap electric power is available. Today's electricity prices are based on supplying a fluctuating load, but the capability of hydrogen storage would even out the load and might reduce the price of electricity somewhat.

Although the cost of hydrogen produced from electricity must always be higher than the cost of the electricity, it is the lower transmission and distribution cost of hydrogen compared with electricity that makes it advantageous to the user. The latest economic figures published by the gas and electrical industries can be used to derive the production, transmission and distribution shares of average prices, charged to all types of customers, for gas and electricity, and these data can be compared in turn with corresponding figures for hydrogen made by electrolysis [see illustration on page 227]. The figures for hydrogen are derived from the hypothetical assumption that all the electricity generated in the U.S. in 1970 was converted to hydrogen, which was sent through the existing natural-gas transmission network (for an average distance of 1,000 miles) and was delivered to customers as a gaseous fuel. The electrolysis charge of 56 cents per million B.t.u. is derived from AEC estimates of the cost of building advanced electrolyzer cells. The hydrogen transmission and distribution costs are based on natural-gas costs, adjusted to take account of the different physical properties and safety factors for handling hydrogen.

Two things are obvious from such a comparison. One is that today it is far cheaper for the average customer to buy energy in the form of natural gas than it is in the form of electricity. The other is that it should already be possible to sell hydrogen energy to the gas user at a lower price than he now pays for electricity. Clearly, however, this hydrogen will find no markets while natural gas is as cheap as it is.

Looking to the future, we see that natural-gas prices, together with all fossil-fuel prices, will increase rapidly. These rises are brought about by their short supply, by the influence of pollution regulations and by such social pressures as land conservation and employee welfare applied to the mining industry. In contrast, the price of nuclear energy, although apparently rising fast now, can be expected to stabilize somewhat in the breeder-reactor era because there will then be no severe supply limit.

It is not possible at this time to fore-

cast accurately what the cost of hydrogen energy is likely to be, but one can certainly look forward to considerably increased prices for all forms of energy. Even so, in the long run delivered hydrogen will be cheaper than delivered natural gas and very probably also cheaper than delivered electricity.

When hydrogen becomes as universally available as natural gas is today, it will easily perform all the functions of natural gas and others besides. Hydrogen can be used in the home for cooking and heating and in industry for heating; in addition it can serve as a chemical raw material in many industries, including the fertilizer, foodstuffs, petro-

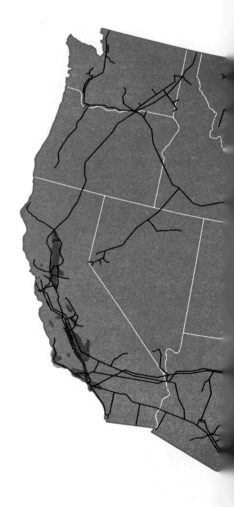

TRUNK PIPELINES extending for 252,000 miles (**black lines**) already exist in the U.S. for transmission of natural gas from areas

chemical and metallurgical industries. Hydrogen can also be used to generate electricity in local power stations.

The combustion properties of hydrogen are considerably different from those of natural gas. Hydrogen burns with a faster, hotter flame, and mixtures of hydrogen with air are flammable over wider limits of mixture. These factors mean that burners of hydrogen must be designed differently from those of natural gas and that modification of every burner will be necessary on changeover. Such widespread modification is not without precedent. A similar operation was carried out when the U.S. changed from manufactured gas (about 50 percent hydrogen) to natural gas; several European countries have recently undertaken the same conversion.

Hydrogen, because it burns without noxious exhaust products, can be used in an unvented appliance without hazard. Hence it is possible to conceive of a home heating furnace operating without a flue, thereby saving the cost of a chimney and adding as much as 30 percent to the efficiency of a gas-fired home heating system. More radical changes are possible, moreover, because without the need for a flue the concept of central heating itself is no longer necessary. Each room can have its heat supplied by unflued peripheral heating devices operating on hydrogen independently of one another. Indeed, the vented water vapor would provide beneficial humidification. Another radical change is the potential use of catalytic heaters. Since hydrogen is an ideal fuel for catalytic combustion, true "flameless" gas heating is possible, with the catalytic bed being maintained at any desired temperature, even as low as 100 degrees C. This prospect promises to revolutionize domestic heating and cooking techniques in the future. With such low temperatures it is virtually impossible to produce nitrogen oxides, thus eliminating the only possible pollutant from a hydrogen system.

Hydrogen is also the ideal fuel for fuel cells. The technological problems that have faced the development of practical, commercially economical fuel cells for

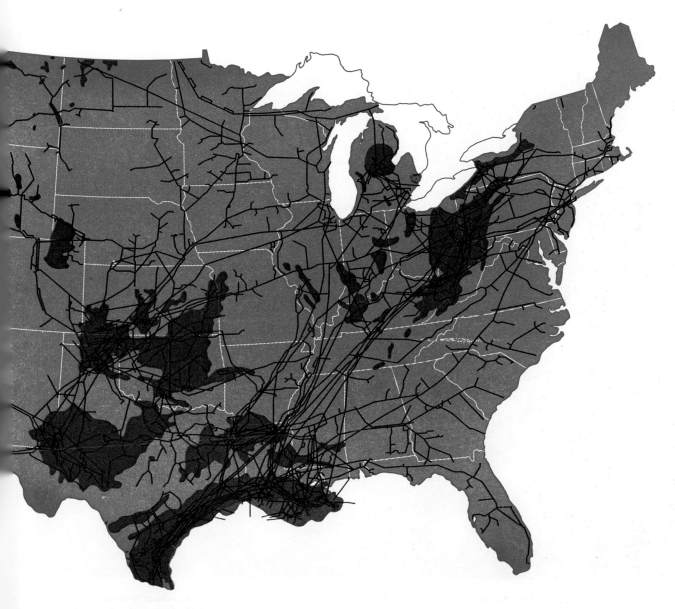

where the gas is produced (*gray*) to areas where it is consumed. The system, which is constructed almost entirely of welded steel pipe, carries approximately 61.4 billion cubic feet (or 1.5 million tons) of natural gas per day. Similar networks of underground hydrogen-gas pipelines would enable the giant nuclear (or solar) power stations of the future to be located far from the load centers.

more than a decade are very much reduced if hydrogen can be used as fuel. Fuel-cell electricity generators operating on hydrogen should be at least 70 percent efficient and can realistically be expected to find a place in the home, in commercial and industrial buildings and in industry. Larger, urban electrical generating stations could be fuel-cell systems or could be hydrogen-fueled steam stations. An earlier concept of operating a closed-cycle steam-turbine system on a hydrogen-oxygen fuel supply could become practical through the use of rocket-engine technology. Workers at the Massachusetts Institute of Technology have proposed such a system for submarines; it has been reported that an overall efficiency of 55 percent can be anticipated from it.

Hydrogen is an excellent fuel for gas-turbine engines and has been proposed as a fuel for supersonic jet transports.

For this kind of use fuel storage and tankage as liquid hydrogen are practical. Although the large volume required may make its use less attractive for subsonic aircraft, the very considerable saving in weight over an equivalent fuel load of kerosene gives hydrogen a distinct advantage. Conventional internal-combustion engines will also operate on hydrogen if they are suitably modified or redesigned. R. J. Schoeppel of Oklahoma State University and others have shown that if hydrogen is injected into the engine through a valve in a manner similar to the way fuel is injected into a diesel engine, the preignition characteristics of hydrogen are overcome. Others, including Marc Newkirk of the International Materials Corporation and Morris Klein of the Pollution Free Power Corporation, have reported satisfactory operation of conventional automobile engines on hydrogen using carburetor and manifold

modifications. Meanwhile William J. D. Escher of Escher Technology Associates has proposed a radically different approach to automobile engine design, using a steam system fueled by both hydrogen and oxygen. The use of liquid hydrogen as a routine private-automobile fuel is questionable on the ground of safety, although it is probably applicable to fleet users, such as bus lines and taxicab fleets.

Richard H. Wiswall, Jr., and James J. Reilly of the Brookhaven National Laboratory have proposed the use of metallic hydrides to store hydrogen as a fuel for vehicles. A magnesium-alloy hydride will store hydrogen energy as efficiently (on a weight basis) as a tank of liquid hydrogen, but some technical problems must still be overcome. At present there seems to be no single, obvious way in which automobiles can be operated on hydrogen fuel, but considerable work is

LARGE ELECTROLYZER PLANT for the production of hydrogen by the electrical decomposition of water is operated by Cominco, Ltd., in British Columbia. The 3,200 electrolytic cells, which cover more than two acres, consume about 90 megawatts of power and produce about 36 tons of hydrogen per day for synthesis into ammonia. By-product oxygen is used in metallurgical processes.

going on to investigate the various options available. If one has to synthesize a suitable liquid fuel for automobiles and aircraft, the starting material for the fuel must be hydrogen in any case.

One of the main criticisms of the hydrogen-economy concept is that hydrogen is too dangerous for use in this way. Undoubtedly hydrogen is a hazardous material and must be handled with all due precautions. If it is handled properly, however, in equipment designed to ensure its safety, anyone should be able to use it without hazard.

In the days of manufactured gas (gas made from coal), which consisted of up to 50 percent hydrogen and contained about 7 percent carbon monoxide, people managed to live with the fire and explosion hazards of hydrogen as well as the toxic hazards of carbon monoxide. Of course, it takes only one major disaster to alert everyone to a hazard. The most famous hydrogen accident, the *Hindenburg* airship disaster of 1937, is still remembered with awe. Indeed, the almost universal fear of hydrogen has been described as the "*Hindenburg* syndrome." Spectacular as it was, however, that fire was almost over within two minutes, and of the 97 persons on board, 62 survived.

Very strict codes are enforced for the use of natural gas today; even stricter ones are applied to industry for the use of hydrogen. Most of these codes are realistically based on reducing the chances of accidents. Just as we have designed apparatus and procedures to enable us to fill our automobile tanks with gasoline and carry the resulting 20-gallon "fire bomb" at speeds of up to 70 miles per hour along a crowded highway and park it overnight right inside our homes, we can surely devise safe practices for handling hydrogen.

Hydrogen cannot be detected by the senses, so that a leak of pure hydrogen is particularly hazardous. Odorants are routinely used to make natural-gas leaks obvious, however, and no doubt the same can be done with hydrogen. Hydrogen flames are also almost invisible and are therefore dangerous on this score. Hence an illuminant may have to be added to the gas to make the flame visible. The flammability limits of hydrogen mixed with air are very wide, from 4 to 75 percent. It is the lower limit, almost the same as that for methane (5 percent in air), that causes the fire hazard with a gas leak. On the benefit side, however, since hydrogen is so much lighter than air and diffuses away at a

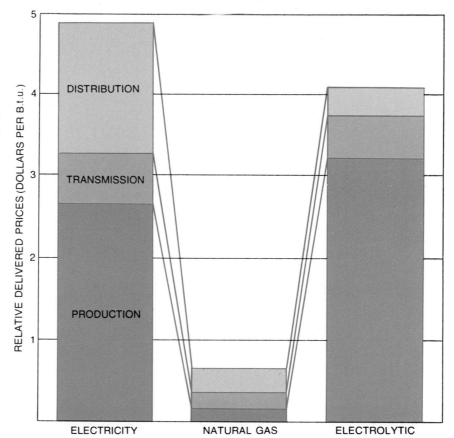

RELATIVE DELIVERED PRICES of various forms of energy are broken down in this bar chart into the shares represented by production (*solid color*), transmission (*intermediate color*) and distribution (*light color*). The comparison reveals that at present it is much cheaper to buy energy in the form of natural gas than in the form of electricity. Moreover, the breakdown shows that although the cost of hydrogen produced from electricity must always be higher than the cost of the electricity, the lower transmission and distribution costs of hydrogen already make it possible to sell hydrogen energy to the gas user at a delivered price lower than what he now pays for electricity. It is expected that natural-gas prices, together with all fossil-fuel prices, will increase rapidly in the future.

far greater rate than methane, a hydrogen leak could actually be less hazardous than a natural-gas leak. The most significant hazardous property of hydrogen is the extremely low energy required to ignite a flammable mixture: only a tenth of the energy required to ignite a gasoline-air mixture or a methane-air mixture and well within the energy levels of a spark of static electricity (a probable cause of the *Hindenburg* fire, which occurred just after a thunderstorm). Thus safety practices will have to be based on the assumption that if a hydrogen fire can occur, it will! Huge quantities of hydrogen are handled in industry quite safely and without accident precisely because proper precautions are taken.

To recapitulate briefly, our recoverable fossil-fuel supplies will sooner or later become exhausted; we are already feeling the effects of the limited supply by having to pay more for fossil-based energy. Within the next 50 years we

must be prepared to pay considerably more for energy from all sources, particularly for fossil fuels. One way of handling nuclear and other energy sources is to use them to convert water to hydrogen in large central plants and then to use hydrogen as a clean, nonpolluting fuel. Technically this is already feasible; only relatively simple developments have to be made, not approaching the magnitude of the technical tasks of developing the alternative energy sources— breeder reactors and solar engines— themselves. Economics and safety are the two obstacles to developing such a hydrogen economy. A combination of technical development and the expected adjustment in relative energy prices can justify the economics, and proper practices and design can ensure safety. If and when we move into a hydrogen economy, the world will undoubtedly be a far cleaner place to live in than it is today.

IV

THE LEGACY OF ENERGY USE

III

THE LEGACY OF ENERGY USE

INTRODUCTION

In 1970, the people of the United States in one way or another consumed energy resources amounting to nearly 68×10^{15} BTUs, or 17×10^{15} kilocalories. About six percent was devoted to material uses, in such forms as chemical feedstocks, lubricants, and paving materials. The remaining 64×10^{15} BTUs represent (except for a small contribution from hydropower) the heat released from fuel materials for conversion to work in our behalf. Of that, approximately 61×10^{15} BTUs were produced by the combustion of fossil fuels, and thus could be said to derive from one general chemical reaction, which can be written as

$$C_xH_y + \left(x + \frac{y}{4}\right) O_2 \rightarrow xCO_2 + \frac{y}{2} H_2O + \text{heat}$$

The equation is deceiving in its simplicity, showing only the net outcome of a complicated series of reactions that take place in a flowing system whose temperature and composition profile are markedly inhomogeneous. It is the kinetic complexity of the burning process and its sensitivity to many variables that makes it so difficult to carry out the oxidation reaction cleanly and completely. "High Temperatures: Flame," by Bernard Lewis, gives some feeling for these factors as it reviews the characteristics of luminous combustion. Of particular interest is the premature quenching of intermediate reactions, which allows escape of hydrocarbon fragments and carbon monoxide from the combustion zone. Internal-combustion engines are especially prone to this problem, and contribute heavily to the burden of atmospheric pollutants.

Another complication arises from the fact that the composition of the reactants is not limited to carbon, hydrogen, and oxygen. Coal and petroleum contain many elements, metallic and nonmetallic, from trace amounts up to several percent by weight. Some of these, notably nitrogen and sulfur, are oxidized, and enter the atmosphere as gases; the nonvolatile inorganic components remain as ash and contribute to formation of atmospheric particulates. Oxygen comes into the system diluted with molecular nitrogen, with which it reacts at flame temperatures. It is true that the by-products of combustion are formed in small amounts relative to the release of carbon dioxide, water, and energy—a power plant boiler without emission controls will produce 0.4 to 0.7 pounds of nitrogen oxides for each million BTUs of fuel it consumes (a million BTUs corresponds to about 80 pounds of coal); gasoline engines in 1970 averaged about three pounds of emitted hydrocarbons for an input of a million BTUs; an uncontrolled industrial boiler burning coal containing one percent sulfur introduces a pound and a half of sulfur oxides per million BTUs. The fact remains, however, that a small fraction of a staggering number can still be a staggering number, and consumption of quadrillions of BTUs worth of fuel must lead to production of very large amounts of pollutants.

The effects of emissions from fossil fuel burning tend to be aggravated because they are concentrated in small areas. When the concentrations happen to occur in regions with plentiful sunshine (southern California is the classical example) the situation is doubly aggravated by the photochemical reactivity of the compounds constituting the mixture and by the fact that their reaction products are even more noxious than many of the primary pollutants. Some basic principles of photochemistry are covered in "The Chemical Effects of Light" by Gerald Oster: how electromagnetic radiation of appropriate wavelengths breaks bonds and how the resulting free radicals go on to react with other molecular species present. ("Free Radicals in Biological Systems," in Section V, provides a more detailed review of radical reactions in general.) A qualitative description of the main reactions that go into the formation of photochemical smog is included in A. J. Haagen-Smit's essay "The Control of Air Pollution," in the context of a somewhat historical discussion of institutional and technological measures for alleviating the problem.

Since Haagen-Smit wrote that article, almost ten years ago, the augmentation of our knowledge of air pollution and our attempts to exert greater control over the factors contributing to smog formation have proceeded as a series of single steps. For a time automotive hydrocarbon emissions were reduced at the expense of increased production of nitrogen oxides, and the situation became worse in many areas. Controls on nitrogen oxides followed; federal legislation placing even stricter limitation on emissions of all atmospheric pollutants will come into effect in 1975. While each step was taking place, the energy use that was producing the pollution that we were trying to curb was steadily increasing, wiping out part of our progress at each stage. Even now many of the tactics that were mentioned in the article remain undeveloped and unimplemented. How much better off we would have been if all the control measures described by Haagen-Smit, including the ones requiring a cutback in pollution-generating activities, had been instituted when the article was written. He made the point then, though, and it still holds today, that our attempts to solve the air pollution problem face more road-blocks than those presented by lack of knowledge and technological skill. There are political and social ones as well, and they might just be the more difficult to overcome.

Some progress is being made, however, both in the public's awareness of what must be done and in their willingness to do it. People have finally come to realize that, depending on the surrounding geography and prevailing meteorology, a given body of air can accept only so much contaminant and no more without becoming not only unpleasant but dangerous. One consequence of this realization has been the tendency for communities where pollution is already severe to try to remove large sources of emissions and locate them in less troubled spots. This procedure has led to construction of large power plants in the Southwest desert that threaten to degrade the air there in order to serve distant metropolitan centers. Another tactic involves dispersion of effluents in such a way that high concentrations are not felt locally. Great success has been achieved in Great Britain in reducing local exposure to sulfur dioxide by discharging flue gases through very high smokestacks. It may or may not be pertinent that complaints about very acidic rainfall are said to be arising in the Scandinavian countries.

The problem is that we have very little idea what our planet's overall capacity for dealing with airborne contamination really is. In "The Global Circulation of Atmospheric Pollutants" Reginald Newell confronts that question at least in a preliminary way, comparing the amounts of "contaminating" substances that we introduce into the air with the amounts of the same materials that occur naturally, and analyzing the airflow that carries them around the world. The human contribution of some of these appears to be small in comparison with the natural; for others we seem to be providing quite a substantial portion of the total turnover. The author emphasizes, however, how little data these initial judgments are based on and how much room there is for improvement of our knowledge of the subject.

Looking to the future, what about nuclear energy? Today 95 percent of our country's fuel requirements are met by burning fossil hydro-carbons, resulting in massive production of air pollutants, and less than one percent of the total need is satisfied by nuclear reactions. All projections indicate that the contribution made by nuclear generation of electricity will grow dramatically; the most optimistic suggest that it will account for more than 17 percent of our energy needs by 1985. Extrapolating current growth trends, that would amount to nearly 300,000 megawatts (electrical) of nuclear generating capacity operating at 80 percent of its full potential during that year.

Although nuclear fission power reactors do not emit the atmospheric pollutants characteristic of fossil fuel-fired plants, they do produce their own brand of potentially hazardous materials—radioactive fission products. Tritium, krypton 85, strontium 90, and cesium 137 are major long-lived products that will require careful waste disposal. Other isotopes, including iodine 131, are generated at very high levels within a reactor, but have half-lives sufficiently short that they would present a hazard only in the case of a catastrophic accident.

No *Scientific American* articles are available that deal directly with the generation and handling of nuclear wastes, or the general question of nuclear safety. Two papers do contribute some understanding about what might be termed the environmental chemistry of radionuclides, and are included here even though they are somewhat outdated. In "Radioactive Poisons," Jack Schubert deals with the physiological effects of radioisotopes (see "Free Radicals in Biological Systems" for discussion of some mechanisms of radiation damage), and in "The Circulation of Radioactive Isotopes" James R. Arnold and E. A. Martell provide details on the behavior of those substances in the biosphere. The authors of the latter place strong emphasis on the consequences of atmospheric weapons testing, and although this is no longer a matter of primary concern, the article is included because it provides useful information on global cycling. In the years since both articles were written, more has been learned about the biological pathways that concentrate radioactive contaminants. The trend during that time has been to lower repeatedly the concentrations of radionuclides in air and water that are regarded as "maximum permissible," in response to advances in knowledge. Readers interested in a current discussion of the subject are referred to Earl Cook, "Ionizing Radiation," in *Environment: Resources, Pollution and Society,* William M. Murdoch, Editor, Sinauer Associates, Incorporated, Stamford, Connecticut, 1971.

24

HIGH TEMPERATURES: FLAME

BERNARD LEWIS

September 1954

Our principal source of intense heat is the luminous combustion of gases. The phenomenon is intensively studied to increase both the temperature of flames and our knowledge of matter and energy

Fire has always been one of the most fascinating of nature's phenomena. It is a subject of inexhaustible mystery: the more intimately man studies it, the more it leads him on to new discoveries and to new questions. From the ancients' concept of fire as one of the four fundamental elements grew the alchemy of the Middle Ages. Later it was the investigations of combustion by Antoine Lavoisier and others that gave rise to modern chemistry. Today the study of flames is leading deep into questions of quantum mechanics and the fundamental nature of our physical world.

The province of this article is the range of temperature from a few hundred to a few thousand degrees—the familiar flames that cook our food, drive our engines, smelt our ores, do our work. To approach an understanding of fire we must first define combustion. Combustion is a heat-producing chemical process that may take place at almost any rate or temperature: it may be as slow and mild as the rusting of iron or as violent as the explosion of hydrogen with oxygen, which gives a temperature of 3,000 degrees centigrade. It does not necessarily involve oxygen, for some

metals may burn in nitrogen, and certain substances such as hydrazine (N_2H_4), hydrogen peroxide (H_2O_2) and ozone (O_3) can burn in the absence of any medium except themselves; that is, at a sufficiently high temperature they decompose and give off heat without combining with another substance. Hydrazine and hydrogen peroxide are well-known rocket fuels. Ozone is particularly interesting because its burning gives only a single product: two molecules of ozone yield three molecules of oxygen plus heat.

The simplest definition of fire is that

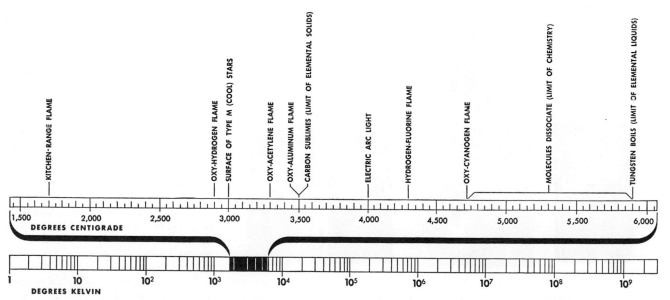

HIGH TEMPERATURES in chemical terms range up to the hottest flames, the dissociation of molecules and the evaporation of elemental liquids. Above the high temperatures are the very high temperatures, and above them are the ultrahigh temperatures.

OXY-ALUMINUM TORCH burns through concrete in the High Temperature Laboratory of Temple University. Fed by powdered aluminum, this flame is 3,500 degrees centigrade.

MATERIALS BURNED by the oxy-aluminum torch are, from left to right: graphite, concrete, red brick and refractory brick. Melted material cooled on the surface of the last two.

it is any combustion intense enough to emit light. It may be a quietly burning flame or a climactic explosion. It grows and sustains itself in the reacting medium by the heat it produces, for the heat raises the temperature and the temperature in turn raises the reaction rate until heat is produced as fast or faster than it is lost to the surroundings. But heat is not always the sole nor even the principal agent that initiates flames and explosions. Another agent is a chemical process known as branching of reaction chains.

The phenomenon of reaction chains, particularly the difference between branched and unbranched chains, is illustrated in the two diagrams on the next page. The first shows the reaction of hydrogen and chlorine. These elements have such an affinity for each other that a hydrogen atom will detach a chlorine atom from the chlorine molecule and *vice versa*. Therefore an atom of hydrogen seizes one atom of chlorine and releases the other, which in turn frees an atom of hydrogen from a hydrogen molecule and so on. The chain of reactions is straight, *i.e.*, "unbranched." On the other hand, in a mixture of hydrogen and oxygen an atom of hydrogen takes an atom of oxygen from an oxygen molecule to form a free radical OH and release an oxygen atom, which in turn takes a hydrogen atom to form another OH radical and free a hydrogen atom. The two OH radicals produce two free hydrogen atoms. The net effect is that from the action of one hydrogen atom two new hydrogen atoms emerge, each capable of propagating its own reaction chain. The chain, in other words, is "branched." The branches may multiply without limit and proceed rapidly to an explosion. This type of chain reaction is comparatively rare in chemistry, but it has been made familiar by the atomic bomb, where each fission of a uranium atom releases neutrons which may start branches.

A typical combustion process is the burning of gasoline in an engine. Here the octane fuel, mixed with air, is compressed and ignited by a spark. Two molecules of octane (C_8H_{18}) react with 25 molecules of oxygen in the air to form 16 molecules of carbon dioxide and 18 molecules of water, with a heat yield of 2,632,000 calories. The hot carbon dioxide and water vapor form the working fluid that exerts pressure in the cylinder and drives the engine.

The spark starts the chemical reaction in a small zone of the fuel mixture; as this flames up, heat flows into the adjacent layer of unburned gas, and so on. In this way a propagating zone of intense burning called a combustion wave is established. If an explosive mixture flows continuously from an orifice, a combustion wave can, under suitable conditions, propagate against the stream at a rate that matches the flow velocity of the stream. This is exemplified in the familiar stationary flame of a kitchen-range burner.

Within the combustion wave itself the temperature rises sharply from the unburned side to the burned side. The wave thickness and temperature gradient vary greatly, depending on the fuel: in a mixture of hydrogen and fluorine the temperature rises about 4,500 degrees C. in the short space of about one thousandth of an inch. Let us follow the progress of a combustion wave through a small zone of gas. The zone at first merely absorbs heat from the low-temperature front of the wave. When a hot enough part of the wave reaches it, the gas in the zone breaks into rapid chemical reaction, or flame. Now the burning gas itself generates heat, first at a rising rate, then at a declining rate as the fuel is used up. While it is generating heat, it passes forward to the advancing wave front as much heat as it absorbed before it began to liberate heat. In this way a wave continuously "borrows" and "repays" heat out of a revolving fund which travels with the wave and is referred to as the "excess enthalpy" (excess heat content) of the wave. The velocity of propagation of the wave is called the burning velocity. It ranges from a few inches per second for weak hydrocarbon-air mixtures to several hundred times this value for mixtures such as hydrogen and fluorine.

Across the combustion wave the chemical population of the reacting zone of gas changes radically: it becomes a mélange of pristine fuel molecules, final products and intermediate dissociated atoms and fragments of molecules known as free radicals [see "Free Radicals," by Paul D. Bartlett; SCIENTIFIC AMERICAN, December, 1953]. There exist "cool" flames which are sustained by a chain-branching process rather than by the production of heat. These waves leave in their wake a residue of intermediate combustion products such as aldehydes and peroxides. By chilling a flame through contact with a cold solid surface, or better still, by diluting the mixture so that intense burning is avoided, it is possible to recover substantial quantities of such intermediate products.

A combustion wave adjusts itself auto-

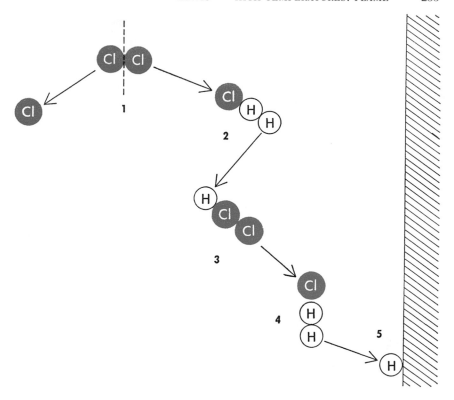

CHAIN REACTION of chlorine and hydrogen begins when a molecule of chlorine splits. One chlorine atom then hits a hydrogen molecule, combining with one hydrogen atom and freeing the other. The reaction continues until an atom is adsorbed on the vessel wall.

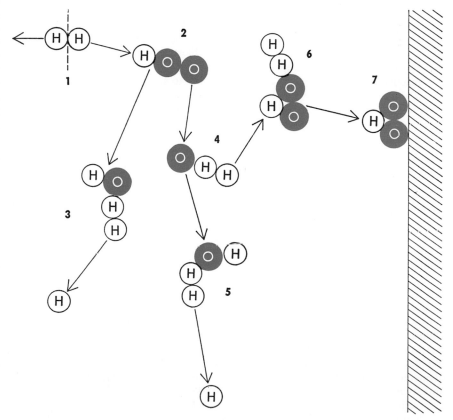

BRANCHED-CHAIN REACTION of hydrogen and oxygen frees three hydrogen atoms to replace the original one. The reaction stops when a majority of the hydrogen atoms form the radical HO_2. This unreactive radical migrates to the vessel wall and is adsorbed there.

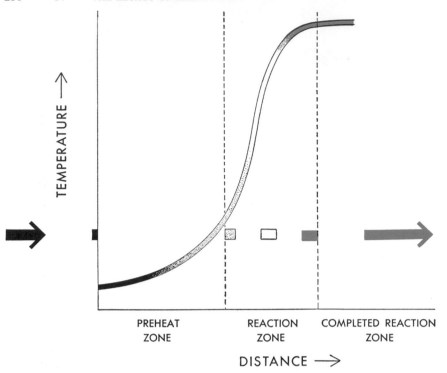

TEMPERATURE →

PREHEAT ZONE   REACTION ZONE   COMPLETED REACTION ZONE

DISTANCE →

COMBUSTION WAVE is divided into three zones. In the preheat zone the temperature of the fuel and air is raised by the "excess enthalpy" of combustion. In the reaction zone the mixture is further heated by the combustion of the fuel. In the completed reaction zone the fuel is completely burned. The small rectangles suggest change in volume with burning.

matically to a steady state in which the reacting zone passes on to the next zone exactly the amount of excess enthalpy it has received. Such a wave possesses the ability to restore itself when disturbed. But if the burning mixture is progressively diluted so that the temperatures and reaction rates in the wave are reduced, the resistance of the steady state to perturbations decreases and finally vanishes. The combustion wave then disintegrates. Any fuel mixture has upper and lower limits of inflammability which are governed by the relative proportions of fuel and oxygen. Methane, for instance, will burn in air at room temperature only if the percentage of methane is between about 5 and 15 per cent. An increase in temperature or pressure will widen the range of inflammability. Information on the limits of inflammability is important to industries that produce, store, handle and transport potentially explosive materials.

Combustion waves lose heat to solid bodies with which they come in contact. A solid therefore quenches burning in a gas for some distance from it. If the diameter of a duct is made small enough, an explosive mixture cannot burn in it. The critical quenching diameter depends upon the composition of the fuel mixture, the pressure, the temperature and the shape of the duct. A mixture of hydrocarbons and air at very low pres-

sure will not burn in a duct as much as several inches in diameter, but a mixture of oxygen with hydrogen or acetylene can propagate a flame in a fine tube with a bore of only a little more than one thousandth of an inch. Bundles of narrow ducts are often used as flame "traps" to arrest flashbacks, just as, conversely, a wide duct is employed to promote flashover in a kitchen range or a jet engine. For high-altitude flight the diameters of engine ducts and chambers are especially critical, because the flame-quenching distance increases markedly with decreasing pressure.

From an ignition source such as an electric spark a combustion wave propagates in all directions. At any instant the combustion wave thus forms a thin shell around a core of burned gas. As the shell grows, it must obtain additional heat or "excess enthalpy" to satisfy the demand created by the enlargement of its area. This heat is furnished by the core of burned gas. Whereas in later stages of the process, when the shell has grown large, the heat thus taken from the burned gas is negligible, it is relatively large in the early stages and leads to a significant decrease of temperature in the wave crest. In fact, unless the wave is given a sufficient initial boost of heat by the ignition source, the temperature drops so much that the wave

peters out. This explains why a low-energy spark may pass through an explosive mixture without igniting it, even when the temperature in the path of the spark is of the order of several thousand degrees. The minimum spark energy required for ignition depends upon the composition of the explosive mixture, the pressure and the temperature. Certain weak mixtures may require as much as a calorie, while hydrogen and oxygen in proper ratios can be touched off by less than one millionth of a calorie—a spark far less energetic than the static electricity a human being generates by walking on a carpet on a dry day.

The spherical way in which a combustion wave develops can be demonstrated by igniting a gas mixture with a spark in the center of a glass sphere; the upper photograph on the opposite page shows such a wave at successive moments. In a cylindrical vessel the wave follows the pattern illustrated in the lower photograph.

A combustion wave, in expanding the gas within it, produces a thrust which raises the pressure in the unburned gas ahead of it. Although this force is generally small, it may produce important effects. It accentuates any turbulence in a gas stream. In a fast-burning mixture

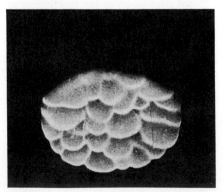

CELLULAR FLAMES in a six-inch glass tube were photographed from below by George H. Markstein of the Cornell Aeronautical Laboratory. This flame aberration is believed due to the preferential diffusion of oxygen in a mixture of butane and air.

confined within a duct the thrust and burning velocity, reinforcing each other, may produce a shock front. The consequent sharp rise in pressure and temperature in turn causes a detonation wave. Such waves travel with high velocities, of the order of several miles per second. They are maintained by the energy released in the chemical reaction and attain a constant velocity which is the sum of the velocities of sound and of the flow in the burned medium. In high explosives such as TNT or nitroglycerin the pressure in the shock front

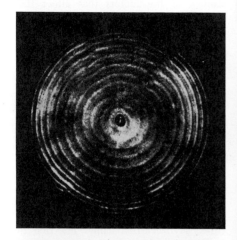

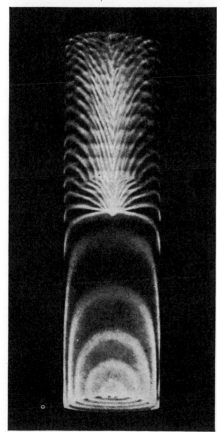

FLAME PROPAGATION from a spark was photographed at regular short intervals. At the top the flame is shown in a spherical glass vessel; at bottom, in a cylindrical one.

UPRIGHT FLAME CONE is stabilized on a ring placed in a stream of inflammable gas flowing from the glass tube at the bottom.

INVERTED FLAME CONE is stabilized on a rod. Without the ring or the rod the flame would simply blow away from tube.

is of the order of 100,000 to 200,000 atmospheres.

The most familiar flame in our civilization is, of course, the gas burner, known to us variously as the burner on the kitchen stove or the oil furnace, the welder's torch, the airplane's jet engine. In the kitchen burner a jet of fuel gas, after entraining air, flows at a given pressure from numerous orifices in the burner head. The flame is stabilized by means of the flow distribution and quenching effects. The combustion wave anchors itself at an equilibrium position whose distance from the burner rim depends on the rate of gas flow. When the velocity of this flow is reduced to less than the burning velocity, the flame flashes back into the tube. When it is increased so that it exceeds the burning

velocity everywhere in the stream, the flame blows off.

When one fuel gas is substituted for another, the critical flow velocity for stabilizing the flame changes. This situation constitutes an important problem in the gas industry, where it is frequently desired to bring in other available fuel gases to meet peak demands. The burner design and flow conditions must be arranged to provide a stable position for each fuel.

A ring or some other obstacle in the gas stream may stabilize a flame under conditions in which it might otherwise blow off [see photograph at left, above]. The gas flow is retarded around the ring, and the combustion wave assumes the shape of an upright

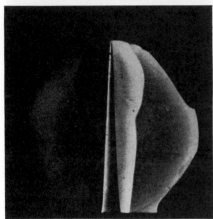

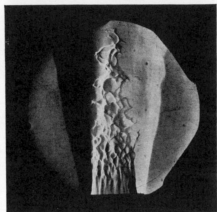

LAMINAR FLAME is photographed by its own light (*top*) and by the schlieren method. The latter shows sharp combustion wave and its rounded envelope of hot gas.

TURBULENT FLAME is similarly photographed by its own light (*top*) and by the schlieren method (*bottom*). The broad flame reflects turbulent combustion wave.

the combustion wave becomes wrinkled. The wrinkles and irregularities are not noticeable to the naked eye, but they show up in photographs made by the schlieren technique, which registers density differences and thus makes visible the unburned gas with its envelope of burned gas [*see the bottom photograph, far left*].

The flame that man has known ever since he discovered fire—the burning of wood, a candle flame—is known as diffusion flame. Scientific research on such flames has been confined essentially to studies of the factors that affect the shape and length of the flame. When wood or coal burn, their hydrocarbons break down to incandescent carbon, which emits intense radiation. Such flames find considerable use as sources of radiant heat in industry.

Not the least interesting aspect of the study of flames is their colors. In the spectral analysis of their light radiation the physical chemist has found a tool to identify the presence of atoms and radicals in the flame and to develop a picture of the kinetics of the chemical processes that occur at high temperatures. Hydrocarbon flames not rich in fuel have a blue color, whose source is the C-H radical; hydrocarbon flames rich in fuel show a green light emission known as the Swan bands, whose source is the C-C radical. OH and other radicals give no visible light; they must be photographed in the ultraviolet or infrared.

After thousands of years we are at last learning some of the inner mysteries of fire, one of man's earliest tools, still his most abundant source of energy, and lately reaching its most advanced expression in jets, flame throwers and atomic fireballs.

cone. A rod placed upright in the stream makes the flame take the form of an inverted cone [*photograph on preceding page*]. The obstacle method of stabilizing flames is finding application in combustors for jet engines, which demand very high flow velocities.

In the foregoing illustrations the gas flow is laminar; that is, the streamlines form a regular pattern and the combustion wave is smooth and steady. When turbulence is introduced into the stream,

25

THE CHEMICAL EFFECTS OF LIGHT

GERALD OSTER

September 1968

Visible light triggers few chemical reactions (except in living cells), but the photons of ultraviolet radiation readily break chemical bonds and produce short-lived molecular fragments with unusual properties

Our everyday world endures because most substances, organic as well as inorganic, are stable in the presence of visible light. Only a few complex molecules produced by living organisms have the specific property of responding to light in such a way as to initiate or participate in chemical reactions ["How Light Interacts with Living Matter," by S. B. Hendricks; SCIENTIFIC AMERICAN, Sept., 1968]. Outside of living systems only a few kinds of molecules are sufficiently activated by visible light to be of interest to the photochemist.

The number of reactive molecules increases sharply, however, if the wavelength of the radiant energy is shifted slightly into the ultraviolet part of the spectrum. To the photochemist that is where the action is. Thus he is primarily concerned with chemical events that are triggered by ultraviolet radiation in the range between 180 and 400 nanometers. These events usually happen so swiftly that ingenious techniques have had to be devised to follow the molecular transformations that take place. It is now routine, for example, to identify molecular species that exist for less than a millisecond. Species with lifetimes measured in microseconds are being studied, and new techniques using laser pulses are pushing into the realm where lifetimes can be measured in nanoseconds and perhaps even picoseconds.

The photochemist is interested in such short-lived species not simply for their own sake but because he suspects that many, if not most, chemical reactions proceed by way of short-lived intermediaries. Only by following chemical reactions step by step in fine detail can he develop plausible models of how chemical reactions proceed in general. From such studies it is often only a short step to the development of chemical processes and products of practical value.

When a quantum of light is absorbed by a molecule, one of the electrons of the molecule is raised to some higher excited state. The excited molecule is then in an unstable condition and will try to rid itself of this excess energy by one means or another. Usually the electronic excitation is converted into vibrational energy (vibration of the atoms of the molecule), which is then passed on to the surroundings as heat. Such is the case, for example, with a tar roof on a sunny day. An alternative pathway is for the excited molecule to fluoresce, that is, to emit radiation whose wavelength is slightly longer than that of the exciting radiation. The bluish appearance of quinine water in the sunlight is an example of fluorescence; the excitation is produced by the invisible ultraviolet radiation of the sun.

The third way an electronically excited molecule can rid itself of energy is the one of principal interest to the photochemist: the excited molecule can undergo a chemical transformation. It is the task of the photochemist to determine the nature of the products made, the amount of product made per quantum absorbed (the quantum yield) and how these results depend on the concentrations of the starting materials. His next step is to combine these data with the known spectroscopic and thermodynamic properties of the molecules involved to make a coherent picture. It must be admitted, however, that only the simplest photochemical reactions are understood in detail.

There is also a fourth way an excited molecule can dissipate its energy: the molecule may be torn apart. This is called photolysis. As might be expected, photolysis occurs only if the energy of the absorbed quantum exceeds the energy of the chemical bonds that hold the molecule together. The energy required to photolyse most simple molecules corresponds to light that lies in the ultraviolet region [*see illustation on page 241*]. For example, the chlorine molecule is colored and thus absorbs light in the visible range (at 425 nanometers), but it has a low quantum yield of photolysis when exposed to visible light. When it is exposed to ultraviolet radiation at 330 nanometers, on the other hand, the quantum yield is close to unity: each quantum of radiation absorbed ruptures one molecule.

Albert Einstein proposed in 1905 that one quantum of absorbed light leads to the photolysis of one molecule, but it required the development of quantum mechanics in the late 1920's to explain why the quantum yield should depend on the wavelength of the exciting light. James Franck and Edward U. Condon, who carefully analyzed molecular excitation, pointed out that when a molecule makes a transition from a ground state to an electronically excited state, the transition takes place so rapidly that the interatomic distances in the molecule do not have time to change. The reason is that the time required for transition is much shorter than the period of vibration of the atoms in the molecule.

To understand what happens when a molecule is excited by light it will be helpful to refer to the illustration on the following page. The lower curve represents the potential energy of a vibrating diatomic molecule in the ground state. The upper curve represents the potential energy of the excited molecule, which is also vibrating. The horizontal lines in the lower portion of each curve indicate the energy of discrete vibrational levels. If the interatomic distance

240

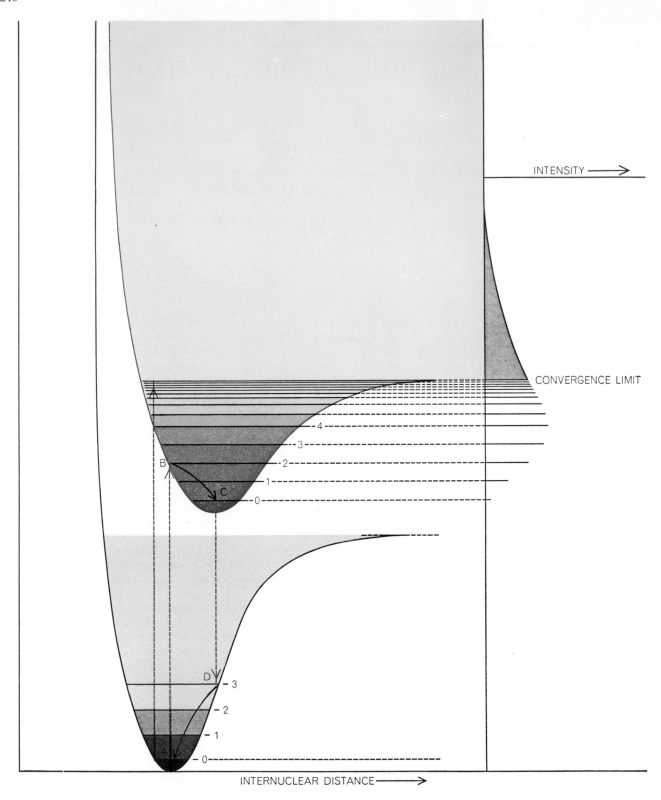

INTENSITY ⟶

CONVERGENCE LIMIT

4

3

2

1

0

3

2

1

0

INTERNUCLEAR DISTANCE ⟶

RESPONSE OF SIMPLE MOLECULES TO PHOTONS can be followed with the help of potential-energy curves. The lower curve represents the potential energy of a typical diatomic molecule in the ground state; the upper curve represents its potential energy in the first excited electronic state. Because the two atoms of the molecule are constantly vibrating, thus changing the distance between atomic nuclei, the molecule can occupy different but discrete energy levels (*horizontal lines*) within each electronic state. The molecule in the lowest ground state can be dissociated, or photolysed, if it absorbs a photon with an energy equal to or greater than ΔE_1. This is the energy required to carry the molecule to or beyond the "convergence limit." The length of the horizontal lines at the right below that limit represents the probability of transition from the ground electronic state to a particular vibrational level in the excited electronic state. Thus a photon with an energy of ΔE_2 will raise the molecule to the second level (B) of that state. There it will vibrate, ultimately lose energy to surrounding molecules and fall to C. It can now emit a photon with somewhat less energy than ΔE_2 and fall to D. This is called fluorescence. After losing vibrational energy molecule will return to A.

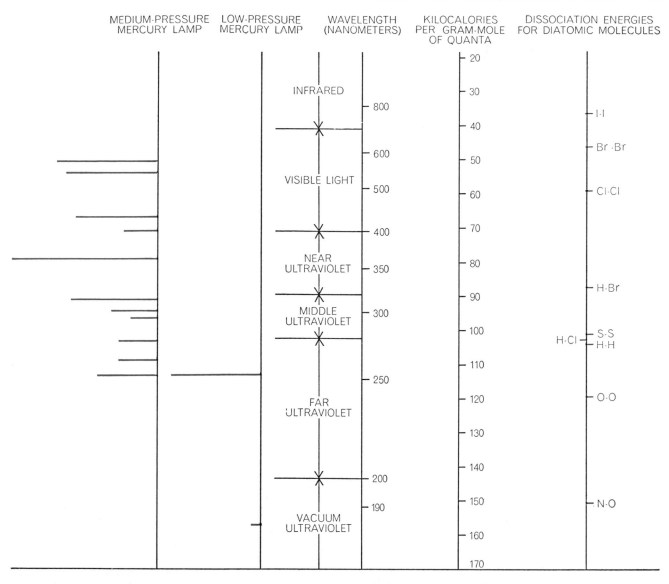

DISSOCIATION ENERGIES of most common diatomic molecules are so high that the energy can be supplied only by radiation of ultraviolet wavelengths. The principal exceptions are molecules of chlorine, bromine and iodine, all of which are strongly colored, indicating that they absorb light. The energy carried by a quantum of radiation, or photon, is directly proportional to its frequency, or inversely proportional to its wavelength. There are 6.06×10^{23} photons in a gram-mole of quanta. This is the number required to dissociate a gram-mole of diatomic molecules (6.06×10^{23} molecules) if the quantum yield is unity. A gram-mole is the weight in grams equal to the molecular weight of a molecule, thus a gram-mole of oxygen (O_2) is 32 grams. The principal emission wavelengths of two commonly used types of mercury lamp are identified at the left. Lengths of the bars are proportional to intensity.

becomes large enough in the ground state, the molecule can come apart without ever entering the excited state. The curve for the excited state is displaced to the right of the curve for the ground state, indicating that the average interatomic distance (the minimum in each curve) is somewhat greater in the excited state than it is in the ground state. That is, the excited molecule is somewhat "looser."

The molecule can pass from the ground state to one of the levels of the excited state by absorbing radiation whose photon energy is equal to the energy difference between the ground state and one of the levels of the excited state. Provided that the quantum of radiation is not too energetic the molecule will remain intact and continue to vibrate. After a brief interval it will emit a quantum of fluorescent radiation and drop back to the ground state. Because the emission occurs when the excited molecule is at the lowest vibrational level, the emitted energy is less than the absorbed energy, hence the wavelength of the fluorescent radiation is greater than that of the absorbed radiation.

When the absorbed radiation exceeds a certain threshold value, the molecule comes apart; it is photolysed. At this point the absorption spectrum, shown at the right side of the illustration, becomes continuous, because the molecule is no longer vibrating at discrete energy levels. As long as the molecule is intact only discrete wavelengths of light can be absorbed.

It is possible for the excited state to pass to the ground state without releasing a quantum of radiation, in which case the electronic energy is dissipated as heat. Franck and Condon explained that this was accomplished by an overlapping, or crossing, of the two potential-energy curves, so that the excited molecule slides over, so to speak, to the ground state, leaving the molecule in an abnormally high state of vibration. This vibrational energy is then readily transferred to surrounding molecules.

As far as life on the earth is concerned, the most important photolytic reaction in nature is the one that creates a canopy of ozone in the upper atmosphere. Ozone is a faintly bluish gas whose molecules consist of three atoms of oxygen; ordinary oxygen molecules contain two atoms. Ozone absorbs broadly in the middle- and far-ultraviolet regions with a maximum at 255 nanometers. Fortunately ozone filters out just those wavelengths that are fatal to living organisms.

Ozone production begins with the photolysis of oxygen molecules (O_2), which occurs when oxygen strongly absorbs ultraviolet radiation with a wavelength of 190 nanometers. The oxygen atoms released by photolysis may simply recombine or they may react with other oxygen molecules to produce ozone (O_3). When ozone, in turn, absorbs ultraviolet radiation from the sun, it is either photolysed (yielding O_2 and O) or it contributes to the heating of the atmosphere. A dynamic equilibrium is reached in which ozone photolysis balances ozone synthesis.

Early in this century physical chemists were presented with a photolytic puzzle. It was observed that when pure chlorine and hydrogen are exposed to ultraviolet radiation, the quantum yield approaches one million, that is, nearly a million molecules of hydrogen chloride (HCl) are produced for each quantum of radiation absorbed. This seemed to contradict Einstein's postulate that the quantum yield should be unity. In 1912 Max Bodenstein explained the puzzle by proposing that a chain reaction is involved [see upper illustration at left].

The chain reaction proceeds by means of two reactions, following the initial photolysis of chlorine (Cl_2). The first reaction, which involves the breaking of the fairly strong H-H bond, creates a small energy deficit. The second reaction, which involves the breaking of the weaker Cl-Cl bond, makes up the deficit with energy to spare. Breaking the H-H bond requires 104 kilocalories per gram-mole (the equivalent in grams of the molecular weight of the reactants, in this case H_2). Breaking the Cl-Cl bond requires only 58 kilocalories per gram-mole. In both of the reactions that break these bonds HCl is produced, yielding 103 kilocalories per gram-mole. Consequently the first reaction has a deficit of one kilocalorie per gram-mole and the second a surplus of 45 (103 − 58) kilocalories per gram-mole. The two reactions together provide a net of 44 kilocalories per gram-mole. Thus the chain reaction is fueled, once ultraviolet radiation provides the initial breaking of Cl-Cl bonds.

The chain continues until two chlorine atoms happen to encounter each other to form chlorine molecules. This takes place mainly at the walls of the reaction vessel, which can dissipate some of the excess electronic excitation energy of the chlorine atoms and allow chlorine mole-

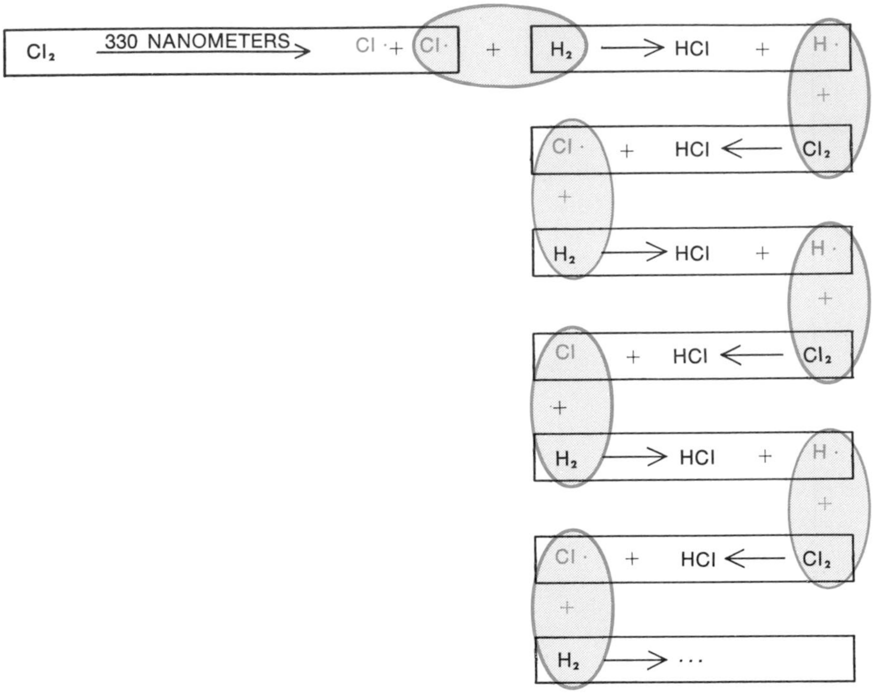

CHAIN REACTION is produced when pure chlorine and hydrogen are exposed to ultraviolet radiation. A wavelength of 330 nanometers is particularly effective. Such radiation is energetic enough to dissociate chlorine molecules, which requires only 58 kilocalories per gram-mole, but it is too weak to dissociate hydrogen molecules, which requires 104 kilocalories per gram-mole. The formation of HCl in the subsequent reactions provides 103 kilocalories per gram-mole. Since 104 kilocalories are needed for breaking the H-H bond, the reaction of atomic chlorine (Cl·) and H_2 involves a net deficit of one kilocalorie per gram-mole. However, the next reaction in the chain, involving H· and Cl_2, provides a surplus of 45 kilocalories (103 − 58). This energy surplus keeps the chain reaction going.

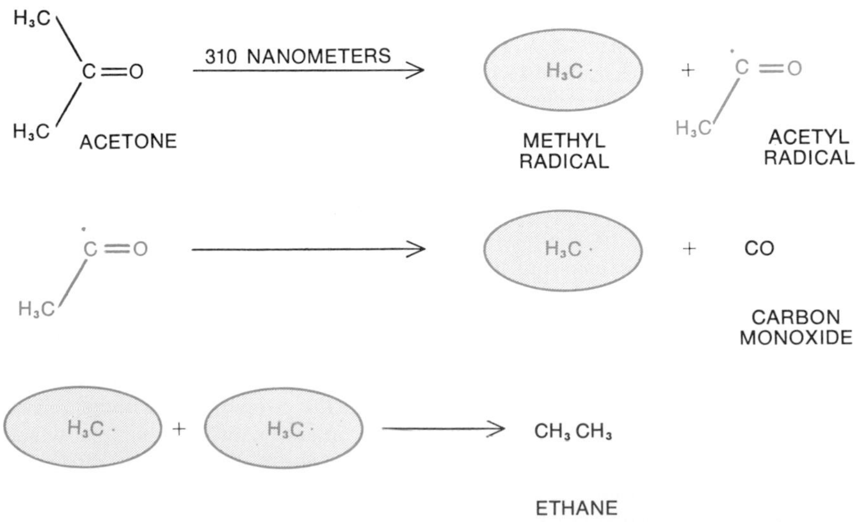

PHOTOLYSIS OF ACETONE, which yields primarily ethane and carbon monoxide, is a much studied photochemical reaction. It was finally understood by postulating the existence of short-lived free radicals, fragments that contain unsatisfied valence electrons.

cules to form. The free atoms may also be removed by impurities in the system.

Bromine molecules will likewise undergo a photochemical reaction with hydrogen to yield hydrogen bromide. The quantum yield is lower than in the chlorine-hydrogen reaction because atomic bromine reacts less vigorously with hydrogen than atomic chlorine does. Bromine atoms react readily, however, with olefins (linear or branched hydrocarbon molecules that contain one double bond). Each double bond is replaced by two bromine atoms. This is the basis of the industrial photobromination of hydrocarbons. Bromination can also be carried out by heating the reactants in the presence of a catalyst, but the product itself may be decomposed by such treatment. The advantage of the photochemical process is that the products formed are not affected by ultraviolet radiation.

An important industrial photochlorination process has been developed by the B. F. Goodrich Company. There it was discovered that when polyvinyl chloride is exposed to chlorine in the presence of ultraviolet radiation, the resulting plastic withstands a heat-distortion temperature 50 degrees Celsius higher than the untreated plastic does. As a result this inexpensive plastic can now be used as piping for hot-water plumbing systems.

A much studied photolytic reaction is one involving acetone (C_2H_6CO). When it is exposed to ultraviolet radiation, acetone gives rise to ethane (C_2H_6) with a quantum yield near unity, together with carbon monoxide and a variety of minor products, depending on the wavelength of excitation. The results can be explained by schemes that involve free radicals—fragments of molecules that have unsatisfied valence electrons. Photolysis of acetone produces the methyl radical (CH_3) and the acetyl radical (CH_3CO). Two methyl radicals combine to form ethane [see lower illustration on page 242].

W. A. Noyes, Jr., of the University of Rochester and others assumed the existence of these free radicals in order to explain the end products of the photolysis. Because the lifetime of free radicals may be only a ten-thousandth of a second, they cannot be isolated for study. Since the end of World War II, however, the technique of flash spectroscopy has been developed for recording their existence during their brief lifetime.

Flash spectroscopy was devised at the University of Cambridge by R. G. W. Norrish and his student George Porter, who is now director of the Royal Institution. They designed an apparatus [see illustration, page 244] in which a sample is illuminated with an intense burst of ultraviolet to create the photolytic products. A small fraction of a second later weaker light is beamed into the reaction chamber; at the far end of the chamber the light enters a spectrograph, which records whatever wavelengths have not been absorbed. The absorbed wavelengths provide clues to the nature of the short-lived species produced by photolysis. In 1967 Norrish and Porter shared the Nobel prize in chemistry with Manfred Eigen of the University of Göttingen, who had also developed techniques for studying fast reactions.

Flash spectroscopy has greatly increased chemists' knowledge about the "triplet state," an excited state that involves the pairs of electrons that form chemical bonds in organic molecules. Normally the spins of the paired electrons are antiparallel, or opposite to each other. When exposed to ultraviolet radiation, the molecules are raised to the first excited state and then undergo a nonradiative transition to an intermediate state in which the spins of two electrons in the same state are parallel to each other. This is the triplet state. If it is again exposed to ultraviolet or visible radiation, the triplet state exhibits its own absorption spectrum, which lies at a longer wavelength than the absorption spectrum of the normal ground state, or state of lowest energy [see illustration at left].

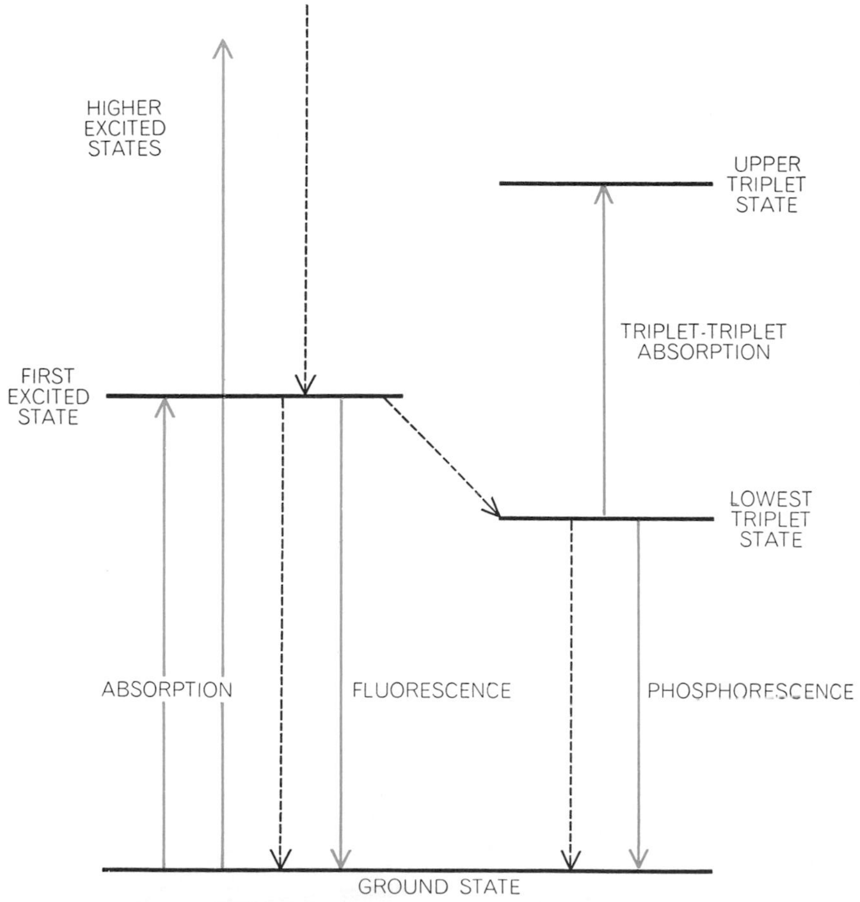

The concept of the triplet state in organic molecules is due mainly to the work of G. N. Lewis and his collaborators at the University of California at Berkeley in the late 1930's and early 1940's. These workers found that when dyes (notably fluorescein) are dissolved

TRIPLET STATE has become an important concept in understanding the photochemical reactions of many organic molecules. Like all molecules, they can be raised to an excited state by absorption of radiation. They can also return to the ground state by normal fluorescence: reemission of a photon. Alternatively, they can drop to the triplet state without emission of radiation. (Broken lines indicate nonradiative transitions.) The existence of this state can be inferred from the wavelength of the radiation it is then able to absorb in passing to a higher triplet state. The triplet state arises when the spins of paired electrons point in the same direction rather than in the opposite direction, as they ordinarily do.

in a rigid medium such as glass and are exposed to a strong light, the dyes change color. When the light is removed, the dyes revert to their normal color after a second or so. This general phenomenon is called photochromism. Lewis deduced the existence of the triplet state and ascribed its fairly long duration to the time required for the parallel-spin electrons to become uncoupled and to revert to their normal antiparallel arrangement.

In 1952 Porter and M. W. Windsor used flash spectroscopy to search for the triplet state in the spectra of organic molecules in ordinary fluid solvents. They were almost immediately successful. They found that under such conditions the triplet state has a lifetime of about a millisecond.

In his Nobel prize lecture Porter said: "Any discussion of mechanism in organic photochemistry immediately involves the triplet state, and questions about this state are most directly answered by means of flash photolysis. It is now known that many of the most important photochemical reactions in solution, such as those of ketones and quinones, proceed almost exclusively via the triplet state, and the properties of this state therefore become of prime importance."

While studying the photochemistry of dyes in solution, my student Albert H. Adelman and I, working at the Polytechnic Institute of Brooklyn, demonstrated that the chemically reactive

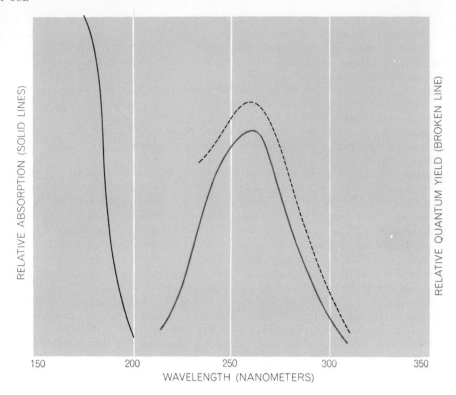

ABSORPTION SPECTRUM OF OZONE (*solid curve in color*) peaks at about 250 nanometers in the ultraviolet. As a happy consequence, the canopy of ozone in the upper atmosphere removes the portion of the sun's radiation that would be most harmful to life. The biocidal effectiveness of ultraviolet radiation is shown by the broken black line. The solid black curve is the absorption spectrum of molecular oxygen. For reasons not well understood, ultraviolet radiation of 200 nanometers does not penetrate the atmosphere.

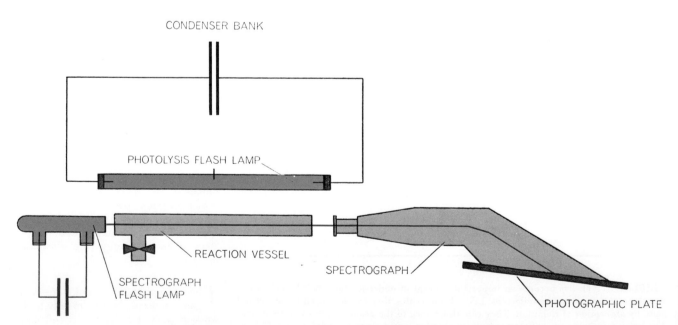

APPARATUS FOR FLASH PHOTOLYSIS was devised by R. G. W. Norrish and George Porter at the University of Cambridge. With it they discovered the short-lived triplet state that follows the photolysis of various kinds of molecules, organic as well as inorganic. The initial dissociation is triggered by the photolysis flash lamp, which produces an intense burst of ultraviolet radiation. A millisecond or less later another flash lamp sends a beam of ultraviolet radiation through the reaction vessel. Free radicals in the triplet state absorb various wavelengths ("triplet-triplet" absorption) and the resulting spectrum is recorded by the spectrograph.

species is the triplet state of the dye. Specifically, when certain dyes are excited by light in the presence of electron-donating substances, the dyes are rapidly changed into the colorless ("reduced") form. Our studies showed that the reactive state of the dye—the triplet state—has a lifetime of about a tenth of a millisecond. The dye is now a powerful reducing agent and will donate electrons to other substances, with the dye being returned to its oxidized state [see illustration on page 246]. In other words, the dye is a photosensitizer for chemical reductions; visible light provides the energy for getting the reaction started.

In the course of these studies I discovered that free radicals are created when dyes are photoreduced. The free radicals make their presence known by causing vinyl monomers to link up into polymers. The use of free radicals for bringing about polymerization of monomers is well known in industry. It occurred to me that adding suitable dyes to monomer solutions would provide the basis for a new kind of photography. In such a solution the concentration of free radicals would be proportional to the intensity of the visible light and thus the degree of polymerization would be controlled by light. It has turned out that very accurate three-dimensional topographical maps can be produced in plastic by this method.

The use of dyes as photosensitizing agents is, of course, fundamental to photography. In 1873 Hermann Wilhelm Vogel found that by adding dyes to silver halide emulsions he could make photographic plates that were sensitive to visible light. At first such plates responded only to light at the blue end of the spectrum. Later new dyes were found that extended the sensitivity farther and farther toward the red end of the spectrum, making possible panchromatic emulsions. Photographic firms continue to synthesize new dyes in a search for sensitizers that will act efficiently in the infrared part of the spectrum. The nature of the action of sensitizers in silver halide photography is still obscure, nearly 100 years after the effect was first demonstrated. The effect seems to depend on the state of aggregation of the dye absorbed to the silver halide crystals.

The reverse of photoreduction—photooxidation—can also be mediated by dyes, as we have found in our laboratory. Here again the reactive species of the dye is the dye in the triplet state. We have found that the only dyes that will serve as sensitizers for photooxidation are those that can be reduced in the presence of light.

The oxidized dye—the dye peroxide—is a powerful oxidizing agent. In the process of oxidizing other substances the dye is regenerated [see illustration on page 246]. My student Judith S. Bellin and I have demonstrated this phenomenon, and we have employed dye-sensitized photooxidation to inactivate some biological systems. These systems include viruses, DNA and ascites tumor cells. That dyes are visible-light sensitizers for biological inactivation was first demonstrated in 1900 by O. Raab, who observed that a dye that did not kill a culture of protozoa did so when the culture was placed near a window.

The inactivation that results from dye sensitization is different from the inactivation that results when biological sys-

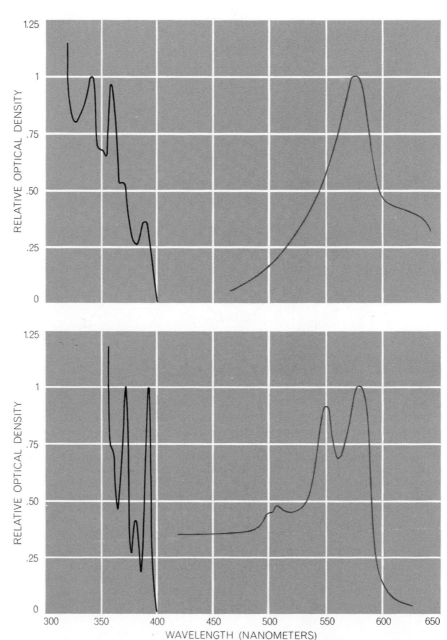

TRIPLET-TRIPLET ABSORPTION OF VISIBLE LIGHT has been observed in the author's laboratory at the Polytechnic Institute of Brooklyn. His equipment sends a beam of ultraviolet radiation into samples embedded in a plastic matrix in one direction and visible light at right angles to the ultraviolet radiation. The visible absorption spectra are then recorded in the presence of ultraviolet radiation. The black curves at the left in these two examples show the absorption of the electronic ground state. The colored curves at the right show the absorption of visible wavelengths that raises the excited molecule from the lowest triplet state to upper triplet states. The top spectra were produced by chrysene, the lower spectra by 1,2,5,6-dibenzanthracene. Both are aromatic coal-tar hydrocarbons.

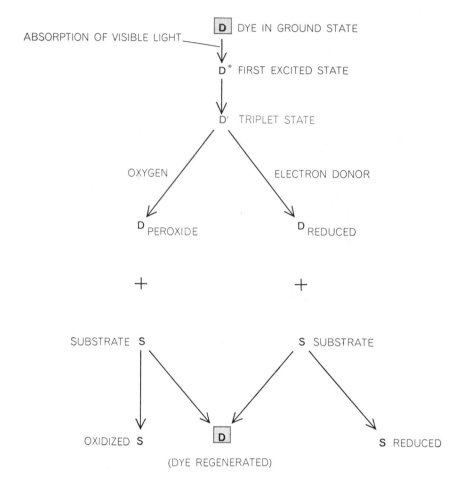

ABSORPTION OF VISIBLE LIGHT

D DYE IN GROUND STATE

D* FIRST EXCITED STATE

D' TRIPLET STATE

OXYGEN

ELECTRON DONOR

D PEROXIDE

D REDUCED

+

+

SUBSTRATE S

S SUBSTRATE

OXIDIZED S

D

S REDUCED

(DYE REGENERATED)

UNUSUAL PROPERTIES OF TRIPLET STATE have been explored by the author. Certain dyes in the triplet state can act either as strong oxidizing or as strong reducing agents, depending on the conditions to which the triplet state itself is exposed. In the presence of a substance that donates electrons (i.e., a reducing agent), the dye is reduced and can then donate electrons to some other substance (*substrate S*). In the presence of an oxidizing agent, the dye becomes highly oxidized and can then oxidize, or remove electrons from, a substrate. In both cases the dye is regenerated and returns to its normal state. The author's studies show that the reactive state of the dye lives only about .1 millisecond.

tems are exposed to ultraviolet radiation. Here the inactivation often seems to result from the production of dimers: the cross-linking of two identical or similar chemical subunits. Photodimerization is implicated, for example, in the bactericidal action of ultraviolet radiation. It has long been known that the bactericidal action spectrum (the extent of killing as a function of wavelength) closely parallels the absorption spectrum of DNA, the genetic material. If dried-down films of DNA are irradiated with ultraviolet, they become cross-linked. According to one view the cross-linking occurs by means of the dimerization of thymine, one of the constituent groups of DNA.

Although this may well be the mode of action of ultraviolet radiation, my own feeling is that insufficient consideration has been given to the photolysis of the disulfide bonds of the proteins in bacteria. This bond is readily cleaved by ultraviolet radiation and has an absorption spectrum resembling that of DNA. Disulfide bonds are vital in maintaining the structure and activity of proteins; their destruction by ultraviolet radiation could also account for the death of bacteria.

In using dyes as sensitizers for initiating chemical reactions we are taking our first tentative steps into a realm where nature has learned to work with consummate finesse. Carbon dioxide and water are completely stable in the presence of visible light. Inside the leaves of plants, however, the green dye chlorophyll, when acted on by light, mediates a sequence of chemical reactions that dissociates carbon dioxide and water and reassembles their constituents into sugars and starches. A dream of photochemists is to find a dye, or sensitizer, that will bring about the same reactions in a nonliving system. There is reason to hope that such a system could be a good deal simpler than a living cell.

26

THE CONTROL OF AIR POLLUTION

A. J. HAAGEN-SMIT

January 1964

It is now clear that smog is not only annoying but also injurious to health. Los Angeles is a leading example of a city that has analyzed the sources of its smog and taken steps to bring them under control

The past decade has seen a change in the public's attitude toward air pollution. Formerly the tendency was to deplore smog but to regard it as one of the inescapable adjuncts of urban life. Now there is a growing realization that smog, beyond being a vexatious nuisance, may indeed present hazards to health, and that in any case the pollution of the air will inevitably grow worse unless something is done about it. As a result many communities have created agencies to deal with air pollution and have, with varying degrees of effectiveness, backed the agencies with laws.

Going considerably beyond these efforts is the program in Los Angeles, a city rather widely regarded as the smog capital of the U.S. There the authorities have adopted the attitude that it is not enough to know smog exists; they have undertaken extensive studies to ascertain its components and to understand something of the complex processes by which it is created. Moreover, with help from the state they have taken pioneering steps toward curbing the emissions of the automobile, which is both a major cause of air pollution and a far more difficult source to control than such stationary installations as petroleum refineries and electric power plants. As a result of California's activities a device to control the emissions from the crankcases of automobiles is now standard equipment on all new cars in the U.S. The state is also working toward a program that will result in a measure of control over emissions from the automobile exhaust.

Complaints about polluted air go far back in time. As long ago as 1661 the English diarist John Evelyn declared in a tract entitled *Fumifugium, or the inconvenience of the Aer and Smoak of London* that the city "resembles the face Rather of Mount Aetna, the Court of Vulcan, Stromboli, or the Suburbs of Hell than an Assembly of Rational Creatures and the Imperial seat of our Incomparable Monarch." Air pollution has drawn similar complaints in many cities over the centuries.

For a long time, however, these complaints were like voices in the wilderness. Among the few exceptions in the U.S. were St. Louis and Pittsburgh, where the residents decided at last that they had inhaled enough soot and chemicals and took steps several years ago to reduce air pollution, primarily by regulating the use of coal. These, however, were isolated cases that did not deeply penetrate the consciousness of people in other parts of the country.

It was probably the recurrence of crises over smog in Los Angeles that awakened more of the nation to the possibility that the same thing could happen elsewhere and to the realization that air, like water, should be considered a precious resource that cannot be used indiscriminately as a dump for waste materials. By the time residents of Washington, D.C., complained of eye irritation and neighboring tobacco growers suffered extensive crop damage, it was clear that Los Angeles smog was not just a subject for jokes but a serious problem requiring diligent efforts at control. As a result the pace of antipollution activity has quickened at all levels of government. In addition to the community efforts already mentioned, a national air-sampling network now exists to assemble data on the extent of air pollution, and extensive studies of the effects of smog on health and the economy are under way.

Still, these efforts seem modest when viewed against the size of the problem. Surgeon General Luther L. Terry spoke at the second National Conference on Air Pollution late in 1962 of "how far we have to go." He said: "Approximately 90 per cent of the urban population live in localities with air-pollution problems—a total of about 6,000 communities. But only half of this population is served by local control programs with full-time staffs. There are now about 100 such programs, serving 342 local political jurisdictions. The median annual expenditure is about 10 cents per capita, an amount clearly inadequate to do the job that is necessary."

Enough has been done, however, to demonstrate that a concerted attack on the smog problem can produce a clearing of the air. Los Angeles, which Terry has called "the area in the United States that's devoting more money and more effort toward combating the problem than any other city," provides an example of the possibilities, the difficulties

LOS ANGELES SMOG, shown in photograph on opposite page, casts thick pall over city. Persistence and severity of smogs led the city to undertake pioneering and extensive programs to curb air pollution.

and the potential of such an attack.

Los Angeles certainly qualifies as a community where air pollution has created an annoying and at times dangerous situation. Two-thirds of the year smog is evident through eye irritation, peculiar bleachlike odors and a decrease in visibility that coincides with the appearance of a brownish haze. According to the California Department of Public Health, 80 per cent of the population in Los Angeles County is affected to some extent.

The city's decision to attack the smog problem dates from a report made in 1947 by Raymond R. Tucker, who as an investigator of air-pollution problems played a major role in the St. Louis smog battle and is now the mayor of that city. His report on Los Angeles enumerated the sources of pollution attributable to industry and to individuals through the use of automobiles and the burning of trash. The report recommended immediate control of known sources of pollution and a research program to determine if there were any other things in the air that should be controlled.

Largely on the basis of the Tucker report, *The Los Angeles Times* started with the aid of civic groups a campaign to inform and arouse the public about smog. As a result the state legislature in 1948 passed a law permitting the formation of air-pollution control districts empowered to formulate rules for curbing smog and endowed with the necessary police power for enforcement of the rules. Los Angeles County created such a district the same year.

The district began by limiting the dust and fumes emitted by steel factories, refineries and hundreds of smaller industries. It terminated the use of a million home incinerators and forbade the widespread practice of burning in public dumps. These moves reduced dustfall, which in some areas had been as much as 100 tons per square mile per month, by two-thirds, bringing it back to about the level that existed in 1940 before smog became a serious problem in the community. That achievement should be measured against the fact that since 1940 the population of Los Angeles and the number of industries in the city have doubled.

Although the attack on dustfall produced a considerable improvement in visibility, the typical smog symptoms of eye irritation and plant damage remained. The district therefore undertook a research program to ascertain the origin and nature of the substances that caused the symptoms. One significant finding was that the Los Angeles atmosphere differs radically from that of most other heavily polluted communities. Ordinarily polluted air is made strongly reducing by sulfur dioxide, a product of the combustion of coal and heavy oil.

Los Angeles air, on the other hand, is often strongly oxidizing. The oxidant is mostly ozone, with smaller contributions from oxides of nitrogen and organic peroxides.

During smog attacks the ozone content of the Los Angeles air reaches a level 10 to 20 times higher than that elsewhere. Concentrations of half a part of ozone per million of air have repeatedly been measured during heavy smogs. To establish such a concentration directly would require the dispersal of about

SMOG CURTAIN falls over the view from the campus of the California Institute of Technology. At top is the scene on a weekday morning; at bottom, the same scene that afternoon.

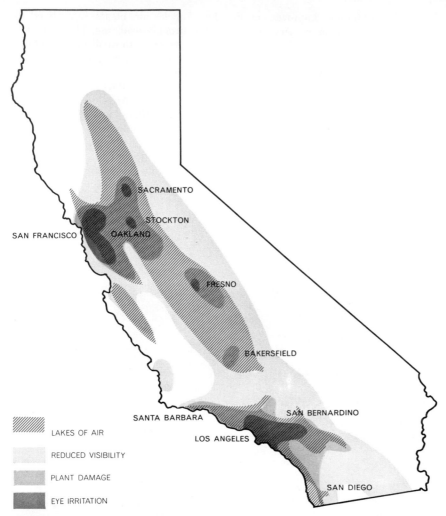

LAKES OF AIR

REDUCED VISIBILITY

PLANT DAMAGE

EYE IRRITATION

EXTENT OF AIR POLLUTION in California is indicated by gray areas on this map. Colored areas show the main natural airsheds, or lakes of air, into which pollutants flow. Sunlight acting on pollutants produces substances that irritate eyes and damage plants.

1,000 tons of ozone in the Los Angeles basin. No industry releases significant amounts of ozone; discharges from electric power lines are also negligible, amounting to less than a ton a day. A considerable amount of ozone is formed in the upper atmosphere by the action of short ultraviolet rays, but that ozone does not descend to earth during smog conditions because of the very temperature inversion that intensifies smog. In such an inversion warm air lies atop the cold air near the ground; this stable system forms a barrier not only to the rise of pollutants but also to the descent of ozone.

Exclusion of these possibilities leaves sunlight as the only suspect in the creation of the Los Angeles ozone. The cause cannot be direct formation of ozone by sunlight at the earth's surface because that requires radiation of wavelengths shorter than 2,000 angstrom units, which does not penetrate the atmosphere to ground level. There was a

compelling reason, however, to look for an indirect connection between smog and the action of sunlight: high oxidant or ozone values are found only during daylight hours. Apparently a photochemical reaction was taking place when one or more ingredients of smog were exposed to sunlight—which is of course abundant in the Los Angeles area.

In order for a substance to be affected by light it has to absorb the light, and the energy of the light quanta has to be sufficiently high to rupture the chemical bonds of the substance. A likely candidate for such a photochemical reaction in smog is nitrogen dioxide. This dioxide is formed from nitrogen oxide, which originates in all high-temperature combustion through a combining of the nitrogen and oxygen of the air. Nitrogen dioxide has a brownish color and absorbs light in the region of the spectrum from the blue to the near ultraviolet. Radiation from the sun can readily dissociate nitrogen dioxide into nitric oxide

and atomic oxygen. This reactive oxygen attacks organic material, of which there is much in the unburned hydrocarbons remaining in automobile exhaust. The result is the formation of ozone and various other oxidation products. Some of these products, notably peracylnitrates and formaldehyde, are eye irritants. Peracylnitrates and ozone also cause plant damage. Moreover, the oxidation reactions are usually accompanied by the formation of aerosols, or hazes, and this combination aggravates the effects of the individual components in the smog complex.

The answer to the puzzle of the oxidizing smog of the Los Angeles area thus lay in the combination of heavy automobile traffic and copious sunlight. Similar photochemical reactions can of course occur in other cities, and the large-scale phenomenon appears to be spreading.

The more or less temporary effects of smog alone would make a good case for air-pollution control; there is in addition the strong likelihood that smog has adverse long-range effects on human health [see "Air Pollution and Public Health," by Walsh McDermott; SCIENTIFIC AMERICAN Offprint 612]. Workers of the U.S. Public Health Service and Vanderbilt University reported to the American Public Health Association in November that a study they have been conducting in Nashville, Tenn., has established clear evidence that deaths from respiratory diseases rise in proportion to the degree of air pollution.

For the control of air pollution it is of central importance to know that organic substances—olefins, unsaturated hydrocarbons, aromatic hydrocarbons and the derivatives of these various kinds of molecules—can give rise to ozone and one or more of the other typical manifestations of smog. Control measures must be directed against the release of these volatile substances and of the other component of the smog reaction: the oxides of nitrogen. The organic substances originate with the evaporation or incomplete combustion of gasoline in motor vehicles, with the evaporative losses of the petroleum industry and with the use of solvents. A survey by the Los Angeles Air Pollution Control District in 1951 showed that losses at the refineries were more than 400 tons a day; these have since been reduced to an estimated 85 tons.

This reduction of one source was offset, however, by an increase in the emissions from motor vehicles. In 1940 there were about 1.2 million vehicles in the Los Angeles area; in 1950 there were two million; today there are 3.5 million. These vehicles burn about seven million gallons, or 21,500 tons, of gasoline a day. They emit 1,800 tons of unburned hydrocarbons, 500 tons of oxides of nitrogen and 9,000 tons of carbon monoxide daily. These emissions outweigh those from all other sources.

When motor vehicles emerged as a major source of air pollution, it was evident that state rather than local government could best cope with these moving sources. As a first step, and a pioneering one for the U.S., the California Department of Public Health adopted community standards for the quality of the air [see top illustration on page 254].

The adoption of these standards provided a sound basis for a program of controlling automobile emissions. Of special importance for that program was the establishment of the figure of .15 part per million by volume as the harmful level of oxidant. Years of observation have demonstrated that when the oxidant goes above .15 part per million, a significant segment of the population complains of eye irritation, and plant damage is readily noticeable. The standards also set the harmful level for carbon monoxide at 30 parts per million by volume for eight hours, on the basis of observations that under those conditions 5 per cent of the human body's hemoglobin is inactivated. A further stipulation of the standards was that these oxidant and carbon monoxide levels should not be reached on more than four days a year. To attain such a goal in Los Angeles by 1970 would require the reduction of hydrocarbons and carbon monoxide by 80 and 60 per cent.

On the basis of these standards the California legislature in 1960 adopted the nation's first law designed to require control devices on motor vehicles. The law created a Motor Vehicle Pollution Control Board to set specifications and test the resulting devices. In its work the board has been concerned with two kinds of vehicular emission: that from the engine and that from the exhaust.

About 30 per cent of the total emission of the car, or 2 per cent of the supplied fuel, escapes from the engine. This "blowby" loss results from seepage of gasoline past piston rings into the crankcase; it occurs even in new cars. Evaporation from the carburetor and even from the fuel tank is substantial, particularly on hot days. Until recently crankcase emissions were vented to the outside through a tube. California's Motor Vehicle Polution Control Board began in 1960 a process leading to a

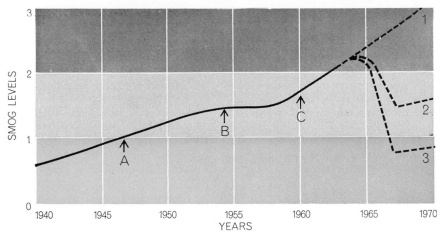

POLLUTION LEVELS in Los Angeles are plotted on scale (left) where 1 is 1947 level, 2 double and 3 triple that. A represents state pollution control law; B, control over refineries; C, motor vehicle controls. Broken lines indicate smog potential without new controls (1), with hydrocarbon controls (2) and with both hydrocarbon and nitrogen oxide controls (3).

requirement that all new cars sold in the state have by 1963 a device that carries the emissions back into the engine for recombustion. The automobile industry thereupon installed the blowby devices in all 1963 models, so that gradually crankcase emissions will come under control throughout the U.S. California is going a step further: blowby devices will have to be installed soon on certain used cars and commercial vehicles.

Two-thirds of the total automobile emission, or 5.4 per cent of the supplied fuel, leaves through the tail pipe as a result of incomplete combustion. For complete combustion, which would produce harmless gases, the air-fuel ratio should be about 15 to 1. Most cars are built to operate on a richer mixture, con-

taining more gasoline, for smoother operation and maximum power; consequently not all the gasoline can be burned in the various driving cycles.

The exhaust gases consist mainly of nitrogen, oxygen, carbon dioxide and water vapor. In addition there are lesser quantities of carbon monoxide, partially oxidized hydrocarbons and their oxidation products, and oxides of nitrogen and sulfur. Most proposals for control of these gases rely on the addition of an afterburner to the muffler. Two approaches appear most promising. The direct-flame approach uses a spark plug or pilot light to ignite the unburned gases. The catalytic type passes them through a catalyst bed that burns them at lower temperatures than are possible

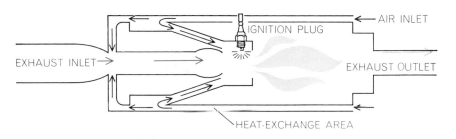

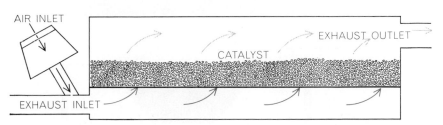

SECOND TYPE of afterburner involves leading exhaust gases through a catalyst bed; they can then be burned at lower temperatures than are possible in a direct-flame afterburner.

with direct-flame burners [*see bottom illustrations on preceding page*].

Building a successful afterburner presents several problems. The high temperatures require more costly materials, thereby increasing initial and replacement costs. Complications in operation arise from the burning of a mixture of gases and air of highly variable concentration. During deceleration the mixture may be so rich that without a bypass ceramics and catalysts will melt. In other cycles of operation there may not be enough fuel to keep the flame going. Moreover, the California law on exhaust-control devices stipulates that they must not be a fire hazard, make excessive noise or adversely affect the operation of the engine by back pressure.

Nine makes of afterburner—six catalytic and three direct-flame—are now under test by the California Motor Vehicle Pollution Control Board. Much testing and modification will be necessary before they are ready for the rough treatment to which they will be subjected when they are attached to all cars. Even after they have been installed a rigorous inspection program will be necessary to make certain that they are properly maintained and periodically replaced.

A preferable method of controlling hydrocarbon emissions from automobile tail pipes would be better combustion in the engine. Automobile engineers have indicated that engines of greater combustion efficiency will appear in the next few years. How efficient these engines will be remains to be seen; so does the effect of the prospective changes on emissions of oxides of nitrogen.

From all the emissions of an automobile the total loss in fuel energy is about 15 per cent; in the U.S. that represents a loss of about $3 billion annually. It is remarkable that the automobile industry, which has a reputation for efficiency, allows such fuel waste. Perhaps pressure for greater efficiency and for control of air pollution will eventually produce a relatively smogless car.

In any case it appears that the proposed 80 per cent control over motor vehicle emissions is a long way off. An alternative is to accept temporary controls at lower levels of effectiveness. It is possible to reduce unburned hydrocarbons and carbon monoxide by modification of the carburetor in order to limit the flow of fuel during deceleration, and by changing the timing of the ignition spark. Proper maintenance can reduce emissions by 25 to 50 per cent, depending on the condition of the car.

Accepting more practical but less efficient means of curbing vehicular emissions requires making up the deficiency in the smog control program some other way. This can be done by control of the other smog ingredient: oxides of nitrogen. At one time it was thought that control of these oxides would be very difficult, and that was why the California law concentrated on curbing emissions of hydrocarbons. It has now been shown, however, that control of oxides of nitrogen, from stationary sources as well as from motor vehicles, is feasible. Oil-burning electric power plants have reduced their contribution by about 50 per cent through the use of a special two-phase combustion system. Research on automobiles has shown that a substantial reduction of oxides of nitrogen is feasible with a relatively simple method of recirculating some of the exhaust gases through the engine.

To arrive at an acceptable quality of air through the limitation of hydrocar-

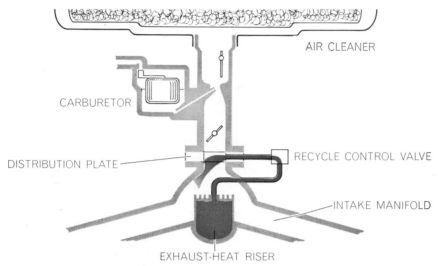

OXIDES OF NITROGEN emitted by automobiles may be curbed by this system, which takes exhaust gases before they leave the engine and recycles them through the combustion process.

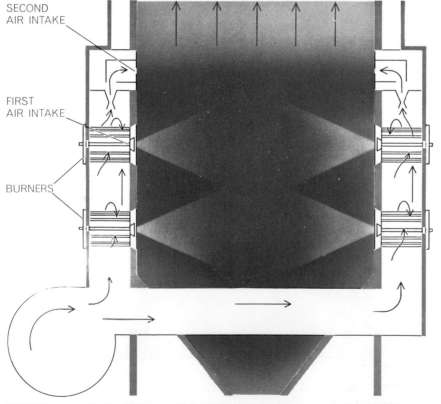

INDUSTRIAL FURNACES have curbed emissions of oxides of nitrogen by two-phase combustion. It lowers temperatures by introducing air at two stages of the burning process.

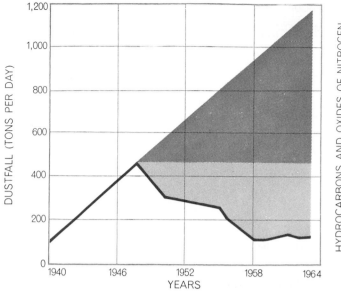

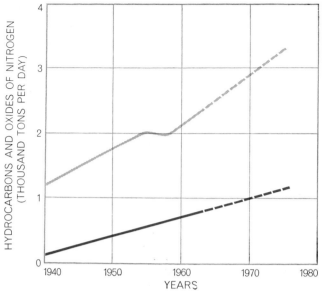

MAJOR POLLUTANTS in Los Angeles County are charted. Dust-fall (*left*) has been visibly reduced (*light color*) by control measures; potential without controls is indicated by darker color. At right, light line shows actual and potential levels of hydrocarbons; dark line similarly represents oxides of nitrogen. Rises in spite of controls reflect growth of population and number of vehicles.

bons alone would require a reduction in the hydrocarbons of about 80 per cent, which could be achieved only with rigorous and efficient controls. The plateau of clean air can also be reached, however, by dealing with both hydrocarbons and oxides of nitrogen. The advantage of such an approach is that each one of the reductions would have to be less complete. An over-all reduction of the two major smog components by half would achieve the desired air quality [*see bottom illustration on following page*].

This combined approach offers the only practically feasible way to return to a reasonably smog-free atmosphere in Los Angeles as well as in other metropolitan areas plagued by photochemical smog. The California Department of Public Health is now considering the expansion of the smog control program to include curbs on emission of oxides of nitrogen. For such a program to succeed, however, there would have to be regular inspection of motor vehicles, control of carburetor and fuel tank losses, stringent additional controls over industry and the co-operation of citizens. Moreover, these efforts must be organized in such a way that they take into account the area's rapid population growth, which will mean proportionate rises in motor vehicle and industrial emissions.

Beyond the efforts to control industries and vehicles lie some other possibilities, all of which would have the broad objective of reducing the amount of gasoline burned in the area. They include electric propulsion, economy cars, increased use of public transportation and improvement of traffic flow. A strong argument for resorting to some of or all these possibilities can be found in an examination of the carbon monoxide readings at a monitoring station in downtown Los Angeles. The readings show clear peaks resulting from commuter traffic. The carbon monoxide increase during a rush period is about 200 tons, representing the emission of about 100,-000 cars. That figure agrees well with vehicular counts made during the hours of heavy commuting.

Greater use of public transportation would produce a considerable reduction of peak pollution levels. So would improved traffic flow, both on the main commuter arteries and on the roads that connect with them. Reduction of the frequent idling, acceleration and deceleration characteristic of stop-and-go driving

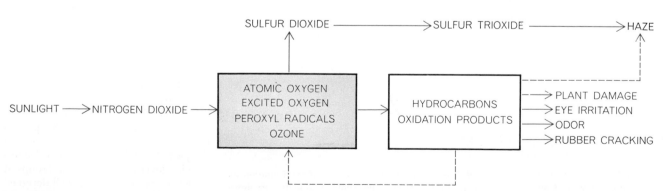

PHOTOCHEMICAL REACTION playing a major role in smog formation begins with sunlight acting on nitrogen dioxide, a product of combustion, to yield oxidants (*gray box*). They attack hydro-carbons, which come mainly from automobile exhausts, to produce irritating materials. Oxidants also attack sulfur dioxide, a product of coal and oil burning. Broken lines indicate interactions.

POLLUTANT	PARTS PER MILLION FOR ONE HOUR		
	"ADVERSE" LEVEL	"SERIOUS" LEVEL	"EMERGENCY" LEVEL
CARBON MONOXIDE		120	240
ETHYLENE	.5		
HYDROGEN SULFIDE	.1	5	
SULFUR DIOXIDE	1	5	10
HYDROCARBONS			
NITROGEN DIOXIDE			
OXIDANT	.15 ON "OXIDANT INDEX"	NOT ESTABLISHED	NOT ESTABLISHED
OZONE			
AEROSOLS			

AIR-QUALITY STANDARDS adopted by California set three levels of pollution: "adverse," at which sensory irritation and damage to vegetation occur; "serious," where there is danger of altered bodily function or chronic disease; "emergency," where acute sickness or death may occur in groups of sensitive persons. Blanks mean "not applicable." Pollutants listed in colored type are involved in or are the products of photochemical reaction. These standards, the first adopted by any state, provided a basis for pollution control measures.

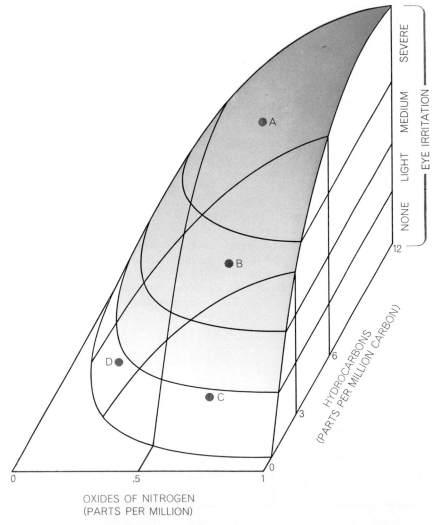

OXIDES OF NITROGEN
(PARTS PER MILLION)

CONTROL POTENTIALS are depicted. Los Angeles is at *A* in degree of eye irritation on a day of heavy smog. Controls reducing hydrocarbons by 50 per cent would bring city down the slope to *B*, still not in clear zone shown in white. Hydrocarbon controls to *C* are impractical; control of both hydrocarbons and oxides of nitrogen would attain clear zone at *D*.

—the very cycles that produce the most hydrocarbons and oxides of nitrogen—could curb vehicular emissions by 50 per cent or more over a given distance. Detroit has a system of computing the optimum speed on certain freeways according to the density and flow of traffic; the speed is then indicated on large lighted signs. The result is a smoother flow. More techniques of this kind, more imaginative thinking about transportation in general, are necessary for a successful attack on smog.

There can be no doubt that the smogs of Los Angeles represent an extreme manifestation of a problem that is growing in every heavily populated area. Similarly, the control steps taken by Los Angeles will have to be duplicated to some degree in other cities. In those cities, as in Los Angeles, there will be difficulties. One is the cost of air-pollution control for communities that already find their budgets stretched; the Detroit City Council annually votes down an ordinance to ban the burning of leaves because it believes the city cannot afford the estimated cost of $500,-000 for carting the leaves off to dumps. Industry also may balk at smog controls out of concern for maintaining a competitive position. There is a related problem of co-ordination: industries are reluctant to install devices for curbing smoke while the city burns trash in open dumps.

Another problem involves mobilizing the public behind air-pollution control programs. Even though smog looks unpleasant, is occasionally offensive to the smell and irritating to the eye, and sometimes precipitates a public health disaster (as in Donora, Pa., in 1948 and in London in 1952), it nonetheless tends to be regarded as a fact of urban life and something that communities can live with if they must. Moreover, so many political jurisdictions must be involved in an effective attack on air pollution that any one community attempting a cleanup may find its efforts vitiated by another community's smog.

Nevertheless, a growing segment of the public is alert to the dangers of air pollution and determined to do something about it. If anything effective is to be done, however, it will require intelligent planning, aggressive public-education programs and resoluteness on the part of public officials. Then leadership by government and civic groups at all levels, united behind well-designed plans, could generate progress toward the goal of cleaner air.

27

THE GLOBAL CIRCULATION
OF ATMOSPHERIC POLLUTANTS

REGINALD E. NEWELL
January 1971

*Worldwide wind and temperature patterns and the
behavior of trace substances are studied in an effort
to learn what effect changes in the atmosphere
caused by man may have on the earth's climate*

Pollution is more than a plume of smoke rising above a factory or a yellowish haze hanging over a city. The foreign substances man introduces into the air spread all over the globe and rise into the upper atmosphere. It is therefore important to learn how each of the major pollutants enters the atmosphere, the speed and extent of its spread and the ways in which it may alter the atmosphere and thus affect temperature and precipitation both locally and worldwide. In this article I shall discuss the movement of the various pollutants and review what is known about the effect on temperature of foreign substances in the atmosphere.

Of the total of 164 million metric tons of pollutants emitted each year in the U.S., about half comes from automobiles. Of the main component, carbon monoxide, 77 percent is from automobiles; so are most of the hydrocarbons and much of the oxides of nitrogen. The oxides of sulfur come mainly from electric power plants, small particles mainly from power plants and industry. In addition to these more obvious pollutants, vast quantities of water and carbon dioxide are produced by the burning of fossil fuels. Other pollutants are lead from automotive gasoline and ozone, which is produced by the action of sunlight on automobile exhaust. Radioactive substances introduced by man's activity include fission products from weapons tests, such as strontium 90, cesium 137 and iodine 131, and neutron-activated casing materials, such as tungsten 185, manganese 54, iron 55, rhodium 102 and cadmium 109. A satellite carrying a portable power plant using plutonium 238 as fuel accidentally burned up in the atmosphere over the southern Indian Ocean in April, 1964, at a height of from 40 to 50 kilometers instead of going into orbit, and the radioactivity from that point source has been tracked over the globe. The reprocessing of nuclear-fuel elements from power plants releases krypton 85, a radioactive gas with a half-life of about 10 years, which is gradually accumulating in the atmosphere.

Trace substances are, of course, present in nature. Carbon dioxide is taken up by plants during their growth cycle and released by the decay of plant material. Sulfur is also involved in plant processes and is abundant in the ocean, from which it is released by sea spray; it is sometimes injected high into the atmosphere in large amounts during volcanic eruptions. Ozone is produced in the upper atmosphere and carried downward toward the earth's surface, where it is destroyed [see "The Circulation of the Upper Atmosphere," by Reginald E. Newell; Scientific American, March, 1964]. As for radioactivity, radon and thoron gas emanate from the soil and decay in the atmosphere, giving rise to a chain of radioactive substances; some of the end products, such as lead 210, are transported up into the stratosphere. Cosmic rays entering the atmosphere collide with air molecules, usually in the stratosphere, to form radioactive nuclides such as beryllium 7, sodium 22 and carbon 14, some of which live long enough to find their way down to the surface.

Whether natural or man-made, some of these trace substances occur as gases, others as aerosols: finely divided liquid droplets or solid particles. Many are involved in phase transformations. For example, water vapor (gas) may be cooled to the point where it changes to water droplets or to ice crystals, forming clouds; sulfur dioxide gas may change in moist air to droplets of sulfuric acid. The gases diffuse and mix quite easily but the aerosols are governed by a number of factors that limit their spread. Large particles with radii of 10 microns (thousandths of a millimeter) or more can be washed out of the air by raindrops or fall out directly. Very small particles can grow by coagulation until they too can be trapped in clouds and be washed out. Particles larger than about .3 micron cannot reach the upper atmosphere under normal circumstances because their fall velocities are greater than the average updraft speeds. Larger particles are nevertheless found in the upper atmosphere, some being introduced directly by volcanic eruptions, others growing there from smaller particles and gases introduced by the air motion. (Incidentally, the electrostatic precipitators on smokestacks are good for trapping particles with radii larger than about a micron; so is the human nose. Smaller particles can penetrate into the lungs, however, and so it is not always the larger particles one sees pouring from chimneys that are the most hazardous to health.)

Trace substances are moved over the globe by wind systems that fluctuate in strength and direction from day to day as cyclones and anticyclones move around the globe. The lowest 10 to 15 kilometers (six to nine miles) of the atmosphere, where the temperature decreases with height, is called the troposphere; above it to about 50 kilometers is the stratosphere, which in some respects resembles a series of stratified layers. Air parcels in both the troposphere and the stratosphere can be tabbed and tracked from one day to another [*see illustration below*]. The prevailing, or mean, wind, which is revealed by averaging over a long time period, blows from west to east over much of the middle latitudes in both hemispheres, but at low latitudes (and in the upper regions in general in summer) the prevailing wind is from the east. The mean wind transfers trace substances fairly rapidly round the globe; for example, the 35-meter-per-second west to east flow at 30 degrees north latitude gives a transit time of about 12 days. Clouds of debris from a nuclear explosion or volcanic eruption can often be identified as they make several circuits of the globe.

While the prevailing wind is along lines of latitude, north-south oscillations occur at the same time; the resulting north-south drift and an accompanying up-and-down motion give rise to an overturning pattern at low latitudes called the Hadley-cell circulation [*see illustration on page 258*]. In the Tropics this large cell is a dominant feature of the circulation, whereas at middle latitudes north-south eddies in the prevail-

ing-wind systems overshadow the mean north-south drifts. Air can be exchanged between the hemispheres by both the mean Hadley circulation and eddies in the upper tropical troposphere. As for vertical exchange, air passes into the stratosphere from the troposphere at low latitudes in the Hadley-cell circulation. Most of the transfer back into the troposphere is thought to occur close to the tropopause level near the middle-latitude jet stream [*see illustration on these two pages*], and some transfer back into the stratosphere may also occur there.

In the troposphere clouds, rain and thunderstorms are evidence of considerable vertical motion. Vertical velocities may reach 10 to 20 meters per second in thunderstorms but are generally no more than 10 centimeters per second in normal middle-latitude cyclones and anticyclones. In the stratosphere it is much harder to push an air parcel up or down because the temperature increases with height, and so vertical motions rarely exceed a few centimeters per second and are often much smaller. The vertical spread of a trace substance in the stratosphere is therefore rather slow, like the

downward migration of a card in a pack of cards that is being shuffled. There is much less shuffling at low latitudes than there is over the polar regions, so that trace constituents can stay in the equatorial stratosphere for several years.

Once the material reaches the bottom of the pack and enters the troposphere it can be mixed vertically rather rapidly. Small particles spend about 30 days in the troposphere before being washed out by rain. Gases spend varying periods there depending on the "sinks" by which each is removed from the atmosphere: incorporation into cloud droplets, reactions with other gases, loss to finely divided liquid or solid particles or the earth's surface and so on. Generally the tropospheric residence times of gases are from about two to four months—provided that there *is* a sink. (Krypton 85, for example, has no known sink and disappears only by radioactive decay.)

The global temperature pattern shows that the coldest air is over the Equator near the tropical tropopause at all seasons and over the winter pole in the stratosphere. The lowest temperatures over the Equator occur in January. Notice that temperature increases with lati-

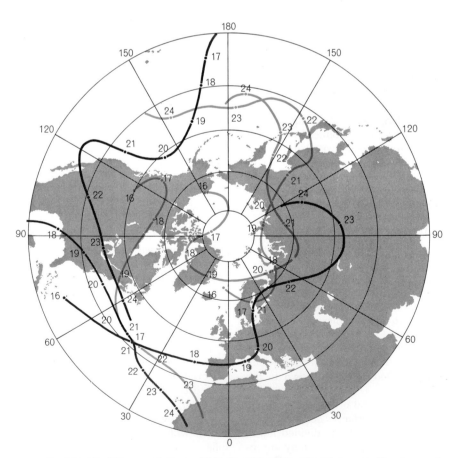

MOVEMENT OF AIR around the world is determined by identifying specific air parcels and tracking them. These tracks were worked out by Edwin Danielsen of the National Center for Atmospheric Research, who identified parcels in the troposphere (*black*) and stratosphere (*color*) and tracked them. The numbers represent successive days in April, 1964.

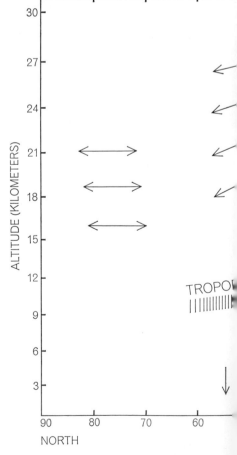

DISTRIBUTION AND MOVEMENTS of ozone (*black*), water (*gray*) and aerosols, or small solid and liquid particles (*color*), are

tude in the lower stratosphere. The temperature pattern is maintained by the liberation of latent heat, by radiation and by the motion of air masses, the sum of which is roughly in balance. When more water is rained out of a given air column than is evaporated into it from the surface, the column gains the latent heat that is liberated. The net effect of latent-heat liberation and radiative processes is that tropospheric air is heated at low latitudes and cooled at high latitudes, generating energy. In the troposphere the net effect of the radiative processes by themselves is to cool the air at all latitudes. Contributions come from absorption of incoming solar radiation, and from absorption and reemission of terrestrial long-wave radiation, by carbon dioxide, water vapor and ozone. Water vapor dominates in the lower layers, producing cooling of up to two degrees per day, with carbon dioxide of secondary importance; in the stratosphere carbon dioxide and ozone dominate [*see top illustration on page 262*]. Horizontal and vertical motions can also produce temperature changes. If air is forced to rise, it cools as it expands; if it is forced to sink, it contracts and be-

comes warmer, as in a bicycle pump. The former effect is thought to maintain the low temperatures over the tropical tropopause, the upward motion being forced from below, where the latent heat is liberated. On the other hand, compression is responsible for the inversion over the Los Angeles basin during a good fraction of the year—the increase of temperature with altitude that inhibits vertical mixing and traps pollutants near the surface.

Carbon dioxide is quite well mixed vertically, whereas water-vapor concentrations decrease with distance from the source (the earth's surface) and ozone decreases away from its source in the middle stratosphere [*see bottom illustration, page 262*]. Aerosols show two regions of high concentration. One is at the source (ground level). The other, in the lower stratosphere, is due to the direct injection of particles and the injection of gases from which particles form, together with a very slow removal rate.

It is fairly clear that if the present atmospheric temperature structure is maintained by a combination of air motions and effects that involve trace substances, it can be altered by changes in

the concentration of trace substances. One therefore needs to know the natural cycles of the atmospheric trace constituents that are important in this context and the changes in the cycles that are being or may in the future be brought about by man.

Ozone (O_3) is produced by the photodissociation of molecular oxygen and the recombination of molecular with atomic oxygen, primarily above about 22 kilometers and at low latitudes. Ozone is transported toward the poles and downward by atmospheric motions between the subtropics and the high latitudes. From the lower stratosphere in the middle latitudes the ozone seeps into the troposphere, and it is eventually destroyed at the earth's surface or in reactions with aerosols in the surface layer. There is a maximum of ozone in the lower stratosphere in the spring and in the troposphere in the late spring; it is caused by an increase in the supply of energy to the stratosphere from the troposphere in this season and a concomitant increase in the large-scale mixing motions. Ozone stays for about four months in the middle-latitude lower

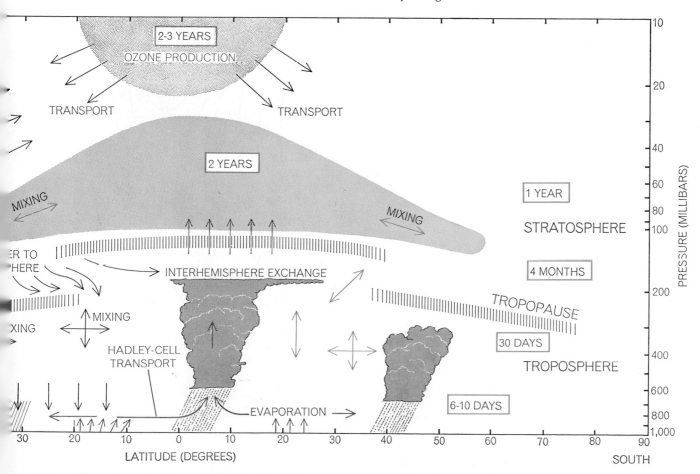

indicated on a schematic diagram drawn along a line of longitude. Effects illustrated for one hemisphere occur in both hemispheres. Boxed figures are residence times for aerosols. Pressure, measured in millibars, is often used by meteorologists as a measure of altitude. The tropopause is the boundary between the troposphere and the stratosphere; its altitude varies with latitude as indicated.

258

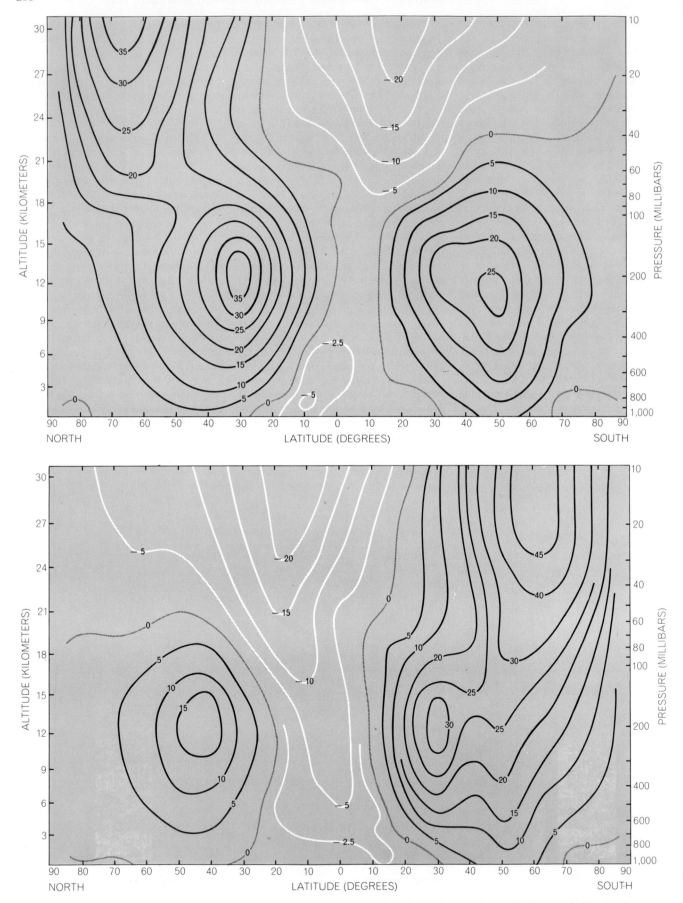

PREVAILING WINDS are revealed by averaging observations over a period of time. Here the mean east-west wind speed is shown as computed for two three-month periods: December–February (*top*) and June–August (*bottom*). The speed is given in meters per second, with positive numbers indicating winds blowing from west to east (*black contour lines*) and negative numbers for east-to-west winds (*white lines*). Note that seasonal changes in the lower atmosphere are larger in Northern Hemisphere than in Southern.

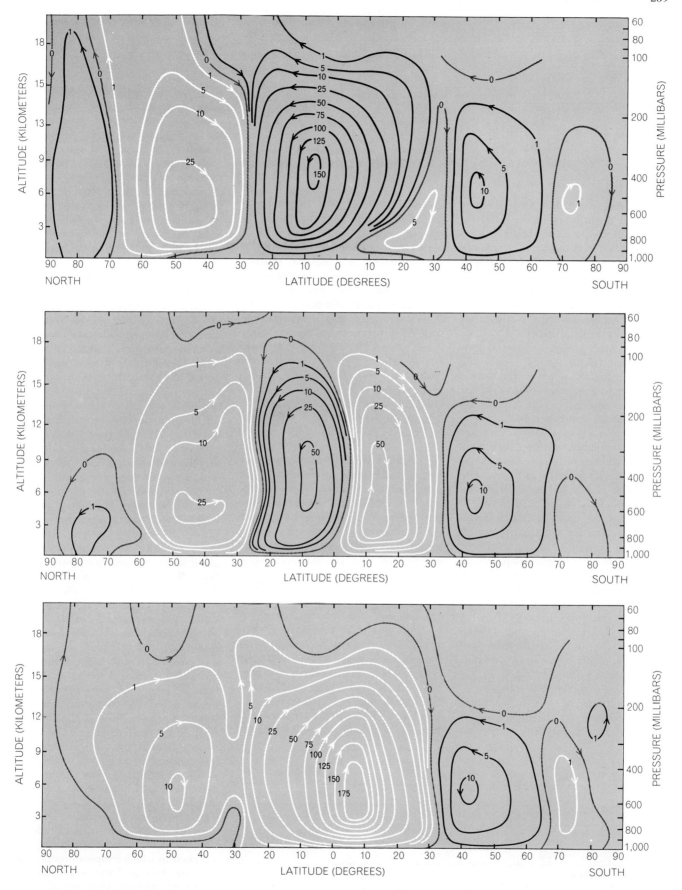

MASS FLUX IN THE TROPICS takes the form of Hadley-cell circulation, the result of a combination of north-south and up-and-down motions. It is shown here for three seasons; December–February (*top*), March–May (*middle*) and June–August (*bottom*). The difference between adjacent contour values gives the flux in millions of metric tons per second based on average values from all longitudes; actually the pattern varies with longitude. The charts are based on work of Dayton G. Vincent and John W. Kidson.

	PARTIC-ULATES	SULFUR OXIDES	NITROGEN OXIDES	CARBON MONOXIDE	HYDRO-CARBONS
POWER AND HEATING	8.1	22.1	9.1	1.7	.6
VEHICLES	1.1	.7	7.3	57.9	15.1
REFUSE DISPOSAL	.9	.1	.5	7.1	1.5
INDUSTRY	6.8	6.6	.2	8.8	4.2
SOLVENT EVAPORATION					3.9
TOTAL	16.9	29.5	17.1	75.5	25.3

POLLUTANTS emitted in U.S. in 1968 are shown. Estimates, in millions of metric tons, are by George B. Morgan and colleagues of National Air Pollution Control Administration.

stratosphere and from two to three months in the troposphere.

Most of the steps of the cycle have been verified in detail by a "dynamical-numerical" model that was developed at the Environmental Science Services Administration's Geophysical Fluid Dynamics Laboratory at Princeton University and has been applied to the ozone problem by Syukuro Manabe and B. G. Hunt. In such a model equations governing temperature, wind speeds and ozone concentration are solved with a computer to trace the evolution of the fields over a period of time. The models have been developed to help in weather forecasting and in understanding the general circulation of the global atmosphere and theories of climatic change. Radiative effects, the earth's surface properties, clouds and even ocean temperatures all have to be included to achieve a good representation of the atmosphere, and the necessary degree of resolution in space and time can only be obtained with the largest computers. As far as can be seen, the influence of man-made ozone (in city smog) is small compared with the natural cycle of production, transport and loss, which totals about two billion tons per year. (This is on a global scale, of course, and is of small comfort to people whose smarting eyes make them quite aware of the ozone in the smog they live with.)

The water-vapor cycle in the troposphere depends on the difference between precipitation and evaporation in a given air column [see illustration on page 263]. In the subtropics there is an excess of evaporation over precipitation, so that these latitudes can be regarded as source regions; the opposite situation occurs over the middle and low latitudes. Water vapor is therefore transported from subtropical latitudes toward the Pole and toward the Equator. Large-scale eddy processes govern the movement toward the Pole and the mean Hadley-cell circulation governs the movement toward the Equator. The average rainfall at a point on the globe is about 100 centimeters per year and the average residence time of a water-vapor molecule in the troposphere is about 10 days.

The water vapor produced by man—by fuel combustion, for example—is not a significant fraction of the natural tropospheric cycle. (Again, people who live near power-plant cooling towers will think otherwise.) Nevertheless, more subtle effects could well be produced by interference with the evaporation-precipitation cycle. (Efforts have already been made to do this on a small scale by spreading a thin film on the water surface of some reservoirs in Australia to prevent excessive evaporation.) One would need a complete dynamical-numerical model to see what effect a given change would have if extensive regions were altered. Less water vapor evaporated would lead to less liberation of latent heat—but also to less cooling by the radiative effect; only a full model can give a proper idea of the interlocking feedbacks and their net effect.

The water-vapor balance of the stratosphere is much more delicate. Alan W. Brewer of the University of Toronto has suggested that most of the water vapor in the stratosphere enters through the region near the tropical tropopause in the rising branch of the Hadley-cell circulation. The temperature is close to −80 degrees Celsius in that region, and so the air can hold only minute amounts of moisture; in the process of passing through, most of the moisture from the troposphere is frozen out and precipitates, staying in the troposphere as cirrus clouds. (The fact that the frost point throughout the lower atmosphere even at middle latitudes is close to the temperature near the tropical tropopause forms the observational basis for Brewer's suggestion.) Henry Mastenbrook of the Naval Research Laboratory has been monitoring the water vapor in the stratosphere since 1964, flying a frost-point device (developed by Brewer) on high-altitude balloons. He finds that the content in the lower stratosphere varies seasonally in phase with the temperature at the tropical tropopause, whereas the content at 30 kilometers varies only slightly, with average values throughout of only two to three micrograms of water vapor per gram of air.

From the mean rising motion and the water-vapor content near the tropical tropopause one can calculate the mass of water entering the stratosphere; it is about seven million grams per second. Now, 500 supersonic transport planes flying at 21 kilometers (70,000 feet) would inject about two million grams of water vapor per second directly into the stratosphere. Since this water is introduced above the cold trap, and since it is introduced at a rate that is of the same order of magnitude as the natural rate, it is clearly going to lead to a significant increase in the water-vapor content at high levels. Wherever rising motion and low temperatures exist together in the stratosphere, clouds may form as expansion and concomitant cooling allow the air to reach the local frost point. Such clouds are occasionally observed near 25 kilometers over Norway and Iceland and over the Antarctic in winter, and also near 80 kilometers at high latitudes in summer and near the tropical tropopause —regions in which temperatures become very low. Increased water-vapor content in the stratosphere, then, would be expected to produce increased cloudiness and therefore a change in the albedo, or reflectivity, of the earth. Again, however, a full dynamical-numerical model

	NATURAL	MAN-MADE
OZONE	2×10^9	SMALL
CARBON DIOXIDE	7×10^{10}	1.5×10^{10}
WATER	5×10^{14}	1×10^{10}
CARBON MONOXIDE	?	2×10^8
SULFUR	1.42×10^8	7.3×10^7
NITROGEN	1.4×10^9	1.5×10^7

NATURAL AND MAN-MADE trace-gas cycles are compared (in metric tons). Sulfur and nitrogen data are from E. Robinson and R. C. Robbins of the Stanford Research Institute. (The man-made carbon dioxide input is not really a cycle; some remains in the atmosphere.)

is required in order to study the possible changes.

The largest pollutant by mass is carbon dioxide, and here there is evidence that man's activities are indeed altering the concentrations formerly controlled by nature. Some 70,000 million tons per year are involved in the natural cycle, corresponding to a fluctuation of nine parts per million by volume in the carbon dioxide concentration of 320 parts per million. Plants and trees start taking up carbon dioxide in the spring and continue to do so until the fall; then the leaves drop, vegetable matter begins to decay and the direction of the net transfer is from matter to air. This seasonal cycle can be monitored on the ground or in the lower troposphere. Superimposed on the seasonal cycle is a long-term increase, produced by the burning of fossil fuels. The carbon dioxide increase expected from the fuel that has been burned is about 1.8 parts per million per year, yet the observations show an increase of only about .7 part per million per year [see the article "The Carbon Cycle," by Bert Bolin, beginning on page 53]. Where does the rest of the carbon dioxide go? Some of it may be incorporated in the biosphere, but it is thought that the largest fraction is dissolved in the oceans, which all together contain about 60 times as much carbon

dioxide as the atmosphere does. The solubility diminishes as the temperature of the water increases. There has been some concern that as carbon dioxide in the air increases, presumably raising the air temperature through the "greenhouse effect," the water temperature will rise also, releasing some of the carbon dioxide from the ocean back into the air. Proponents of this view argue that a runaway effect will occur, with the additional carbon dioxide producing still more atmospheric heating.

It is well to bear in mind, however, that the net radiative contribution of carbon dioxide is to cool the atmosphere. With computer programs developed by Thomas G. Dopplick, my colleagues at the Massachusetts Institute of Technology and I have rerun the radiative-heating computations, assuming a tripled carbon dioxide concentration of 1,000 parts per million and the same temperature and cloudiness distributions. We find that the cooling rate diminishes by only a small fraction of a degree per day in the lower levels and actually increases in the stratosphere. Other things being equal, one could interpret a smaller cooling rate as effective heating in the troposphere. The outgoing infrared flux emitted by carbon dioxide is smaller for higher concentrations; for the atmosphere to radiate the original amount of infrared radiation back to space it would have to

radiate at a higher effective temperature. Other things are not equal, however. If the temperature distribution changed, it is likely that the clouds, albedo and therefore the radiation-stream balance would change also. For example, a slightly higher temperature would mean more evaporation and hence more water-vapor radiative cooling. Although it is tempting to argue from our results that little net temperature change should be expected, it is again clear that the proper way to proceed is to run a complete dynamical-numerical model with higher carbon dioxide levels and see what happens.

Another substance that is copiously produced by combustion is carbon monoxide. Levels of 50 parts per million are not uncommon in city streets, with values up to several hundred parts per million in traffic tunnels and underground garages; weekly average levels of 20 parts per million are sometimes found in the Los Angeles basin. The toxic effect is proportional to the ambient-air concentration and the time of exposure. Carbon monoxide that enters the bloodstream combines with hemoglobin, forming carboxyhemoglobin, and thus reduces its capacity to carry oxygen. Impairment of mental function, as measured by visual performance and ability to discriminate time intervals, occurs when carboxyhemoglobin in the blood

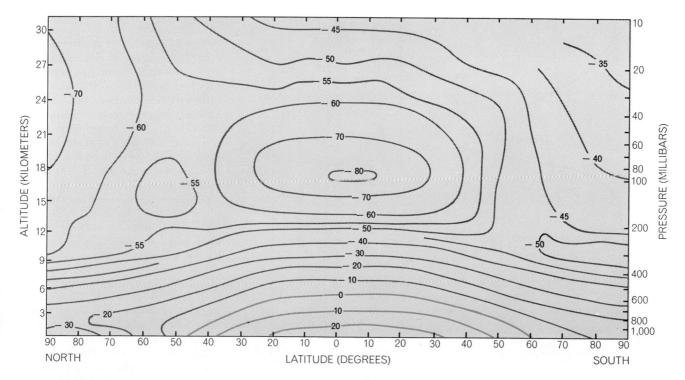

TEMPERATURE PATTERN is given for a north-south cross section of the atmosphere for the three-month period December–February. The isotherms, or contour lines of equal temperature, indicate the temperature in degrees Celsius. The coldest air is over the Equator and near the winter pole. In the latter region the temperature sometimes reaches −85 degrees C. at certain longitudes.

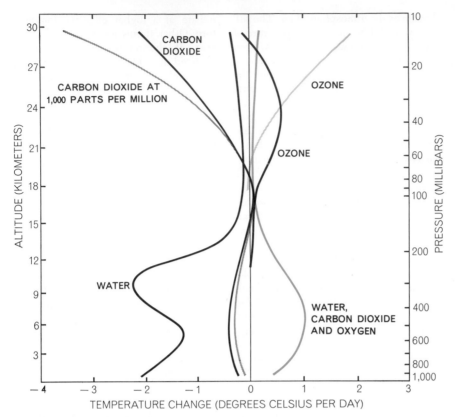

RADIATIVE TEMPERATURE CHANGE is brought about by the absorption of solar visible and ultraviolet (*light color*) and near-infrared (*dark color*) radiation and the absorption and reemission of thermal radiation from the earth (*black*) by various atmospheric gases. The direction and magnitude of each effect vary with altitude, as shown for at the Equator in January. If the carbon dioxide concentration were tripled to 1,000 parts per million, its effects would change as shown (*gray*), according to computations by author's group.

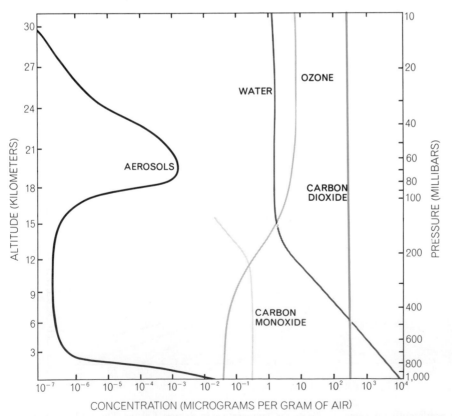

MIXING RATIO, or concentration, of the trace constituents of the atmosphere varies with altitude, decreasing with distance from their sources, except in the case of carbon dioxide.

goes above about 2.5 percent (compared with a normal level of about .5 percent). Such values accompany exposure to 200 parts per million for about 15 minutes or 50 parts per million for about two hours.

The 200 million tons injected each year would correspond to an increase of .03 part per million per year. No such increase has been observed, and so a search for sinks of carbon monoxide is under way.

Christian E. Junge of the University of Mainz and his student Walter Seiler have reported that near sea level there is a boundary close to the Equator with high carbon monoxide values to the north and low values to the south; in the upper troposphere aircraft measurements show no such boundary [*see bottom illustration, page 264*]. Such a carbon monoxide discontinuity is compatible with the fact that there are more automobiles in the Northern Hemisphere and with what is known of the Hadley-cell circulation. At low levels the air just north of the discontinuity is streaming toward the Equator, whereas the air to the south has come from the Southern Hemisphere, with a much smaller population of automobiles. (A similar boundary was frequently revealed by surface-air measurements of strontium 90 when nuclear tests were being held in the Northern Hemisphere.) Higher up in the troposphere eddy mixing eliminates the interhemisphere gradients.

There is still much to be learned about the carbon monoxide sink. Junge's data show a decrease in the stratosphere, and it has been suggested that carbon monoxide combines above the tropopause with hydroxyl radicals (OH) to produce carbon dioxide and hydrogen. There are no direct measurements of the hydroxyl radical in the lower stratosphere and the computed amounts, together with the reaction rate, are just barely sufficient to account for the loss of 200 million tons per year. Furthermore, the rate at which air is transferred from the upper troposphere to the lower stratosphere also seems a shade too low to ascribe the entire loss to this path. The ocean, which has been found to contain carbon monoxide, has been suggested as both a sink and a source; so little is known that it is difficult to say which! Clearly many more observations are needed all over the atmosphere and in the ocean at different latitudes.

Nitrogen constitutes the largest fraction of the atmosphere and is involved in a variety of reactions, including plant growth. The total atmospheric turnover in all forms is about 10 billion tons per year, whereas the amount introduced by man is only about 50 million tons. While this much pollution is very important in some regions (city smogs are produced by the action of sunlight on oxides of nitrogen from automobiles), on a global scale it seems that nature's contributions dominate.

There is no such clear distinction for the sulfur compounds. Sulfur is an abundant constituent of ocean water and is released to the atmosphere by sea spray and by biological decay; it is removed by precipitation, intake by vegetation and direct deposition. The total amount involved in the natural cycle is about 142 million tons per year; the 73 million tons injected by man as a pollutant is therefore a very significant amount. This additional sulfur is thought to end up in the ocean and, as Erik Eriksson of the University of Stockholm has stressed, increases the acidity of terrestrial waters. (Fish cannot live in waters of high acidity, and they already shun some inland waterways in Sweden; man has been given a warning.) Sulfur dioxide and particles together in the air seem to produce respiratory ailments.

A considerable amount—no one knows exactly how much—of sulfur dioxide emitted to the atmosphere ends up as sulfate ions or as ammonium sulfate particles. The time necessary for the sulfur dioxide to disappear as a gas and become incorporated into particles varies from about half an hour to a few days, depending on the air's moisture content and other factors. Measurements made by Junge show that the very small particles called Aitken nuclei (radius about .03 micron) decrease in concentration above the tropopause in a manner consistent with the view that their source is the troposphere. Their composition is unknown. In addition there is a layer of somewhat larger particles (mean radius about .3 micron, with some radii of a micron or more) in the lower stratosphere, composed of ammonium sulfate or sulfuric acid. James P. Friend of New York University and Richard D. Cadle of the National Center for Atmospheric Research have independently verified Junge's finding of this layer of larger particles. Junge has suggested that the large particles in the lower stratosphere may grow on the Aitken nuclei from the gases injected there, and gradually coagulate to form larger particles before they fall, or are transported by exchange, into the troposphere.

When the distribution of sizes of particles of various kinds in the lower troposphere is measured, it is found that most of the mass is accounted for by particles whose radius is between .1 micron and 10 microns. Observations over land away from major cities have shown an increase in worldwide particle mass in the past 10 years, even though over some city areas there has been a decrease. There has been considerable speculation that an increase in the concentration of particles in the troposphere would cause more solar radiation to be scattered back to space and thus contribute to the lowering of terrestrial temperatures. There are no measurements to support this speculation. In fact, George Robinson of the British Meteorological Office has found that solar radiation is absorbed, rather than scattered, by tropospheric aerosols.

The efforts now being made to reduce particulate pollution, coupled with the relatively short washout time for aerosols, provide reason to hope the long-term effect of large man-made aerosols may be small. If particulate material is not removed but is simply more finely divided, however, more will find its way into the stratospheric regions where the residence times are long and the influence of aerosols generated at the surface could be appreciable.

In the case of hydrocarbons, very little is known about the natural cycle and consequently about the worldwide effect of man's interference. Methane (marsh gas) is abundantly produced by nature and finds its way to the stratosphere, but there are no global-scale measurements. Yet hydrocarbons do appear to play a role in smog formation and cannot be ignored at the local level.

Great volcanic eruptions, such as that of Krakatoa in 1883 or Mount Agung on Bali in 1963, increase the layer of stratospheric aerosols, so that colorful sunsets are produced all around the world. After Krakatoa, investigators suggested that the sunsets were due to the injection of sulfuric acid as well as volcanic dust, basing this suggestion on the fact that there is a strong odor of sulfur near active volcanoes. Samples taken in the lower stratosphere over Australia after the Bali eruption showed that this was indeed the case.

As the concentration of particles (and perhaps their size too) increases, solar radiation is intercepted and less arrives at the ground. Such blocking has often been proposed as a major cause of climatic change [see "Volcanoes and World Climate," by Harry Wexler; SCIENTIFIC AMERICAN Offprint 843]. We examined stratospheric temperature patterns after the Bali eruption in March, 1963, and found there was an immediate rise in stratospheric temperatures, with values

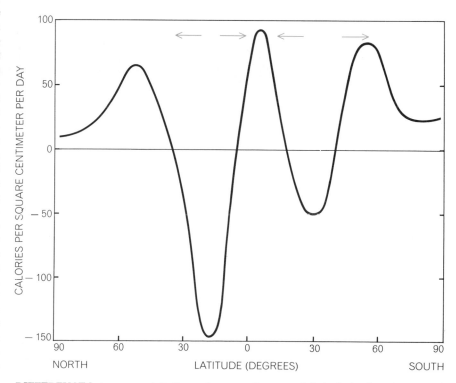

DIFFERENCE between precipitation and evaporation at each latitude is given in terms of the amount of heat lost through evaporation or gained through precipitation in each column of air. (About 600 calories are required to evaporate a column of water one centimeter high covering one square centimeter.) The curve, for December–February, is based on three sets of data. Imbalances of precipitation (*top*) and evaporation (*bottom*) are redressed by air motions that transport water toward Equator and toward poles (*arrows*).

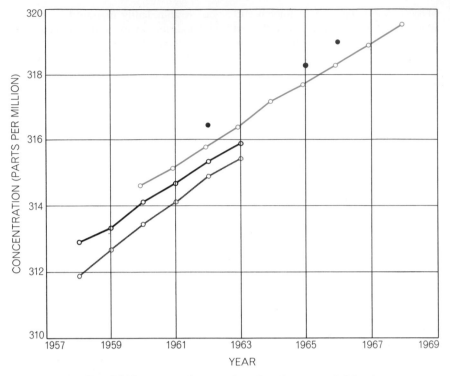

LONG-TERM INCREASE in atmospheric carbon dioxide is revealed by four investigations. The data are from measurements at Barrow, Alaska, by John Kelley of the University of Washington (*black dots*), aircraft observations by Bert Bolin and Walter Bischof of the University of Stockholm (*color*) and measurements at Mauna Loa in Hawaii (*black curve*) and in the Antarctic (*gray*) by Charles Keeling of the Scripps Institution of Oceanography.

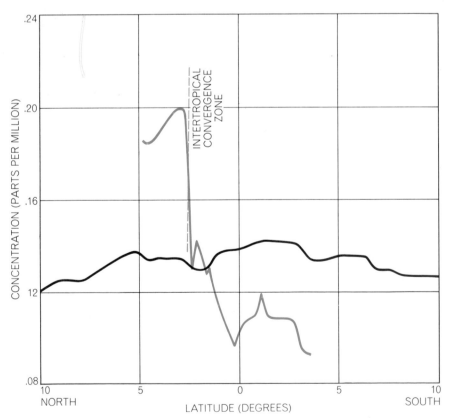

CARBON MONOXIDE measurements made from a ship at sea (*color*) and from an aircraft at 10 kilometers (*black*) by Christian E. Junge and Walter Seiler of the University of Mainz reveal a change in concentration at sea level but not in the upper troposphere. The boundary is explained by the larger number of automobiles in the Northern Hemisphere and the rising motion of the Hadley-cell circulation at the Intertropical Convergence Zone (**broken line**). Higher up eddy currents mix the air, eliminating the interhemisphere gradient.

as high as eight degrees C. above average being recorded; the rise was global in extent [*see illustrations on next page*]. It was mainly concentrated in the region above 15 kilometers and there were concomitant changes in the wind field. Presumably the small particles intercepted and absorbed solar radiation. (Their size is of the same order as the wavelength of light.)

The variation of the temperature change with latitude and height was very similar to the patterns that have been found for clouds of radioactive debris in the stratosphere. Such clouds persist longest over the Equator and slope downward toward the poles. The volcanic cloud from Bali did likewise—and the same slope was discussed in a report on Krakatoa published in 1888. The 1963 temperature increase is the largest climatic change ever observed by man. (There were no balloon observations of the stratosphere after Krakatoa.) It serves to warn us that we should watch the sulfur cycle carefully and try to learn soon whether the increased amounts injected into the troposphere by man can find their way into the stratosphere.

Investigators are able to base deductions about atmospheric motions on a network of 800 balloon sounding stations, at many of which temperature is also measured; global maps of temperature constructed from satellite observations are also becoming available. Yet the pollutants in the air we breathe are measured at very few places, and there are no systematic measurements above the surface layers. Moreover, most of the surface measurements are made near cities. It is perfectly well established that if a city has oil-fired or coal-fired power plants, there will be sulfur in the air; if it has automobiles on crowded streets, there will be carbon monoxide and oxides of nitrogen. The most intensive monitoring efforts near cities cannot provide information concerning the global buildup of pollutants.

Water vapor and ozone have been measured above the surface because of their interest as natural trace substances. Moisture sensors are included on the same balloon flights that collect wind and temperature information, but the sensors only operate up to about six to eight kilometers. Mastenbrook's special frost-point hygrometer, which records moisture content up to about 30 kilometers, makes only one flight a month from Washington, D.C.—pitifully inadequate for a study of the global distribution of stratospheric moisture. About six balloon stations measure ozone up to about

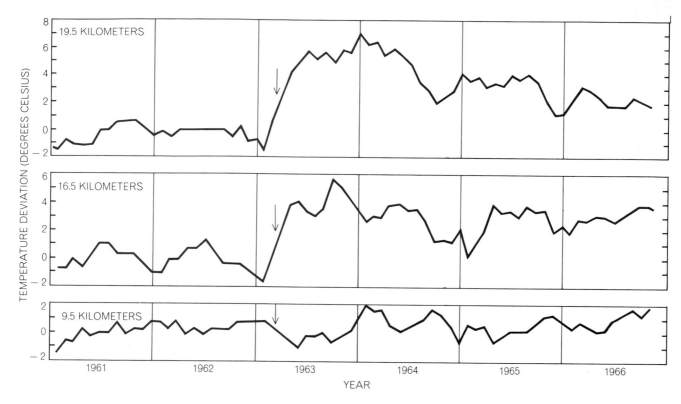

TEMPERATURE CHANGES above Port Hedland, Australia, show the heating effect (first noted by James G. Sparrow of the University of Adelaide) of particles from the 1963 eruption of Mount Agung on Bali. Monthly means were calculated for five years before the eruption (*arrows*); deviations from the means were computed and three-month running averages were plotted.

30 kilometers once a week. The Atomic Energy Commission launches balloons from four sites to obtain samples from up to 35 kilometers, mainly for analysis of radionuclides, although carbon dioxide has sometimes been included. High-altitude sampling aircraft such as the U-2 have also been used, but as nuclear tests in the atmosphere have decreased these sampling programs have naturally been pared. Practically all the published data on carbon monoxide measured away from the surface is represented in Junge's results. (The instrument he used was made portable and carried on a commercial jet flight; it occupied one seat, the observer another.) A few spot measurements of sulfur dioxide and sulfate above the surface layers have been made by Hans Georgii of the University of Frankfurt; again no extensive coverage is available.

There are many opportunities to collect data on atmospheric trace substances by taking advantage of commercial airline flights and regular ocean voyages as well as special oceanographic cruises, the unique AEC balloon network, the surface-air sampling networks established to monitor global radioactivity levels, mountaintop observatories and so on. Now that some important trace substances are being introduced into the atmosphere by man at rates comparable to those in the natural cycle, it seems appropriate to start making these measurements.

I must mention in closing that one cannot take the global view of pollutants without feeling some concern about the rate of use of natural resources and the rate of generation of pollutants. Both could be slowed considerably by a serious effort to use every pound of fuel in the most efficient manner possible.

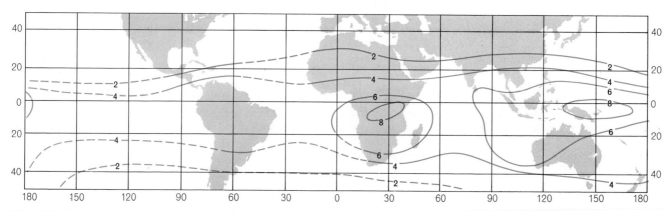

HEATING EFFECT after Mount Agung is mapped. The isotherms give the increase in mean temperature at about 19 kilometers, in degrees Celsius, from January before the eruption to one year later. (Broken lines indicate uncertainty due to scarcity of observations.)

28

RADIOACTIVE POISONS

JACK SCHUBERT
August 1955

The term refers to those radioactive atoms and molecules which, when they enter the human body, tend to accumulate in certain organs and subject them to damaging radiation

Man was created in a radioactive world. We are bombarded constantly by cosmic rays from outer space and by radiations originating within the earth and ocean. The food we eat and the air we breathe are laden with minute but measurable amounts of radioactive isotopes. As a result we ourselves contain naturally radioactive atoms in every cell of our bodies. What has protected the human race up to now is that the total exposure to natural radioactivity during a lifetime is very small. In a sense, then, our present concern about radioactive "poisoning" arises from the fact that the age of man-made radioactivity has raised the exposure to a higher level. Unfortunately radioactive poisons are far more difficult to cope with than the ordinary chemical poisons we have known. They are incomparably more potent (the toxic dose is usually so small it cannot even be weighed), and they may produce a slow, insidious disease of which the victim is not aware until many years after the exposure.

Consider these facts. Sodium fluoride, one of the most potent chemical poisons, may be lethal to a man at a dose of one gram. But as little as one half-millionth of a gram of radium in the body has been known to kill a human being. Amounts of this order may not kill immediately but may cause bone cancer, which develops years after the radium has entered the body.

Radioactivity is measured in curies: one curie is equal to the radioactivity from one gram of radium (37 billion atoms disintegrating per second). Thus a millionth of a curie, acting on the body over a period, is a dangerous dose. From the discovery of radium in 1898 until the present, the total production of radium has been about 1,500 grams (1,500 curies). But today our chain-reacting piles produce fission products whose total radioactivity runs into hundreds of millions of curies.

Man was slow to realize the hazard of radioactivity. Probably the first human death from acute radioactive poisoning was reported at a meeting of the Berlin Medical Society in 1912. A 58-year-old woman suffering from arthritis had been treated for the disease with frequent injections, for 16 days, of thorium X, a short-lived isotope of radium. Within a month afterward she died, showing symptoms now recognized as those of radiation sickness, including hemorrhages and diarrhea. Eight years earlier Pierre Curie had observed that laboratory animals died within hours after breathing the radioactive gas radon, emitted from radium during its decay.

But these danger signals went unheeded. Between 1915 and 1930 thousands of people in the U. S. actually ate or drank radium. Patients wealthy enough to afford this "cure" took radium water or injections of radium salts for all sorts of diseases. One physician alone administered radium salts to about 5,000 patients. An emaciated 52-year-old patient, admitted to a hospital, related that he had drunk a two-ounce bottle of water containing two micrograms of radium each day for about five years. All in all he had consumed 1,400 bottles! Post-mortem examination disclosed that his skeleton contained 74 micrograms of radium. Few persons anywhere had any notion of the deadliness of this novel substance. In Europe a candy firm marketed radium-containing chocolate bars. We may laugh at that age of innocence, but we cannot be too smug even today. As late as 1953 a company in the U. S. sold contraceptive jelly incorporating nearly a microgram of radium in each two-ounce tube!

The tragic case of the New Jersey watch-dial painters finally aroused the world to the necessity of examining the biological effects of radioactivity. For eight years girls in a New Jersey factory painting luminous dials with radium had followed the practice of pointing their brushes with their lips. The first indication that they had been poisoned by the radium was discovered in 1924 by a dentist, Theodor Blum, who treated many of the girls for severe jaw infections caused by bone destruction. By 1929 15 of the girls had died. Meanwhile they had been subjects of the first intensive and systematic study of chronic radioactive poisoning in human beings, carried out by Harrison Martland, the Medical Examiner of Essex County.

Many other studies followed. One of the obvious areas for investigation of radiation hazards was the uranium mines, the source of radium. It had been known even before the discovery of radioactivity that in the uranium mines of Joachimsthal in Bohemia more than half of the miners died of lung cancer. The opening of a particularly rich vein of pitchblende was always followed by an increased death rate some years later. Measurements of radioactivity in the mines showed that the level was some 30 times what is now considered the tolerance dose.

The newer uranium mines now being worked in the U. S., on the Colorado plateau, have about the same radon concentrations as those in Germany, but U. S. Government agencies have taken steps, particularly improved ventilation, to reduce the hazard.

Man can learn to live even with the tremendous amounts of radioactivity

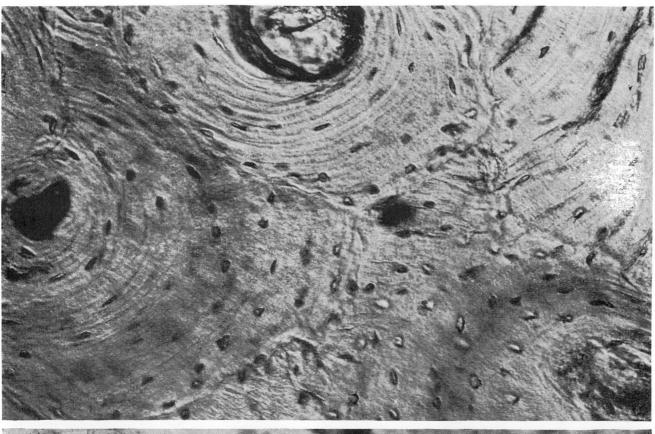

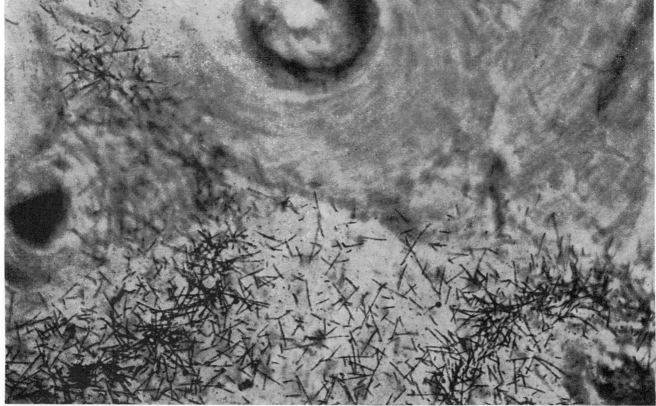

RADIUM POISONING is illustrated by these two photomicrographs. The top photomicrograph (magnification: 300 diameters) shows a section of bone from the upper arm of a 48-year-old woman who had died in 1951 as a result of a tumor induced by radium. She had been given radium water 22 years previously for the treatment of an arthritic condition. In the bottom photomicrograph a thin sheet of photographic film has been affixed to the same section of bone and developed. The tracks of alpha particles from the decaying radium atoms in the bone are visible as short, needlelike streaks in the film. The photomicrographs were furnished by W. B. Looney and Lois A. Woodruff of the Argonne National Laboratory. They were made by A. S. Tracy of its Biological-Medical Division.

	ELEMENT	TOTAL RADIOACTIVITY (MICROCURIES)	ATOMIC DISINTEGRATIONS (PER MINUTE)
MAN		(TOTAL IN BODY)	
	POTASSIUM 40	.1	220,000
	CARBON 14	.06	130,000
	RADIUM 226	.0001	200
OCEANS		(PER KILOGRAM)	
	POTASSIUM 40	.00025	560
	CARBON 14	.00000013	.3
	RADIUM 226	.0000001	.2
	URANIUM 238	.0000005	1.2
	RUBIDIUM 87	.0000036	8
SOIL		(PER KILOGRAM)	
	RADIUM 226	.0001-.001	200-2,000
	ALL OTHER RADIOISOTOPES FROM URANIUM, THORIUM, POTASSIUM	.001-.01	2,000-20,000
FOOD		(PER KILOGRAM)	
	RADIUM 226	.000001-.000005	2-10
WELL WATER		(PER LITER)	
	RADIUM 226	.000005	11
SURFACE WATER		(PER LITER)	
	RADIUM 226	.00000003	.07
ATMOSPHERE		(PER LITER)	
	RADON 222	.0000002	.5
	RADON 220 (THORON)	.00000001	.02

NATURAL BURDEN of radioactive isotopes is tabulated. The amount of radium 226 in newborn infants is only .00000003 microcurie. The well water for which a figure is given is from an area in Illinois. The atmosphere also contains tiny amounts of other isotopes formed by reactions involving cosmic rays: carbon 14, hydrogen 3 (tritium) and beryllium 7.

RADIOISOTOPE	HALF-LIFE	PRINCIPAL RADIATION EMITTED	MAXIMUM PERMISSIBLE AMOUNTS (MICROCURIES)		
			TOTAL BODY	WATER (PER LITER)	AIR (PER LITER)
PLUTONIUM 239	24,400 YEARS	ALPHA	.04	.0015	.000000002
RADIUM 226	1,620 YEARS	ALPHA	.1	.00004	.000000008
POLONIUM 210	138 DAYS	ALPHA	.04	.003	.0000004
STRONTIUM 90	19.9 YEARS	BETA	1	.0008	.0000002
CALCIUM 45	152 DAYS	BETA	13	.25	.000006
CARBON 14	5,600 YEARS	BETA	250	3	.001
PHOSPHORUS 32	14.3 DAYS	BETA	10	.2	.0001
IRON 59	45.1 DAYS	BETA	11	.1	.000015
IODINE 131	8.1 DAYS	BETA	.3	.03	.000003

NINE RADIOACTIVE ISOTOPES are among the potentially more hazardous. The maximum permissible amounts are for continuous exposure to a soluble form of each isotope. The maximum permissible amounts for carbon 14 apply to its intake as carbon dioxide.

that have been released by the discovery of nuclear fission. The best evidence of this is the excellent safety record of the U. S. atomic energy enterprise. In 10 years there has not been a single radiation injury at Hanford, where some 9,000 men and women have worked on the production of plutonium, perhaps the most dangerous of all the new radioactive poisons. But as time goes on, and radioactive materials accumulate on our planet, the problem will become less and less simple. Safety so far has been bought at the price of the strictest possible precautions against exposure, and it has been achieved only because mankind has been extremely gingerly and limited in its use of radioisotopes. To handle them with complete safety and confidence we shall need to learn much more precisely than we know now what the permissible limits of exposure are, how the various radioactive poisons affect the body and how such poisoning can be prevented or cured.

Radioactive substances emit two kinds of radiations which particularly concern us: alpha and beta rays. Alpha particles are harmless when they strike the outside of the body; since they cannot penetrate more than about 50 microns into tissue, they are absorbed by the dead outer layers of the skin and do not reach living cells. Beta particles, which can travel several millimeters through tissue, may burn the skin, but they do not penetrate as far as the vital inner organs. It is when they get inside the body, through breathing, swallowing or entry into the bloodstream, that the radioactive poisons are most dangerous. Within the body the radioactive substance comes directly into contact with living cells, and even the short-ranged alpha particle may pass through five cells before it reaches the end of its trail. Here the massive alpha particle is far more damaging than the beta particle. Either type of particle, however, may injure the cells. Fortunately the cells have a remarkable capacity for self-repair, but if they are subjected to insult or injury for a long enough time, they will finally die or give rise to cancer-producing cells.

The most dangerous radioisotopes are those that stay in the body, instead of being quickly excreted, and have a long enough half-life to keep bombarding the cells with particles for months or years. One of the most hazardous, for instance, is plutonium: it tends to lodge in the bones, is excreted very slowly, and has a radioactive half-life of 24,000 years,

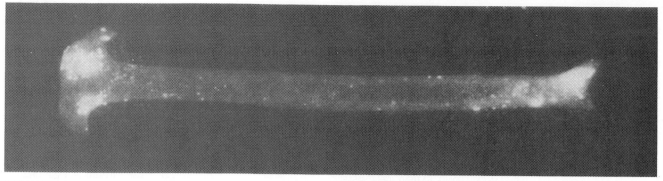

LEG BONE OF A DOG which eight and a half years earlier had been injected with the radioactive isotope strontium 90 was mounted in a block of plastic and sliced down its long axis. The cut section of the bone was then placed in contact with a sheet of special photographic film. When the film was developed, it showed both diffuse exposure and "hot spots" that were due to the radioactive strontium. This radioautograph was furnished by Miriam P. Finkel and Juanita Lestina of the Argonne National Laboratory.

emitting destructive alpha particles. On the other hand, radioactive iodine (the isotope iodine 131) may be kept in the body for a long time but is not very dangerous because its radioactive half-life is only eight days; within two months almost all of it will have decayed and ceased to be radioactive even if none is excreted. In some cases the tenure of the poison in the body depends on the form in which it is introduced. For example, if carbon 14 (radioactive half-life: 6,000 years) is injected into the bloodstream in the form of carbonate, most of it is eliminated within a matter of minutes, but if the radiocarbon is incorporated in a molecule which the body does not break down readily (e.g., certain dyes), it may take years to eliminate all the carbon 14. So the hazard of a given radioisotope depends basically on a composite quantity called the "biological half-life"—a measure of the duration of its activity within the body.

A radioactive poison entering the body may at first be excreted very rapidly in the feces and urine. But after about a week the excretion rate falls, and the body may then take years to rid itself of as much of the radioisotope as it did during the first week. Most fission products are known as "bone-seekers": they tend to concentrate in the skeleton. Radium and strontium 90 have this tendency because they resemble calcium in chemical behavior. Plutonium, which is insoluble in body fluids, is stored not only in bone but also in the liver, spleen and other soft tissues. A particularly malignant feature of some of the radioisotopes is their tendency to concentrate in "hot spots" instead of distributing themselves evenly through the bone or other tissue they invade [see photograph above]. Plutonium, which is two and a half times as toxic as radium, may owe its greater toxicity to the circumstance

that it forms more intense hot spots or concentrates in more sensitive tissue centers.

Radioactive dust breathed into the lungs is a great hazard. Fortunately the lungs have some defense: the upper respiratory tract efficiently eliminates most foreign particles by propelling them into the mouth so that the particles are swallowed and excreted. English coal miners, who during their average 39-year lifetime of work in the mines inhale about seven pounds of dust, were found to have less than one ounce of dust retained in their bodies at death. However, insoluble dust particles of certain sizes—particularly those about one micron in diameter—do tend to collect and stay in the lungs.

To learn how to live and work comfortably in a world in which radioisotopes are being produced in great quantities, we must begin by determining how much exposure the body can safely tolerate—i.e., the maximum permissible amount (MPA) for each radioisotope. How much can be allowed in our air, water and food? We cannot avoid exposure altogether, for if we set the tolerance levels too low, it would be impossible to produce or use radioactive substances at all.

The estimation of the MPA is a complicated business. It involves the uptake and retention by the body of various radioisotopes in various chemical forms, their relative concentration in different parts of the body, the comparative sensitivity of different tissues, the energies and damaging effects of various kinds of radiation, and so on. As if all this were not enough, we also must consider the age of the exposed person. The bone-seeking radioisotopes related to calcium, such as strontium, barium and radium, concentrate to a greater extent in grow-

ing bone. Young, growing tissues are generally more radiosensitive. And the hazard is also greater for younger people because it may take 20 years or more of exposure for the body to develop cancer.

The present permissible levels have been fixed by the National Committee on Radiation Protection, which is sponsored by the National Bureau of Standards in cooperation with radiological organizations in the U. S. and abroad [see lower table on opposite page]. Some information has been gained from experiments on animals, but the standards are based mainly on the past 50 years of experience in exposure of people to X-rays, gamma rays and other ionizing radiation and on accidental cases of radioisotope intake.

For example, one tenth of a microcurie of radium 226, distributed throughout the skeleton, is stated to be the maximum permissible amount of that isotope because no individual with this amount of radium in his body has ever suffered detectable harm. One person with 3.5 microcuries of pure radium 226 in his body died of leukemia with accompanying bone damage. Hence the MPA for radium in effect at present apparently provides a "safety factor" against serious damage of 35. If in the future as little as two tenths of a microcurie is found to have injured an individual, or if a large number of people is found to be unaffected by as much as one microcurie of radium, the MPA may very well be lowered or raised, as the case may be. Generally a safety factor of at least 10 is applied. It must be emphasized that an individual who has somewhat more than the MPA of a radioisotope in his body is not necessarily in danger. Besides the tenfold safety factor, there are great variations in individual susceptibility: one patient with 23 microcuries of radium stored in his body was in better health

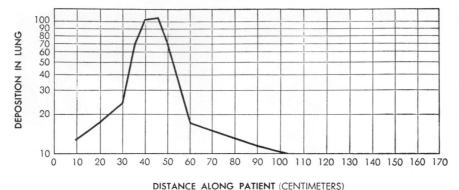

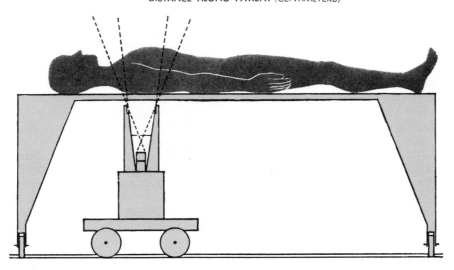

SCINTILLATION COUNTER mounted on wheels is used to detect the concentration of radioactivity in various organs of a patient. In the case illustrated by the curve at the top the patient had inhaled an insoluble salt of radium. The concentration of radium in the lung is shown by the increased counts as the counter traveled along the length of the patient.

than another patient who had 1.5 microcuries.

How does one measure the amount of radioisotope in a person's body? It is easy enough when the patient has died: all one need do is cremate the body and analyze a sample of the ashes by standard radiochemical procedures. This method has obvious disadvantages for application to a living person. The job can be done indirectly, however, by analysis of the urine, feces and blood. One of the earliest and best of these studies was made in Berlin nearly 45 years ago by J. Plesch. He injected small amounts of the short-lived radium isotope thorium X into two people and measured the amounts eliminated in the urine and feces during the next few days. He even went so far as to measure the radium excreted in their sweat: he had them wear long woolen underwear for 24 hours and then burned the perspiration-soaked underwear and analyzed the residue for radium. Apparently he recovered about one tenth of one per cent of the injected radium in the sweat secreted by a person in one day.

One of the most reliable methods for estimating the amount of radium in the body was introduced many years ago by Herman Schlundt, a chemistry professor at the University of Missouri. He found that the amount of radon in a person's breath is a good measure of the radium fixed in the body. Radon, an alpha-emitting daughter of radium, is easily measured by its radioactivity.

In the case of plutonium the usual measuring method is to analyze the urine: it is known that a few months after exposure the amount of plutonium present in a 24-hour sample of a person's urine is one 10,000th of the amount fixed in the body. At the Los Alamos Scientific Laboratory the urine of persons working with plutonium is analyzed periodically to make sure that they have not accumulated more than three hundredths of a microcurie.

The intake of radioactive dust into the lungs can be measured roughly by analysis of the feces, because the solid particles are coughed up, swallowed and passed through the stomach and intestines. Contamination of the air by radio-

isotopes can even be detected by analyzing swabs from the dust-filtering hairs in the nose.

When the radioisotope in the body is one that emits gamma rays, it can be measured directly by a Geiger counter or scintillation counter. The counter is simply passed over the patient's prone body, and it not only measures the amount of radioisotope in the body but also locates it [see illustration at left]. The method is valuable for detecting radium, because radium is a gamma-ray emitter. Moreover, it is so sensitive that it can identify different radioisotopes, if more than one is present, by the different energies of gamma radiation.

Much research has been done on the problem of removing radioactive poisons from the body. Promptness and speed are important, both to prevent the beginning of damage to the cells and to try to catch the isotope while it is still in the stomach or the bloodstream and has not yet migrated to a permanent lodging place in the bones or other tissues.

A person who has swallowed radium or strontium may be able to get rid of a large part of it in the feces by taking epsom salts or certain other substances which will react with the soluble radioisotope in the digestive tract and form an insoluble precipitate with it. Another possible and certainly more palatable approach is to eat rhubarb or spinach, which contain large amounts of oxalate—a good precipitating agent.

In experiments on animals Marcia White Rosenthal and the writer have been able to divert plutonium in the bloodstream from bone by injecting salts of certain metals, notably zirconium, titanium or aluminum, into the blood. The salts decompose, releasing insoluble particles of the metal hydroxides, and these soak up plutonium in the bloodstream much as a blotter soaks up ink. If the treatment is applied within an hour after the plutonium gets into the bloodstream, about half of it is removed in the urine within a day. Even if the circulating plutonium is not excreted, it is diverted from bone to other tissues.

After radioisotopes have left the bloodstream and become deposited in tissues, the situation is still not completely hopeless. Because radium behaves like calcium, one of the treatments that has been tried is a low calcium diet, or a drug, which hastens the removal of calcium (and radium) from bone. However, this decalcification therapy does not remove radium fast enough, and it cannot be continued indefinitely. A very promising newer approach, which unfortunately does not work for radium but may be effective in removing plutonium and rare-earth fission products, is the use of chelating agents [see the article "Chelation," by Harold F. Walton, beginning on page 344]. These metal-grasping compounds have been used successfully many times to remove mercury, arsenic and other metallic poisons, and they have now been found applicable to plutonium. The powerful chelating agent known as EDTA has a high affinity for plutonium and the rare earths, and when administered early it has removed as much as 25 per cent of the plutonium in a patient's body. Unfortunately it is not effective when the radioisotope has lodged in bone.

Perhaps the most effective treatment so far is a combination of zirconium salt and EDTA—the first to divert the radioisotope from bone, the second to remove the diverted poison from the soft tissues to which it has been sidetracked. One problem is that, since EDTA and the zirconium salt must be given intravenously, it does not appear practical to treat large numbers of people with these drugs for indefinite periods, as would be necessary in case of an emergency. One new approach that we are exploring is to induce the body to form its own chelating agents to capture radioisotopes. This may be done by alteration of the body's metabolism. It has been found possible to induce various tissues, including the spleen, blood and bone, to produce extra amounts of a fairly good chelating agent, citric acid.

On the whole, it is still the best policy to be supercautious in the use of radioisotopes. Their use in medicine is developing into a good-sized business, but it is questionable whether they should be employed as freely as they are for routine diagnostic tests of blood volume, liver and kidney function and so on, particularly where other tests are available. There are many reasons for caution. Certain radioisotopes such as iron concentrate in what are equivalent to hot spots in restricted regions of an organ. The injection of radioisotopes for routine diagnosis in pregnant women and babies seems particularly unwise, in view of the high radiosensitivity of embryonic tissue.

The administration of radioisotopes also involves a long-range, genetic risk. We have no information on the genetic effects in mammals of radioisotopes stored in the body. Here is an important area which needs to be explored.

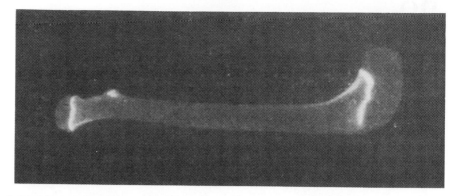

RADIUM was injected into a dog four days before this radioautograph of its leg bone was made. The radium has concentrated in the growing areas of the bone, especially the ends.

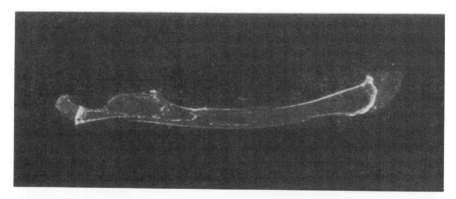

PLUTONIUM in the bone of a rabbit is shown by this autograph. The top two photographs were made by W. P. Norris and Lois A. Woodruff of the Argonne National Laboratory.

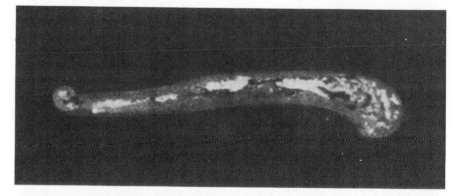

SODIUM BICARBONATE labeled with carbon 14 was injected into a rat. This radioautograph of a bone shows that the carbon 14 accumulated both in the bone and in its marrow.

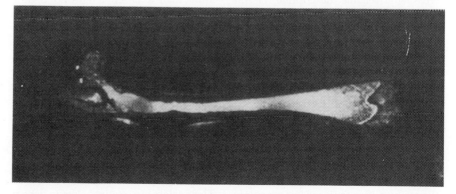

DYE bearing carbon 14 in the central part of its molecule accumulated in the rat's bone marrow. The bottom photographs were made by A. Lindenbaum of the Argonne Laboratory.

29

THE CIRCULATION OF RADIOACTIVE ISOTOPES

JAMES R. ARNOLD AND E. A. MARTELL

September 1959

Man's experiments with nuclear explosions have stimulated him to study the distribution of radioisotopes in his environment. These investigations have clarified many geophysical processes

Life has evolved on earth in a haven relatively free of ionizing radiation. The fast-moving particles and hard electromagnetic radiation that impinge upon our planet from the sun and from elsewhere in our galaxy barely trickle through the sheltering atmosphere. Radiation from space accounts for only a quarter of the natural "background" radiation at the earth's surface. The balance comes from radioactive isotopes in rocks, soils, waters and the atmosphere.

The study of radioactivity in our environment has acquired new significance since man began to contribute measurable amounts to the total through his experiments with nuclear fission and fusion. Fortunately for us radioactive atoms are still rare, so that sensitive methods of detection are usually required. The high energy associated with nuclear processes makes it possible, however, to detect events on the scale of single nuclei. Moreover, the virtually unchangeable rate of decay (half-life) of each nuclear species allows high precision in the interpretation of data. Thus it has been possible to develop an increasingly reliable description of the world-wide transport and distribution, via the atmosphere and waters of the earth, of minute quantities of natural and man-made radioisotopes. These same studies have enhanced our understanding of the circulation of the atmosphere and the oceans.

Considerations of human welfare have prompted investigation of still a third major system of transport. This is the biosphere, with its intricate food chains that tie together the existence of living organisms. Radioactive zinc taken up in the sea by microscopic marine organisms may appear on the table in the tissues of a fish; radiostrontium from the soil of a meadow may similarly be present in a glass of milk.

Until 1932, when Frédéric and Irène Joliot-Curie produced the first artificial radioisotopes, the natural variety stood alone. Not until the first nuclear explosion in 1945 did man-made transmutations begin to influence the environment at large. Even today natural radioisotopes irradiate human beings far more intensely than man-made fallout.

Since radioisotopes are unstable by definition and thus are constantly decaying, why do they occur in nature at all? There can be only two reasons: either they were made long ago and their half-lives are long enough to permit them to persist, or they are still being made. Isotopes of both types in fact exist.

The long-lived "old" isotopes so far discovered range from neodymium 144, which has a half-life of about 5,000 trillion years, to uranium 235, which has a half-life of 710 million years. No radioisotope with a half-life of less than 710 million years occurs in this group. This fact, among others, enables us to date the formation of the earth's elements at between five and 10 billion years ago. More than a dozen long-lived natural radioisotopes have been discovered, and the list continues to grow as our methods for detecting them become more sensitive. However, three of them—uranium 238, thorium 232 and potassium 40—generate the overwhelming bulk of natural radiation in our environment.

The location of these isotopes within the earth depends not on their nuclear properties but on their chemistry. All three are easily oxidized metals. Their oxides are relatively low in density and thus occur in the crust of the earth rather than in its dense mantle and metallic core. Perhaps half of the radioactivity in the earth lies within 40 miles of the surface. This concentration significantly increases the radiation to which all living things are exposed—and makes uranium and thorium available to man's uses.

The radiation from isotopes in the crust is rapidly attenuated as it passes through rock and soil; most of the radiation that reaches living organisms originates less than six inches below the surface. The average surface radioactivity is on the order of two curies per square mile, though values in different places vary by a factor as large as five. In general, the more acid rocks, such as granites, are more radioactive than the alkaline basalts; limestones show especially low radiation levels.

The radioactivity of potassium 40 ends with its transformation into stable calcium or argon. Uranium and thorium, however, decay into long series of radioactive "daughter" elements with half-lives ranging from several hundred thousand years down to microseconds. In impermeable rocks and other sealed reservoirs parents and daughters remain trapped together; their abundance with respect to one another dates the formation of the rock. Elsewhere they take different paths [*see illustration on pages 274 and 275*]. Thus radon, a gas, diffuses out of soils into the air. Its decay products constitute the chief natural source of atmospheric radioactivity. Radon itself has a four-day half-life and so reaches the upper atmosphere only in minute quantities. Its concentration at various altitudes tells us something about the rate at which air mixes.

As rocks weather away, some of the radioisotopes find their way in solution to the sea. Thorium quickly combines with constituents of sea water to form

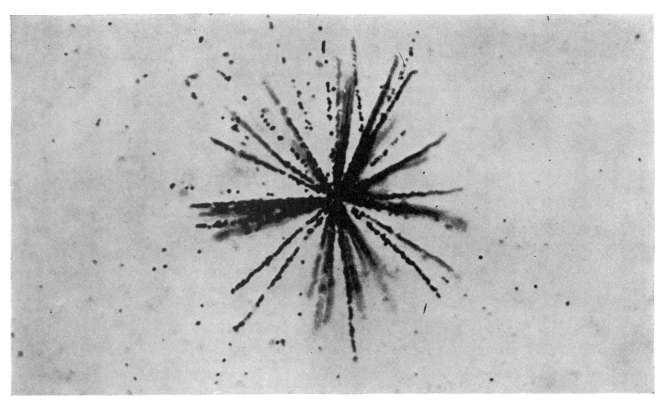

NATURAL RADIOACTIVITY in the atmosphere is shown by this nuclear-emulsion photograph of tracks (enlarged 2,000 diameters) emitted by a grain of dust. The tracks are produced by alpha particles; their length identifies the dust as polonium, a decay product of uranium 238. The photographs on this page were made by Herman Yagoda of the Air Force Cambridge Research Center.

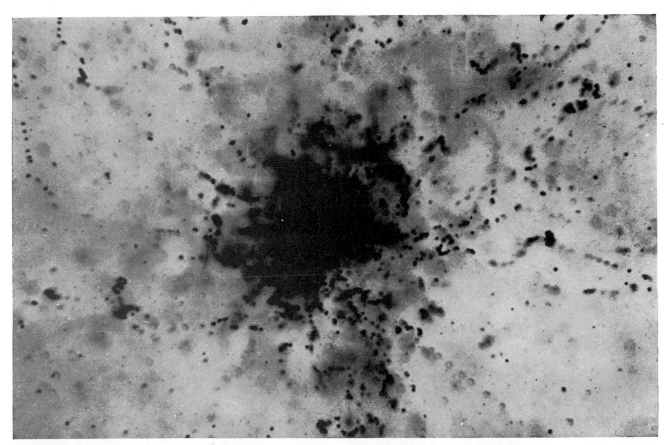

MAN-MADE RADIOACTIVITY in the atmosphere produced this second nuclear-emulsion photograph. The irregular tracery surrounding the dust particle is produced by beta particles. The character of these emissions identifies the substance as a fission product formed in a nuclear explosion, but poor definition makes precise identification impossible. The enlargement here is 1,200 diameters.

insoluble compounds that precipitate onto the ocean floor. Uranium precipitates far more slowly, so that its concentration in sea water reaches the comparatively high value of three parts per billion. Potassium also tends to be carried to the bottom; though virtually all of its compounds are soluble, its ions are adsorbed on particles of clay and other ion-exchange materials. This segregation of thorium and potassium lowers the already low concentration of radioisotopes in sea water; thus fishes suffer less external radiation than men.

In sea water the chemical transformations of decaying uranium lead to some unexpected effects. The second long-lived daughter of uranium 238 is thorium 230 (ionium). Its salts, like those of thorium 232, quickly precipitate into the constantly forming marine sediments. The precipitated ionium, however, eventually decays into radium, part of which dissolves out of the sediments and back into the bottom waters, whence it drifts upward. Its concentration in the ocean, like that of radon in the atmosphere, falls off from the bottom of the ocean to the top and provides an index of the slow mixing of the ocean waters.

Neither uranium 238 nor thorium 232 enters appreciably into the metabolic processes of living organisms, and neither do most of their descendants. The radiation that man receives from these elements is thus largely external. However, radium 226 and lead 210, of the uranium series, and radium 228, of the thorium series, can be taken up by plants and so travel eventually to man. In the human body they concentrate in bone and thus by themselves contribute about half as much radiation to the skeleton as do all their relatives put together. Potassium also enters into biological processes, and distributes itself through the soft tissues. It provides the bulk of the natural radiation that originates within the body.

Just after World War II Willard F. Libby at the University of Chicago opened up a whole new field of natural radioactivity. He began the study of radioisotopes produced by the most energetic radiations known: cosmic rays. These particles crash into the upper atmosphere and there shatter nuclei of nitrogen, oxygen and other atmospheric gases. On the average these collisions

TRANSPORT OF RADIOISOTOPES from soil and rocks to atmosphere, living organisms and ocean is shown in this schematic diagram. Solid-line arrows indicate radioactive decay; broken-line arrows, physical transport of isotopes. Uranium 238 (U^{238}), thorium 232 (Th^{232}) and potassium 40 (K^{40}) are the three chief natural radioisotopes. In soils U^{238} and Th^{232} both release radon (Rn), a decay product, into the atmosphere; K^{40} releases traces of stable argon (A^{40}). Radium (Ra) and lead 210 (Pb^{210}), radio-

transmute about four atoms per gram of air per minute in the region between seven and 15 miles above the ground. Some of the new atoms are stable and so are indistinguishable from "old" atoms of the same nuclear species. Some, however, are radioactive and can be detected by sensitive methods. The most interesting of these is carbon 14, formed by the interaction of a secondary neutron and an atom of nitrogen [see top illustration on next page]. Because neutrons are common products of nuclear reactions, and nitrogen is the principal constituent of the atmosphere, carbon 14 is relatively abundant—perhaps we should say only slightly rare. The carbon 14 produced by cosmic rays accounts for about one atom of atmospheric carbon in a trillion.

The distribution of carbon 14 in the atmosphere and at the surface depends on its chemical properties, which are of course identical with those of ordinary carbon (carbon 12). Most carbon 14 exists in the form of carbon dioxide; thus it is carried in the turnover of carbon dioxide from the atmosphere into the oceans and through the biosphere. Delicate measurements of the differences in radioactive carbon dioxide in different places show that on the average a carbon dioxide molecule spends five years in the atmosphere, five years in the surface layers of the ocean and 1,200 years in the deep waters. Such calculations are made possible by the long half-life of carbon 14 (5,600 years) and the precision with which its concentration can be measured. If the time-scale of the turnover of carbon dioxide were extremely short with respect to the half-life of carbon 14, all parts of the atmosphere-ocean exchange system would maintain approximately the same ratio of carbon 14 to stable carbon 12. However, if carbon 14 atoms are "kept waiting" (as they are at the interface between the atmosphere and the oceans), some of them will decay in the air rather than in the water, giving atmospheric carbon dioxide a slightly higher radioactivity. Measurable differences of this sort actually exist; they enable us to plot the planetary circulation of carbon dioxide.

Since living organisms consist almost entirely of carbon compounds and water, carbon 14 is concentrated as it passes through the biosphere. Its internal emissions in human tissues amount to about 1 per cent of the background radiation.

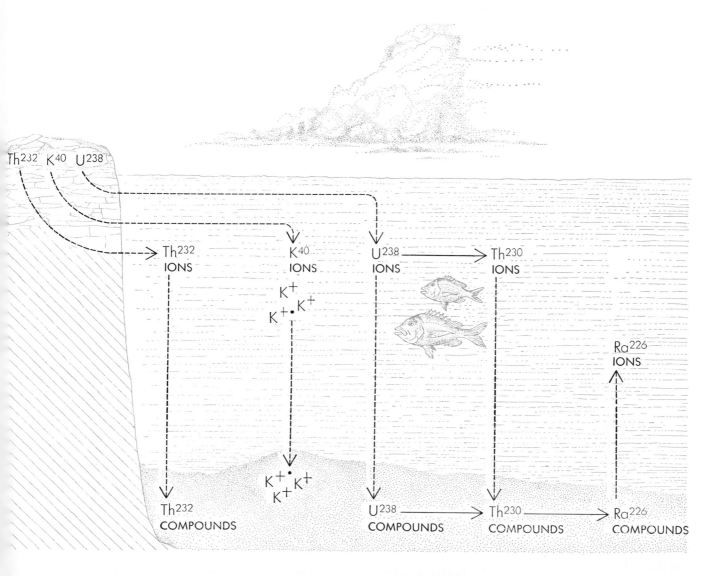

active decay-products of uranium and thorium, can be absorbed by plants and then by animals. Thorium washed into the ocean rapidly precipitates as insoluble compounds. Potassium remains in solution, but some of its ions become attached to particles of clay (black dots) which fall to the bottom. Uranium precipitates more slowly than thorium. Its decay product, thorium 230, precipitates rapidly, but decays into radium, some of which dissolves back into the ocean. Diagram does not show all steps of the decay reactions.

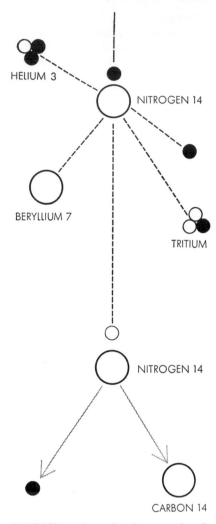

CARBON 14 is formed in the atmosphere by a two-stage process. A high-energy proton (*black sphere*) from space shatters a nitrogen atom; one of the fragments, a neutron (*white sphere*), reacts with another nitrogen atom, yielding carbon 14 and a proton

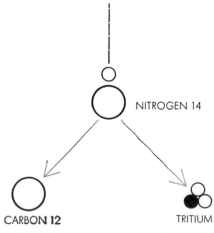

TRITIUM (hydrogen 3) is produced by a similar process; in this case the reaction between the neutron and the nitrogen atom yields tritium and carbon 12. Tritium can also be produced by the shattering of a nitrogen atom, as shown in the drawing at top.

The relative abundance of carbon 14 in organic substances gives the archaeologist a technique for dating wood, bone and other old organic materials.

Man's consumption of fossil fuels during the past 50 years has been decreasing the radioactivity of atmospheric and biological carbon. Because the carbon 14 in coal and oil has long since decayed, the carbon dioxide from the combustion of these substances, amounting to about 15 per cent of the mass of atmospheric carbon dioxide, is nonradioactive and has diluted the carbon 14 content of the atmosphere. As the inert carbon dioxide has mixed into the oceans our consumption of fossil fuels has provided a sort of tracer experiment in reverse.

Another cosmic-ray product of great scientific interest is hydrogen 3 (tritium). Tritium is formed by a number of processes, most often by the transmutation of atmospheric nitrogen atoms. Only about 20 pounds of natural tritium exist on earth, the bulk of it in the oceans. Combining with oxygen to make water, it reaches the earth as rain or snow. With a half-life of only 12.5 years it does not travel nearly so far as carbon 14 before disintegrating. Water samples from different sources show marked differences in tritium content. Because of the vast quantity of normal water in the ocean, water evaporating from the ocean surface has a low tritium content; as the water moves upward in the atmosphere its activity is increased by tritium from above. Water evaporated from land, on the other hand, consists mainly of recent precipitation relatively rich in tritium. As an air mass moves across a continent it is enriched with tritium from below as well as above. The cycle is closed when the rain falls on the ocean, and its load of tritium mixes into the surface water. A significant amount of tritium is also carried into the deep ocean before it decays.

The tritium picture has been badly blurred in recent years by the relatively enormous quantities of the isotope released in the testing of thermonuclear bombs. Some puzzling recent measurements have suggested that even apart from bomb tests the earth contains much more tritium than can be accounted for by cosmic-ray production. Studies now under way may resolve this problem before long.

One does not use a grandfather clock to time a footrace, nor a stop watch to reckon the passing of the seasons. An isotope "clock" must likewise be chosen for its appropriateness to a particular scale of time. The best uranium measure-

ments, for example, are accurate to no more than 20 million years—a small error in measuring the age of the earth but hopelessly too large to resolve the comparatively recent onset of the Ice Age. As our list of cosmic-ray-produced isotopes expands, we are finding that nature has been remarkably accommodating in providing clocks suitable for a variety of measurements. Thus beryllium 10, with a half-life of 2.5 million years, promises to help fill the gap between uranium and carbon-14 dates. Silicon 32, with a half-life of 700 years, may enable us to time processes of oceanic mixing for which the carbon-14 clock is too slow and the tritium clock too fast. Phosphorus 32, beryllium 7 and sulfur 35, with half-lives of a few weeks, serve to clock the much faster processes of atmospheric circulation. Certain man-made isotopes, principally those generated in nuclear explosions, are beginning to be used for the same purpose.

Starting with the first thermonuclear explosions in 1952, the testing of nuclear weapons has been radically altering the relative abundance of radioactive isotopes in the environment. The energy yield of all the tests conducted by the three nuclear powers prior to the tacit agreement to suspend tests in November, 1958, had added up to 174 megatons (equivalent to the yield from 174 million tons of TNT), of which 92 megatons came from the fission of heavy nuclei and 82 megatons from the fusion of light nuclei. From these tests, as their principal radioactive by-product, came more than 4.5 tons of fission products: the direct offspring of the fission of uranium and plutonium, which are produced at the rate of roughly 100 pounds per megaton of fission yield. Thermonuclear explosions in addition yield tritium as a direct product of the fusion reaction; the total output to date is estimated at 100 pounds, about five times the amount of natural tritium. Other isotopes arise as secondary products of the interaction of neutrons with the materials of the bomb and the gases of the atmosphere, and with the substances in the soil and sea if the burst is at the surface. Among these isotopes the most abundant is carbon 14.

Produced at a considerably higher rate over the past seven years than the natural product of cosmic-ray neutrons, the man-made carbon 14 now in circulation around the earth amounts to about one ton and equals about 1 per cent of the total natural abundance. Most of it is still in the atmosphere compounded in carbon dioxide. In living organisms it has

raised the carbon-14 content as much as 10 per cent above normal, a situation that may confuse future archaeologists.

The fission products liberated by the nuclear explosion would be cause for greater concern than they have aroused were it not for the fact that most of them are so short-lived. Roughly speaking, the radioactivity of the fission-product mixture decreases by 90 per cent for each seven-fold increase in time following the explosion that produced it. Thus the level of activity seven hours after the explosions is about a tenth the level after one hour; two weeks (about 7^3 hours) later the activity has dropped to about a thousandth the one-hour figure.

Of the more than 90 different radioisotopes identified among the fission products, two have figured at the center of scientific and public concern: strontium 90 and cesium 137. Both are produced in substantial quantities by nuclear fission, both have relatively long half-lives (about 25 and 30 years respectively) and both become engaged in the metabolism of the human body. Neither element is a normal constituent of biological processes. But the chemistry of strontium resembles that of calcium. Like calcium it concentrates in the bone, where its radioactivity may give rise to leukemia and bone tumors. Strontium 89, a short-lived fission product, follows the same metabolic pathway. Cesium, on the other hand, somewhat resembles potassium and its radioisotope concentrates in the soft tissues of the body, with particular hazard to the genes.

Tests so far have produced about 140 pounds of strontium 90, with a total radioactivity of nearly 9.2 million curies. By the end of 1958 decay had reduced these figures to about 130 pounds and 8.5 million curies; cesium 137 radioactivity at the same time amounted to 15 million curies.

Two other isotopes, iodine 131 and barium 140, have recently attracted attention. Barium, like strontium, concentrates in the skeleton; iodine, in the thyroid gland. Because they have short half-lives, radioiodine and radiobarium have been at near-zero levels since early this year. Other bomb products, such as radioisotopes of zinc and cobalt, may play a minor role in marine ecology, because some oceanic organisms concentrate relatively enormous quantities of these elements in their tissues. However, only minute quantities of these isotopes are likely to reach human beings.

The distribution of strontium 90 and other fission products, and hence their effects on man, depend largely on the type and location of the explosion that produced them. Surface explosions over land vaporize and irradiate great quantities of soil and so yield a heavy local fallout; 80 per cent of it comes down in a matter of hours, spreading downwind from the site of the explosion in an irregular elliptical pattern. Over water the close-in fallout drops to as low as 20 per cent of the total. Although contamination of areas adjacent to testing sites has occurred, local fallout has played only a small role in human affairs. In war, however, local fallout from a megaton weapon would deliver lethal external gamma radiation to unsheltered people in areas of thousands of square miles, and the strontium 90 would render soils unfit for agriculture over even larger areas for generations afterward.

Detonations of 100- to 200-kiloton yield at or near the surface propel their fission products no higher than the troposphere, the turbulent lower layer of the atmosphere. Convection currents in the troposphere rapidly mix air from various altitudes, so that tropospheric fallout comes down in a few months at most. Most of it is scavenged by rain or snow, with the larger particles simply drifting to earth. Since tropospheric winds rarely cross the Equator, most tropospheric fallout remains in the hemisphere where it originated. It accounts for only a small percentage of the fallout from weapons tests.

High-yield nuclear tests thrust their radioactive clouds into the stratosphere, where vertical mixing of air is slow and precipitation nil. The tiny radioactive particles (less than 1/10,000 inch in diameter) remain aloft for periods ranging from months to years, and are spread over large parts of the earth's surface by high-altitude winds. Stratospheric fallout, which accounts for about two thirds of total fallout, is the dominant source of artificial radioactivity that weapons tests have introduced into the environment.

The first observations of stratospheric fallout indicated that it remained aloft for five to 10 years—long enough to become evenly distributed between the Northern and Southern hemispheres. This comported with the idea that the stratosphere was a continuous, stable envelope enclosing the turbulent air masses of the lower atmosphere. Recently, however, the record revealed that the fallout has tended to concentrate in the North Temperate Zone and to come down at a maximum rate in the spring. Some investigators proposed that seasonal and latitudinal variations in the transport of air from stratosphere to troposphere might be responsible. However, this hypothesis did not explain the absence of comparable concentrations in the Southern Hemisphere.

By measuring the ratios of long-lived to short-lived fallout isotopes in rainwater, investigators have been able to establish the time and place of the explosions that generated the fallout. Since the initial proportion of one fission product to another is about the same in each nuclear explosion, such measurements make it possible to extrapolate backward from the observed ratios and thus to estimate the dates of the original explosions. Recent interpretations of these measurements have revealed the existence of stratospheric fallout much "younger" than five years. During the summer of 1958, a separate study of rainfall from the tropical air masses that move up over the U. S. from the Gulf of Mexico showed that the extrapolated time of origin of the fallout delivered by this rainfall matched the dates of certain U. S. medium-yield tests conducted several months previously. Fallout in rains from "polar" air masses during the same period had apparently originated in high-yield Soviet tests conducted a

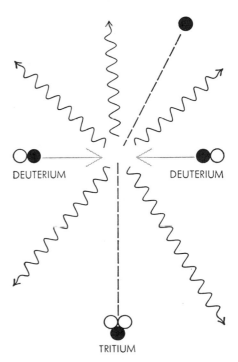

MAN-MADE TRITIUM is a by-product of some thermonuclear (fusion) reactions, such as that between two deuterium (hydrogen 2) atoms; proton (*black sphere*) is another by-product. The reaction also yields energy in the form of gamma rays (*wavy arrows*).

little earlier. The highest concentrations of strontium 90 were invariably associated with the Soviet tests.

The unexpected rapidity of the stratospheric fallout from Soviet tests focused attention on a peculiarity of the tropopause, the boundary between stratosphere and troposphere. Over the Equator, the tropopause lies about 10 miles up. Over the middle latitudes it dips sharply to about six miles. Here the jet streams—high-altitude winds that blow from west to east parallel to this dip—create a gap in the tropopause. Several meteorologists had already called attention to the possibility that the stratosphere and the troposphere exchange air through the gap.

A little calculation showed that Soviet fallout injected into the lower stratosphere could have been carried by the movement of stratospheric air through the tropopause gap into the troposphere, where vertical mixing would bring it fairly rapidly to earth [*see illustration on page 281*]. Fallout from U. S. and British high-yield tests, all of which have taken place near the Equator, has been evenly distributed over the world. Soviet tests, on the other hand, have been carried out north of a latitude of 50 degrees. Substantially all the fallout they have

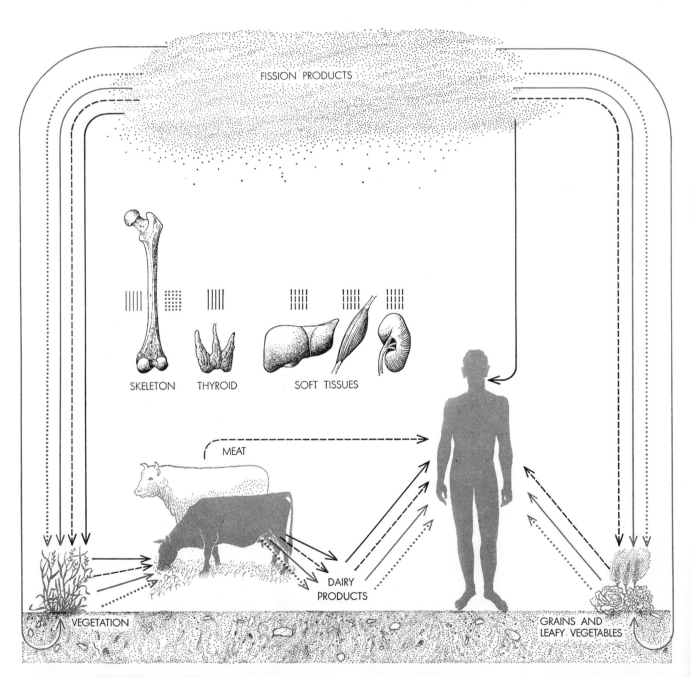

FISSION PRODUCTS

SKELETON   THYROID   SOFT TISSUES

MEAT

DAIRY PRODUCTS

VEGETATION   GRAINS AND LEAFY VEGETABLES

PATHS OF ENTRY by which fallout reaches human beings are shown schematically for the five fission products of greatest biological importance. Strontium 89 and 90 (*solid colored arrows*) reach man through the soil and, more directly, by contaminating surfaces of plants. Cesium 137 (*broken black arrows*) forms insoluble compounds in soil and thus reaches us only by the more direct paths. Iodine 131 (*solid black arrows*) and barium 140 (*dotted arrows*), because of their short half-lives, reach man only as surface contaminants and (in the case of iodine) by inhalation. Once absorbed, the isotopes concentrate in different parts of the body.

produced has remained in the Northern Hemisphere. Most of it has reached or will reach the ground in the densely populated North Temperate Zone. The spring peaks in fallout evidently stem from the timing of the Soviet tests, which have been conducted in the autumn and winter.

The tests in the U.S.S.R. last October gave strong confirmation to this theory: they produced a record peak in the fallout of strontium 90 over the Northern Hemisphere this past spring. Short-lived fission products appeared in this fallout in high concentrations, while throughout the Southern Hemisphere they remained near zero.

These findings with respect to manmade fallout suggest that a similar pattern may be found in the distribution of "fallout" from cosmic-ray action. Because the magnetic field of the earth funnels cosmic rays toward the poles, the major production of natural radioisotopes occurs in those latitudes and would presumably come to the surface via the gap in the tropopause.

The revised picture of atmospheric circulation makes it possible to forecast with some assurance the present and future distribution of strontium 90. From the nine million curies of this isotope produced to date, local fallout on U. S. and British test sites or in the Pacific Ocean subtracted three million curies at the outset. The Soviet tests produced little local fallout, since few or none of them were surface shots. Thus essentially all the remaining six million curies were injected into the stratosphere.

On the basis of a recent world-wide survey of strontium in soils, made by L. T. Alexander of the U. S. Department of Agriculture, it appears that 3.2 million of these six million curies had already come down to earth by October, 1958: 2.4 million in the Northern Hemisphere and .8 million in the Southern. The strontium reached a peak concentration, averaging about 33 millicuries per square mile, between 40 and 50 degrees north latitude. As for the 2.8-million-curie balance still in the stratosphere, the atmospheric circulation data indicate that most of it will have fallen out by 1965. Coming down in the same geographic pattern, it will approximately double the radioactivity of the soil produced by strontium 90 as of October, 1958. The fallout of cesium 137 should follow a similar pattern, and will contribute a somewhat higher radioactivity to the soil.

So long as these man-made radioisotopes remain on the ground, their

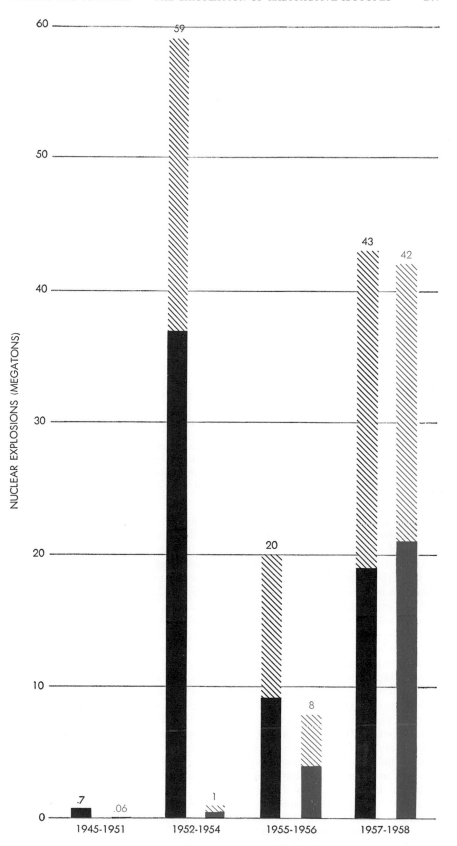

INCREASE IN NUCLEAR EXPLOSIONS since 1945 is depicted in this chart. Black bars indicate U. S.-British explosions; colored bars, Soviet tests. Fission explosions, indicated by solid parts of the bars, are the chief source of man-made environmental radioactivity, yielding about 100 pounds of fission products per megaton. Fusion reactions, shown by hatching, produce almost no fallout. The proportion of fission to fusion in Soviet tests is estimated.

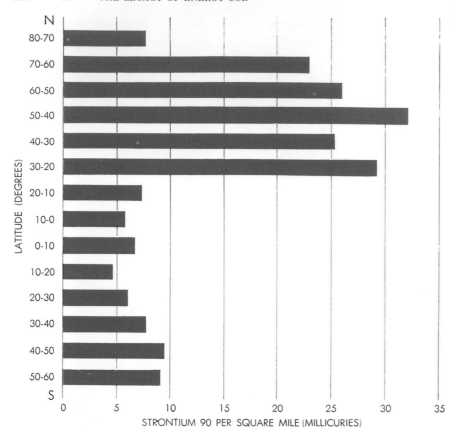

CONCENTRATION OF FALLOUT in the Northern Hemisphere is shown on this chart, based on a world-wide study of strontium in soils by L. T. Alexander of the U. S. Department of Agriculture. The concentration results from the peculiarities of atmospheric circulation shown on the opposite page. Fallout still aloft will roughly double these figures by 1965.

activity constitutes only a small addition to the external background-radiation to which living things are exposed. But once they are entrained in the chemistry of the biosphere their effective activity is heightened by concentration, and they irradiate living cells and tissues from within. Not many fission products find a place in biochemistry. The few that do follow various pathways in the food chain and some of them ultimately arrive in the tissues of human beings.

Strontium 90, for example, can be picked up from the soil in the metabolism of a plant, ingested by a dairy cow and delivered to the table in milk or butter. It may, however, skip one or more of these links in the food chain and arrive as a mere surface contaminant adhering to leafy vegetables and grain. Cesium 137, on the other hand, tends to form insoluble compounds in the soil and thus does not enter into plant metabolism. It is, however, ingested by beef or dairy cattle as a contaminant of grasses and passed on in meat or milk.

At each stage in the food chain the nutritional processes of the plant or animal tend to discriminate against strontium 90 in favor of its chemical analogue, calcium. Thus the ratio of radiostrontium to calcium will decrease as the two move together up the chain. At the first stage it seems that the strontium-calcium ratio in the plant varies not only with the ratio in the soil, but also with the character of the soil and ground cover, the method of cultivation and the type of plant. For example, in a soil matted with roots the radiostrontium coming down from above will not be so quickly diluted with calcium and so will be absorbed by the plant in a higher ratio to calcium. In animals, according to C. L. Comar of Cornell University, the chemical balance of metabolism regulates the strontium-calcium ratio more rigorously; the animal preferentially incorporates calcium into its tissues and rejects strontium. The ratio in a cow's milk is about 11 per cent of the ratio in the animal's fodder. In the human body the ratio diminishes still further; Comar estimates that the overall decline in the ratio from plant to man will be 90 per cent in the case of the average U. S. citizen, who obtains most of his calcium from dairy products. Where people live on a cereal diet and get most of their calcium from plants, without the intervening dis-

crimination by animal metabolism, the drop is smaller. In any case the introduction of the radioisotopes into the tissues raises the effective biological activity far above the contribution that the same quantity of radioisotopes makes to external background radiation. Clearly the subject of biological transport of fallout requires much more research.

Some recent tracer experiments promise to refine our knowledge of the fallout of natural and man-made radioactivity and of its ultimate distribution over the earth. During the U. S. nuclear tests of last summer, rhodium and tungsten were incorporated into some of the bombs. Neutrons from the explosions partially transformed them into the radioisotopes tungsten 185 and rhodium 102. The travel of these isotopes in the atmosphere can be followed easily. Tungsten 185, with a half-life of 74 days, was injected by several medium-yield explosions into the lower stratosphere. Several laboratories are sampling stratospheric and surface air and analyzing precipitation at a number of points in order to follow its history; the findings of this study should be published by the end of this year. Rhodium 102, with a longer half-life of 210 days, was chosen to study the longer-term processes at higher altitudes, and so was injected chiefly into the upper stratosphere. By comparing its history with that of the tungsten we should obtain some useful measurements of the effect of altitude on residence time and distribution of fallout.

The waste products of nuclear reactors constitute an additional source of man-made radioisotopes. These wastes, mainly fission products, pose many of the same biological problems as bomb debris. Indeed in the long run—barring a nuclear war—radioactive wastes will far outweigh the products of nuclear explosions. Fortunately man can decide within wide limits when and where he will release these substances into the environment. Our expanding knowledge of the atmosphere and ocean should enable us to do so in such a way as to minimize their effect on living organisms.

As this article is written, international discussions of a ban on the testing of nuclear weapons are continuing at Geneva. Such a ban would bring an end to this peculiar form of atmospheric contamination. As a consequence it would certainly improve our opportunity to investigate the fallout of both natural and artificial radioactivity and thus to learn more about the circulation of the oceans and the atmosphere.

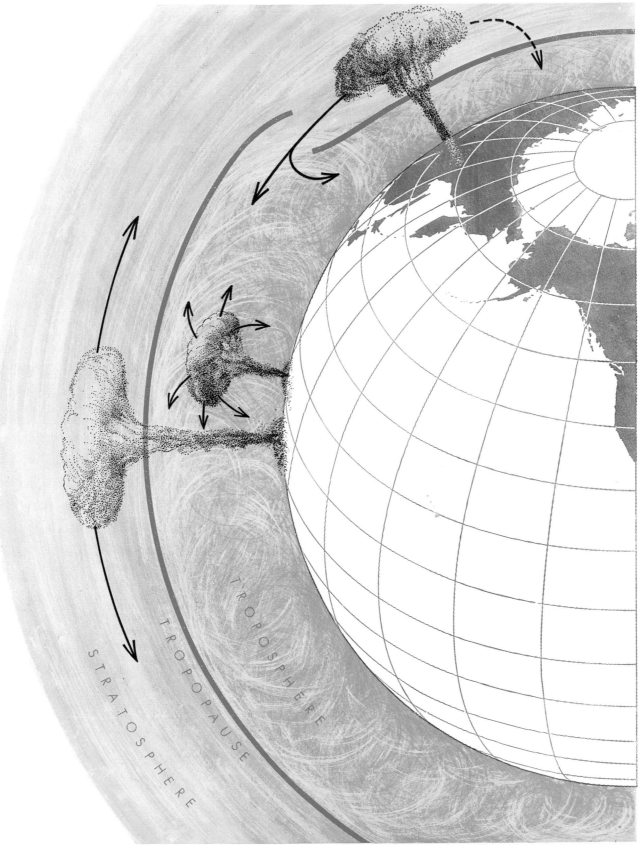

DISTRIBUTION OF FALLOUT depends on the size and location of the original explosion. High-yield test near the Equator (*left*) thrusts its fission products into the stratosphere, where horizontal circulation slowly spreads it over both hemispheres. Low-yield test (*left center*) does not reach the stratosphere; vertical mixing in the troposphere brings down its debris into one hemisphere in a matter of weeks. Debris from high-yield Soviet test at high latitudes (*top*) will be carried by horizontal circulation through tropopause gap and to some extent through the tropopause itself at higher latitudes (*broken arrow*). Most of it will come down over the North Temperate Zone. The scale of the atmosphere and the explosions is greatly exaggerated for purposes of clarity.

V

IMPLICATIONS OF MATERIAL WEALTH

V

IMPLICATIONS OF MATERIAL WEALTH

INTRODUCTION

Today's technological societies might be regarded as the most recent stage in man's continuing attempts to synthesize an environment in which he can live insulated from the vagaries of nature. One need only ponder the present size of the world's population to appreciate the survival value provided by that insulation. Previous sections have dealt with the energy costs, both direct and indirect, that go into the development and maintenance of the trappings that surround us in our so-called civilized state. The contribution from materials is equally important. Indeed, the two factors cannot be separated from one another; exploitation of energy resources and exploitation of materials go hand in hand, and neither class of technology could have reached its present heights without the other.

It is in the handling of materials that the growth of chemical knowledge has had some of its most profound effects. Our lives today are carried on in a milieu made up of an almost uncountable number of substances fulfilling at least as many purposes, and most of those materials are synthesized by people, starting with atomic or molecular precursors. Even a casual attempt to list the chemical products an average city-dweller might encounter in a day of normal business will quickly numb the imagination. Startled by an alarm clock ringing in its plastic case, then throwing off a blanket made of synthetic fiber, and ignoring the dyes and pigments in fabrics, carpets, towels, wallpaper, paint, and toiletpaper, we are, in our waking minutes, likely to be further assaulted with the mixtures of chemicals in toilet soap, shaving lotions, shampoos, deodorants, hair sprays, cosmetics and what not — and we haven't arrived in the kitchen for breakfast yet!

Even the breakfast table offers no relief (an egg cooked in a teflon-lined pan with partially hydrogenated, artificially flavored vegetable oil, perhaps?). "Food Additives," by G. O. Kermode, begins to give some clue to how absolutely inundated with chemicals we can be. In his excellent summary of the sorts of compounds that are added to our foodstuffs and the purposes for which they are used the author also brings up the difficult question: under what circumstances is the incorporation of an additive justified? Certainly no one could question the benefits of some preservatives in a world where the consumers of food are usually far from the areas in which it is produced. Some part of today's highly processed food products might well be necessary to meet the sheer volume of remote demand. Extensive processing for the purpose of convenience alone, however, seems a highly questionable practice — particularly at the price of chronic exposure to substances whose potentially adverse effects, or lack of them, are as overwhelmingly difficult to assess as the article implies.

All questions of necessity aside, the fact remains that our society consumes chemicals, either by themselves or formulated into various products, as well as materials manufactured in processes that use and then discard chemicals. The widespread distribution of the fantastic variety of compounds and mixtures that results cannot fail to have implications for the health and safety of humans, as individuals and occupants of an ecosystem. As pointed out in "Poisons," by Elijah Adams, any substance administered in sufficient quantities and the proper manner can be lethal to an organism, its toxicity being determined by the amount required to produce fatal effects. It is not surprising, then, that some of our chemical output does represent a potential threat, whether intended or not. Sometimes the threat is recognized, and sometimes it is not. Adam's article and the following one, "Radiation-Imitating Chemicals," by Peter Alexander, in their analyses of the modes of action of several classes of compounds that are known to act as poisons, outline some molecular characteristics that can be expected to render a substance harmful. The former article deals with chemicals whose action is immediate (not all of which are man-made, by the way); the latter covers those that are just as likely to exhibit delayed effects. As the title implies, the symptoms of exposure to these are very similar to those caused by radiation; it seems the radiomimetic compounds act by alkylating either proteins or nucleic acids, or both.

A third article documenting our personal vulnerability to chemical disruption is included in this group because, while it does not deal with poisonous materials and their effects as such, it does present an excellent up-to-date discussion of basic chemical principles and parallel mechanisms for creation of the same effects observed with the radiation-imitating substances. The radical reactions described in "Free Radicals in Biological Systems," by William A. Pryor, account nicely for the process of radiation damage — though they need not be the only route initiated by ionizing radiation — and seem to be implicated in aging. Chemical initiators, of course, and less energetic radiation such as ultraviolet light are also capable of starting similar reaction series.

Our exposure to hazardous chemical species is not limited to organic compounds, although their variety and ubiquity make them a major part of the problem. Almost every element is called upon to take some part in fabricating the things produced by technological societies, and consequently such activities provide a fertile source of a wide range of more or less exotic toxicants. "Beryllium and Berylliosis" comprises an interesting case history of the difficulty that was met in tracing the severe physiological effects of that rare metal to their real source. Author Jack Schubert presents a warning that should never be forgotten in this era of swift technological innovation. "Advances in technology now develop so rapidly that the rare material of yesterday becomes the widely used material of today . . . the harmlessness of a material cannot be taken for granted."

The numerous incidents involving mishandling of heavy metals that have been so highly publicized in recent years raise grave doubts about how well we are heeding that warning. It is one thing to be caught unawares by unexpected toxic properties, as with beryllium. It is quite another to engage in the indiscriminate release of metals whose toxicity has been recognized for centuries, such as mercury and lead. Perhaps we are too easily lulled into a false sense of security by partial knowledge. This seems to be true of our dealings with mercury, as surveyed by Leonard J. Goldwater in "Mercury in the Environment." Here it took a major outbreak of poisonings to lead us to the discovery that biological pathways exist for converting the relatively harmless element to its toxic methyl derivatives and concentrating them in a food chain that we participate in. With that knowledge now in hand, industries are taking steps to curtail introduction of mercury into the environment and investigating ways to alleviate the hazard posed by previous releases. The rate at which we can assimilate such lessons is apparently limited, though, as is well illustrated in "Lead Poisoning," by J. Julian Chisolm, Jr., as he documents repetitive "mistakes" with that element that have allowed overexposure to its toxic effects. We still know little if anything about the results of chronic exposure to marginal quantities of lead and its compounds, even though few people, especially those in cities, can escape such exposure today.

This section closes with "Chelation," a much-cited article by Harold

F. Walton. It is referred to in several of the other articles in this reader, usually in connection with the treatment of acute metals poisoning (including radioactive metals). The basic chemistry covered therein has implications that go beyond the therapeutic value of chelating agents, however. The power of chelating compounds to mobilize (or immobilize) metal ions gives them great potential for affecting the distribution of toxic metals in the environment. A case in point was nitrilotriacetic acid (NTA), an apparently nontoxic biodegradable substance considered as a replacement for polyphosphates in the formulation of detergents after phosphates had been implicated in the accelerated eutrophication of bodies of water. The plans to use NTA, a well-characterized complexing agent, were dropped after it was observed that it solubilized lead.

There can be no question that we live in a threatening chemical environment, full of pitfalls for those who ignore its hazardous nature. It must also be recognized that wholesale rejection of all things related to any dangerous material is as foolish as blind acceptance of a careless technology's mismanagement. Either course would be devastating to our continuing efforts to inhabit the planet in comfort and in health. What mankind needs to deal with the dangers, as well as the benefits, arising from manipulation of materials is a constant hardheaded assessment of the pros and cons of every action, and a good memory for the mistakes and surprises encountered in the past.

30

FOOD ADDITIVES

G. O. KERMODE
March 1972

*Perhaps as many as 2,500 substances are currently
being added to foods for flavoring, coloring,
preservation and other purposes. How are the necessity
and safety of these substances determined?*

Men have added nonfood substances to their food throughout recorded history, but in recent decades they have become concerned about such practices because of the large number of substances and motivations that have become involved. The questions at issue for any food additive are whether or not it is necessary and, if so, whether or not it is safe. For many years the United Nations (through the Food and Agriculture Organization and the World Health Organization) and many governments have kept watch on additives with these questions in view. The questions must be faced whenever a new additive is proposed; sometimes, as the recent case of cyclamate additives in the U.S. showed, they must be reconsidered when new evidence puts the safety of an old additive in doubt.

The distinction between food ingredients and food additives is somewhat imprecise. Sugar, being a natural product, is usually regarded as an ingredient, whereas saccharin, being an artificial sweetener, is likely to come under the heading of an additive. Perhaps the best method of classification is by function. Additives are employed for such purposes as enhancing flavor, improving color, extending shelf life and protecting the nutritional value of a food. They are, in short, valuable but not always essential items in the manufacture of food products.

Whatever one's views on additives may be, it is true that without additives many food products could not be offered for sale in their present form. This is exemplified in particular by the many convenience foods that have become popular in North America and in western Europe. Moreover, if food production is to increase enough to keep pace with population growth and the effort to improve nutrition generally in undernourished areas, chemicals that are not normally part of food will inevitably play an increasingly important role.

From the earliest times foods were preserved with incidental additives that resulted from cooking. Food was also preserved extensively in ancient times by heating, drying, salting, pickling, fermenting and smoking. Food colors were used in ancient Egypt. In China kerosene was burned to ripen bananas and peas; the reason the method succeeded, although the Chinese did not know it, was that the combustion produced the ripening agents ethylene and propylene. Flavoring and seasoning were arts in many ancient civilizations, with the result that spices and condiments were important items in commerce.

Additives have not invariably been employed with beneficial aims. The adulteration of food, in order to pass an inferior product off as a good one, is as old as trade. Expensive items such as tea, coffee, sugar, spices and essential oils were often adulterated. Common adulterants included coloring substances and burned or roasted vegetable material,

which was mixed with flour. Bread, beer, and wine were widely adulterated.

Eventually such practices led the authorities of the time and place to try to suppress them. The earliest food laws were often designed to control the more obvious forms of adulteration and fraud. In addition to these efforts the merchant guilds tried to protect the genuineness and the reputation of their products. The means at hand for testing foods were limited; checking the appearance, taste and smell of a food was about all one could do. The basis of knowledge making possible the national food laws that are common today was not established until about the middle of the 19th century. In the latter part of the century pure-food laws were enacted in country after country to control the composition of food and regulate the use of additives.

These developments coincided with a number of discoveries, mainly in organic chemistry, that led to the production of several of the important food additives now in use. For example, discoveries that resulted in the development of aniline and the coal-tar dyes led eventually to many of the synthetic colors now added to food. The active principles of odor and flavor were isolated from vegetables and other organic materials, leading first to alcoholic solutions of those materials as flavors and later to synthetic flavors, some of which did not appear to be present in natural edible aromatic substances used to flavor foods and some of which proved to have more flavoring

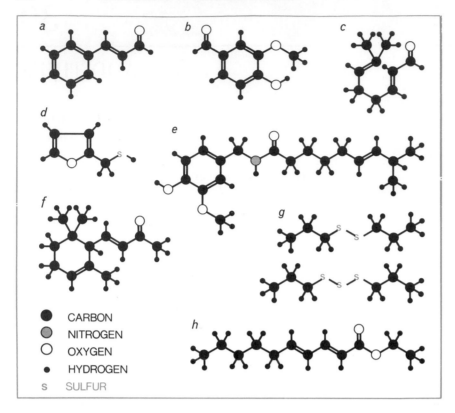

TYPICAL FLAVORING COMPOUNDS have chemical structures shown here. Cinnamaldehyde (*a*) supplies a cinnamon flavor; vanillin (*b*), a vanilla flavor; citral (*c*), lemon; furfuryl mercaptan (*d*), coffee; capsaicin (*e*), the pungent ingredient of red pepper; alphaionone (*f*), a principal component of strawberry and raspberry flavors; propyl disulfide (*g*), onion, and ethyl *trans*-2,*cis*-4-decadienoate (*h*), pear. Some 1,400 flavorings are in use.

power than the analogous natural flavors. By 1900 the flavorings in use were nearly all artificial and, except for vanilla, lemon, orange, peppermint and wintergreen, were being made with synthetic substances.

More than 40 functions now served by food additives can be listed. In this discussion, however, I shall group additives broadly in five categories: flavors, colors, preservatives, texture agents and a miscellaneous group.

Flavors constitute the largest class of food additives; estimates of the number of natural and synthetic flavors available range from 1,100 to 1,400. It is probably fair to say that flavors pose the largest regulatory task, not only because there are so many of them but also because of insufficient toxicological data, rapid changes in the field and many other factors. In general little is known about the toxicological aspects of flavors. Part of the problem is that many natural flavors have been used for centuries, and fully evaluating them all for safety would be an immense task. It is often argued that doing so would divert a large part of the effort that is needed to investigate the safety of more important and potentially more dangerous additives.

Over the past 30 years the use of flavorings has grown tremendously, paralleling the expansion in new types of food, new food-processing techniques and new methods of distribution. Governments have approached the question of controlling flavors from various directions. Some publish lists of permitted and prohibited flavors; some have a short list of prohibited flavors, many of which are natural, and others allow flavorings (both natural and synthetic) that are found only in the aromatic oils of edible plants.

Related to flavors are the additives known as flavor-enhancers. The commonest of them is monosodium glutamate (MSG), which is the monosodium salt of glutamic acid, one of the amino acids. A good deal of research is under way to find other flavor-enhancers, particularly in the group of substances known as the 5'-nucleotides. Similar work is being done on enhancers for fruit flavors. In recent years the use of maltol, which can intensify or modify the flavor of preserves, desserts, fruit, soft drinks and foods generally high in carbohydrates, has expanded greatly.

Manufacturers are also doing considerable research to find flavors that are cheaper or more effective than existing flavoring agents and flavor-enhancers. It is probably in this field that the greatest need for new additives will arise in the future, particularly for additives that can be put in simulated food products to imitate the complex flavor properties of traditional foods. At present the most widely sold simulated products are meat substitutes made from spun soybean proteins or proteins from other vegetables. With the addition of flavors, colors, vitamins, emulsifiers, acidifying agents and preservatives these proteins are sold as "vegetable steaks," "soya chicken breast" and "vegetable bacon" or are included in compounded products that normally have meat as a major ingredient. Other simulated foods are substitutes for dairy products. Flavored drinks, made so as to simulate the properties of genuine fruit juices, are also on the list. Because additive flavorings are expected to play such an expanding role in these products, they represent the field where ways of protecting the consumer's interest will need close attention, particularly with regard to designation and labeling of simulated foods.

Colors are put in food mainly to give it an appetizing appearance, on the tested assumption that the way food looks has an effect on its palatability. Foods are also colored to enhance the appreciation of flavor. Many people have become accustomed to the standardized color of a food product and would not accept the product if the color were substantially changed, even though nothing else had been done to the food. One need think only of blue or red butter to recognize the importance of accepted colors.

Much research has gone into food coloring. The colors most used in the food industry are synthetic dyestuffs. They are notably pure. Since they also have strong coloring power, little coloring is needed to achieve the desired result in a food product.

The manufacturer needs a color that not only produces the desired appearance but also will remain stable under certain conditions of manufacture, storage and cooking. Color put into candy, cakes and biscuits must be stable both to high temperature and to the action of carbon dioxide. Other colors must be able to withstand high processing temperatures and the action of acids.

Color regulation, like flavor regulation, varies from country to country. Many countries have fairly short lists of permitted food colors. The regulations specify purity and identity for the permitted colors and also restrict the number of foods to which color can be added. Since most of the lists are based on the toxicological evaluation of the dyes, one might expect a reasonable degree of uniformity among the lists. It is not so, however, and therefore one of the most troublesome problems facing a food manufacturer who wants to export his products is the need to vary the color according to the different regulations of the importing countries.

The World Health Organization has evaluated more than 140 kinds of coloring matter, declaring a number to be unsafe and publishing a fairly short list of colors deemed to be safe. In some countries the food industry manages quite well with a choice of no more than a dozen dyes. Other countries allow more dyes. The difference is illustrated by a

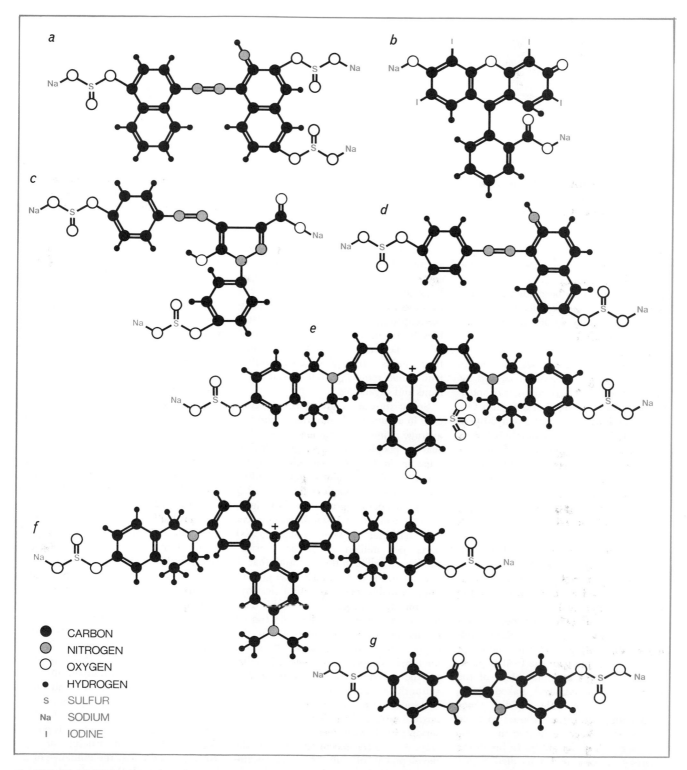

CERTAIN COLORS employed as food additives are portrayed according to their chemical structure. The colors have both numerical and descriptive names: (a) red 2, amaranth; (b) red 3, erythrosine; (c) yellow 5, tartrazine; (d) yellow 6, sunset yellow; (e) green 3, fast green; (f) violet 1, benzylviolet, and (g) blue 2, indigotine. The characteristic ring structure evident in the seven diagrams is more likely than an aliphatic, or open-chain, structure to produce color because of the way it absorbs and reflects light.

a   *b*   *c*

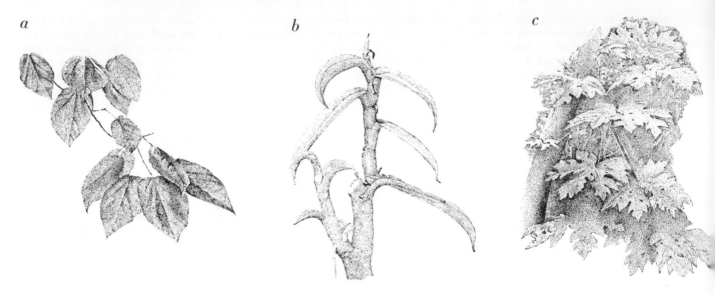

BOTANICAL SOURCES of four natural additives are depicted. The annatto (*a*) is a tropical tree, *Bixa orellana*, that produces a yellowish-red coloring agent made from the pulp around the seeds. Natural vanilla extract comes from the pods of several species of orchid, chiefly *Vanilla planifolia* (*b*). Sap from the papaya tree (*c*), *Carica papaya*, is the source of the enzyme papain, which is

problem that the British will face when their country becomes a member of the European Economic Community: kippers will no longer be golden and sausages will no longer be "nicely pink" unless the Community's list of permitted colors is extended.

Preservative additives are one means of deterring food spoilage caused by microorganisms. The seriousness of spoilage is shown by the World Health Organization's estimate that about 20 percent of the world's food supply is lost in this way. Indeed, shortages of food in many parts of the world could be alleviated with the wider use of preservatives.

Spoilage can be prevented or retarded not only with additives but also with physical and biological processes such as heating, refrigeration, drying, freezing, souring, fermenting and curing. Some of these processes, however, achieve only partial preservation. Additives therefore have a role in prolonging a food's keeping qualities.

A number of different types of preservative have to be employed, depending on the kind of food, the method of manufacture, the way the food is packaged and stored and the nature of the microorganisms involved. Baked goods, for example, go stale rapidly. Once made, they are often exposed to mold spores that become active in warm weather or high humidity. In bread the spores produce a condition called "rope." Sodium diacetate, acetic acid, lactic acid, monocalcium phosphate, sodium propionate and calcium propionate are all effective in preventing rope. Sorbic

acid and its salts have many uses, such as preventing mold in cheese, syrup and confections containing fruit or sugar. Benzoic acid and sodium benzoate serve as preservatives in margarine, fruit-juice concentrates, juices and pickled vegetables. Sulfur dioxide is widely used to inhibit mold and discoloration in wine, fruit pulps, fruit-juice concentrates, fruit drinks requiring dilution and dried fruits and vegetables.

Sulfur dioxide is giving rise to concern in a number of countries where the average wine consumption is so high that the people who drink a good deal of wine are in danger of exceeding the acceptable average daily intake of sulfur dioxide. The acceptable daily level is 1.5 milligrams per kilogram of body weight, which means about 100 milligrams a day or a half-liter of wine containing 200 parts per million of sulfur dioxide. Studies on sulfite in the rat found that .1 percent in the diet inhibited the growth rate, probably because sulfite destroys vitamin B_1. The significance of this finding in man, whose diet does not consist exclusively of sulfited food as in the experiments with rats, is questionable; nonetheless, more work is needed to dispel the uncertainty about the toxicity of sulfur dioxide and sulfites.

As a result of such uncertainties serious attempts to find alternatives to preserving foods with chemicals are under way. Among the recent advances is the development of antibiotics as antimicrobial additives. Antibiotics commonly have a more transitory effect than the traditional preservatives and are more selective. These advantages are sig-

nificant when antibiotics are directed against known food pathogens and when their action is required only during the manufacturing stage. Antibiotics can be said to have a major disadvantage, however, in that by changing the normal spoilage pattern of certain foods they may result in unfamiliar forms of spoilage that consumers cannot recognize.

A number of countries have permitted such antibiotics as tetracyclines, nystatin, nisin and pimaricin as direct or indirect additives to chilled or raw fish, meat, poultry, cheese and bananas. The applications are strictly limited. Many other countries, although they recognize the efficacy of antibiotic additives, have taken the view that it would be unwise to approve them widely for food, since the antibiotics are important in medicine and their liberal use in food might produce resistant strains of pathogens that could affect humans.

Another development that has attracted interest as an alternative way of protecting food is the experimental work wherein ionizing radiation is employed to destroy the microorganisms and insects that cause food spoilage. An advantage of irradiation is that it produces little or no rise in the temperature of the food during treatment. A disadvantage at present is the possibility that irradiation will have an effect on the food and leave residues. For example, it is possible for the extranuclear structure of atoms to be excited under the influence of ionizing radiation. If the atoms are constituents of a molecule, the molecule as a whole may be excited, which may lead to rupture of one or more chemical

d

employed as a meat tenderizer. One source of wintergreen flavor is the leaves of the evergreen plant *Gaultheria procumbens* (d).

bonds, giving rise to free radicals. The free radicals may be capable of starting chemical chain reactions.

Nonetheless, work with irradiation has advanced to a point where in a small number of countries the sale of certain irradiated foodstuffs is now allowed. Irradiation raises the possibility that perishable foodstuffs could be more widely distributed in a fresh or nearly fresh condition. It is likely to take several years, however, for the irradiation of food to become a widespread practice because of the effort that must still be devoted to developing procedures for testing irradiated foods. The International Atomic Energy Agency, in conjunction with the Food and Agriculture Organization and the World Health Organization, has indicated a number of possible treatments by ionizing radiation to achieve long-term preservation, without refrigeration or chemical preservatives, of perishable foods and also to prevent food poisoning by destroying microorganisms such as salmonella.

Many traditional preservatives—notably salt, vinegar and sugar—still play an important role in homes and factories. It can be argued that recent improvements in food processing, coupled with improved standards of hygiene, should reduce the need for chemical preservatives. On the other hand, developments in making ready-to-use foods and the widespread resort to prepackaging have tended to increase the need for preservatives.

Related to preservatives are antioxidants, which are added to fatty foods primarily to prevent rancidity. Typical products containing these additives are margarine, cooking oils, biscuits, potato chips, cereals, salted nuts, soup mixes and precooked meals containing fish, poultry or meat. Certain foods, such as virgin olive oil, contain their own natural antioxidants in the form of tocopherols and therefore do not need the addition of antioxidants. If such foods are heated in a manufacturing process, however, they tend to lose their natural antioxidants, which should be restored if the product is to have a reasonable shelf life.

The most widely added antioxidants are butylated hydroxyanisole, butylated hydroxytoluene, propyl, octyl and dodecyl gallates and natural or synthetic tocopherols singly or in combination. Certain acids (ascorbic, citric and phosphoric) combined with antioxidants increase the antioxidant effect.

Preventing rancidity is not the only problem with a number of foods. The growing practice of using transparent wrapping for food presents its own problems by exposing the product to light and increasing the likelihood of discoloration. Ascorbic and isoascorbic acid have proved effective in preventing discoloration in certain fruit juices, soft drinks, canned vegetables, frozen fruits and cooked cured meat such as ham.

Often more than one antioxidant is put into a food, producing a synergistic action that allows more effective control of the product. Many countries authorize several antioxidants as food additives, but a number of others will allow only the so-called natural antioxidants, such as ascorbic acid (vitamin C) and the tocopherols (vitamin E). Much research is in progress to find compounds that are more potent than the present antioxidants. The search is particularly keen for antioxidants that are less likely than the present ones to impart odor, flavor or color to foods. Another quest is for antioxidants with a required solubility in both water and oil. With the development of simulated foods the search for antioxidants that are more effective in extending shelf life will gain further impetus.

In the class of texture agents I have included emulsifiers, stabilizers and thickening agents. In terms of quantity consumed they probably constitute the largest class of additives, being employed extensively in preparing bread, pastry, ice cream, frozen desserts, whipped products, margarine, candy and certain soft drinks and milk products. Many of the newer convenience foods have only become practicable as a result of the development of new and improved emulsifiers and stabilizers.

Among other things, the texture agents permit oil to be dispersed in water, produce a smooth and even texture and supply the desired body and consistency of many food products.

The first emulsifiers were few in number and were either natural substances such as gums, alginates and soaps or synthetic substances of fairly simple composition. Their action was often variable. Progress in chemical synthesis has now made available a large number of new texture agents with characteristics suitable for almost any requirement. Among the most common emulsifiers and stabilizers, aside from the natural ones, are stearyl tartrate, complete glycerol esters, partial glycerol esters, partial polyglycerol esters, propylene glycol esters, monostearin sodium sulfoacetate, sorbitan esters of fatty acids and their polyoxyethylene derivatives, cellulose ethers and sodium carboxymethyl cellulose. Thickeners include natural products such as agar, alginates, celluloses, starches, vegetable gums, dextrins and pectin; modified celluloses such as methyl cellulose, and starches modified by bleaching, oxidizing and phosphating.

The miscellaneous group of additives is so numerous that I can only indicate a few of the functions they serve. Acids, alkalis, buffers and neutralizing agents are added to many processed foods where the degree of acidity or alkalinity is important; manufacturers of baked goods, soft drinks, chocolate and processed cheese employ these additives extensively. The baking industry also makes heavy use of bleaching and maturing agents, which render flour whiter and bring it to maturity sooner. Sequestrants are added to food to bind trace metals and thus prevent any oxidative activity the metals in an ionized state might have on the food; in shortening, for example, unsequestered metals could catalyze processes leading to rancidity. Humectants, which are hygroscopic, offset changes in the humidity of the environment to which food is exposed, so that a desired level of moisture can be maintained in a food product such as shredded coconut. Anticaking agents keep many salts and powders free-flowing. Glazing agents make certain food surfaces shiny and in some cases protect the product from spoiling. Firming and crisping agents prevent flaccidity in processed fruits and vegetables and also aid the coagulation of certain cheeses. Release agents help food to separate from surfaces it touches during manufacture or transport. Foaming agents coupled with propellants make whipped

ANTICAKING AGENTS
Aluminum calcium silicate
Calcium silicate
Magnesium silicate
Sodium aluminosilicate
Sodium calcium aluminosilicate
Tricalcium silicate

CHEMICAL PRESERVATIVES
Ascorbic acid
Ascorbyl palmitate
Benzoic acid
Butylated hydroxyanisole
Butylated hydroxytoluene
Calcium ascorbate
Calcium propionate
Calcium sorbate
Caprylic acid
Dilauryl thiodipropionate
Erythorbic acid
Gum guaiac
Methylparaben
Potassium bisulfite
Potassium metabisulfite
Potassium sorbate
Propionic acid
Propyl gallate
Propylparaben
Sodium ascorbate
Sodium benzoate
Sodium bisulfite
Sodium metabisulfite
Sodium propionate
Sodium sorbate
Sodium sulfite
Sorbic acid
Stannous chloride
Sulfur dioxide
Thiodipropionic acid
Tocopherols

EMULSIFYING AGENTS
Cholic acid
Desoxycholic acid
Diacetyl tartaric acid esters
  of mono- and diglycerides
Glycocholic acid
Mono- and diglycerides
Monosodium phosphate
  derivatives of above
Propylene glycol
Ox bile extract
Taurocholic acid

NUTRIENTS AND DIETARY SUPPLEMENTS
Alanine
Arginine
Ascorbic acid
Aspartic acid
Biotin
Calcium carbonate
Calcium citrate
Calcium glycerophosphate
Calcium oxide
Calcium pantothenate
Calcium phosphate
Calcium pyrophosphate
Calcium sulfate
Carotene
Choline bitartrate
Choline chloride
Copper gluconate
Cuprous iodide

Cysteine
Cystine
Ferric phosphate
Ferric pyrophosphate
Ferric sodium pyrophosphate
Ferrous gluconate
Ferrous lactate
Ferrous sulfate
Glycine
Histidine
Inositol
Iron, reduced
Isoleucine
Leucine
Linoleic acid
Lysine
Magnesium oxide
Magnesium phosphate
Magnesium sulfate
Manganese chloride
Manganese citrate
Manganese gluconate
Manganese glycerophosphate
Manganese hypophosphite
Manganese sulfate
Manganous oxide
Mannitol
Methionine
Methionine hydroxy analogue
Niacin
Niacinamide
D-pantothenyl alcohol
Phenylalanine
Potassium chloride
Potassium glycerophosphate
Potassium iodide
Proline
Pyridoxine hydrochloride
Riboflavin
Riboflavin-5-phosphate
Serine
Sodium pantothenate
Sodium phosphate
Sorbitol
Thiamine hydrochloride
Thiamine mononitrate
Threonine
Tocopherols
Tocopherol acetate
Tryptophane
Tyrosine
Valine
Vitamin A
Vitamin A acetate
Vitamin A palmitate
Vitamin B$_{12}$
Vitamin D$_2$
Vitamin D$_3$
Zinc sulfate
Zinc gluconate
Zinc chloride
Zinc oxide
Zinc stearate

SEQUESTRANTS
Calcium acetate
Calcium chloride
Calcium citrate
Calcium diacetate
Calcium gluconate
Calcium hexametaphosphate
Calcium phosphate, monobasic
Calcium phytate
Citric acid

Dipotassium phosphate
Disodium phosphate
Isopropyl citrate
Monoisopropyl citrate
Potassium citrate
Sodium acid phosphate
Sodium citrate
Sodium diacetate
Sodium gluconate
Sodium hexametaphosphate
Sodium metaphosphate
Sodium phosphate
Sodium potassium tartrate
Sodium pyrophosphate
Sodium pyrophosphate, tetra
Sodium tartrate
Sodium thiosulfate
Sodium tripolyphosphate
Stearyl citrate
Tartaric acid

STABILIZERS
Acacia (gum arabic)
Agar-agar
Ammonium alginate
Calcium alginate
Carob bean gum
Chondrus extract
Ghatti gum
Guar gum
Potassium alginate
Sodium alginate
Sterculia (or karaya) gum
Tragacanth

MISCELLANEOUS ADDITIVES
Acetic acid
Adipic acid
Aluminum ammonium sulfate
Aluminum potassium sulfate
Aluminum sodium sulfate
Aluminum sulfate
Ammonium bicarbonate
Ammonium carbonate
Ammonium hydroxide
Ammonium phosphate
Ammonium sulfate
Beeswax
Bentonite
Butane
Caffeine
Calcium carbonate
Calcium chloride
Calcium citrate
Calcium gluconate
Calcium hydroxide
Calcium lactate
Calcium oxide
Calcium phosphate
Caramel
Carbon dioxide
Carnauba wax
Citric acid
Dextrans
Ethyl formate
Glutamic acid
Glutamic acid hydrochloride
Glycerin
Glyceryl monostearate
Helium
Hydrochloric acid
Hydrogen peroxide
Lactic acid
Lecithin

Magnesium carbonate
Magnesium hydroxide
Magnesium oxide
Magnesium stearate
Malic acid
Methylcellulose
Monoammonium glutamate
Monopotassium glutamate
Nitrogen
Nitrous oxide
Papain
Phosphoric acid
Potassium acid tartrate
Potassium bicarbonate
Potassium carbonate
Potassium citrate
Potassium hydroxide
Potassium sulfate
Propane
Propylene glycol
Rennet
Silica aerogel
Sodium acetate
Sodium acid pyrophosphate
Sodium aluminum phosphate
Sodium bicarbonate
Sodium carbonate
Sodium citrate
Sodium carboxy-
  methylcellulose
Sodium caseinate
Sodium citrate
Sodium hydroxide
Sodium pectinate
Sodium phosphate
Sodium potassium tartrate
Sodium sesquicarbonate
Sodium tripolyphosphate
Succinic acid
Sulfuric acid
Tartaric acid
Triacetin
Triethyl citrate

SYNTHETIC FLAVORING SUBSTANCES
Acetaldehyde
Acetoin
Aconitic acid
Anethole
Benzaldehyde
N-butyric acid
d- or l-carvone
Cinnamaldehyde
Citral
Decanal
Diacetyl
Ethyl acetate
Ethyl butyrate
Ethyl vanillin
Eugenol
Geraniol
Geranyl acetate
Glycerol tributyrate
Limonene
Linalool
Linalyl acetate
1-malic acid
Methyl anthranilate
3-Methyl-3-phenyl
  glycidic acid ethyl ester
Piperonal
Vanillin

GROUP OF ADDITIVES included in the U.S. Food and Drug Administration's list of additives "generally recognized as safe" is given, except for large groups of natural flavors and oils. To be on this list an additive must have been in use before 1958 and have met certain specifications of safety. Additives brought into use since 1958 must be approved individually. Occasionally substances are removed from the list by the FDA in the light of new evidence; recent examples include the cyclamate sweeteners and saccharin.

toppings come out of their containers as a foam, whereas foam inhibitors have an opposite role, where a tendency to foam, as with pineapple juice, makes filling a container difficult. Clarifying agents remove small particles of minerals from liquids such as vinegar, which might otherwise turn cloudy. Solvents serve as carriers for flavors, colors and other additives, and solvent extraction is the method whereby oil is obtained from oilseeds, coffee is decaffeinated and a number of instant beverages are prepared.

Additives have become a public issue because of recurrent episodes bringing into question the safety of certain additives that have been used for some time. Cyclamates were tested extensively in the U.S. before they were put on the market as artificial sweeteners, but in 1969 it was reported that large doses had caused bladder cancer in rats. The U.S. Government ordered cyclamates off the market. Subsequently it was reported that rats fed with cyclamate and saccharin at a sixth of the dose that led to the original ban also developed bladder cancer. As a result saccharin is now being critically reviewed in the U.S. and other countries. Sodium nitrite, which fixes a red color in frankfurters, sausages and hams, is currently under review in many countries because of the possibility that it may form a cancer-producing agent during digestion and storage. Laboratory evidence has linked monosodium glutamate with the "Chinese restaurant syndrome" (more precisely Kwok's disease), a tightening of the muscles of the face and neck, occasionally accompanied by headache, nausea and giddiness, experienced by some people who have eaten in restaurants where monosodium glutamate has been used in large amounts. Many countries have therefore restricted the use of monosodium glutamate or required its presence in food to be prominently stated on the label.

Food additives, unlike the chemicals put in pesticide preparations, are not designed to be toxic, and most of them would have to be ingested in large single doses to produce acute toxic symptoms. Many additives by nature are of extremely low potential toxicity. It is therefore difficult to determine their possible hazards to man, even after exhaustive testing. It is probably true to say that there will always be an area of doubt concerning the possible effects of ingesting small amounts of additives over the course of a lifetime. One cannot be fully sure of the safety of an additive until it has been consumed by people of all ages in specified amounts over a long period

of time and has been shown conclusively, by careful toxicological examination, to have no harmful effects.

Since humans cannot be used for testing by exposing them to unknown chemicals for a substantial period of time, tests are made on rats and other animals such as mice and dogs. Test animals are fed quantities of the additive that far exceed the amount likely to be found in food. Tests are made both for short periods and over the animal's lifetime and are often continued into succeeding generations. Any change in growth, body function, tissue and reproduction is reported, as is the incidence of tumors.

The largest dose that appears to produce no effects in animals is taken, and a safety factor reducing that dose by about 100 is applied in most countries in order to arrive at an acceptable dose for humans. The "acceptable daily intake" thus calculated is the daily intake that for an entire lifetime appears to be without appreciable risk on the basis of all known facts at the time. It is expressed in terms of milligrams of the additive per kilogram of body weight. One must then calculate how much of the additive a person might be expected to ingest in a day from all dietary sources and compare this figure with the acceptable daily intake in order to decide whether the applications of the additive should be permitted and whether the specific tolerances or maximum limits required for it by good manufacturing practices in individual foods are safe to the health of the consumer.

Many national authorities publish information on the tests they require for proposed additives. International guidelines have been published by the Joint Expert Committee on Food Additives of the Food and Agriculture Organization and the World Health Organization. They require a comprehensive series of tests on laboratory animals, including short- and long-term studies covering acute toxicity, metabolism of the additive and carcinogenic effects, among others.

Assessing safety on the basis of toxicological tests calls for expert judgment of all the evidence available. The judgment may have to be modified in the light of further experiments and experience with the additive in food for humans. The tests required to obtain approval of an additive can cost upward of $100,000.

It is therefore the hope of the government and the food industry in many countries that the work of such international bodies as the Codex Alimentarius Commission of the Food and Agriculture

Organization and the World Health Organization will lead to a greater exchange of toxicological data and to the evaluation of the safety of more additives as soon as possible. The commission has published a list of six general principles on the use of food additives. The first is that the use of an additive is justified only when it has the purpose of maintaining a food's nutritional quality, enhancing its keeping quality or stability, making the food attractive, providing aid in processing, packing, transporting or storing food or providing essential components for foods for special diets, and that an additive is not justified if the proposed level of use constitutes a hazard to the consumer's health, if the additive causes a substantial reduction in the nutritive value of a food, if it disguises faulty quality or the use of processing and handling techniques that are not allowed, if it deceives the customer or if the desired effect can be obtained by other manufacturing processes that are economically and technologically satisfactory.

The second principle is that the amount of additive should not exceed the level reasonably required to achieve the desired effect under good manufacturing practice. The third principle calls for additives to conform with an approved standard of purity; the fourth holds that all additives, in use or proposed, should be subjected to adequate toxicological evaluation and that permitted additives should be kept under observation for possible deleterious effects; the fifth states that approval of an additive should be limited to specific foods for specific purposes and under specific conditions, and the sixth relates to the use of additives in foods consumed mainly by special groups in the community. In this case the intake of the food by the group should be taken into account before authorizing the use of the additive.

Food additives have become part of everyday life and undoubtedly will play an increasing role with advances in food technology. The prospect is not necessarily bad, because properly used additives can bring the consumer significant benefits. Moreover, provided that in each case sound justification for the additive is demonstrated, that government and manufacturers exercise the utmost care to ensure that the additive entails no appreciable risks to health and that clear labeling informs the consumer of the nature and composition of the product he is buying, consumers should be reasonably assured as to the safety of officially authorized food additives.

31

POISONS

ELIJAH ADAMS
November 1959

*What is the molecular mechanism by which a toxic
substance produces its effect? In seeking the answer
for various poisons, investigators have found
invaluable tools for the study of normal cell physiology*

"Poisons can be employed as agents of life's destruction or as means for relief of disease, but in addition to these universally recognized uses there is a third that particularly interests the physiologist. For him the poison becomes an instrument that dissociates and analyzes the most delicate phenomena of the living machine, and by studying attentively the mechanism of death in diverse types of poisoning he can learn indirectly much about the physiological processes of life."

After almost a century these words of the eminent French physiologist Claude Bernard still summarize concisely the scientific significance of poisons. Curare, the specific agent that he was discussing, remains one of the best examples of the tripartite nature of this loose category of substances. When curare first came to Bernard's attention, Europeans knew it only as an arrow poison used by Indians in South America to kill game and enemies. Bernard himself first applied it as a physiological scalpel in his pioneering studies of nerve and muscle function. More recently curare has become a "means for the relief of disease": as an adjunct to surgery it relaxes the muscles and thereby obviates dangerously deep levels of anesthesia.

A poison is difficult to define with legalistic rigor. Even distilled water is toxic when it is consumed by the gallon, and considerably smaller quantities of water can cause death when they are inhaled. The accepted pharmacological definition follows popular usage: A poison is any substance that in relatively small quantities can cause death or illness in living organisms by chemical action. The last clause rules out such mechanically lethal effects as those produced by a small quantity of lead entering the body at high velocity.

The list of poisons furnished readymade by nature has been extended mightily in recent years by human ingenuity and the chemical industry. No reasonable definition of poisons can exclude the thousands of substances which in small doses produce physiological changes but which, being used customarily for the treatment of disease, are identified as drugs. Another deceptive category is that of the pesticides, especially those designed to kill our close physiological relatives the rodents. It is not surprising that poisons represent a major public hazard, particularly to children. In 1955 about 8,000 Americans died of poisons (exclusive of ethyl alcohol). The fatalities were about equally divided between accidents and suicides; homicides accounted for only .5 per cent of the total. Non-fatal poisonings are estimated at a million or more per year, about 25 per cent of them in children under five.

The term "poison" (akin to "potion") originally included medicinal as well as lethal draughts. For example, digitoxin —the medication of choice in certain cardiac conditions—is among the most toxic substances known; its use as an industrial chemical would necessitate elaborate safety precautions. Courses in pharmacology properly emphasize the broad overlap between drugs and poisons, and physicians, nurses and pharmacists dare not forget it.

To the physiologist, on the other hand, poisons are no more interesting than any other physiologically active compound. Like all such substances, they challenge him to interpret their gross effects in molecular terms—to elucidate the chemical mechanisms by which they derange or destroy cells and thereby induce more obvious disturbances of the entire organism. So far the attempt to reduce toxicity to molecular mechanisms has succeeded in only a few cases. As Bernard anticipated, however, it has meanwhile taught us a good deal about the chemistry of living matter.

The corrosive poisons—strong acids and alkalis—have the most obvious effects. At first glance their mode of action seems simple enough. By massive destruction of cells they can produce death from shock, hemorrhage or incapacitation of some vital organ. Thus the corrosive gas phosgene, used in chemical warfare during World War I, reacts with water in the lungs to produce hydrochloric acid. This destroys lung tissue and by its irritant action fills the lungs with fluid. Death ultimately results from asphyxiation. With most corrosive poisons, however, the mode of action is not so clearly discernible. Concentrated sulfuric acid, for example, does so many things to the body that its specifically lethal activities are hard to isolate. Moreover, its catastrophic effects on tissue leave little for the physiologist to examine. Some corrosive poisons may produce death simply by shifting the delicate acid-alkaline balance of the body to the point where vital chemical reactions can no longer occur.

More subtle in their effects are the metabolic poisons, a group that includes most of the drugs in the pharmacopoeia as well as many other substances that have no place in medicine. These compounds do not destroy tissues; many of them produce no visible tissue-change whatever. Instead they accomplish their lethal work by disrupting one or another of the intricate chemical reactions upon which life depends.

The effects of two common metabolic poisons were explained many years ago

295

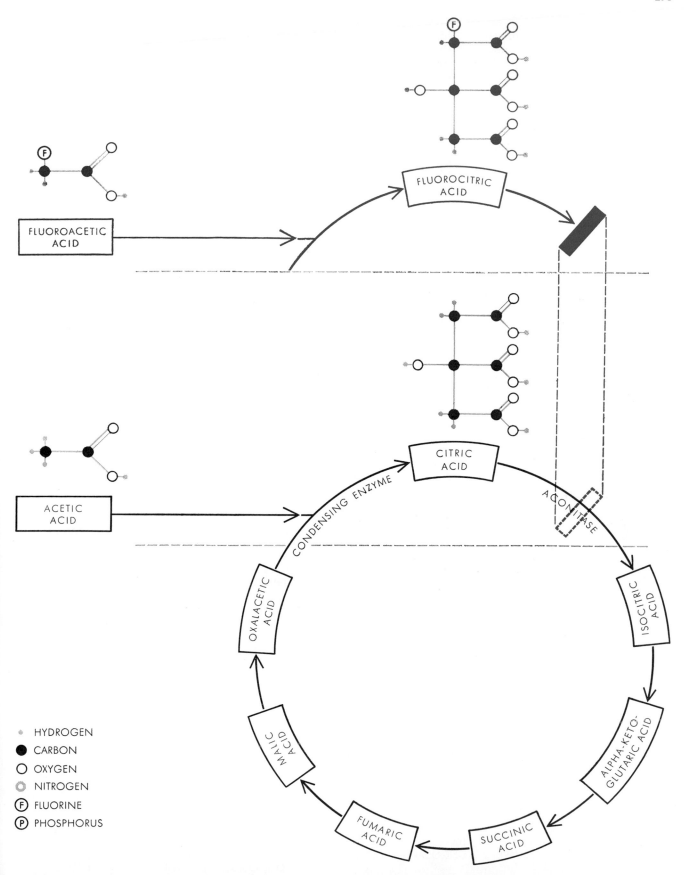

HYDROGEN
CARBON
OXYGEN
NITROGEN
FLUORINE
PHOSPHORUS

FLUOROCITRIC ACID
FLUOROACETIC ACID
CITRIC ACID
ACETIC ACID
CONDENSING ENZYME
ACONITASE
OXALACETIC ACID
ISOCITRIC ACID
ALPHA-KETO-GLUTARIC ACID
MALIC ACID
FUMARIC ACID
SUCCINIC ACID

CITRIC ACID CYCLE, the basic energy-producing mechanism of the body, is blocked by the poison fluoroacetic acid. In the normal cycle (*black*) partially broken-down foodstuffs in the form of acetic acid combine with oxalacetic acid to yield citric acid. This in turn undergoes a series of transformations, most of which yield energy. In the lethal cycle (*color*) fluoroacetic acid substitutes for acetic acid, forming fluorocitric acid. This substance cannot undergo further metabolism; instead it blocks the normal cycle by inactivating the enzyme aconitase. At lower left is the key to structural diagrams here and on subsequent pages in this article.

in biochemical terms. One is carbon monoxide (CO), a major constituent of automobile-exhaust gas and currently the leading chemical agent of suicide in the U. S. Its lethal effects seem largely attributable to its strong affinity for hemoglobin, the protein in red blood cells that transports oxygen from the lungs to the tissues. Carbon monoxide combines with hemoglobin 200 to 300 times more readily than does oxygen. Thus a carbon monoxide concentration in the air only .5 per cent of the oxygen concentration will convert half the hemoglobin in the blood into carboxyhemoglobin, a compound that cannot transport oxygen. Moreover, the presence of carboxyhemoglobin alters the properties of normal hemoglobin so that even the remaining fraction functions with reduced efficiency. The victim undergoes a sort of internal asphyxiation.

The treatment of acute carbon monoxide poisoning is based on the fact that the association between carbon monoxide and hemoglobin is reversible. If the victim is removed from the contaminated atmosphere and given pure oxygen to breathe, the carboxyhemoglobin gradually gives up its carbon monoxide

to yield active hemoglobin. Its "half life" in the blood is about 30 minutes. Unfortunately a few persons who recover from carbon monoxide poisoning suffer damage to vision or to the nervous system that may be permanent.

Cyanide (CN^-), a homicidal and suicidal agent notorious in both medical and fictional literature, resembles carbon monoxide in that it cuts down the body's supply of oxygen. It does so, however, not by hindering the intake or transport of oxygen outside the cells but by interfering with the utilization of oxygen in the cells. The crucial step blocked by cyanide is thought to be the energy-yielding reaction catalyzed by the enzyme cytochrome oxidase. This protein, somewhat similar to hemoglobin, plays an almost ubiquitous role in cellular respiration. The cyanide ion, introduced into the body in the form of hydrocyanic acid or its salts, is believed to inactivate cytochrome oxidase somewhat as carbon monoxide inactivates hemoglobin. This reaction, though it is possibly not the only significant one, adequately explains the deadly effects of small quantities of cyanide (the lethal dose for man is about a millionth of the body weight).

Since the process postulates the almost instantaneous choking-off of cellular respiration, it also explains the notoriously rapid action of cyanide. A few tenths of 1 per cent by volume of hydrocyanic acid gas in air can cause death within minutes. The gas facilitates its own lethal action by provoking deep breathing, thus ensuring the rapid absorption of even low concentrations. These properties make it a hypothetically humanitarian agent for gas-chamber execution in several states and aggravate the problem of therapy after acute accidental exposure to large doses.

Current treatment of cyanide poisoning is based upon suggestions made by K. K. Chen of Eli Lilly & Company. Chen took advantage of the fact that sulfur-containing compounds in the body can transfer their sulfur to cyanide ions, converting them to relatively harmless thiocyanate (SCN^-) ions. His treatment neutralizes the cyanide by administering thiosulfate ($S_2O_3^{--}$), the sodium salt of which ("hypo") is used in photographic development.

Subsequent investigations have shown that the transfer of sulfur to cyanide is catalyzed by rhodanese, an enzyme of

TOXICITY RATING	PRACTICALLY NONTOXIC	SLIGHTLY TOXIC	MODERATELY TOXIC	VERY TOXIC	EXTREMELY TOXIC	SUPERTOXIC
EXAMPLES	GLYCERIN, WATER, GRAPHITE, LANOLIN	ETHYL ALCOHOL, LYSOL, CASTOR OIL, SOAPS	METHYL (WOOD) ALCOHOL, KEROSENE, ETHER	TOBACCO, ASPIRIN, BORIC ACID, PHENOL, CARBON TETRACHLORIDE	MORPHINE, BICHLORIDE OF MERCURY	POTASSIUM CYANIDE, HEROIN, ATROPINE
PROBABLE LETHAL DOSE (MILLIGRAMS PER KILOGRAM)	MORE THAN 15,000	5,000 TO 15,000	500 TO 5,000	50 TO 500	5 TO 50	LESS THAN 5
PROBABLE LETHAL DOSE FOR A 70 KILOGRAM (155 POUND) MAN	MORE THAN ONE QUART	ONE PINT TO ONE QUART	ONE OUNCE TO ONE PINT	ONE TEASPOON TO ONE OUNCE	SEVEN DROPS TO ONE TEASPOON	A TASTE (LESS THAN SEVEN DROPS)

SCALE OF TOXICITY rates substances according to the size of the probable lethal dose. As the examples indicate, many drugs fall into the highly toxic categories. Drawings suggest the fatal doses of water, whiskey, ether, aspirin, morphine and cyanide, the last three depicted in terms of aspirin-sized tablets. The scale was suggested by Marion N. Gleason, Robert E. Gosselin and Harold C. Hodge.

obscure function found in many mammalian tissues. The recent availability of purified rhodanese has suggested the administration of the enzyme along with the thiosulfate. Animal experiments in Sweden by Bo Sörbo and his colleagues indicate that the combined treatment is more effective than the thiosulfate alone. Chen also introduced a parallel therapeutic measure: the administration of compounds such as sodium nitrite that oxidize part of the blood's hemoglobin to methemoglobin. The latter compound combines avidly with cyanide and thus protects the vital cytochrome oxidase at the cost of a little hemoglobin.

These antidotes have proven their efficiency in animal experiments, but are generally most effective when given before or simultaneously with the poison. Few victims of massive cyanide poisoning survive long enough to be treated.

The nervous system, so vital to the regulation and coordination of the body's activities, is the target of a wide spectrum of toxic agents, ranging from classical poisons such as strychnine and atropine to the "nerve gases" developed during World War II. All of these neurotoxins in one way or another disrupt the microchemical mechanisms that transmit nerve impulses. Though their exact mode of operation is in most cases still unknown, many of them have advanced our knowledge of neurophysiology.

Nerve poisons have helped to establish, for example, that the synthesis and breakdown of the compound acetylcholine plays a central role in the body's internal communications. Studies extending over 40 years have identified this substance as the chemical messenger that transmits impulses between motor nerves and voluntary muscles. The list of "cholinergic" nerves also includes important portions of the autonomic nervous system, which controls respiration, digestion and the rest of the body's involuntary activities [see illustration on next page].

A nerve impulse reaching the end of a cholinergic nerve rapidly liberates a minute quantity (less than 10^{-9} gram) of acetylcholine. This activates a receptor on the adjacent nerve fiber, muscle fiber or gland, causing the nerve to fire, the muscle to contract or the gland to secrete. To leave the receptor free to receive further impulses the acetylcholine must then be inactivated. The enzyme cholinesterase, present in all nerve tissue, performs this essential task by cleaving the acetylcholine molecule into choline and acetic acid. Other enzymes

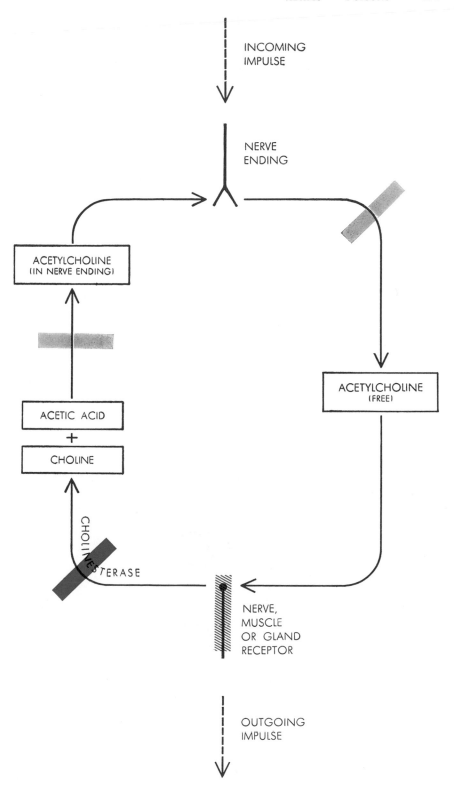

ACETYLCHOLINE CYCLE, a basic mechanism in the transmission of nerve impulses, is the target of many poisons. In the normal cycle an impulse reaching a nerve ending liberates acetylcholine, which stimulates a receptor. To free the receptor for further impulses, the enzyme cholinesterase breaks down acetylcholine into acetic acid and choline which other enzymes resynthesize into new acetylcholine. The "anticholinesterase" poisons prevent the breakdown of acetylcholine by inactivating cholinesterase (*solid color*). Botulinus and dinoflagellate toxins hinder the synthesis or the release of acetylcholine (*light color*). Curare and atropine make the receptor less sensitive to the chemical stimulus (*hatched color*).

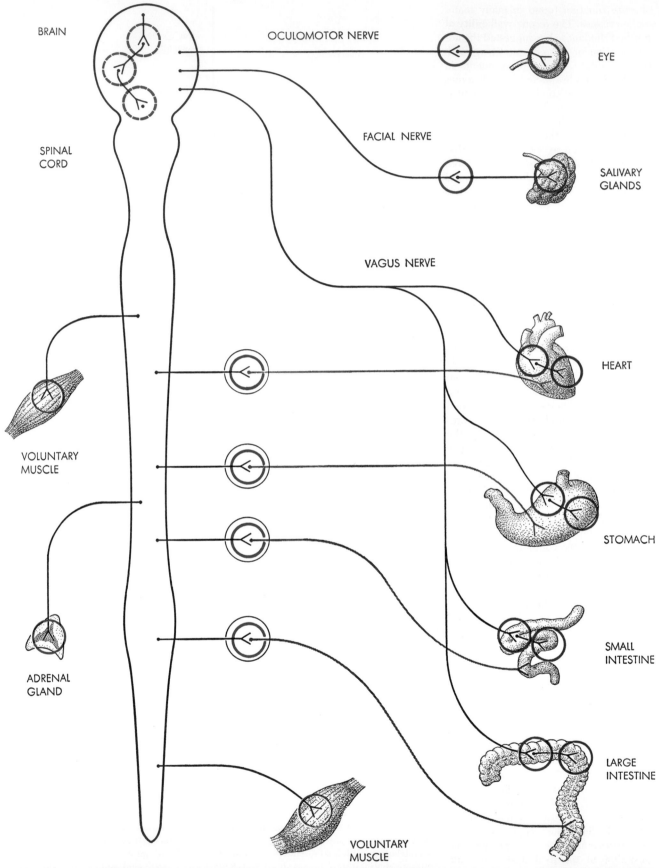

BRAIN

OCULOMOTOR NERVE

EYE

SPINAL CORD

FACIAL NERVE

SALIVARY GLANDS

VAGUS NERVE

HEART

VOLUNTARY MUSCLE

STOMACH

ADRENAL GLAND

SMALL INTESTINE

LARGE INTESTINE

VOLUNTARY MUSCLE

"CHOLINERGIC" NERVES, which transmit impulses by means of acetylcholine, include nerves controlling both voluntary and involuntary activities. Exceptions are parts (*gray*) of the "sympathetic" nervous system that utilize norepinephrine instead of acetylcholine. Sites of acetylcholine secretion are circled in color; poisons that disrupt the acetylcholine cycle can derange the body's communications at any of these points. The role of acetylcholine in the brain is uncertain, as is indicated by the broken circles.

then reconstitute the two substances into acetylcholine, thus completing the cycle [see illustration on page 297].

The nerve poisons that are best understood are those that block the action of cholinesterase. This group includes the alkaloid physostigmine, its synthetic relative neostigmine and the nerve gases. By inactivating the enzyme they prevent the breakdown of acetylcholine. The resulting excess of acetylcholine hyperstimulates nerves, glands and muscles, producing convulsions, choking, heart irregularities and other distressing symptoms. If the victim survives these disorders, the rising level of acetylcholine ultimately brings on flaccid paralysis and death.

Physostigmine (sometimes called eserine), a poison once used in West African ordeal trials, has in recent years worked its way up to clinical respectability. It can temporarily restore muscular strength to an individual suffering from the disease myasthenia gravis and can relieve pressure within the eyeball in the serious eye disorder glaucoma [see "Glaucoma," by Sidney Lerman; SCIENTIFIC AMERICAN, August, 1959]. Neostigmine, an improved synthetic substitute, has found additional uses in slowing the rapidly beating heart and in relieving the intestinal paralysis that sometimes follows abdominal surgery.

Both substances are classed as reversible anticholinesterase poisons. Similar in structure to acetylcholine [see illustration on next page], they attach themselves semipermanently to the cholinesterase molecule. Their high affinity for it—about 10,000 times that of acetylcholine—ensures that even minute doses can inactivate a dangerously high proportion of the enzyme. If their concentration in the body does not reach fatal levels, however, they slowly dissociate from cholinesterase and are eliminated.

The nerve gases, including DFP (diisopropylfluorophosphate), TEPP (tetraethylpyrophosphate), Sarin and tabun, are termed irreversible poisons. They too attach themselves to cholinesterase, but so tightly that for practical purposes the restoration of normal nerve function must await the formation by the body of new cholinesterase molecules free of the poison.

Their toxic action, theatrical enough to satisfy the most demanding detective-story addict, has been well documented in a number of laboratory accidents and in their field use as insecticides. Exposure to even traces of the vapor produces rapid contraction of the pupils and usually transient constriction of the res-

piratory passages. Larger doses of vapor bring death in a matter of minutes from asphyxia, heart failure or shock. Some of the nerve gases are as lethal as cyanide and more easily absorbed. Because they are soluble in fat, they can enter the body through the unbroken skin. In one case a five-minute fingertip contact with a few drops of Sarin produced unpleasant symptoms for weeks.

The original treatment for both reversible and irreversible anticholinesterases involved the administration of two counter-poisons: atropine and curare. These substances combat the effects of excess acetylcholine by making nerve, muscle and gland receptors less sensitive to it. The atropine-curare treatment produces only symptomatic relief, however; full recovery must await the elimination of the poisons and (in the case of the nerve gases) the gradual manufacture of unpoisoned cholinesterase. More recently Irwin B. Wilson and David Nachmansohn of the Columbia University College of Physicians and Surgeons have produced an antidote for the nerve gases based on a new principle.

The new treatment is the fruit of fundamental research into the mechanism by which cholinesterase breaks down acetylcholine. Investigators have hypothesized that the enzyme has two "active sites" which interact with different parts of the acetylcholine molecule. By attaching itself to these two sites, the molecule splits into free choline and an acetyl group (CH_3CO^-) that remains bound to the enzyme. The acetyl group then reacts with water, forming free acetic acid and liberating the enzyme for further work [see illustration on page 301].

The nerve gases, probably by means of the electron-seeking phosphorus atoms which all of them contain, tenaciously preempt the site on the cholinesterase molecule that normally accepts the acetyl group. However, Wilson and his colleagues were able to devise a compound (pyridine aldoxime methiodide) that can pry the nerve gas loose from the enzyme by attaching itself at one end to the phosphorus group in the nerve-gas molecule and at the other to the choline-accepting site of the enzyme [see illustration on page 302]. As anticipated, this compound not only reactivates inhibited cholinesterase in the test tube but also protects experimental animals against otherwise lethal doses of nerve gas.

Though interest in the nerve gases originally stemmed from their possibili-

ties in chemical warfare, the compounds have yielded an abundant nonmilitary harvest. For example, DFP has helped to unravel the fine structure of several enzymes which, like cholinesterase, bind it tightly at their active sites. Molecules of trypsin and chymotrypsin (protein-digesting enzymes secreted by the pancreas) can be labeled with DFP containing radioactive phosphorus atoms and then be degraded into small fragments; the labeled fragments identify the active segment of the molecule. In medicine the nerve gases serve much the same therapeutic purposes as physostigmine and neostigmine; their use as insecticides represents another nonmilitary application.

The toxic protein secreted by the botulinus bacillus, occasionally a cause of human food-poisoning, disrupts the acetylcholine cycle no less efficiently than do the anticholinesterase poisons, but is believed to work the opposite side of the biochemical cycle. Instead of raising the concentration of acetylcholine, it lowers it to paralytic levels by preventing its synthesis or its release from nerve endings. Botulinus toxin has the distinction of being the most poisonous substance yet discovered. As methods for purifying it have improved, the apparent minimum lethal dose has correspondingly decreased. According to a recent estimate the lethal dose for a mouse may be as little as 1,000 molecules—less than 10^{-13} gram per kilogram of mouse! Similar in action to botulinus toxin and almost as deadly is the toxin secreted by the poisonous dinoflagellates. These microscopic marine organisms can kill fish by the millions when they multiply explosively in "red tides"; the concentration of their toxin in the digestive tracts of mollusks accounts for many cases of shellfish poisoning in man.

Another potent metabolic poison that is fairly well understood is fluoroacetic acid. This simple compound constitutes the active ingredient in the gifblaar plant, notorious in South Africa as a cattle killer. Pharmacological studies place it, with cyanide and the nerve gases, in the supertoxic class. Dogs, which seem especially sensitive to its effects, may succumb to a dose of .05 milligram per kilogram of body weight. Curiously the fatal concentration of this poison is about 100 times greater for rats and 10,000 times greater for toads. The lethal dose for man, rather unreliably estimated from a few cases of accidental poisoning, is about 2.5 milligrams per kilogram of body weight.

Fluoroacetic acid acts relatively slowly; even when it is injected directly into the bloodstream the symptoms often take several hours to develop fully. They vary in different animals, but involve chiefly the heart and central nervous system. As with many metabolic poisons, the damage is purely functional; examination of the tissues shows no distinctive changes.

A decade ago Sir Rudolph Peters of the University of Oxford and the German biochemist Carl Martius independently suggested that fluoroacetic acid interferes with the citric acid cycle.

This is the basic biochemical mechanism by which foodstuffs are oxidized to yield energy. The intermediate breakdown-products of sugar and fatty acids enter the cycle as a form of acetic acid, which combines with oxalacetic acid in the next step of the cycle to yield citric acid.

ACETYLCHOLINE

DIISOPROPYLFLUOROPHOSPHATE (DFP)

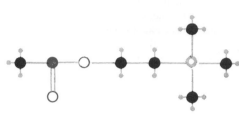

NEOSTIGMINE

SARIN

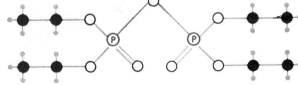

TETRAETHYLPYROPHOSPHATE (TEPP)

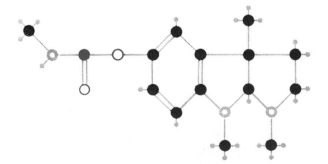

PHYSOSTIGMINE

PARAOXON

STRUCTURES OF ANTICHOLINESTERASE POISONS help to explain their toxic properties. Acetylcholine attaches itself to the cholinesterase molecule by means of an ester bond and a positively charged nitrogen atom, both shown in color. The poisons neostigmine and physostigmine are structurally analogous (the nitrogen atoms in physostigmine become positively charged in an acid environment). They combine similarly with cholinesterase, but more tenaciously. The "nerve gases" at right all contain a phosphorus atom with an easily broken bond (*also shown in color*). These poisons inactivate cholinesterase as shown in illustration, page 302.

Peters and Martius proposed that fluoro-acetic acid, which is structurally simi-lar to acetic acid, can enter the cycle at the same point, forming fluorocitric acid. This is a lethal error, because the fluorocitric acid powerfully inhibits the next enzyme in the sequence, which catalyzes the reorganization of citric acid into isocitric acid [*see illustration on page 295*].

Peters and his colleagues have but-tressed this hypothesis with many ob-servations. In test tube preparations of liver cells, fluoroacetic acid and oxala-cetic acid yield fluorocitric acid as pre-dicted. Citric acid accumulates in the tissues of poisoned animals, indicating that its conversion into isocitric acid is indeed blocked. At present, however, there are still some inconsistencies that need explaining. It may be that other toxic biosynthetic products of fluoro-acetic acid, such as fluorinated amino acids, play a significant part in the chem-ical pathology.

There is a rich variety of ways to wreck the machinery of a cell, and a corresponding diversity of poisons. Un-doubtedly the future will uncover many new poisons, both by accident and de-sign. It seems currently impossible to classify poisons in consistent and logical categories. Such a system might identify each poison by the metabolic reaction which is its target, assuming that toxic mechanisms prove to be that simple when we understand them. Apart from the examples already cited, however, the metabolic targets remain uncertain. Thus the classic poison arsenic, like sev-eral other metallic poisons, is known to inactivate enzymes by attaching itself to sulfhydryl (SH⁻) groups in their struc-ture, but we can as yet only guess which enzymes are the crucial ones.

The animal venoms alone constitute a remarkably varied group of poisons. All animal phyla, from single-celled proto-zoans to mammals, include species that produce toxic compounds. Snakebite causes thousands of deaths in tropical re-gions every year, yet we know little about the chemistry of snake venoms. They are known to contain many enzymes, some of which can break open red blood cells and cleave protein molecules. Rattle-snake venom has become a commercial source of enzymes used to study the structures of proteins and nucleic acids. The truly potent constituents of many venoms, however, appear to be nerve poisons whose specific mechanisms still elude us. Some of these neurotoxins are extraordinarily powerful: the fresh

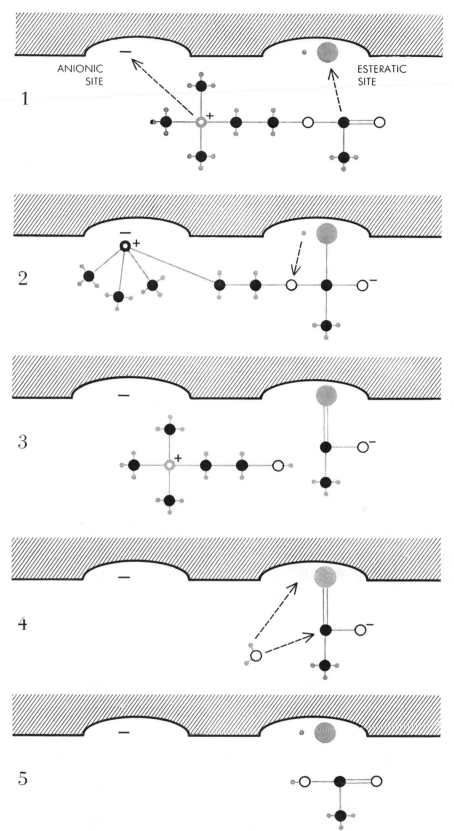

BREAKDOWN OF ACETYLCHOLINE probably takes place as shown in these diagrams. The cholinesterase molecule (*hatched*) is thought to include a negatively charged anionic site, which attracts one end of the acetylcholine molecule, and an esteratic site whose "nu-cleophilic" group (*large gray sphere*) attracts the other end. When acetylcholine attaches itself to these sites (1, 2), it breaks into free choline and an attached acetyl group (3). This group then reacts with water (4), forming free acetic acid and returning the enzyme to its active state (5). This scheme was proposed by Irwin B. Wilson and David Nachmansohn.

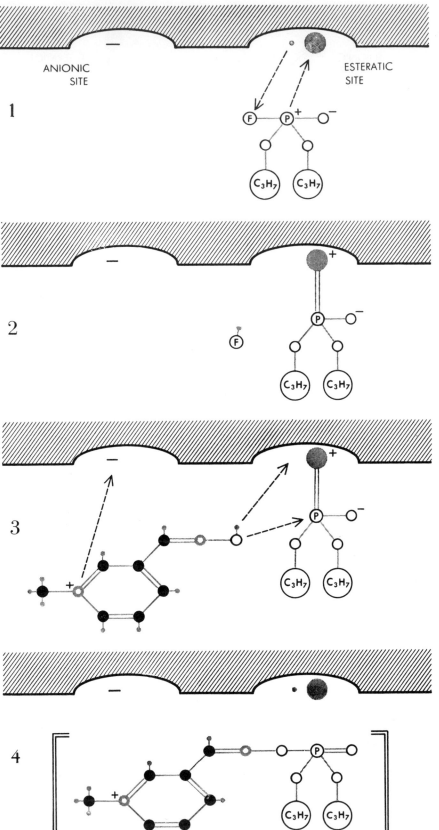

INACTIVATION OF CHOLINESTERASE by a nerve gas, according to Wilson and Nachmansohn, is illustrated above. The phosphorus atom in the nerve gas binds itself tightly to the nucleophilic group in the enzyme, liberating hydrofluoric acid (1, 2). The compound pyridine-2-aldoxime methiodide (iodide ion is not shown) reactivates the enzyme. By attaching itself to the anionic site and the phosphorus atom (3) it pries the nerve gas loose from the nucleophilic group, presumably forming the unstable molecule shown in brackets (4).

venom of the Australian taipan can kill guinea pigs at a concentration of .025 milligram per kilogram of body weight, even though a considerable fraction of the crude venom must be relatively inert.

The alkaloids furnish another inexhaustible list of poisons, many of which double as drugs. In crude form, indeed, these plant compounds represent both the earliest drugs and the earliest poisons of mankind; the opium poppy is said to have been cultivated in the Stone Age. The alkaloids include, in addition to atropine and curare, such well-known drugpoisons as quinine, morphine, cocaine, strychnine and nicotine. Among the many mysteries surrounding these compounds is why plants produce them [see "Alkaloids," by Trevor Robinson; SCIENTIFIC AMERICAN Offprint 1082].

The profusion of toxic substances in common circulation and use in this country has created a growing problem for the physician. The pediatrician in particular may at any moment receive a telephone call from a frantic mother whose child has swallowed one of the dozens of household cleaning preparations or one of the 80,000-odd formulations of pesticides. Many of these substances are relatively innocuous, in which case reassurance is the chief therapy, but some represent true pharmacological emergencies requiring prompt and specific treatment.

No physician, of course, can possibly identify by memory the ingredients of these thousands of products, let alone recall the prognosis and therapy appropriate to each. Accordingly health departments in a number of U. S. cities have organized poison-control centers which provide around-the-clock information to physicians and parents on the ingredients of brand-name products and on emergency therapy. Since the establishment of the first of these centers six years ago in Chicago, their number has grown to more than 200. Some have accumulated extensive reference-libraries and have even gone into toxicological research. In 1956 the U. S. Public Health Service organized a national clearinghouse for poison information.

The wide realm of poisons thus includes the savage tending a curare pot, the organic analyst at his microbalance, the pediatrician on the telephone and the industrial physician calculating the tolerable maximum of a toxic vapor. Last and by no means least, it encompasses the biochemist, who is learning to employ an almost infinitely varied set of probes for sounding the recesses of cell physiology.

32

RADIATION-IMITATING CHEMICALS

PETER ALEXANDER
January 1960

Nitrogen mustards and certain other compounds produce the same apparent biological effects as high-energy radiation. These chemicals have some use in cancer therapy and great value in basic research

As a result of the development of atomic energy it is now widely appreciated that radiation causes profound changes in living matter. The changes have been studied for 50 years, but they are many and varied and still far from completely understood. Taken together they make up the well-defined group of radiation effects whose potential for good and ill has now become the subject of such wide concern. What is not generally realized is that these ef-fects are not unique to radiation. There is a considerable number of chemicals that produce, so far as we can tell, the same biological results.

The knowledge that substances can be "radiomimetic" is largely a product of World War II research in poison gases. For a few years after the war these sub-stances were expected to open the way to a cure for cancer. Today it is clear that they are no cure, but they are therapeu-tically useful in many types of malignant disease. More important, however, is their role in fundamental research, where studies of their mode of action are throwing new light on cell biology. Al-though the radiomimetic chemicals are themselves not the answer to cancer, they are helping to achieve the funda-mental knowledge in which the answer lies.

What are the characteristic effects of radiation? On the cellular level they can be divided into four categories. First,

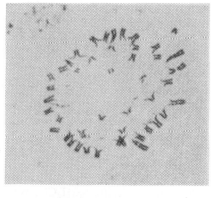
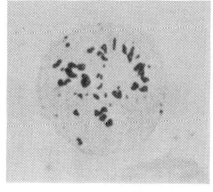
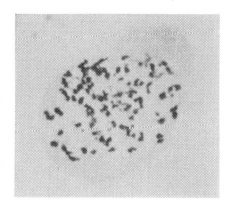
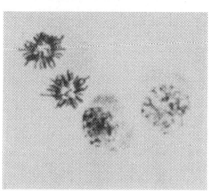
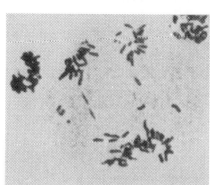
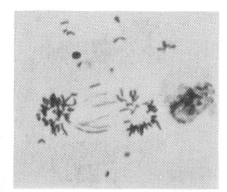

CHROMOSOME ABNORMALITIES are seen in cells treated with nitrogen mustard (*center and right*). At metaphase (*top*), when the chromosomes are about to separate, the treated ones show fragmentation. At anaphase (*bottom*) treated cells have bridges between chromosome halves. These photographs were made by P. C. Koller of the Chester Beatty Research Institute in London.

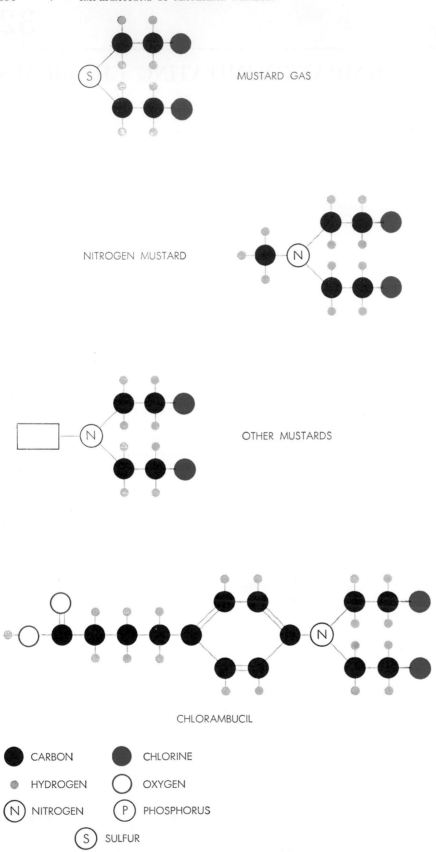

MUSTARD GAS

NITROGEN MUSTARD

OTHER MUSTARDS

CHLORAMBUCIL

● CARBON   ● CHLORINE

• HYDROGEN   ○ OXYGEN

Ⓝ NITROGEN   Ⓟ PHOSPHORUS

Ⓢ SULFUR

MUSTARD MOLECULES are diagrammed schematically. Common to all are two chains containing two carbon atoms and a chlorine. In mustard gas (*top*) the chains are attached to a sulfur atom. In original nitrogen mustard (*second from top*) sulfur is replaced by nitrogen. Other nitrogen mustards (*third from top*) are formed by attaching any of various chemical groups (*rectangle*) to nitrogen atom. One such mustard is chlorambucil (*bottom*).

radiation can cause genetic affects: gene mutations and permanent changes in the structure of chromosomes. Second, it interferes with some of the processes of cell division; this interference may lead eventually to the death of the affected cells, although they usually multiply a few times after the exposure. Third, it kills certain types of cell outright; radiation-sensitive cells include white blood cells, mammalian egg-cells and certain pigment-producing cells. Fourth, it induces cancer and leukemia, the disease usually appearing years after the exposure. An additional strange effect on the whole organism is a nonspecific shortening of life-span. Animals that have been exposed to radiation and have apparently recovered completely from the immediate damage live less long, on the average, than unexposed animals even if they do not develop malignant diseases.

Substances have been known for many years that produce some of the effects of radiation. In this article I shall deal with chemicals that are radiomimetic in the strictest sense, that is, chemicals that produce all the effects listed above.

The story begins with two separate lines of research, started under wartime secrecy, that converged only after security restrictions were lifted. Among the chemical-warfare agents being studied in a number of laboratories was mustard gas, a substance that produces blisters on human skin (but curiously not on the skin of most other animals). It was discovered that a relatively minor change in the mustard-gas molecule—the substitution of a nitrogen atom for a sulfur—greatly enhanced the effectiveness of the gas. The new agent was called nitrogen mustard [*see illustration at left*]. Subsequently a number of different compounds, the molecules of which incorporated the chloroethylamine group ($-NHCH_3CH_2Cl$), were found to have the same activity, and these compounds are now called nitrogen mustards also. While investigating their pharmacology, Alfred Gilman at the Yale University Medical School was impressed by the fact that they destroyed lymphoid tissue and organs containing rapidly dividing cells. This suggested that the nitrogen mustards might be useful for controlling cancer and leukemia. A group at Yale pursued the idea from 1943 onward, though they could not publish their results until after the war ended.

Meanwhile in Britain, Alexander Haddow and his colleagues at the Chester

Beatty Research Institute of the Royal Cancer Hospital in London had been experimenting with entirely different chemicals for the control of cancer. Learning of the U. S. work, they attempted to combine the active groups in one of their substances with the active group of the nitrogen mustards. The biological activity of these new compounds turned out to be dominated by the nitrogen-mustard groups. Therefore they undertook an intensive program to prepare new derivatives that might show improved biological properties. Clinically their efforts have had a limited success. Many different compounds are now used in treatment, but the advantages over the parent compound are relatively small. Scientifically, however, the research by Haddow's group was extremely valuable, because it revealed the two key properties of the nitrogen mustards that account for their growth-inhibiting activity.

First, the reactivity of the chlorine atom proved all-important. When nitrogen mustard is dissolved in body fluids, the chlorine atom splits off, leaving a reactive intermediate that readily combines with many of the molecules of the cells [see illustration below]. The result is the attachment to these molecules of a

straight-chain group, a reaction known as alkylation. On the theory that alkylation was responsible for the capacity to kill cells, Haddow's group tested other compounds, chemically quite different from the nitrogen mustards but capable of the same type of reaction. These substances included epoxides, ethylene imines and methane sulfonates [see illustration on next page]. All were growth inhibitors. Having no other feature in common, their activity seems inescapably to be due to their alkylating ability under physiological conditions.

As so often happens in science, these discoveries were subsequently found to have been partially anticipated many years earlier. In 1898 Paul Ehrlich, the father of chemotherapy, had recognized the unusual pharmacological properties of ethylene imine and of the simplest of the epoxides: ethylene oxide. He noted that they differed from all the hundreds of other substances he had studied in causing intense cell destruction in tissues containing rapidly dividing cells. However, his report came to the general attention of scientific workers only in 1956, with the publication of his collected papers. Had it not been overlooked in the wealth of Ehrlich's researches, we should not have had

to wait for a world war to focus attention on the potentialities of alkylating agents.

The second critical property of the growth-inhibiting chemicals to emerge from the studies of Haddow and his colleagues was the presence of two alkylating centers in each molecule. The original mustards—nitrogen or sulfur—have two such groups. Of all the other derivatives tested, only those with two reactive chlorine atoms proved effective in arresting the development of tumors in experimental animals. (Single-group, or "one-armed," compounds do inhibit growth in experiments on single cells, but only in concentrations too high to be tolerated by an animal.) The same condition applied, in general, to the other biological alkylating agents; for example, a compound having two epoxy groups is very much more effective than a mono-epoxide.

We shall presently consider the significance of these findings, but first let us turn to a second radiation-like feature of the alkylating agents: their genetic effects. In 1928 H. J. Muller showed that X-rays produce mutations in genes, the hypothetical units of heredity that are strung out along the chromo-

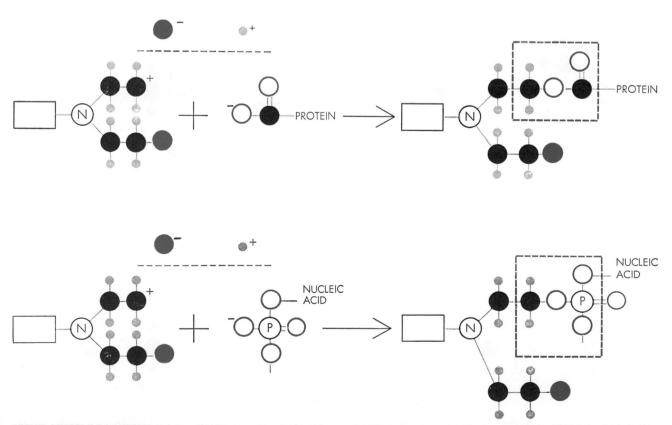

ALKYLATION REACTIONS involve attachment of active group on mustard molecule to a receptor such as an acid group on a protein (*top*) or a phosphate group on a nucleic acid (*bottom*). Broken rectangles enclose region where reaction has taken place.

somes and passed down from generation to generation. Subsequently it was found that all types of high-energy radiation cause mutations in all forms of life that have been tested. These radiations can also damage the structure of chromosomes, the changes becoming apparent when cells divide. For many years geneticists looked for chemicals that would have the same actions, but none was found. Radiation seemed to be unique in these respects.

Then during the war P. C. Koller, working at the University of Edinburgh, observed that mustard gas produces permanent chromosome abnormalities that are indistinguishable from those induced by X-rays. Charlotte Auerbach and William Robson, also at the University of Edinburgh, had set out to test whether this compound would also cause mutations in the fruit fly. It did; they had found the first chemical mutagen. Quite independently of the British work, the publication of which was again delayed by security requirements, I. A. Rapoport, a geneticist in the U.S.S.R., also discovered the mutagenic action of an alkylating agent: diethyl sulfate. He published his results in 1947, though his studies had not been stimulated by military research.

It soon developed that the ability to cause mutation and chromosome damage, like the ability to inhibit tumors, is shared by all the biological alkylating agents. Mutagenic action, however, requires only one alkylating center per molecule and is not, in general, greatly enhanced by the presence of a second reactive group. On the other hand, compounds with two centers are much more active than one-armed substances in producing abnormalities in chromosomes.

Like X-rays, the alkylating agents cause mutations in the genetic systems of organisms ranging from bacterial viruses to mammals. Many other types of

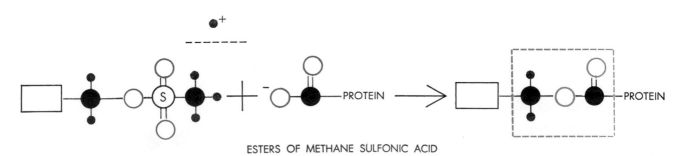

OTHER ALKYLATING AGENTS are diagrammed schematically at left. Although they are chemically unlike the mustard molecule, they react with cell molecules such as proteins in the same way. All these substances produce the detectable effects of radiation.

chemical substance have since been found that cause mutations and chromosome damage in certain cells, but none of these seems to be so universal in its action. Apparently genetic changes can be induced by a variety of different mechanisms, some of which work rather indirectly, possibly by altering metabolism. The fact that the alkylating agents affect the genetic system wherever it is found indicates that they directly attack the fundamental genetic material, common to all forms of life. It is this that makes them true radiomimetics.

One of the many unsolved problems in cancer research is the connection between these genetic effects and the ability of alkylating agents to kill cells, which is the basis of the agents' therapeutic use. Undoubtedly chromosome abnormalities can be fatal to isolated cells, although the effect is delayed for some generations. The cells usually pass through a few divisions after sustaining this damage. Eventually unequal sharing of genetic material between daughter cells, or mechanical interference with the division process, will bring reproduction to a halt. This type of destruction may be called mitotic death, since it is bound up with the process of cell division, or mitosis.

The role that mitotic death plays in the whole animal is not certain. It may contribute to damage in organs containing rapidly dividing cells, such as the bone marrow, which turns out a continuous supply of blood cells. But what is called interphase death is more important in the treatment of many types of

tumor. Here the cells die some hours after exposure without an intervening division (*i.e.*, during interphase), so that there seems to be no role for chromosome damage. Some cells are very resistant to direct killing, by radiation or chemicals, while others, notably the mammalian egg, the mature sperm and the white blood cells known as lymphocytes, are extremely sensitive. In any case the influence, if any, of genetic effects in this type of cell destruction is obscure.

Let us turn now to the question of how the radiomimetic chemicals produce their effects. Knowing the chemical reaction responsible for their biological activity, one might expect that the rest of the problem would be relatively easy to solve. Unfortunately alkylating agents react with nearly all the important components of cells. Every protein has many points to which they can attach. So do nucleic acids; so do nearly all the different vitamins. Thus when alkylating agents enter a cell, they are dissipated in a large number of different reactions. Many of these reactions, if they involved a significant proportion of the available molecules, would no doubt lead to the death of the cell. But the concentrations necessary to produce radiation-like effects are so low that most of the possible receptor molecules must be left untouched. The problem, then, is not to find the kinds of molecule that are alkylated, but to eliminate the trivial reactions and to determine those vital sites where a few alkylations can cause deep-seated and permanent damage.

The chief chemical groups with which alkylating agents react in living material are the sulfhydryl group (SH), the amino group (NH$_2$) and the acid group (COOH). Many other substances also combine with sulfhydryl and amino groups, but none exhibits true radiomimetic properties. Hence the process of elimination points to acid groups as the important sites. Every chemical that can alkylate an acid group under conditions existing in living cells has proved to be radiomimetic. The reasoning is by no means conclusive, and if we accept it we are still left with a bewildering number of possibilities. Every protein and every nucleic acid contains many eligible acid groups. Furthermore, the different biological end-effects are not necessarily initiated by the same primary chemical reaction.

Our next guidepost is the fact that compounds having two or more alkylating groups per molecule are very much more effective than one-armed compounds in inhibiting growth and damaging chromosomes. This suggests that such injuries may be produced when the chemical agent reacts with two of the cell molecules, joining them together. Specifically, if the two threads of a chromosome were cross-linked before a cell divided, the two could not separate properly during division and abnormalities would result. Although there is little evidence for cross-linking of chromosome threads, the general notion of cross-linking has proved most fruitful. This reaction is quite plausible. Cross-linking will almost surely put a giant

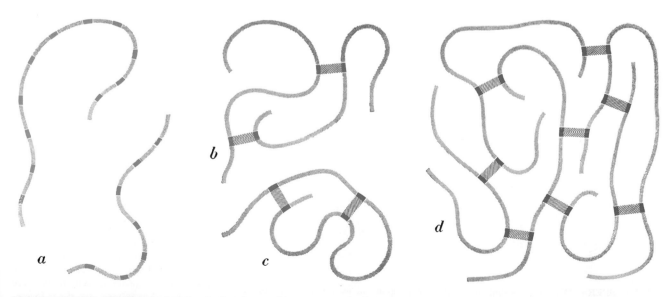

CROSS-LINKING OF DNA molecules (*colored strands*) occurs when an alkylating agent with two active groups per molecule attaches to receptors on DNA chains (*dark segments*). Linking may be between different molecules (*b*) or between parts of the same molecule (*c*). When enough different molecules are joined together (*d*), this portion of the DNA is not soluble but gel-like.

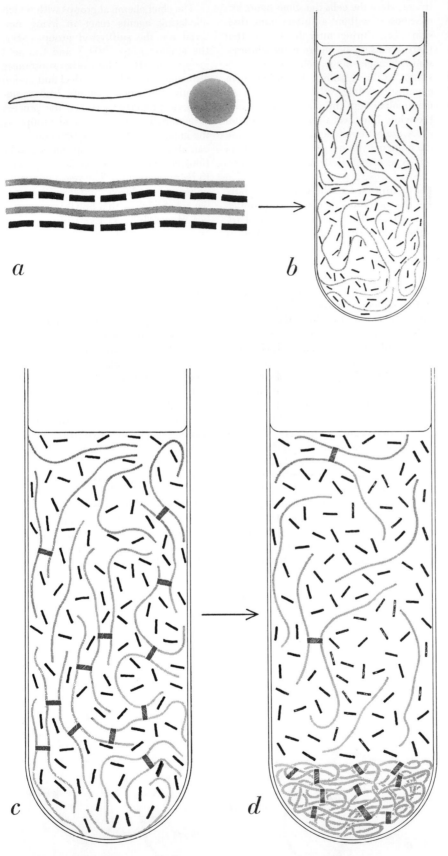

SALMON-SPERM DNA is cross-linked by alkylating agents in experiment diagrammed here. Intact nucleus contains an aggregate (*a*) of DNA (*colored strands*) and protein (*black segments*). When nuclei are dissolved in strong salt solution (*b*), the DNA and protein separate. If cells have been treated, part of the DNA is cross-linked (*c*). Upon centrifugation (*d*) the part of the DNA that is most heavily cross-linked separates in form of a gel.

molecule out of action, no matter where the linking takes place. The simple attachment of a small side-group will have no effect if it takes place at a biologically unimportant site. For example, many enzymes, which are proteins, can be extensively alkylated without losing their activity.

Assume, then, that radiomimetic chemicals cross-link the giant molecules of cells by attaching themselves to groups that can be alkylated. The question remains: What molecules are involved? We know that the chemicals do cross-link proteins. However, the universality of their genetic effects suggests that, in this case, they act directly on the genetic material: deoxyribonucleic acid (DNA). Another reason for choosing DNA is that many substances cross-link proteins under physiological conditions, but do not show radiomimetic properties. These biologically inactive cross-linking agents do not combine readily with isolated DNA, while the biological alkylating agents do. Moreover, the ability of the alkylating chemicals to cross-link DNA has been demonstrated directly [*see illustration at left*].

Does cross-linking explain both types of cell-killing? It is certainly not hard to see how it applies to mitotic death. Joining DNA molecules together in an unnatural configuration must lead to trouble during cell division, when the DNA has to be shared between the daughter cells. Once the genetic material has been altered, the change will be perpetuated and eventually give rise to nonviable cells.

As to interphase cell-death, the picture is less clear. The cells most sensitive to death without intervening division are the lymphocytes, which have a large nucleus and very little of the surrounding cytoplasm. There are indications that damage to the nucleus kills them, but we do not know whether cross-linking of DNA has anything to do with it.

In the case of mutation it is obvious that the cross-linking mechanism is not involved, because one-armed molecules are just as effective as those with double groups. As readers of this magazine are well aware, current genetic theory holds that hereditary information is stored in the DNA molecules in the form of a code. DNA is composed of four (sometimes five) basic units called nucleotides, and the order in which they are strung together determines the information in the molecule. In reproduction the dividing cell makes identical copies of its DNA which it transmits to its descendants. Occasionally there is an error in copying, the sequence of a few nucleotides is changed, and a spontaneous mutation has occurred.

How can the alkylating agents increase the number of mutations? Since they do this in the genetic system of organisms ranging from the bacterial virus to the mouse, it is most probable that they work by direct reaction with DNA. Apparently they interfere in some way with the replication process, increasing the chance of copying-errors. It seems likely that the alkylation of one of the many reactive groups in the DNA molecule would have such an effect. The alkylated DNA is not itself the "mutant," since it is a structure that cannot be copied by the normal metabolic processes. We suggest that alkylation of an isolated group slightly complicates the synthetic process, thereby increasing the likelihood of a copying-error.

The final radiation-like property of the alkylating agents is their capacity to cause malignant growths years after the organism is exposed to them. All the different classes of chemicals we have mentioned have been found to be carcinogenic in animals. Whether their cell-killing or their mutagenic properties or both are responsible is not clear. According to one school, the cancer results from a mutation that robs a cell and its descendants of the ability to respond to the restraining influence of the host in which they grow. If so, we should expect a close correlation between the mutagenic and cancer-producing properties of a compound. An alternative view is that the carcinogenic agent damages tissue, making it possible for the occasional cancer cell that arises by chance mutation in this tissue to establish itself and develop into a malignant growth. On the latter theory the important reaction is not with the parent cancer cell but with the neighboring cells. The two ideas are not mutually exclusive. Some cancers are almost surely produced by the indirect mechanism and others seem to result from mutation, though this has never been unambiguously proved. Since radiation causes both tissue damage and mutations, it does not provide a means of deciding between the two possibilities. As we have seen, some of the alkylating agents produce mutations, but do not, in general, damage tissues by killing cells. Thus we have high hopes that research with these chemicals may furnish valuable information about the nature of the cancer-inducing process.

The remarkable similarity between

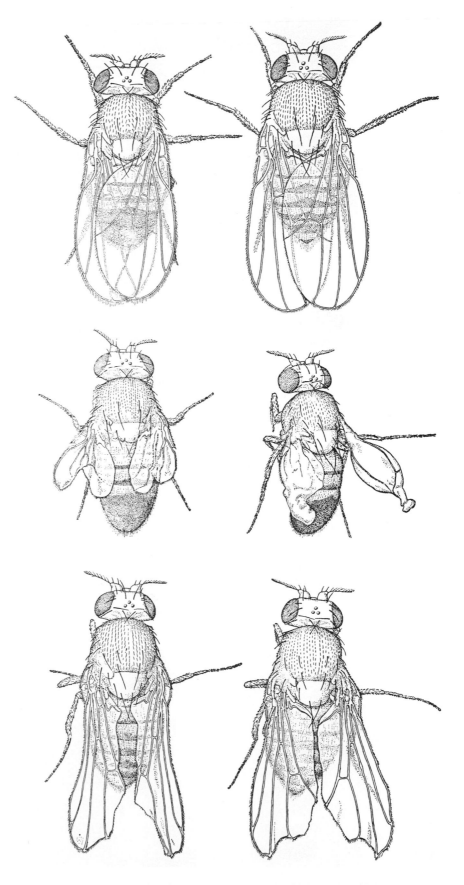

WING MUTATIONS in fruit flies are produced by injecting alkylating agents into the testes. At top are the normal male (*left*) and female (*right*). The drawings of male (*center*) and female (*bottom*) progeny of treated fruit flies illustrate four of the wing mutations produced. This study was done by O. Fahmy of the Chester Beatty Research Institute.

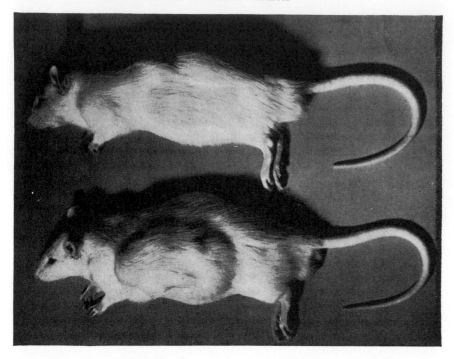

INHIBITION OF TUMOR GROWTH was produced by injecting an alkylating agent. Pictures show 14 days' tumor growth in a treated animal (*top*) and an untreated one (*bottom*). Photographs were furnished by Alexander Haddow of the University of London.

the detectable effects of radiation and of alkylating agents naturally raises the question of whether the ends are reached by the same biochemical route. We know that several stages intervene between the primary action of radiation—damage to an important center in a molecule of the cell—and the appearance of biochemical disturbances [see the article "Radiation and the Cell," by Alexander Hollaender and George E. Stapleton; SCIENTIFIC AMERICAN Offprint 57]. Metabolism develops the injury just as a developer brings

out an image on a photographic plate. Is the initial reaction of the alkylating agent comparable to the primary chemical lesion of radiation? It is tempting to think so, and to suppose that the biological pathway from then on is the same. But the evidence is not clear.

Radiation can, under certain conditions, cross-link the DNA in the nucleus of the cell. Unfortunately, however, there are many reasons for believing that this reaction is not the starting point for radiation injury of cells (though it

may well be responsible for the inactivation of viruses by atomic radiations). Thus the train of events set in motion by alkylating agents and by radiation may meet only at the stage of the detectable biochemical disturbances. It seems unlikely, however, that there is no common pathway at all, and that the many identical end-effects are reached in quite different ways.

Ten years of intensive research on alkylating agents has failed to produce a cure for cancer. Thousands of compounds have been synthesized and tested, but none shows any fundamental advantage over the original nitrogen mustards. Some of the newer substances do have fewer toxic side-effects, making them more useful in therapy. All share the basic defect that they do not specifically seek out the malignant cell, but attack all rapidly dividing cells. For example, they destroy the normal cells in bone marrow as readily as the cells that give rise to leukemia. A dose large enough to kill all the leukemic cells would therefore be fatal to the patient. So the chemicals produce remissions, but not cures. Because their action is not localized they cannot be used to eradicate a local tumor, as radiation often can.

Probably the greatest benefit that has been, and will be, derived from the radiomimetic substances is the increase in basic knowledge. Interfering as they do with the most fundamental cellular processes, they offer one of the best ways of learning more about these processes.

Finally it seems worthwhile to consider these agents from another point of view. The dangers of atomic radiations continue to be a matter of lively public concern, and rightly so. What about the chemicals that so faithfully imitate radiation? They may well represent a greater danger. We add chemicals to food, spread them widely in pest control, inject them into the air we breathe. In setting standards of safety it is usual to consider only the quantities necessary to cause acute toxic effects. If the level of exposure is well below these limits, it is considered safe. We test not at all for genetic effects, and for cancer risk only in the case of food additives.

The facts reviewed in this article ought to put us on our guard. We are trying very hard to assess the subtle and long-term risks of radiation. Should we not pay equal attention to the chemicals, new and not so new, that we encounter in our everyday lives? The effort might pay a great dividend.

 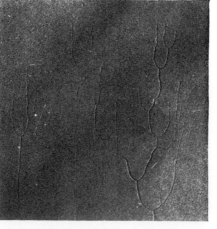

DNA MOLECULES show visible alterations when treated with nitrogen mustard. Normal molecules (*left*) are long and flexible. They are stretched straight by surface tension in being prepared for the electron microscope. Treated molecules (*right*) have a different form. They are not similarly straightened but remain more coiled. These electron micrographs were made by M. S. C. Birbeck and K. A. Stacey of the Chester Beatty Institute.

33

FREE RADICALS IN BIOLOGICAL SYSTEMS

WILLIAM A. PRYOR

August 1970

*Short-lived and highly reactive, free radicals are
essential intermediates in many chemical processes.
In living systems radicals play important roles in
radiation damage and in aging*

In early 1969 an exquisite creature named Vanessa showed a mouse called Mimi to Simon Templar, hero of the television program "The Saint." Vanessa explained that the mouse's life expectancy had been increased by more than 40 percent as the result of an experiment conducted by Vanessa's father: he had fed Mimi a diet including "butylated hydroxytoluene," or BHT. The episode was actually based on a scientific report: Denham Harman of the University of Nebraska has found that BHT and several other chemicals of widely varying types appear to increase the average life-span of laboratory animals [see "Science and the Citizen," SCIENTIFIC AMERICAN, March, 1969]. The most obvious similarity among these life-lengthening chemicals is that they all interfere with or entirely stop the reactions of ephemeral chemical entities called free radicals.

A free radical is a chemical compound that has an odd number of electrons and is therefore generally highly reactive and unstable and cannot be isolated by ordinary methods. In contrast, most chemical compounds have an even number of electrons and are stable. As we shall see, chemical bonds are made up of a pair of electrons, and the high reactivity of free radicals stems directly from the fact that they have an odd electron. Any species with an odd electron seeks another odd-electron species, and the two proceed to pair their odd electrons to unite and form a bond.

Free radicals are known to be key intermediates in many laboratory, industrial and biochemical processes. Most of the reactions of oxygen involve free radicals, including the slow degradation of organic materials in air, burning and the drying of paints. Many plastics are made by processes that involve free radicals as transient intermediates. Radicals can also be detected in most animal and plant cells, and it is clear that radical chemistry plays a vital role in life processes. Radicals appear to be involved in the production of at least some types of cancer, and the concentration of radicals is different in cancerous cells from what it is in normal cells. Some of the reactions that mediate respiration by living organisms involve radicals. Fats in foods become rancid on oxidation, and so inhibitors of radical reactions, such as BHT, are added to lengthen the storage life of foods. Radiation damage to living systems occurs partly through free radicals. Finally, aging itself has been postulated to involve random destructive reactions by radicals present in the body.

To understand the great reactivity of free radicals it is necessary to understand first why electrons pair to form the normal two-electron bond. Electrons have a magnetic moment and are small magnets; as such they can be considered to have the property termed spin, which can be either "pointing up" or "pointing down" for any electron. Chemical compounds are assigned a "multiplicity" depending on the arrangement of the spins of their electrons. If a molecule has an even number of electrons entirely arranged in pairs with opposed spins, the molecular species is said to be a singlet. If a molecule has an odd number of electrons, then the odd electron must have an unpaired spin; the molecule is called a doublet and is a radical. If a molecule has an even number of electrons but the electrons of one pair have parallel spins, the species is said to be a triplet. Bonds are formed only between electrons that are paired and have antiparallel spins; clearly, then, most molecules will be singlets.

When an organic molecule containing a normal electron-pair bond is heated above a certain temperature (for example, when a hydrocarbon molecule such as is found in petroleum is heated to a temperature between 700 and 800 degrees Celsius), the weakest bond in the compound breaks and the two fragments fly apart. In this process of bond thermolysis, or thermal bond scission, the two electrons in the bond divide, one going with each fragment. The process can be symbolized as $A:B \rightarrow A\cdot + B\cdot$, where AB is an ordinary molecule and the two dots represent the two electrons in the A-B bond that hold the molecule together. A process such as this one, in which the bonding pair of electrons divides symmetrically as the bond breaks, is called homolysis, or homolytic bond scission. The two electrons in the A-B bond must be paired and have antiparallel spins, and so the odd electrons in A⋅ and B⋅ initially must have antiparallel spins. If these two radicals come together again, they can re-form the A-B

METHANE is a simple molecule consisting of a carbon atom bonded to four hydrogen atoms (*left*). The methyl free radical (*center*) lacks one hydrogen. A chemical bond consists of two paired electrons; the methyl radical has an odd electron (*dot*) and so it is highly reactive. Two methyl radicals can combine to form an ethane molecule (*right*).

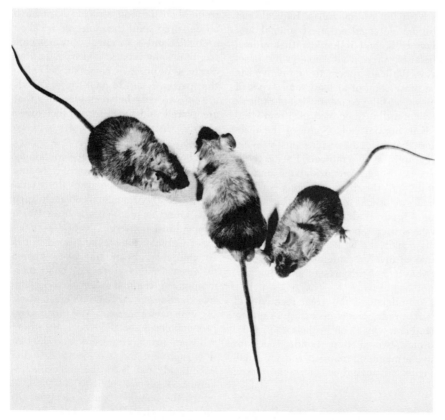

FREE RADICALS appear to be implicated in the process of aging and in damage to tissues from radiation. The connection between radiation and aging is demonstrated in these photographs, made by Howard J. Curtis of the Brookhaven National Laboratory, of two groups of 14-month-old mice. Originally there were nine mice in each group. One group received a large but nonlethal dose of radiation as young adults. The untreated mice are healthy (*top*). Of the irradiated mice only three survive (*bottom*) and they are senile and gray.

bond. This bond-making process is extremely fast for almost all radicals, so that radicals exist only in very low concentrations. As soon as their concentrations build up, A· and B· collide with each other more frequently, and stable A-B molecules are produced.

Suppose, however, that the A· and B· radicals have not come from the same A-B molecule. If a random pair of A· and B· radicals collide in solution, the spins of the odd electrons in the two radicals are randomly oriented, and the chance that a pair of radicals that have odd electrons with antiparallel spins will collide will be only one in four. This is the case because the triplet state is three times as probable as the singlet, and three in four of the collisions lead to a triplet multiplicity, which cannot form a bond.

There are two possibilities for these triplet pairs: either one of the electrons could "flip" its spin to the other direction, converting the triplet to a singlet pair so that the A-B bond could form, or diffusion could occur faster than this conversion, and the two radicals would simply separate without reacting. Experimentally it is found that triplet radical pairs usually diffuse apart and do not combine to form a bond. Diffusive separation of a pair of radicals occurs in solvents of ordinary viscosity in a time period of only about 10^{-10} second; apparently the triplet-to-singlet conversion requires more time than that. In summary, an A-B molecule will be re-formed every time A· and B· radicals with antiparallel spins encounter each other in solution. However, A· and B· radicals that do not have correlated spins will re-form A-B only one time in four; three in four of the encounters form triplet pairs of radicals, which cannot form bonds.

Since radicals are extremely reactive substances that normally exist only in very dilute solutions, it is not surprising that most practical radical reactions are chain reactions. Radicals, formed in an initiation phase, react in a cyclic propagation sequence in which a product molecule is produced at the same time that another radical is formed to carry on the chain. The cycle is ended by a termination phase, in which radicals recombine. For example, the reaction of a hydrocarbon with chlorine has a chain length of 1,000 or more: every primary radical produced eventually leads to the formation of 1,000 or more molecules of product.

The chain reaction can be initiated by irradiation with light or X rays, by simply heating the system to a high temperature or by the use of an "initiator," a compound with an unusually weak bond that breaks to form radicals at a convenient temperature. The cracking of petroleum is initiated by the scission of one of the carbon-carbon bonds in the petroleum hydrocarbon molecules themselves. A simple hydrocarbon such as ethane has a carbon-carbon bond with a bond strength of 85 kilocalories per mole. The rate of breaking of such a bond becomes appreciable only at temperatures near 700 degrees C. The compounds called peroxides, on the other hand, contain the oxygen-oxygen bond, which is unusually weak. For example, the bond-dissociation energy of the central bond in hydrogen peroxide, HO—OH, is 48 kilocalories per mole. Different organic peroxides dissociate at temperatures ranging from 50 to 200 degrees, so that one can generally find a commercially available peroxide initiator that decomposes to produce radicals at a convenient rate at any desired temperature.

The decomposition of an initiator is not always as simple a process in solution as it is in the gas phase. In the gas phase the A· and B· fragments fly apart and each A-B molecule yields two radicals. In solution, however, it is often observed that each initiator does not produce the two free-radical fragments expected. The reason is that when an initiator undergoes bond scission in solution, the A· and B· fragments are held together very briefly by the "cage" of surrounding solvent molecules. The two fragments strike these solvent molecules as they try to separate and are reflected back toward each other. Consider the decomposition of an azo compound of the type R—N=N—R, which decomposes by splitting out a nitrogen molecule, N≡N, and forming two R· radicals. Since these two radicals have opposed spins and can immediately couple, the formation of the stable molecule R—R in the cage can compete with the diffusion apart of the two R· radicals. This reduces the number of free R· radicals that can initiate chemical reactions, and azo initiators therefore range from about 50 to 100 percent in efficiency.

One of the striking features of radical reactions is that only two general types of propagation reaction are commonly observed: atom abstraction and addition. In the abstraction reaction a free radical attacks another molecule to pull off an atom with one valence electron, usually a hydrogen atom. This reaction can be symbolized as M· + RH → MH + R·, where M· is any free radical and

INITIATION

a   Cl—Cl $\xrightarrow{\text{LIGHT}}$ 2 Cl·

b   INITIATOR $\xrightarrow{\text{HEAT}}$ R·

R· + Cl$_2$ $\longrightarrow$ R—Cl + Cl·

PROPAGATION

c   Cl· + CH$_4$ $\longrightarrow$ HCl + $\overset{\bullet}{C}$H$_3$

d $\overset{\bullet}{C}$H$_3$ + Cl$_2$ $\longrightarrow$ Cl—CH$_3$ + Cl·

SUM CH$_4$ + Cl$_2$ $\longrightarrow$ Cl—CH$_3$ + HCl

TERMINATION

2 Cl· $\longrightarrow$ Cl$_2$

Cl· + CH$_3$· $\longrightarrow$ Cl—CH$_3$

2CH$_3$· $\longrightarrow$ CH$_3$—CH$_3$

SIMPLE CHAIN PROCESS, the chlorination of methane, begins with an initiation step in which chlorine atoms (free radicals) are produced (a) by light, which dissociates chlorine molecules, or (b) by other free radicals (R·) provided by the decomposition of a chemical initiator. Two propagation steps (c, d) make up the chain sequence, during which the number of radicals is conserved; the steps can be summed to give the overall chemical change. The chain reaction can be terminated by the coupling of any two radicals. A propagation reaction such as c or d, involving the transfer of an atom from a molecule to a radical, is an atom abstraction, one of the two common types of free-radical propagation reactions.

RH is any hydrogen-containing molecule. Notice that in this reaction, as in all propagation reactions of radicals, the number of radicals is not reduced; one radical is used and another is made.

In the addition reaction a radical adds to a material that contains a double bond: M· + CH$_2$=CHR → M—CH$_2$—CHR. Notice that two of the electrons in the double bond unpair; one joins with the odd electron of the free radical to form a new bond and one becomes localized on the adjacent carbon atom to form a new radical center. Materials such as polyethylene, polyvinyl chloride and polystyrene are produced by processes that involve this reaction. A monomer

such as ethylene, vinyl chloride or styrene is mixed with an initiator and the mixture is heated to a temperature at which the initiator decomposes to form free radicals [see illustration below].

A dramatic advance in the study of radical reactions was made in 1945 with the invention of an instrument that detects and identifies radicals by the magnetic properties of their odd electron. In this technique, called electron-spin resonance, or ESR, a strong magnetic field is applied to the sample and the energy absorption is measured when the odd electrons flip their spins from being aligned in the same direction as the field to being aligned in the opposite direction. The

INITIATION

CH$_2$=CH$_2$ $\xrightarrow{\text{HOT AIR AND HIGH PRESSURE}}$ ROO·

ROO· + CH$_2$=CH$_2$ $\longrightarrow$ ROO—CH$_2$—$\overset{\bullet}{C}$H$_2$

PROPAGATION

ROO—(CH$_2$—CH$_2$)$_n$—CH$_2$—$\overset{\bullet}{C}$H$_2$ + CH$_2$=CH$_2$ $\longrightarrow$

ROO—(CH$_2$—CH$_2$)$_n$—CH$_2$—CH$_2$—CH$_2$—$\overset{\bullet}{C}$H$_2$

TERMINATION

ROO—(CH$_2$—CH$_2$)$_n$—CH$_2$—$\overset{\bullet}{C}$H$_2$ + ROO· $\longrightarrow$

ROO—(CH$_2$—CH$_2$)$_n$—CH$_2$—CH$_2$—OOR

ADDITION REACTION is another common free-radical propagation reaction. In it a radical adds to a compound containing a double bond. For example, polyethylene is made from ethylene by using air as an initiator. The air oxidizes the ethylene to produce peroxide radicals (ROO·), which initiate the polymerization (molecular chain-building) process.

amplitude of the resulting energy peaks and the field strengths at which they occur give information on the concentration of radicals and their nature. Often, particularly if the radicals can be studied either in the liquid phase or in a single crystal, the detailed structure of the radical can be identified from its ESR spectrum. This technique has had great impact on the study of radicals in biochemical systems and in tissues, since the radicals can be studied even when the chemistry of the system is incompletely understood.

A living organism is a machine that requires vast amounts of energy for the chemical and physical work it must perform. Organisms obtain their energy by the oxidation of biological materials—by burning food as fuel. The burning does not proceed in the random and inefficient way it would in a furnace; rather, enzymes act as catalysts and the oxidations take place in a controlled sequence of small steps in which more nearly the maximum obtainable energy is liberated.

Oxidation is defined as the loss of electrons; the reverse process, reduction, is the gaining of electrons. Oxidation can proceed through loss of electrons from a substance either in pairs or one at a time. For example, a reaction in which a substance is oxidized by the overall loss of two electrons to form a stable product might occur in one step, in which both electrons are transferred, or in two steps, by the transfer of one electron at a time with a free radical as a transient intermediate. If this intermediate were very unstable, it would rapidly lose the second electron to form the ultimate product and would exist only at extremely low concentrations and be very difficult or even impossible to detect. The difference between oxidations that involve radicals and proceed one electron at a time and those that proceed in two-electron steps is therefore not always obvious just from the products that are formed.

In the early 1930's Leonor Michaelis of the Rockefeller Institute initiated a series of investigations to prove that biological oxidations might involve free radicals. In 1946 he published his highly provocative statement that "all oxidations of organic molecules, although they are bivalent, proceed in two successive univalent [one-electron] steps, the intermediate being a free radical." We now know that this theory is incorrect: there are two-electron oxidations in biochemistry that proceed pairwise and do not involve radicals as intermediates. Nevertheless, Michaelis' views inspired research that is continuing today, providing insight into the nature of the processes by which living organisms obtain energy. Studies by numerous workers in the 1940's and 1950's indicated that an intermediate could sometimes be identified in enzymatic oxidation-reduction reactions of biological molecules, but in spite of Michaelis' conviction there was no proof that the intermediate was a free radical. Starting in 1954, however, and using the newly developed techniques of ESR, Barry Commoner and George E. Pake of Washington University, Helmut Beinert of the University of Wisconsin, Anders Ehrenberg of the Nobel Institute of Sweden and others were able to show that a paramagnetic intermediate can be detected in some enzyme-substrate systems. Commoner in 1956, Melvin Calvin in 1957 and later other investigators showed that ESR signals are produced in plant systems during photosynthesis.

Early ESR instruments were insensitive. They could detect radicals only in biological materials that had been freeze-dried, killing most living systems, and so it was difficult to correlate concentrations of radicals with biological activity. The possibility existed that the radicals being detected were artifacts not directly involved in the biochemical reactions under investigation. Because water absorbs microwave energy very near the frequency used for most commercial ESR instruments, it was particularly difficult to study biological samples in aqueous solution in their natural state.

Around 1957 various workers developed new ESR methods that made it possible to study aqueous samples, and as the ESR instruments have improved so has the observed correlation between radical concentration and biological activity. It now appears that radicals are important intermediates in a number of biological processes. In some systems a correlation between biological activity and the concentration of radicals can be established, in others the structure of the radicals can be identified and in still others the rate of reaction of the radicals can be followed. To date, however, all three of these things have been done for very few systems. Two areas in which ESR techniques have been successfully applied are enzymatic oxidations and the mechanisms by which radiation damages organic materials.

Enzymatic oxidations usually proceed through the removal of electrons from the substrate by an enzyme and their transfer to a coenzyme. These are steps in an elaborate electron-transport system that carries the electrons from the substrate to oxygen and effects the reduction of oxygen to water, and they occur in the cigar-shaped organelles of the cell called mitochondria. It is probable that all living cells contain some radicals, but ESR signals can be detected only in certain cells. In general stronger signals are detected in cells with high concentrations of mitochondria. The greatest success in correlating ESR signals with biological activity has come from studies of the flavin coenzymes and coenzyme Q. Several workers have shown that relatively stable radicals are formed in these enzyme systems and that radicals are directly involved as intermediates in the oxidations.

Although most enzymes do not transfer electrons from the substrate directly to oxygen, the oxidases and some of the flavins are able to do so. They generally

PEROXIDE	STRUCTURE	TEMPERATURE (DEGREES CELSIUS)
t-BUTYL HYDROPEROXIDE	$(CH_3)_3CO-OH$	172
CUMENE HYDROPEROXIDE	$C_6H_5C(CH_3)_2O-OH$	158
t-BUTYL PEROXIDE	$(CH_3)_3CO-OC(CH_3)_3$	127
t-BUTYL PERBENZOATE	$C_6H_5CO_2-OC(CH_3)_3$	105
t-BUTYL PERACETATE	$CH_3CO_2-OC(CH_3)_3$	100
BENZOYL PEROXIDE	$C_6H_5CO_2-O_2CC_6H_5$	76
ACETYL PEROXIDE	$CH_3CO_2-O_2CCH_3$	67
t-BUTYL PHENYLPERACETATE	$C_6H_5CH_2CO_2-OC(CH_3)_3$	66
t-BUTYL TRIPHENYLPERACETATE	$(C_6H_5)_3CCO_2-OC(CH_3)_3$	11

INITIATORS can be selected from among a number of organic compounds that dissociate to produce free radicals at a wide range of temperatures. Most such compounds are peroxides, which contain the readily broken oxygen-oxygen bond. The table lists some peroxide initiators and gives the temperature at which each of them has a half-life of 10 hours.

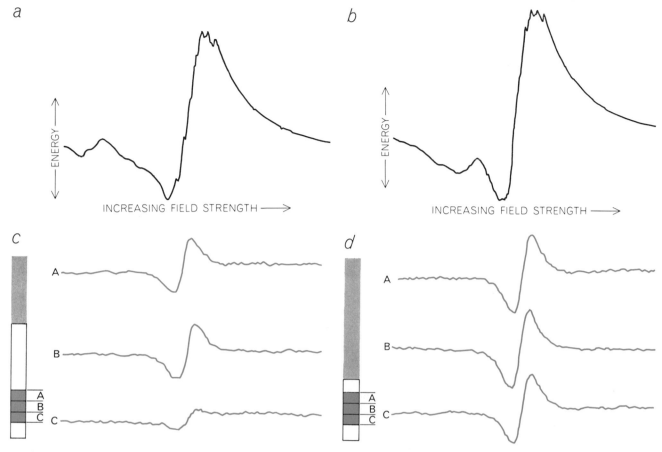

ELECTRON-SPIN RESONANCE (ESR) detects unpaired electrons and thus measures the number and properties of free radicals in a sample placed in a magnetic field. An ESR curve traces the changes in energy associated with the flipping of electron spins as the magnetic field varies; the amplitude of a peak and its location are characteristic of the concentrations and the chemical sites of free radicals. These ESR spectra from the Japan Electron Optics Company are those of the tobacco in an unsmoked cigarette (a) and in a smoked butt (b) and of various levels (A, B and C) of the filters of half-smoked (c) and nine-tenths-smoked (d) cigarettes.

reduce oxygen to hydrogen peroxide, which is subsequently reduced to water by another class of enzymes called catalases or peroxidases. Hydrogen peroxide is produced in all cells through the reduction of oxygen, and the peroxidase enzymes are important in keeping peroxide concentrations down to levels that are not damaging to the cell. Lawrence H. Piette of the University of Hawaii and others have shown that peroxidase-substrate systems give ESR signals, and the activity of the enzyme has been correlated with the strength of the signal.

The processes by which radiation interacts with organic materials are quite complex, and we shall consider only the simplest scheme here. Ionizing radiation can consist of electromagnetic radiation such as very-short-wavelength light or X rays, or of highly energetic particles such as high-speed electrons or alpha particles. Radiation causes ionization by knocking electrons away from the molecules to which they belong, thus producing free electrons and positively charged ions, or cation radicals. The ions and electrons may recombine to produce neutral molecules in an excited state, which can undergo homolysis to produce free radicals; radicals can also be produced by reactions between the cation radicals and the neutral molecules.

Cells are from 60 to 80 percent water and when a plant or an animal is irradiated, most of the energy is deposited in the aqueous phase; less often will a primary ionization occur in an organic biomolecule. A portion of the damage to living systems therefore results from reactive particles that are formed in the water phase and diffuse to an organic molecule in the cell, causing secondary reactions. The chief radical species produced in the radiolysis of water and implicated in radiation damage are solvated electrons (electrons associated with water molecules), hydrogen atoms and hydroxyl radicals (HO·).

Cells are extremely sensitive to radiation. Calculations show that radiation that destroys perhaps only one molecule in 100 million can have profound biological consequences and can even kill the cell. The most reasonable explanation of this biological magnification is that the energy from radiation can be transferred to critical polymer molecules in the nucleus of the cell. Aside from certain effects on cellular membranes, which will be discussed later, it is probable that the cell nucleus and its chromosomes are critical in determining radiation sensitivity. About half of the dry mass of the nucleus consists of chromosomes and of this about half is deoxyribonucleic acid (DNA) and about half is associated protein material. Two types of reactions are therefore particularly important in terms of the damage they can cause. The first is the reaction of radicals from water with DNA and the second is their reaction with proteins.

Each strand of the double helix of DNA is a chain of subunits called nucleotides, arranged in an ordered sequence to spell out genetic instructions according to the genetic code. The distinctive element of each nucleotide is one of four nitrogenous bases: thymine,

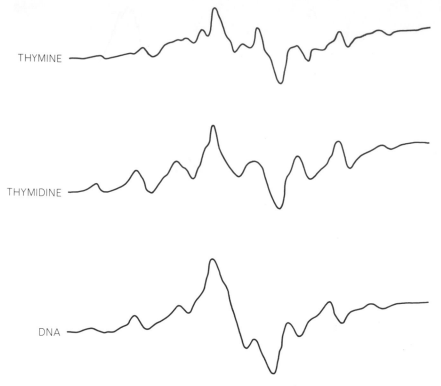

THYMINE

THYMIDINE

DNA

THYMINE, one of the four bases in DNA, is sensitive to radiation, probably at least in part because hydrogen atoms add to thymine to form radicals, which can react to alter the DNA. The top ESR spectrum, from an experiment conducted by Anders Ehrenberg and his colleagues, shows the pattern typical of the radical, produced when thymine was subjected to gamma radiation. The other two spectra show that the same radical is present in thymidine (a part of the DNA molecule that contains thymine) and in a purified DNA sample.

adenine, guanine and cytosine. All three of the radicals produced in water radiolysis react with these bases at a high rate. For example, hydrogen atoms add to thymine to produce a radical that can be identified by its characteristic ESR spectrum. The same spectrum can be observed when hydrogen atoms react with the thymine nucleotide, thymidylic acid, or when either solid DNA or an aqueous solution of DNA is irradiated [*see illustration above*]. The thymine radicals are not stable and they react further to alter the DNA molecule and interfere with the coding of genetic information. Such a change could be nonlethal, since the cell has mechanisms for excising and repairing damage to its DNA, but it would be expected to often alter a gene in such a way that it would kill the cell.

Dov Elad and his co-workers at the Weizmann Institute of Science have re-

cently shown that certain organic compounds can add to the bases in DNA. These reactions proceed in the same way whether they are induced by light, high-energy radiation or normal free-radical initiators, and they unquestionably involve free radicals as intermediates.

Another reaction that may cause mutations has been identified by H. J. Rhaese of the National Institutes of Health. He has shown that hydrogen peroxide reacts with adenine and modifies its structure. This reaction occurs when adenine is irradiated with X rays or when it is simply treated with hydrogen peroxide and ferric ions, a mixture known to produce hydroxyl radicals. Although it has not yet been established that this slight modification of adenine causes mutations in organisms, there is some evidence that it may contribute to the weak mutagenic effect of X rays. For example, there appears to be a higher incidence of chromosome aberrations in cells with low concentrations of catalase and consequently with higher than normal concentrations of hydrogen peroxide.

Proteins, which play an important role in the chemistry of all plant and animal cells, are made up of many amino acid molecules joined together in a peptide chain. Proteins are held in a particular configuration by molecular forces, and the shape of each protein is critical to its exact functioning. One amino acid, cysteine, contains a sulfur-hydrogen group, and two of these groups can be converted to produce a disulfide bond between two cysteine residues in the protein chain. This S—S link helps to hold the protein molecules in their active configuration. (A macroscopic example of this is the permanent-waving of hair. Hair is made of the protein keratin, which contains disulfide cross-links that help the hair to hold its shape. In permanent-waving these bonds are first reduced to S—H groups; then the molecules are arranged in the desired conformation by rolling the hair on rods and the S—S bonds are then remade by oxidation with peroxide.)

In 1955 Walter Gordy of Duke University examined the ESR spectra of irradiated proteins and proposed that two types of radical are produced. One gives a two-peak spectrum that varies slightly from one protein to another. It is now known that this spectrum results from an odd electron localized on a carbon atom of the peptide-chain backbone. The second type of radical gives an ESR absorption at the low-field end of the spectrum and is easy to identify

SPERM HEADS   SPERM HEADS PLUS MEA

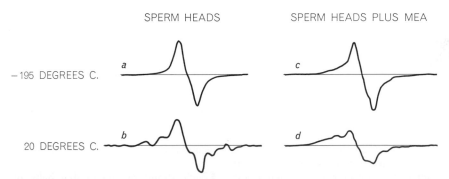

−195 DEGREES C.

20 DEGREES C.

EFFECTS of temperature and of a sulfur-containing drug on radical formation were demonstrated by Peter Alexander. Fish-sperm heads, which are rich in DNA, were cooled to —195 degrees Celsius and irradiated. The frozen sample shows a typical low-temperature spectrum (*a*). When the sample is warmed, the DNA spectrum appears as the radicals become mobile (*b*). Addition of MEA, containing thiol (S—H) groups, has little effect on the frozen sample (*c*). When the sample is warmed, the radicals transfer from DNA to thiols and a new plateau characteristic of sulfur radicals appears at the left in the spectrum (*d*).

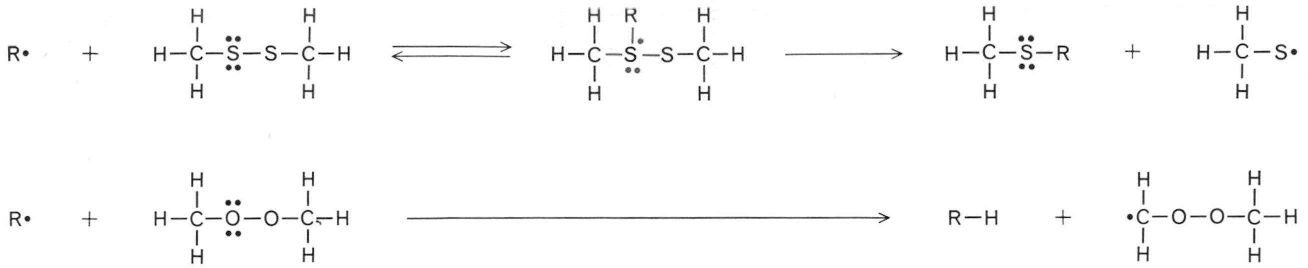

SULFUR COMPOUNDS protect against radiation because they react readily with radicals, which attack the sulfur-sulfur bond of
disulfides faster than they attack the oxygen-oxygen bond of peroxides, according to the author and his colleagues. It may be that
an attack by a radical on methyl disulfide, for example (*top*), can
form a relatively stable intermediate because the sulfur atom can
accommodate nine electrons (*color*). An attack by a radical on
methyl peroxide (*bottom*), on the other hand, cannot involve this
path, since the oxygen atom can only accommodate eight electrons.
As a result the reaction proceeds by hydrogen abstraction instead.

as a sulfur radical, RS·; it is observed in
enzymes that contain S—H or disulfide
groups and in mixtures of enzymes with
added sulfur compounds. Thormod Hen-
riksen in Norway, Peter Alexander and
M. G. Ormerod in England and Harold
C. Box and several other workers in this
country have shown that irradiation of
proteins at low temperatures produces
ESR signals of the carbon-radical type,
but that heating the material gradually
converts at least some of these carbon
radicals to sulfur radicals. Clearly protein molecules possess a mechanism for
transferring the site of the destruction
from one part of the molecule to another
or to nearby molecules.

Although the sulfur-sulfur bond is critical in determining the shape of
many proteins, it is cleaved remarkably
easily. My research group at Louisiana
State University has studied the cleavage
of the disulfide bond by radicals and we
believe it may be faster than analogous
reactions of oxygen compounds because
of the ability of sulfur to react with radicals to form a relatively stable intermediate [*see illustration above*]. There is evidence from other laboratories that some
of the reactions of ions with sulfur-sulfur
bonds may also proceed through a related intermediate complex.

Recently several workers in this country, Gabriel Stein in Israel and H. Jung
in Germany have shown that the hydrogen atom attacks the sulfur-sulfur bond
so efficiently that each hydrogen atom
produced in a model system leads to the
inactivation of an entire enzyme molecule. Since enzymes are among the
largest molecules known and the hydrogen atom is the smallest atom, this is a
bit like killing an elephant with a BB
pellet. When a cell is irradiated, very
few of the hydrogen atoms from the
aqueous phase are likely to collide with
an enzyme molecule, but apparently
those that do can cause the total destruction of the biological activity of the
enzyme.

Since sulfur groups react readily with
radicals, it is not surprising that sulfur
compounds act as drugs that protect
against radiation. There are a number of
mechanisms by which a molecule might
protect a cell from radiation damage.
For example, compounds containing
S—H groups (thiols) can protect important biological molecules through a repair process. A number of workers have
shown that radiation can result in the
removal of a hydrogen atom from a biological polymer, leaving a radical. The
thiol can then repair this damage by a
hydrogen-transfer reaction, donating a

hydrogen atom to the polymer radical
and creating a less lethal thiyl radical,
R—S·. These reactions are thought to be
partly responsible for the significant radiation-protection activity of thiols. Most
compounds containing S—H or S—S
bonds act as protective agents in laboratory systems, but in the body problems
of solubility, diffusion and toxicity arise
and so only certain sulfur compounds
are effective radiation-protection drugs.

It is interesting that the potent radiation-protection drug beta-mercapto-
ethylamine (MEA), or cysteamine, also
has been found by Harman to be effective in lengthening the mean life-span of
mice. There are radicals in the body and
it is clear that they can damage biochemical systems in cells; radiation involves
radical reactions and also effects aging.
These facts have suggested to many investigators that aging itself must be at
least partly due to damage caused by
radical reactions within the body.

In an organism such as man, aging can
be expected to result from many chemical reactions, and there is no reason to
anticipate a single cause or mechanism.
On the contrary, it is likely that several
different and perhaps complex mechanisms make significant contributions to
the total changes that occur in our bodies

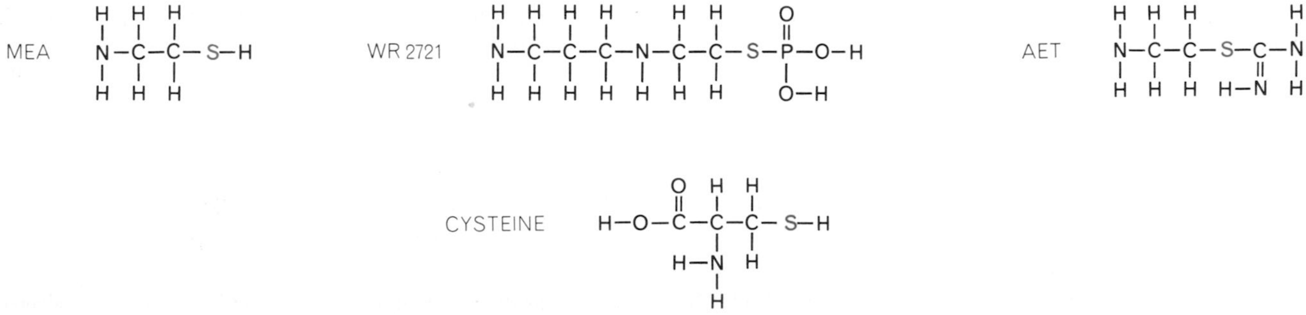

PROTECTION AGAINST RADIATION is afforded by a number of compounds, most of which contain sulfur. The structures of
four of the most effective sulfur-containing drugs are shown here:
MEA, or cysteamine, WR 2721, AET and the amino acid cysteine.

COLLAGEN, a protein connective tissue, becomes less flexible with age because its fibers become cross-linked. In the speculative scheme shown here the cross-linking is caused by free radicals. Carbon atoms, which for the sake of simplicity are shown here with two hydrogen atoms (*left*), are attacked by radicals; they lose a hydrogen atom, giving rise to radical centers on the peptide chain. Pairs of radical centers could couple with each other, producing cross-links that bind the fibers to one another, stiffening them.

with time. There is increasing evidence that at least some of these mechanisms involve the reactions of free radicals.

One theory suggests that aging may be partly due to changes in the connective tissues collagen, elastin and reticulin. These are the structural materials of the body, largely protein in composition, that give tissues shape, plasticity, resilience and elasticity. The biological role of collagen depends on its high plasticity and its ability to bear stress and maintain shape and form; it is present throughout the body but it occurs in particularly high concentrations in flexible organs such as the lungs, blood vessels, skin and muscles. With age, collagen fibers become denser, stiffer, thicker and less plastic. They also become insoluble in organic solvents, indicating that they have become cross-linked by chemical bonds. It is not unreasonable to suggest that some of this cross-linking could result from free-radical reactions.

Most collagen occurs in organs with high concentrations of blood serum, in which radicals are known to be present, and ESR studies show that free radicals attack proteins, producing radical centers on the protein chains. It might be expected that the subsequent combination of these protein radicals could sometimes lead to cross-linking between the protein collagen fibers, and thus to stiffness and increased density [see *illustration on this page*]. It may not be the free radicals themselves that cause cross-linking but rather some product of their reactions. It is known that under most conditions radiation does not cross-link collagen but rather stimulates collagen synthesis at an increased rate, producing some of the same symptoms as cross-linking. This is one of the ways in which radiation mimics the effects of aging, as William F. Forbes of the University of

Waterloo in Canada has pointed out, rather than truly accelerating natural aging.

A theory favored by Howard J. Curtis of the Brookhaven National Laboratory and others suggests that aging in mammals results from mutations in the animals' somatic cells (the body cells, as distinguished from the germ cells involved in reproduction). The theory is conceptually quite simple: DNA directs the synthesis of ribonucleic acid (RNA), which in turn directs the synthesis of all the proteins produced by the cell. If errors gradually accumulate in the genetic information of the DNA or RNA, the cell begins to malfunction and may die. As in other aging theories, the concept is simple but the proof is not. The DNA molecule is enormously complex and there is no direct chemical evidence that it changes with time. Furthermore, it is not possible to measure mutations in somatic cells directly and indirect techniques must be found. In support of the mutation theory, it has been shown that certain radiation effects on chromosome aberrations and on life expectancy are similar, that short-lived animals develop chromosome aberrations faster than long-lived ones and that mouse strains with a longer lifespan also have a lower sensitivity to radiation. There remain several serious

difficulties with this theory, however, and it is not universally accepted.

A theory that aging is due at least in part to the peroxidation of lipids explicitly implicates radical reactions. The lipids, which include the fats, constitute the most concentrated source of energy available to the organism. They are oxidized in the cell, chiefly in the mitochondria, in a series of reactions that normally proceed in carefully controlled enzyme-regulated steps. Like all organic materials, however, lipids can also react with oxygen in a nonenzymatic, free-radical pathway. Those that contain reactive hydrogen atoms called allylic hydrogens are particularly prone to undergo the radical reaction. This type of hydrogen occurs in the polyunsaturated fatty acids, which account for about 13 percent of the caloric intake of the human diet.

The peroxidation of lipids is a typical free-radical chain process involving both hydrogen transfer reactions and addition reactions [see *illustration below*]. The importance of lipid peroxidation to aging rests on the belief that damage to cells accumulated over the lifetime of the organism gradually reduces the efficiency with which the cell carries out its functions. For example, A. L. Tappel of the University of California at Davis has shown that when enzymes are present in systems in which fatty acids are

ABSTRACTION	$P{-}H$	$\xrightarrow{ROO\cdot}$	$P\cdot$		
ABSTRACTION	$P{-}H$ + O_2	$\longrightarrow$	$POO\cdot$		
ABSTRACTION	$P{-}H$ + $POO\cdot$	$\longrightarrow$	$P\cdot$ +	$POOH$

PEROXIDATION OF LIPIDS involves both abstraction and addition. A lipid polymer ($P{-}H$) first loses a hydrogen atom through abstraction by a peroxide radical, forming a lipid radical ($P\cdot$). Then the lipid adds a molecule of oxygen to form a peroxidized radical ($POO\cdot$), which in turn abstracts a hydrogen, leaving a radical ($P\cdot$) to carry on the chain.

being oxidized, the biological activity of the enzymes is destroyed. The oxygen does not react with the enzymes; rather the enzymes are attacked either by radicals produced through the interaction of the lipids and oxygen or by nonradical molecular products from the oxidation of the fats, or by both. Furthermore, there are numerous similarities between the damage to enzymes produced by lipid peroxidation and the damage from radiation.

The body provides a natural lipid antioxidant in vitamin E, and the products derived from vitamin E show that it functions at least partly as a free-radical inhibitor that is sacrificially oxidized to protect the lipids. It is striking that the effects of diets deficient in vitamin E resemble certain effects of both radiation damage and aging. In all three cases there is evidence for structural damage to various cellular membranes.

The membranes in a cell are the partitions that compartmentalize reaction systems so that they do not interfere with one another. Membranes also appear to contain many of the specific receptor sites for the binding of hormones and drugs. Moreover, they must be permeable to particular chemical species at precisely defined rates. They are therefore highly structured and extremely sensitive components, and any deterioration of a cell's membranes must seriously affect its ability to function and could be responsible for some of the effects of cellular aging. Membranes consist of lipids and proteins in varying proportions; the membranes of mitochondria, for example, are about 27 percent lipid and 73 percent protein. Research in a number of laboratories makes it clear that both radiation and free radicals produce significant structural deterioration in membranes and that peroxidation of lipids leads to products that cause cross-linking, reduced permeability and structural decay in membranes.

A particularly clear case of free-radical damage to membranes has been demonstrated for human red blood cells by Edward M. Kosower of the State University of New York at Stony Brook and his wife Nechama S. Kosower of the Albert Einstein College of Medicine. They used a drug that reacts with glutathione, a thiol in blood cells, inhibiting its protective action. The same drug also produces an intermediate that, under the controlled conditions of their experiments, reacts with oxygen in the cell membrane to form radicals; the radicals cause lipid autoxidation, which destroys the viability of the cell wall and bursts

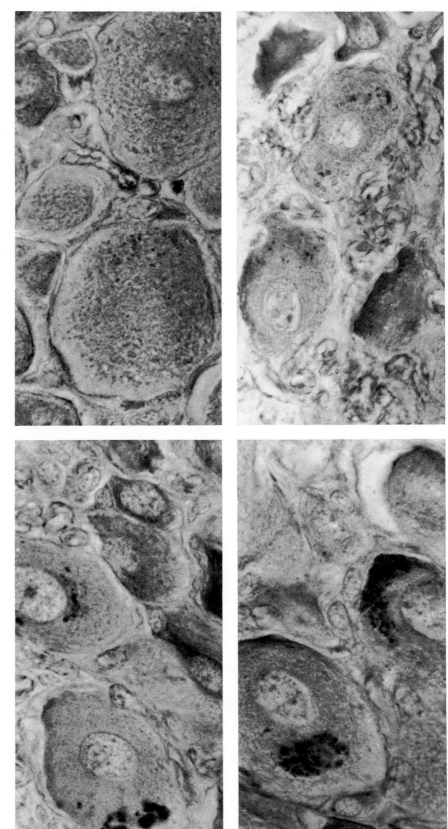

AGE PIGMENTS accumulate in cells with age and there is evidence that they are produced in part by radical reactions. Photomicrographs made by Thaddeus Samorajski and his colleagues at the Cleveland Psychiatric Institute compare nerve cells from the dorsal root ganglia of mice four (*top left*), eight (*top right*), 20 (*bottom left*) and 30 (*bottom right*) months old. The increase in size and concentration of the dark granules is clearly seen.

the cell. This hemolysis occurs even when only a small fraction of the lipid in the cell wall is damaged.

A link between lipid peroxidation and aging is found in the chemistry of the age pigments. These materials, a kind of metabolic debris of the cell, are fluorescent brown compounds that accumulate slowly in cells that are not regularly replaced, such as the nondividing cells in the heart, nervous system and lungs. Age pigments are about 60 percent protein, 25 percent lipid and 15 percent carbohydrate. Several observations indicate that radical reactions are probably important in the production of age pigments. The lipid fragments in them appear to be peroxidized, their proteins are cross-linked in ways suggestive of the cross-links found in proteins exposed to peroxidizing lipids and Tappel has shown that lipid peroxidation of some cellular organelles gives materials that have the characteristic fluorescent spectra of the age pigments.

The color-bearing components of these age pigments appear in many cases to belong to the melanin class of compounds, which are also responsible for the color of the pigments of skin, eyes and hair. Melanins consist of large polymeric networks that have been known for more than 15 years to contain free-radical centers that can give rise to ESR signals. Donald C. Borg of Brookhaven has shown recently that some age pigments from various organs produce the type of ESR signals characteristic of melanins, indicating that age pigments also contain free radicals of the melanin type.

Bernard L. Strehler of the University of Southern California and others have shown that age pigments increase almost linearly with age in human heart tissue, for example, and it is clear that these materials are related to aging. Whether they play an important role in the process or are merely harmless by-products is unknown. The occurrence of peroxidized lipid fragments in these age pigments does prove, however, that lipids are attacked by radicals and are peroxidized in living systems. Furthermore, the fact that lipids are attacked by peroxidizing radicals confirms that reactive free radicals actually are present in living tissue, an assumption that underlies many of the free-radical theories of aging. In this area, as in so many others, it is apparent that the study of radical reactions will continue to provide important new lines of research for chemists and biologists.

34

BERYLLIUM AND BERYLLIOSIS

JACK SCHUBERT

August 1958

*Extracted from beryl, a crystal which is almost
identical with emerald, this light metal has found
glamorous applications in modern technology.
It is also capable of causing a serious disease*

Injury and death to hundreds of American workers during the 1940s might have been averted if the prescient advice given a decade earlier by an Italian physician had been heeded. On the basis of animal experiments and known human cases, Stefano Fabroni gave the name berylliosis to an illness which followed soon after the inhalation of dust and fumes of irritating salts of the relatively obscure metal beryllium. He wrote: "In view of the important practical applications of beryllium in industry . . . which all indicate that the importance of this metal will continue to grow, we feel it is of interest now to call the attention of scientists to the picture of berylliosis . . . so that there will be the chance of timely development of measures to prevent this grave malady of the lungs from being added to the catalogue of industrial diseases."

The story of berylliosis is one of the most fascinating, contradictory, infuriating and controversial episodes in medical history. Some medical people argue even now that beryllium is incapable of causing disease. One writer in *The Lancet*, the esteemed British medical journal, invoked sheer sentimentality: "Beryllium seems to be the Admirable Crichton of metals. . . . To charge such an admirable metal with having poisonous properties is about as distasteful as accusing a trusted butler of stealing the family plate."

When one examines the clinical, biochemical and toxicological evidence, however, one cannot escape the fact that

beryllium has caused at least 500 cases of poisoning in the U. S. alone during the past two decades. Particles of the metal or its compounds, wherever they are deposited, but especially in the lungs and skin, do extensive local damage to the tissues; reduction in respiratory capacity and general systemic debility have terminated in death in 10 to 30 per cent of all the cases.

Beryllium performs heroic functions in hundreds of everyday products even though its total annual production is measured in pounds. A little beryllium goes a long way, especially in the alloys which account for most of its consumption. It is a glamorous newcomer to metallurgy, one of a number of rare metals now brought into·wide use by virtue of highly specific properties which technology has only recently learned to appreciate.

Beryllium has special glamour because it is the element that distinguishes the crystal structure of the precious emerald. The discovery of beryllium in 1798 stemmed from the observation by the French mineralogist René Just Haüy that the optical properties of the emerald were identical with those of the more common mineral beryl. He asked Louis Nicolas Vauquelin to make a chemical analysis. Vauquelin proved that both substances had the same composition and contained a new element. Since the salts of beryllium are sweet-tasting, the new element was first called glucinium. The name beryllium was given by Fried-

rich Wöhler in Germany in 1828, when he succeeded in isolating the pure metal. With an atomic weight of nine, it is one of the lightest elements.

The emerald is a transparent, intensely green variety of beryl. The hexagonal crystals of beryl itself may be greenish, bluish or rosy, but in the field they tend to assume the color of the granitic rocks with which they are generally associated. Occasionally they attain gigantic size, up to two or three feet in diameter and several feet in length, weighing several tons. Some 60-ton crystals have been found.

Beryl is the single industrially significant beryllium-bearing mineral. The small size of the beryllium ion favored its diffusion as a minor constituent in numerous minerals, and discouraged the formation of high-concentration beryllium minerals. The world resources of beryl are estimated at four million tons, the principal deposits being in Brazil. Miners have to move, on the average, about 100 tons of rock in order to hand-pick half a ton of beryl crystals. A ton of beryl, which sells for $400 to $500, yields about 70 pounds of beryllium, worth about $4,200. (Emerald is worth some $25 billion per ton!) For better or worse, the production of beryllium will always be measured in pounds.

As a light element, beryllium should be a great deal more abundant on earth. It appears, however, that the beryllium nucleus is easily destroyed by collision with high-energy protons, as in the sun and other stars. In the cold reaches of

BERYL closely resembles emerald and aquamarine. The large crystal is beryl. At left in foreground is limestone containing emeralds. Second from left is an aquamarine. Third and fourth are emeralds. Specimens are in the American Museum of Natural History.

interstellar space, where such collisions are rare, beryllium occurs in much higher relative concentration. The scarcity of beryllium on earth, then, indicates that our planet passed through a stage of stellar temperature. Curiously the cosmic ray bombardment of the earth's upper atmosphere results in the transformation of nitrogen and oxygen into beryllium 7 and beryllium 10, two radioactive isotopes of the element. The latter has a half-life of 2.5 million years; incorporated in snow and ocean sediments, it fills an important gap in geological age measurement.

The principal uses of beryllium stem from the discovery in the 1920s that the addition of only 2 per cent of beryllium to copper forms an alloy six times stronger than copper. Beryllium-copper alloys stand up at high temperatures, have great hardness, show resistance to corrosion, do not spark and are nonmagnetic. One finds them used in the critical moving-parts of aircraft engines and in the key components of precision instruments, mechanical computers, electrical relays, switches and camera shutters. Springs made of these alloys retain their springiness almost indefinitely. Beryllium-copper hammers, wrenches and other tools are employed in petroleum refineries and other plants where a spark from steel against steel might ignite an explosion or a conflagration.

The peculiar virtues of its alloys brought a sudden expansion of beryllium production during World War II. The needs of war machines were so specialized and urgent that it was easier to expand the supply of this rare element than to look for substitutes. Moreover, it happened that beryllium was destined for a special role in the development of the atomic weapons.

When beryllium atoms are bombarded with alpha particles (e.g., from radium), their nuclei disintegrate and yield a profusion of neutrons. This reaction led James Chadwick to the discovery of the neutron itself in the early 1930s, and a radium-beryllium source provided the neutrons for the historic experiments of Enrico Fermi that led to the discovery of uranium fission. Beryllium is one of the most efficient materials for slowing down the speed of neutrons; it is also an excellent neutron reflector. The Manhattan Project overnight created a big new market for beryllium. Incidentally, one of the physicists working in the program is numbered among the first clearly diagnosed cases of berylliosis in the U. S.

Beryllium remains today an important material in nuclear technology. At least one U. S. atomic submarine went to sea powered by a beryllium-moderated reactor, and a reactor in the U.S.S.R. has a beryllium moderator and reflector. Last year the Atomic Energy Commission invited proposals for the supply of about 100,000 pounds of "reactor-grade" beryllium a year for five years. This amounts to roughly a 30-per-cent increase in production.

Presently the makers of advanced aircraft, missiles and of space ships to come are developing ambitious plans for beryllium. The metal has much higher rigidity and heat stability than the other light metals, qualities that commend it for use in airframes and rocket bodies. The oxide has a higher heat conductivity than any known ceramic and looks like a promising material for re-entry cones. Were it not for its toxicity and scarcity beryllium might make the ideal rocket fuel; it packs more energy per unit volume than any other element.

The general public first became acquainted with beryllium as an ingredient in the phosphors that give fluorescent lamp tubes their glow. Until 1949 nearly all fluorescent tubes were coated with beryllium phosphors. It was this use of beryllium that first exposed large numbers of industrial workers and the general public as well to the hazard of berylliosis. Most of the cases have been traced to exposure to the phosphor dust, either in the manufacturing process or in the disposition and incidental destruction of old fluorescent tubes. By 1949 the list of cases had grown to such alarming proportions that the major manufacturers of fluorescent tubes decided in consultation with officials of the U. S. Public Health Service to discontinue the use of beryllium phosphors.

The ill effects of beryllium compounds on workers who handled them were first observed in the U. S. in 1940 by Howard S. Van Ordstrand and his colleagues at the Cleveland Clinic. The workers were employed in processing beryllium ores; one of them died late in 1940. Not long afterward the first cases involving workers exposed to beryllium phosphors were recorded in Pennsylvania. Suspicion of beryllium was allayed at this early stage, however, by medical literature dealing with similar cases in Europe. Beginning in 1933 there had been reports from Germany, Italy and the U.S.S.R. of an occupational disease peculiar to beryllium workers. Because the victims were exposed to fluorine and other acid-forming compounds of the metal, and because the symptoms so closely resembled the effects of wartime poison gases in which fluorine was the active ingredient, the trouble was laid to the corrosive action of acids, not to beryllium. The clinical picture of the Ohio and Pennsylvania cases fitted precisely the symptoms described in the European literature: a first stage of chills, fever and profuse perspiration, then progressive involvement of the lungs, shortness of breath, a painful cough and an X-ray picture suggesting tuberculosis. Accordingly the U. S. physicians adopted the European diagnosis and exculpated beryllium, even though some of the victims, those working with beryllium phosphors, were exposed to beryllium oxide, which forms no acids.

By 1945 Van Ordstrand had reported on 170 cases of illness in beryllium workers. These patients exhibited damage to the skin as well as to the lungs. The findings were typical: 42 of the patients had severe skin ulcers with the most intense reaction occurring on the hands, arms, face and neck. When exposure to beryllium compounds was terminated, the dermatitis subsided and cleared up. Some ulcers persisted. In these cases a minute crystal of beryllium or a beryllium compound was invariably found entrapped within the skin; healing followed removal of the crystal.

In 1946 the hazard that threatened beryllium workers developed an alarming new aspect. Most of the cases reported up to that time had been acute attacks arising during exposure to beryllium compounds. Two physicians associated with the Massachusetts industrial hygiene office, Harriet L. Hardy and Irving R. Tabershaw, now reported on 17 cases of chronic illness arising long after exposure had ceased. All these patients had been engaged in the manufacture of fluorescent lamps in a plant in Massachusetts, and had worked an average of 17 months in the building where the phosphors were compounded. Their illness did not become noticeable until six months to three years after they had quit working in the building and had ceased to have contact with beryllium.

The beryllium compounds involved in these cases had no fluorine or other acid-forming element to divert concern away from beryllium itself. The recipe for phosphors called for pure beryllium oxide, mixed with silica and the oxides of other metals and fused in a furnace at temperatures in excess of 2,000 degrees Fahrenheit. After firing, the rock-like batch of phosphors had to be pulverized for coating the inside of the lamp tubes. The operation was at times a dusty one and exposed the skin and respiratory system of the workers to beryllium compounds in powdered form.

The cases of delayed or chronic illness clearly implicated beryllium itself as the poison. The principal distinction between the various beryllium compounds now appeared to be merely their relative solubility in the body fluids. The acid-forming compounds are the most soluble; in the Cleveland ore-processing plant they produced immediate illness. The lesions in the lungs and skin in these cases were only incidentally complicated by the action of acids. In its oxide, beryllium appears to be somewhat less soluble, and in its silicate phosphor-compounds even less so. These compounds produced the delayed illness among the fluorescent-lamp workers.

The chronic form of berylliosis is distinguished from the acute chiefly by the delay between the exposure to beryllium and the onset of symptoms. The case of the Manhattan Project physicist mentioned earlier typifies the course of the chronic disease. His exposure began in 1938 at Columbia University, where on several occasions he handled finely powdered pure beryllium metal in making up neutron sources. Later at Chicago,

where he was engaged in assembling the world's first nuclear reactor, he handled beryllium-oxide bricks. The first sign of his illness was not detected until 1946, when a routine chest X-ray yielded the finding "the lung fields are diffusely granular in appearance and look subjectively like silicosis." At that time he felt in good health, and nothing was done about his condition. In the summer of 1948, however, he went to the hospital, suffering general debility and with his weight down 20 pounds.

There are 35 similar cases, in which symptoms were not noted until more than 10 years after exposure. Chronic berylliosis is difficult to diagnose. The symptoms of shortness of breath, lowered vitality, weight loss and reduction in respiratory capacity are typical of many debilitating illnesses. The chest X-ray, which is said at times to give the appearance of a sand- or snow-storm, may be mistaken for tuberculosis and other diseases. Unless the physician has special experience and instruction he may misdiagnose a case of chronic berylliosis such as still turns up occasionally among persons who were exposed a decade ago, before the disease and its cause were adequately recognized.

Just as in the delayed form of the lung disease, the skin lesions caused by

the less-soluble beryllium compounds do not appear until several months after exposure. This was a hazard particularly for the unwary consumer when beryllium was used in fluorescent-lamp phosphors. A typical case was that of a 12-year-old boy who was admitted to a Boston hospital with small painless swellings around the angle of the jaw on the right side. His story was that three months previously he had been playing with some friends at a dump. One of the boys, deciding an old fluorescent lamp tube would make an excellent baseball bat, hit a bottle with it. The tube broke, and pieces of phosphor-coated glass hit the patient on the right side of the neck. About eight weeks later small lumps began to appear beneath the scars on his face and neck. Upon analysis the tissue was found to contain several micrograms of beryllium. With the complete removal of the beryllium from the tissue, the condition cleared up.

The epidemiology of berylliosis assumed a truly bizarre character in the discovery of the so-called "neighborhood" cases of the chronic disease in 1947 in a Midwestern city. Not one of these people had ever worked with beryllium or handled it in any form. All of them, however, had lived within three quarters of a mile of a plant which pro-

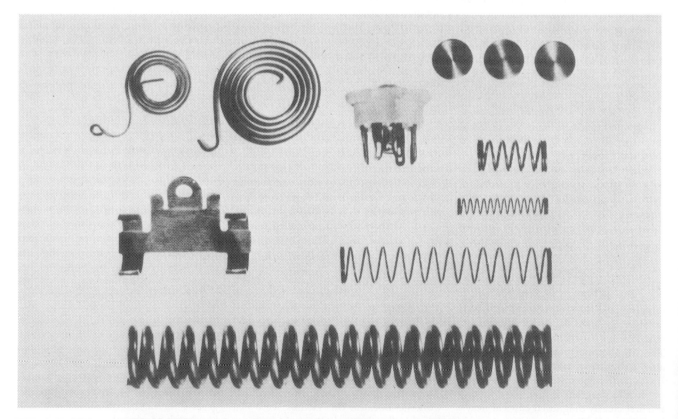

METAL PARTS are made of beryllium and its alloys. At upper right are three small disks of pure beryllium, used as windows for X-ray tubes. All the other metal parts shown here are made of beryllium-copper alloy, which is six times stronger than copper.

duced beryllium compounds from ore. When physicians in the community learned about these neighborhood cases, they soon discovered others in their records of undiagnosed lung disease. The apprehensive plant management instituted a chest X-ray survey; among 10,000 persons examined, two additional cases of berylliosis were discovered. It turned out that traces of beryllium were being emitted from the plant in the stack gases. Altogether more than 40 such neighborhood and nonoccupational cases of beryllium poisoning have been recorded. One woman lived nearly two miles from a beryllium plant; her husband worked in the plant, and she received her exposure from the dust she inhaled while washing his work clothes!

The registry of cases kept by Harriet Hardy at Massachusetts General Hospital in Boston shows that about 40 per cent have been of the chronic type, with a mortality rate of about 30 per cent. About 15 cases have involved individuals working in atomic-energy developments during the war years—an ironic fact in view of the extreme measures taken to protect workers from radioactive poisons. The Atomic Energy Commission has since led the way in the establishment of safe working conditions in the beryllium industry.

While beryllium alloys as commonly used appear safely nontoxic, one laboratory experience is instructive. During World War II a group of surgeons, testing various alloys for repair of massive bone damage, tried one alloy containing 1.6 per cent beryllium. They fixed plates of the alloy in dog skulls with screws made of the same alloy. After six months they were startled to find that the screws had loosened and that the screw holes and tissues in contact with the plate were lined with inflamed lesions.

As yet the biochemistry of beryllium poisoning is little understood. The ion of the metal, dissociated from the oxide or salt in solution with the body fluids, appears to be the active principle. We know also that the chemical reaction of beryllium ions in the body invariably involves a hydroxide group (OH) attached to a benzene ring. Such phenolic hydroxide groupings, as they are called, are found in the amino acid tyrosine, which in turn undergoes metabolic transformation to such compounds as adrenalin which react with beryllium ions.

In the test tube beryllium inhibits the action of many enzymes. Injection of tiny amounts of its compounds in experimental animals causes massive damage

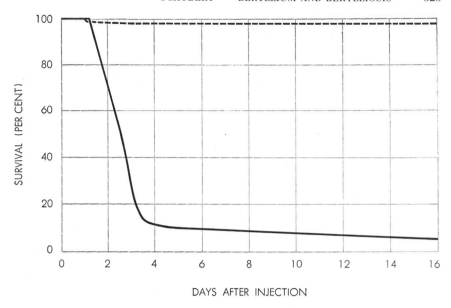

MICE WERE INJECTED with a beryllium salt at Argonne National Laboratory. An hour later half of them were injected with aurintricarboxylic acid (ATA). Broken line shows survival of mice injected with ATA; solid line, survival of those not injected.

to the cells of practically every organ with which the compounds come in contact. Similarly in human victims, upon autopsy, lesions are found scattered throughout the organs of the body as well as in the lungs. Beryllium poisoning is now accepted as a general disease, and it is recognized that the patient is sicker than the lung picture would suggest.

Examination of the lumps and nodules that form in the tissues suggests that beryllium possesses some power to cause the growth and proliferation of cells. This suspicion has been supported by the discovery that beryllium may induce cancer in experimental animals. The late Leroy Gardner of the Saranac Laboratory showed in 1946 that the intravenous injection of beryllium compounds in rabbits led to bone cancers. This finding has since been confirmed and extended to various animals, including guinea pigs and rats. Lung cancers also have been produced repeatedly with beryllium compounds. The investigators conclude: "These studies pose the grave question whether beryllium may not ultimately prove a factor in the genesis of certain cases of human lung cancer."

The question how beryllium acts on the tissue cells leads into the whole general problem of the role of trace metals in the processes of life. M. B. Hoagland at Harvard University has demonstrated that the growth-stimulating effect of beryllium is a general one, acting on algae and higher plants as well as on animal tissues. When he supplied a very small amount of beryllium salt to the

nutrient of a tomato plant that had been stunted by depriving it of magnesium, the growth of the plant immediately showed a marked acceleration.

Ignorant as we are about the toxicity of beryllium, investigators have made one significant finding which already facilitates diagnosis and may lead to the discovery of an effective treatment. Patients invariably exhibit a kind of allergic response to beryllium compounds. It would seem, therefore, that beryllium must combine with protein in the body to form an antigen. The antigen stimulates the formation of beryllium-specific antibodies. As a result, in all cases of active berylliosis a small amount of any beryllium compound placed on the skin produces a local allergic reaction. This "patch test" is now a valuable aid in the diagnosis of beryllium poisoning. The finding that beryllium forms compounds with protein suggested in turn that the logical approach to treatment is to look for a way to tie up beryllium chemically so that it can no longer react with substances in the body.

With this in mind, our group at the Argonne National Laboratory began a search for a specific antidote to beryllium poisoning. At first we studied the possibility of injecting materials into experimental animals which would hasten the elimination of beryllium. However, we soon learned that it is nearly impossible to eliminate the poison once it has been deposited in the tissues.

We then determined to look for a drug

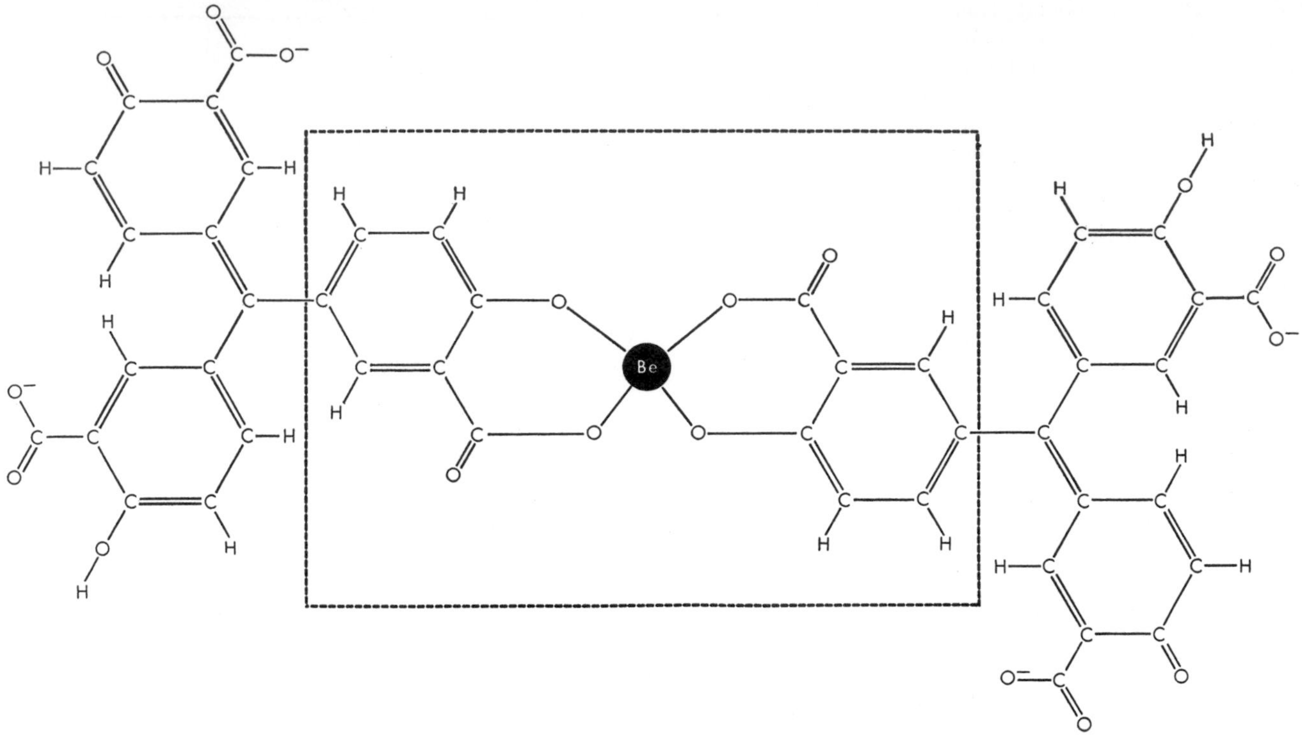

ACTION OF ATA is explained. The molecular structure of ATA is given in the diagram at left. ATA is a chelating agent, *i.e.,* a substance which chelates or sequesters a metal ion within a larger structure of atoms. In this case the metal ion is beryllium (Be). The

which would seek out beryllium deposited in the body and form an insoluble compound, thus inactivating the metal. Such an agent, we thought, might be found among the chelates, an interesting class of dyestuffs which have the power to incorporate various metal ions selectively in their structures [see "Chelation," by Harold F. Walton, beginning on page 344].

We did not have to search long. There is a deep-red dye, well known to analytical chemists, which is used for the analysis of aluminum and to a lesser extent for beryllium and which, on paper, seemed to meet our requirements. This dye is known by the trade name "aluminon," and by the chemical name aurintricarboxylic acid, or simply ATA. In the first test of ATA we injected mice with enough beryllium salt to kill them within a few days. We then injected half the animals with a small dose of ATA and left the others untreated. The results were dramatic: virtually every animal treated with ATA survived and lived on normally, while all of the untreated animals died. We have repeated this experiment with hundreds of animals of different species, with the same high degree of protection.

We checked the conclusions suggested by this experiment by injecting the animals with radioactive beryllium

and with ATA tagged with tracer atoms. The picture that emerged was clear and unambiguous: ATA was found in practically every cell where beryllium was present. Previously damaged cells recovered, and within a few days could not be distinguished from the normal tissue. Nor could any abnormality be detected a year or more later, despite the fact that the beryllium-dye combination remained in the tissues.

We have tested many other compounds and found none as completely effective as ATA. Some of them have demonstrated a high affinity for beryllium in the test tube, yet have failed completely when injected in the experimental animal. Capacity to tie up beryllium does no good if the compound is unable to make contact with the beryllium in the tissues. All the compounds that failed have chemical groups which render them more water-soluble. As a result, they tend to remain in the water of the body and are rapidly excreted.

It is too early to say whether ATA will prove to be successful for the treatment of human beings. Clinical tests are being made. The main problem is to promote contact between ATA and beryllium. One approach involves the use of a special aerosol generator which breaks down the ATA into extremely small particles to insure that the material penetrates deep into the lung spaces. Intravenous injections will be tried along with the aerosol treatment. On the theory that ACTH and cortisone might render the lumps and nodules in the tissues surrounding the beryllium more permeable, these drugs may also be used in combination with ATA. As to the toxicity of ATA, physicians report they "are satisfied beyond any doubt that the ATA preparation is nontoxic, easy to administer and well tolerated by the patient."

Meanwhile, treatment of berylliosis patients with ACTH or cortisone has shown some success. Just why these drugs are helpful remains unknown. Certainly they have no effect on the beryllium itself. They do, however, provide many patients with some or with considerable relief. In the case of the Manhattan Project physicist described earlier, treatment with ACTH restored him to almost full activity.

Can the beryllium hazard be controlled? Is it possible to work with beryllium compounds in safety? The answer is a somewhat qualified "Yes." The problem reduces itself to one of good industrial housekeeping, which means keeping the levels of beryllium low. But how low is low? One important factor to be taken into account is that different forms of the same beryllium compound

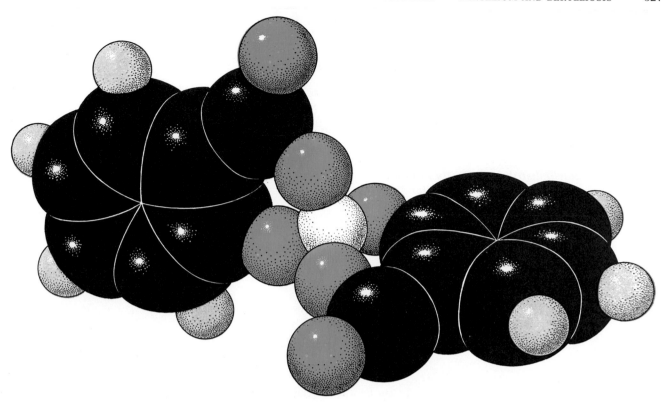

three-dimensional structure of the atoms within the broken rectangle at left is shown in the illustration at right. When ATA is injected into body fluids containing beryllium, it sequesters the beryllium ions so that they cannot exert their poisonous effects.

may have radically different degrees of toxicity, depending upon particle size as well as solubility. On the basis of extensive tests the AEC early in 1950 concluded that all known cases of the acute disease could be attributed to air concentrations of soluble salts in excess of 100 micrograms per cubic meter, and that when the air level exceeded 1,000 micrograms of beryllium, nearly everyone developed acute beryllium poisoning. To minimize the risk of the chronic disease, on the other hand, the AEC found it necessary to recommend a limit of two micrograms per cubic meter. The strictness of this standard can be appreciated when it is realized that the corresponding limits for dusts of other metals such as lead, mercury, arsenic and cadmium range from 100 to 500 micrograms per cubic meter. As to the air in the neighborhood of a beryllium plant, the recommended limit is one hundredth of a microgram per cubic meter. To the best of our knowledge, no cases of beryllium poisoning have appeared in or around plants which have adopted these rigid standards.

These limits are, in fact, so low that they raise the general and sinister question whether the natural occurrence of beryllium in the environment might not at times be sufficient to induce beryllium lung disease in hypersensitive in-dividuals. The tissues of most persons contain negligible amounts of beryllium, actually less than a microgram in the entire lungs, for example. Ordinary outdoor air contains about a thousandth of a microgram or less of beryllium per cubic meter. Beryllium, however, is concentrated by some plants, especially those growing in areas where the soils contain higher-than-average amounts of beryllium. The ashes of wheat straw are reported to contain as much as 2 per cent beryllium. Again, its concentration in certain coal ashes goes as high as 2 per cent. Another site of high concentration is forest litter. Examination of the lungs of coal miners has shown beryllium usually present in amounts far in excess of normal and in many instances greater than in known berylliosis cases. Yet the lungs of these miners had no lesions suggesting the disease. It must be emphasized that the severity of berylliosis does not necessarily reflect the amount of beryllium in the lungs. The total amount is not as important as the fractional amount present in an active form. It seems, therefore, we do not have to worry that we might fall victim to chronic beryllium poisoning merely from breathing outdoor air.

I say "seems" because physicians remain puzzled by some 1,800 cases of sarcoidosis, a lung malady, discovered among troops from the southeastern states during World War II. Their condition resembles berylliosis. The soil of their granitic native countryside has a higher than average beryllium content. Recently investigation has shown that the pollen dust of local evergreens carries lipid compounds similar to lipids found in tubercle bacilli. But the case is not yet closed.

More than 250 years ago Bernardino Ramazzini, the great student of industrial diseases, wrote that when physicians have a working man as a patient they should inquire into all the details of his occupation, because without this information a correct diagnosis cannot be made. The story of beryllium highlights the whole problem of occupational disease in the present era. Advances in technology now develop so rapidly that the rare material of yesterday becomes the widely used material of today. The beryllium mishaps teach the lesson that the harmlessness of a material cannot be taken for granted; a new material must be regarded as harmful until proved otherwise. Industrial medicine must provide safe working conditions before harm results; public health agencies must see that the public is not exposed to fumes from industrial processes until safe tolerances are known.

35

MERCURY IN THE ENVIRONMENT

LEONARD J. GOLDWATER
May 1971

*The metal is widely distributed, mostly in forms
and amounts that do no harm. The question is whether
its concentration by industrial and biological processes
now endangers animals and human beings*

In the early 1950's fishermen and their families around Minamata Bay in Japan were stricken with a mysterious neurological illness. The Minamata disease, as it came to be called, produced progressive weakening of the muscles, loss of vision, impairment of other cerebral functions, eventual paralysis and in some cases coma and death. The victims had suffered structural injury to the brain. It was soon observed that Minamata seabirds and household cats, which like the fisherfolk subsist mainly on fish, showed signs of the same disease. This led to the discovery of high concentrations of mercury compounds in fish and shellfish taken from the bay, and the source of the mercury was traced to the effluent from a factory.

Since then there have been several other alarming incidents. In 1956 and 1960 outbreaks of mercurial poisoning involving hundreds of persons took place in Iraq, where farmers who had received grain seed treated with mercurial fungicides ate the seed instead of planting it. There were similar outbreaks later in Pakistan and in Guatemala. In Sweden, where poisoning of game birds and other wildlife, apparently by mercury-treated seeds, began to be noticed in 1960, the Swedish Medical Board in 1967 banned the sale of fish from about 40 lakes and rivers after it was found that fish caught in those waters contained high concentrations of methyl mercury. In 1970 alarm rose to a dramatic pitch in North America. Following the discovery of mercury concentrations in fish in Lake

Saint Clair by a Norwegian investigator working in Canada, restrictions on fishing and on the sale of fish were imposed in many areas in the U.S. and Canada, and government agencies in both countries began to take action to control the discharge of mercury-containing wastes into lakes and streams.

Suddenly, almost overnight, mankind has become acutely fearful of mercury in the environment. The alarm is understandable. Quicksilver has always been regarded as being magical and somewhat sinister, in part because of its unique property as the only metal that is a liquid at ordinary temperatures. Mercury's peculiarities have been recognized since medieval times, when the alchemists took a keen interest in the element's fascinating properties. Its toxic properties became so well known that some mercury compounds came to be used as agents of suicide and murder. There are indications that Napoleon, Ivan the Terrible and Charles II of England may have died of mercurial poisoning, either accidental or deliberate. (Charles II experimented with mercury in his laboratory.) It has been suggested (incorrectly) that mercury is what made Lewis Carroll's Hatter mad (since it is used in the manufacture of felt hats). And it is authentically recorded that as early as 1700 a citizen of the town of Finale in Italy sought an injunction against a factory making mercuric chloride because its fumes were killing people in the town.

Nevertheless, although the recent incidents give us justifiable concern about

the potential hazards of mercury in the environment, a panicky reaction would be quite inappropriate. Mercury, after all, is a rare element, ranking 16th from the bottom of the list of elements in abundance in the earth and comprising less than 30 billionths of the earth's crust. There are comparatively few places in the world where it occurs naturally in more than trace amounts, and the ore-bearing deposits of commercial value are so limited that a handful of mines scattered over the globe account for most of the world production. The uncompounded element in liquid form is not a poison; a person could swallow up to a pound or more of quicksilver with no significant adverse effects. Nor should it be forgotten that certain compounds of mercury have been used safely for thousands of years, and some are still prescribed, as effective medications for various infections and disorders. We need not shrink from mercury as an unmitigated threat. What is now required is detailed investigation of how mercury is being redistributed and concentrated in the environment by man's activities and in what forms and compounds it may be harmful to life. Extensive research on these questions is under way.

We consider first what might be called the normal distribution of mercury in nature. The element is found in trace amounts throughout the lithosphere (rocks and soil), the hydrosphere, the atmosphere and the biosphere (in tissues of plants and animals). In the rocks and

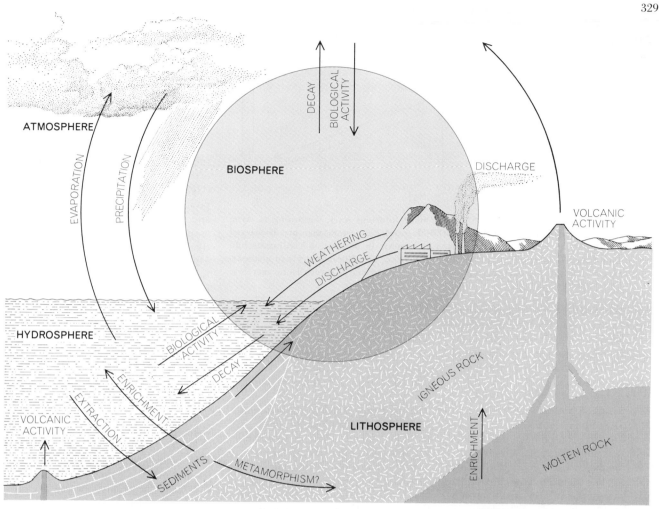

MERCURY CYCLE disperses the metal through the lithosphere, hydrosphere and atmosphere and through the biosphere, which interpenetrates all three. Mercury is present in all spheres in trace amounts, but it tends to be concentrated by biological processes. Man's activities, in particular certain industrial processes, may now present a threat by significantly redistributing the metal.

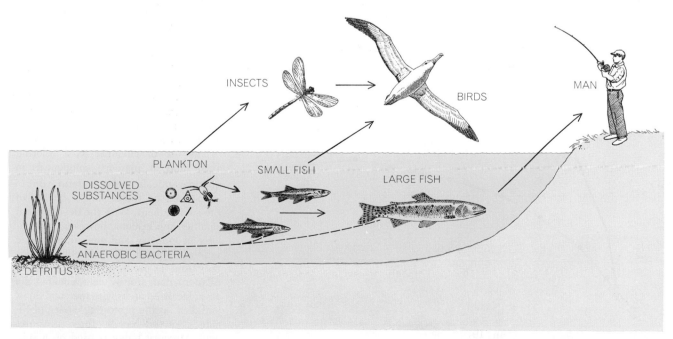

AQUATIC FOOD CHAIN is a primary mechanism by which mercury is concentrated. At each trophic level less mercury is excreted than ingested, so that there is proportionately more mercury in algae than in the water they live in, more still in fish that feed on the algae and so on. Bacteria and the decay chain (*broken arrows*) promote conversion of any mercury present into methyl mercury.

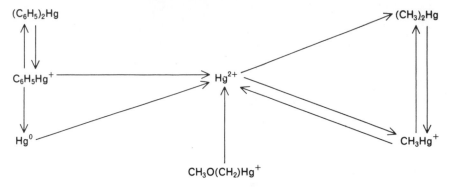

METHYL MERCURY COMPOUNDS are the most injurious ones. According to Arne Jernelöv of the Swedish Water and Air Pollution Research Laboratory, mercury discharged into water in various forms can be converted by bacteria in detritus and sediments into methyl and dimethyl mercury (*right*). Phenyl mercurials, metallic mercury and methoxyethyl mercury (*left and bottom*) are converted into methyls primarily through ionic mercury.

soils (apart from ore concentrations) mercury is measured in fractions of one part per million, except in topsoils rich in humus, where the amount may run as high as two parts per million. In the hydrosphere (the seas and fresh waters) it generally occurs only in parts per billion. In the atmosphere mercury is present both as vapor and in the form of particles. Under natural conditions, however, the amount is so small that extremely sensitive methods are required to detect and measure it; the measurements that have been made at a few locations indicate that this atmospheric "background" amounts to less than one part per billion.

The situation is somewhat different when we come to the biosphere. Plants and animals tend to concentrate mercury; it has been found, for example, that some marine algae contain a concentration more than 100 times higher than that in the seawater in which they live, and one study of fish in the sea showed mercury concentrations of up to 122 parts per billion. There is considerable variation, as we shall see, in the amounts of mercury found in plants and animals, depending on circumstances. Under natural conditions, however, the concentration in the earth's vegetation (aside from cultivated plants) averages no more than

a fraction of one part per million.

Thus the natural cycle of circulation of mercury on the earth [*see top illustration on preceding page*] disperses it widely through the habitable spheres in trace amounts that pose no hazard to life. How seriously has man altered its distribution?

The only ore containing mercury in sufficient concentration for commercial extraction is cinnabar, or mercuric sulfide (HgS). There are minable cinnabar deposits in many regions around the world, and man was attracted to its use as early as prehistoric times. There is evidence that cinnabar was mined in China, Asia Minor, the Cyclades and Peru at least two or three millenniums ago. Cinnabar, a brilliant red mineral, came into use at first as a pigment, but it was not until medieval times that physicians and other investigators became interested in extracting quicksilver from ore to produce medicines and other useful compounds. Hippocrates is believed to have prescribed mercury sulfide as a medication, and this was probably one of the first compounds of a metal to be employed therapeutically.

By the Middle Ages, when alchemists had synthesized chlorides, oxides and various other inorganic compounds and mixtures of mercury, its use in medications began to spread. Calomel (mercurous chloride, or HgCl) came into wide use as a cathartic, and in the 16th century mercury compounds were introduced as a treatment for syphilis. By the 19th century scores of mercurials were being employed in medicine. Many are still in the pharmacopoeias; the most useful ones today are the diuretics. It has been found that even among the organic compounds of mercury there are some that can be used safely. As much as 78.5 grams of mercury in the form of an organic compound has been given to a patient without harmful side effects!

The use of cinnabar as a coloring agent and of mercury compounds in pharmaceuticals under careful control introduced no threat to the quality of the environment. With the development of other applications, however, particularly in industry and agriculture, came serious problems. The extraction of mercury from the ore by heating (that is, distillation) dangerously contaminates the air in localized areas with mercury vapor and dust (as the protesting citizen of Finale observed nearly three centuries ago). Mercury today is used on a substantial scale in chemical industries, in the manufacture of paints and paper and in pesticides and fungicides for agriculture. The world production of mercury

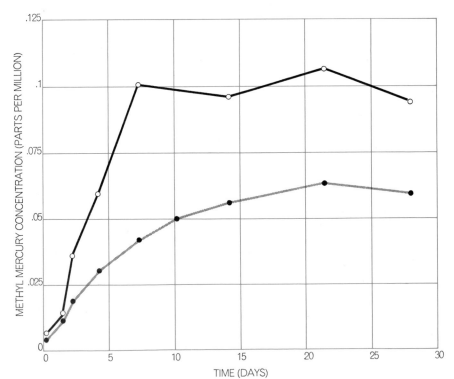

CONVERSION of inorganic into methyl mercury in sediment was measured by Sören Jensen and Jernelöv at intervals after the addition of 10 (*gray*) and 100 (*black*) parts per million of inorganic mercury. At lower mercury concentrations methylation may not occur.

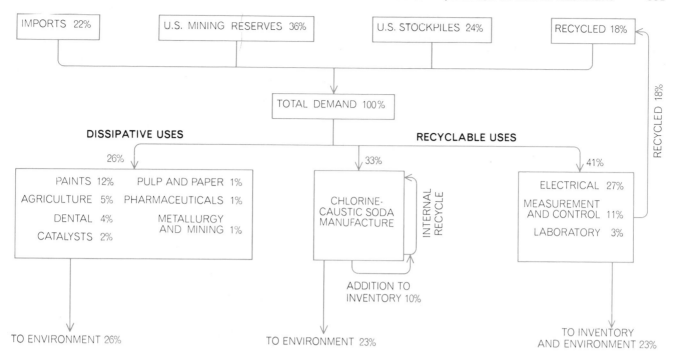

MERCURY FLOW through the U.S. is shown for 1968. The chart is based on one prepared by Robin A. Wallace, William Fulkerson, Wilbur D. Shults and William S. Lyon of the Oak Ridge National Laboratory. Major use of mercury has been as a cathode in the electrolytic preparation of chlorine and caustic soda. In this process a large inventory of mercury is continuously recycled, but in 1968 23 percent of total mercury demand still went to make up what was wasted. Another 10 percent went for start-up of new plants. Since then legislation and lawsuits have required manufacturers to increase recycling sharply, reducing emissions to the environment.

now amounts to about 10,000 tons per year, of which about 3,000 tons are used in the U.S. (The principal producers are Spain, whose mines at Almadén are the richest in the world, Italy, the U.S.S.R., China, Mexico and the U.S.) From the large-scale uses a considerable amount of mercury wastes is flowing into the air, the soil and streams, lakes and bays.

In agriculture, for example, corrosive sublimate ($HgCl_2$) is used to disinfect seeds and to control many diseases of tubers, corms and bulbs (including potatoes). The chlorides of mercury (both mercuric and mercurous) are also employed to protect a number of vegetable crops. In recent decades farmers in Europe and the U.S. have adopted the use of organic compounds of mercury, some of which are highly toxic, principally to prevent fungal diseases in seeds and in growing plants, fruits and vegetables. These chemicals may present a potential threat to health through the ingestion of treated seeds by birds (and people), through concentration in food plants and through percolation or runoff from the fields into surface waters. The U.S. Geological Survey, after analyzing the concentration of mercury in a number of U.S. rivers in 1970, reported that although the mercury content was only one part per 10 billion or less in most rivers and streams, "it may be several

thousand times this concentration in some natural waters."

In order to evaluate the hazards of mercury in the environment we must examine the forms in which it occurs there and the relative toxicity of its various compounds. Liquid mercury itself, as we have already noted, is not ordinarily toxic to man. Inhalation of mercury vapor, however, can be injurious. In acute cases it causes irritation and destruction of the lung tissues, with symptoms including chills, fever, coughing and a tight feeling in the chest, and there have been reports of fatalities from such exposure. The acute exposures, however, usually come about not from the general environment but by accident, such as heating a household mercurial. More common is a chronic form of injury resulting from occupational exposure to mercury vapor, for example among mercury miners and workers in felt hat factories employing mercury nitrate for processing. These exposures, as we have found in examinations of miners, are not necessarily incapacitating; they produce tremors, inflammation of the gums and general irritability.

The soluble inorganic salts of mercury have long been known to be toxic. Mercury bichloride (corrosive sublimate), which has been used on occasion for suicide and homicide, produces corrosion of the intestinal tract (leading to bloody

diarrhea), injury to the kidneys, suppression of urine and ultimately death from kidney failure when it is taken by mouth in a substantial dose. Its former use in moderate doses by mouth for treatment of syphilis did not, however, result in observable poisoning in most cases. Mercurous chloride is less soluble than the mercuric salt and therefore is less dangerous. It is still used medicinally, but some of its uses have been abandoned because it was found to cause painful itching of the hands and feet and other symptoms in children. Among other inorganic mercurials, some of the oxides, such as the red oxide used in antifouling paint for ship bottoms, may be potentially hazardous. In general, however, these inorganic mercurials are not important factors in contamination of the general environment.

What does cause us concern now with regard to the environment is the presence of some of the organic compounds of mercury, specifically the alkyls: the methyl and ethyl compounds. In Minamata Bay the substances that had poisoned the fish and people were identified as methyl mercurials. The grain that caused outbreaks of illness and death among the farmers of Iraq had been treated with ethyl mercury p-toluene sulfonanilide. And alkyls of mercury were similarly incriminated in Sweden and other places.

It has been known for some time that alkyl mercury can cause congenital mental retardation, and recent laboratory studies have shown that it can produce abnormalities of the chromosomes and, through "intoxication" of the fetus in the uterus, can bring about cerebral palsy. The alkyl mercurials attack the brain cells, which are particularly susceptible to injury by this form of mercury. The chemical basis of this effect seems to be mercury's strong affinity for sulfur, particularly for the sulfhydryl groups (S-H)

in proteins (for which arsenic and lead have a similar affinity). Bound to proteins in a cell membrane, the mercury may alter the distribution of ions, change electric potentials and thus interfere with the movement of fluid across the membrane. There are also indications that the binding of mercury to protein disturbs the normal operation of structures such as mitochondria and lysosomes within the cell. Alkyl mercury appears to be especially dangerous because the mercury is firmly bonded to a

carbon atom, so that the molecule is not broken down and may maintain its destructive action for weeks or months. In this respect it differs from the inorganic and phenyl (aryl) mercurials, and that may explain why it produces permanent injury to brain tissue, whereas the injury caused by inorganic and aryl mercurials is almost invariably reversible.

At one time it was thought that aryl mercurials (compounds based on the phenyl group) might act like the alkyls; however, in an extensive series of studies I initiated in 1961 at the Columbia University School of Public Health we found that chemical workers who continually handled phenyl mercurials and experienced exposures far above the supposedly safe limit did not show any evidence of toxic effects.

Having examined the nature of the mercury "threat" and its quantitative presence in the environment, we should now look at the other side of the equation: the extent of man's exposure and his response to this factor up to now.

Without question the major source of man's intake of mercury is his food. Alfred E. Stock in Germany initiated analyses of the mercury concentration in foods in the 1930's, and there have been several follow-up studies since then, including one by our group at Columbia in 1964. (Numerous further investigations are now in progress.) The measured concentrations in samples from various sources are in the range of fractions of one part per million [see illustration on next page], but the concentrations in fish in contaminated waters may run several hundred times higher than that. A joint commission of the Food and Agriculture Organization and the World Health Organization proposed in 1963 that the permissible upper limit for mercury in foods (except fish and shellfish) should be .05 part per million; there is as yet no firm basis, however, for determining what the safe standard ought to be. Perhaps the most significant conclusion that can be drawn from these sets of samplings is that in general the concentration of mercury in foods does not appear to have changed substantially over the past 30 years. The comparisons may not be entirely valid, however, because of differences in analytical methods.

In addition to food, there are other possible everyday sources of exposure to mercury. It is used fairly commonly in antiseptics, paint preservatives, floor waxes, furniture polishes, fabric softeners, air-conditioner filters and laundry preparations for suppression of mildew; no doubt there are other such exposures

ATMOSPHERIC MERCURY LEVELS were measured by S. H. Williston at a station south of San Francisco. The concentration averaged about .0002 microgram of mercury per cubic meter of air when the wind blew from the Pacific (a) and was somewhat higher when the wind was from the generally nonindustrial southeast (b). The average was .008 microgram, with many peaks that went off the record at .02 microgram, when the wind was from the industrial area to the northeast (c). The mercury was often associated with dust particles.

of which we are not aware. In view of all these factors, it is not surprising to find that 20 to 25 percent of the "normal" population—persons who have apparently had no medicinal or occupational exposure to mercury—show easily measurable amounts of mercury in their body fluids. Several studies of this matter have been made, including a fairly extensive international investigation we carried out at Columbia in 1961–1963 as a joint project with the WHO. Analyzing 1,107 specimens of urine collected from "normal" subjects in 15 countries, we found that except in rare instances the mercury content in the urine ran no higher than about 20 to 25 parts per billion. A similar examination of blood samples showed that the highest mercury concentration in the blood among "normal" subjects was 30 to 50 parts per billion. And analyses of human tissues made at autopsy have indicated that similar traces of mercury are present in the body organs.

It is important to consider these findings in the context of the evolutionary relationship of life on our planet to the presence of mercury. Unquestionably the element has been omnipresent in the sea, where life originated, from the beginning, and presumably all plants and animals carry traces of mercury as a heritage from their primordial ancestry. Man as the top of a food chain must have added to that heritage by eating fish and other mercury-concentrating forms of food. Over the millions of years he presumably has built up an increased tolerance for mercury. (The development of tolerance for chemicals is of course well recognized today. It was put to use more

than 2,000 years ago by Mithradates the Great, king of Pontus, who armed himself against poisoners by taking small and increasing doses of toxic agents.) Tolerance for a potent substance not infrequently grows into dependence on it, and it is reasonable to suppose that man, as well as other forms of life, may now be dependent on mercury as a useful trace element. Whether its effects are beneficial or harmful may be influenced decisively by the form in which it is incorporated in tissues, by the dose and probably by other factors. It has been found, for instance, that the highly toxic element arsenic is sometimes present in healthy shrimp in concentrations of close to 200 parts per million (dry weight) in the form of trimethylarsine. Methylation in this case apparently suppresses the element's toxicity. The biochemical behavior of mercury has much in common with that of arsenic, which suggests that there may be harmless forms of methyl mercury as well as toxic ones in fish.

Our concern, then, must be with any disturbance of the environment that alters the natural balance of mercury in relation to other substances or that generates virulent forms of mercurials. In the case of the fish in Minamata Bay apparently both factors were at work. The polluting effluent from the chemical plant itself contained methyl mercury, and elemental mercury in the effluent was methylated by microorganisms in the mud on the bottom of the bay. This conversion was fostered by the enrichment of the water with a high concentration of mercury and organic pollutants that promoted the growth of the methylating bacteria. The result was the accu-

mulation in the fish of concentrations of methyl mercury as high as 50 parts per million (wet weight), which is 100 times the *total* mercury concentration currently accepted as "safe" in the U.S. and Canada. The effect on the fishermen and their families was compounded by the fact that their diet consisted largely of the bay fish and may have been deficient in some essential nutrients; dietary deficiencies are known to enhance the adverse effects of toxic agents.

The current journalistic outcry on the "mercury problem" has produced a state of public alarm approaching hysteria. "Protective" measures are being proposed and applied without basis in established knowledge. Research on mercury poisoning in the past has focused primarily on occupational hazards involving prolonged exposure, principally by way of inhalation. Mercury in the general environment, however, presents an almost entirely different problem. Discharge of mercury into the atmosphere, either as vapor or as particulate matter, is not likely to become a serious general hazard. The main threats to which we shall have to give attention are solid and liquid wastes that may ultimately enter bodies of water, thus threatening fish and eaters of fish, and agricultural uses of mercury that may dangerously contaminate food. We do not yet have enough information to estimate the magnitude of these threats or to establish realistic standards of control.

To begin with, we need a better understanding of what should be considered a toxic level of mercury in the human body. Analysis of the mercury con-

	A	B	C	D	E
MEATS	.001 — .067	.005 — .02	.0008 — .044	.31 — .36	.001 — .15
FISH	.02 — .18	.025 — .18	.0016 — .014	.035 — .54	0 — .06
VEGETABLES (FRESH)	.002 — .044	.005 — .035	0	.03 — .06	0 — .02
VEGETABLES (CANNED)			.005 — .025		.002 — .007
MILK (FRESH)	.0006 — .004	.0006 — .004	.003 — .007	.003 — .007	.008
BUTTER	.002 (FATS)	.07 — .28			.14
CHEESE	.009 — .01				.08
GRAINS	.02 — .036	.025 — .035	.002 — .006	.012 — .048	.002 — .025
FRUITS (FRESH)	.004 — .01	.005 — .035		.018	.004 — .03
EGG WHITE				.08 — .125	.01
EGG YOLK				.33 — .67	.062
EGG (WHOLE)	.002	.002	0		
BEER	.00007 — .0014	.001 — .015			.004

CONCENTRATION IN FOODS was reported by Alfred E. Stock in Germany in 1934 (*A*) and 1938 (*B*), O. S. Gibbs in the U.S. in 1940 (*C*), Y. Fujimura in Japan in 1964 (*D*) and the author's group in the U.S. in 1964 (*E*). A listing of "0" means simply a concentration too low to be detected by the method used. The World Health Organization proposed a permissible upper limit of .05 part per million for foods other than fish; the U.S. Food and Drug Administration has set an upper limit of .5 part per million for fish.

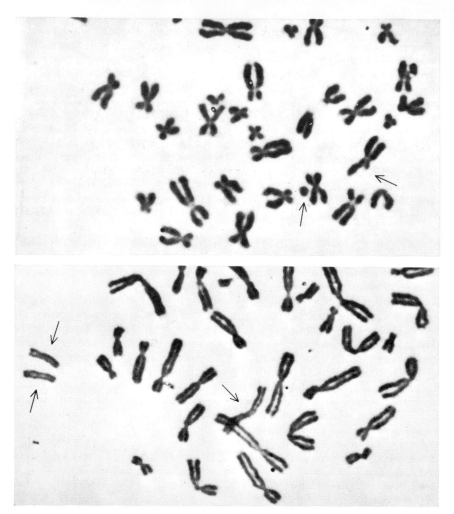

CHROMOSOME DAMAGE is found in persons with high blood levels of mercury after exposure to methyl mercury in fish. Photographs of lymphocytes made by Staffan Skerfving of the Swedish National Institute of Public Health show a broken chromosome and extra fragment (top) and three sister-chromatid fragments that lack centromeres (bottom).

does not apply to the general environment, although direct contact with organic mercurials can cause severe (second degree) local burns of the skin and can result in absorption of measurable quantities of mercury into the body from underclothing and bed linen.

With regard to food, we can identify the most important possible sources of trouble. There is abundant evidence that the inorganic compounds of mercury and the phenyl mercurials are relatively non-toxic compared with the alkyl mercurials. We need to be concerned, however, about the potential conversion of the inorganic forms and the phenyls to methyl mercury, which on the basis of the experience so far must be rated as the prime hazard in the environment. A number of alkyl mercury fungicides have already been eliminated, quite properly, from use on food crops. Government agencies are now beginning to move to ban other organic mercurials as well from any use that might contaminate our food or water.

A calm view of the present state of affairs regarding mercury in the environment suggests that the best way to deal with the problem is to apply the techniques of epidemiology, preventive medicine, public health and industrial hygiene that have been effective in meeting hazards in the past. A system should be set up for frequent monitoring of the environment for the detection of significant increases in mercury contamination. Research should be carried forward to establish measures for the levels and forms of mercurial pollution that signal a threat to health. Techniques for mass screening of the population to detect mercurial poisoning should be developed. Controls should be applied to stop the discharge of potentially harmful mercury wastes at the point of origin. Toxic mercurials in industry and agriculture should be replaced by less toxic substitutes. To implement such a program we shall need, of course, realistic education of the public and legislative action with adequate enforcement. And these measures should be applied to all contaminants that threaten man's environment, not only mercurials.

It would be foolish to declare an all-out war against mercury. The evolutionary evidence suggests that too little mercury in the environment might be as disastrous as too much. In the case of mercury, as in all other aspects of our environment, our wisest course is to try to understand and to maintain the balance of nature in which life on our planet has thrived.

centration in the urine or in the blood has not given much enlightenment on this point. Among workers exposed to mercury in their occupations it has been found that the mercury content in the urine varies greatly from day to day and from one individual to another. As a general rule the body's excretion of mercury does tend to reflect the amount of exposure, but recent studies of workers in the chemical industries have disclosed that individuals who have been exposed to mercury in high concentrations or for long periods often show no sign of adverse effects. Furthermore, high levels in the urine or the blood do not necessarily indicate poisoning; many cases have been observed in which the individual had a mercury concentration in the blood amounting to 10 to 20 times the "normal" upper limit and yet showed no indications of illness or toxic symptoms! All in all there is substantial evidence that host factors may be more important than the amount of exposure, up to a

point, in determining the individual's response to mercury in the environment. In any case, urine or blood analysis is of no value for early diagnosis of mercury poisoning. Other possible indicators, such as disturbances of the blood enzyme system, are being investigated, but no reliable diagnostic test has yet been produced. Nor can we define a precise threshold for the toxic level, either for exposure to or for absorption of mercury.

A good deal of useful information is available, on the other hand, about the sources and avenues of possible danger. Mercury in one form or another can invade the human system by way of the lungs, the skin or the ingestion of food. (Incidentally, recent studies have shown that the mercury in dental fillings is not a hazard; most people with amalgam fillings have negligible amounts of mercury in their urine or blood.) Mercury in the air, as we have noted, is a local problem confined to certain industries. Attack by way of the skin is also a problem that

36

LEAD POISONING

J. JULIAN CHISOLM, JR.
February 1971

Among the natural substances that man concentrates in his immediate environment, lead is one of the most ubiquitous. A principal cause for concern is the effect on children who live in decaying buildings

Lead has been mined and worked by men for millenniums. Its ductility, high resistance to erosion and other properties make it one of the most useful of metals. The inappropriate use of lead has, however, resulted in outbreaks of lead poisoning in humans from time to time since antiquity. The disease, which is sometimes called "plumbism" (from the Latin word for lead) or "saturnism" (from the alchemical term), was first described by the Greek poet-physician Nicander more than 2,000 years ago. Today our concerns about human health and the dissemination of lead into the environment are twofold: (1) there is a need to know whether or not the current level of lead absorption in the general population presents some subtle risk to health; (2) there is an even more urgent need to control this hazard in the several subgroups within the general population that run the risk of clinical plumbism and its known consequences. In the young children of urban slums lead poisoning is a major source of brain damage, mental deficiency and serious behavior problems. Yet it remains an insidious disease: it is difficult to diagnose, it is often unrecognized and until recently it was largely ignored by physicians and public health officials. Now public attention is finally being focused on childhood lead poisoning, although the difficult task of eradicating it has just begun.

Symptomatic lead poisoning is the result of very high levels of lead in the tissues. Is it possible that a content of lead in the body that is insufficient to cause obvious symptoms can nevertheless give rise to slowly evolving and long-lasting adverse effects? The question is at present unanswered but is most pertinent. There is much evidence that lead wastes have been accumulating during the past century, particularly in congested urban areas. Increased exposure to lead has been shown in populations exposed to lead as an air pollutant. Postmortem examinations show a higher lead content in the organs of individuals in highly industrialized societies than in the organs of most individuals in primitive populations. Although no population group is apparently yet being subjected to levels of exposure associated with the symptoms of lead poisoning, it is clear that a continued rise in the pollution of the human environment with lead could eventually produce levels of exposure that could have adverse effects on human health. Efforts to control the dissemination of lead into the environment are therefore indicated.

The more immediate and urgent problem is to control the exposure to lead of well-defined groups that are known to be directly at risk: young children who live in dilapidated housing where they can nibble chips of leaded paint, whiskey drinkers who consume quantities of lead-contaminated moonshine, people who eat or drink from improperly lead-glazed earthenware, workers in certain small-scale industries where exposure to lead is not controlled. Of these the most distressing group is the large group of children between about one and three to five years of age who live in deteriorating buildings and have the habit of eating nonfood substances including peeling paint, plaster and putty containing lead. (This behavior is termed pica, after the Latin word for magpie.) The epidemiological data are still scanty: large-scale screening programs now in progress in Chicago and New York City indicate that between 5 and 10 percent of the children tested show evidence of asymptomatic increased lead absorption and that between 1 and 2 percent have unsuspected plumbism. Small-scale surveys in the worst housing areas of a few other cities reveal even higher percentages.

There is little doubt that childhood lead poisoning is a real problem in many of the older urban areas of the U.S. and perhaps in rural communities as well. Current knowledge about lead poisoning and its long-term effects in children is adequate to form the basis of a rational attack on this particular problem. The ubiquity of lead-pigment paints in older substandard housing and the prevalence of pica in young children indicate, however, that any effective program will require the concerted and sustained effort of each community. Furthermore, the continued use of lead-pigment paints on housing surfaces that are accessible to young children and will at some future date fall into disrepair can only perpetuate the problem.

Traces of lead are almost ubiquitous in nature and minute amounts are found in normal diets. According to the extensive studies of Robert A. Kehoe and his associates during the past 35 years at the Kettering Laboratories of the University of Cincinnati, the usual daily dietary intake of lead in adults averages about .3 milligram. Of this, about 90 percent passes through the intestinal tract and is not absorbed. Kehoe's data indicate that the small amount absorbed is also excreted, so that under "normal" conditions there is no net retention of lead in the body. In addition the usual respiratory intake is estimated at between five and 50 micrograms of lead per day. These findings must be recon-

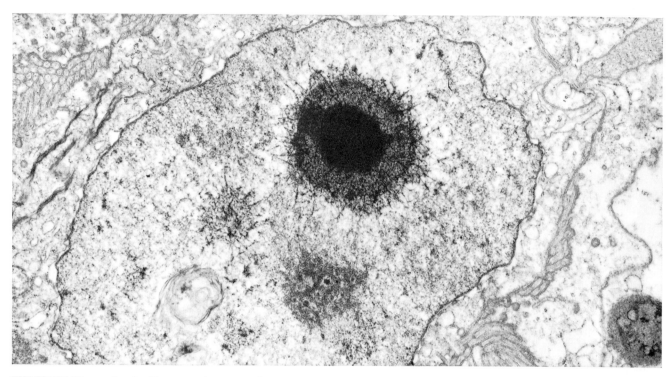

EXCESS LEAD complexed with protein forms inclusion bodies in the nuclei of certain cells in lead-poisoned animals and man. In an electron micrograph made by Robert A. Goyer and his colleagues at the University of North Carolina School of Medicine the nucleus of a cell from a proximal renal tubule of a lead-poisoned rat is enlarged 15,000 diameters. The large structure with a dense core and a filamentous outer zone is an inclusion body; below it to the left is a smaller one. The dark area below the large body is the nucleolus.

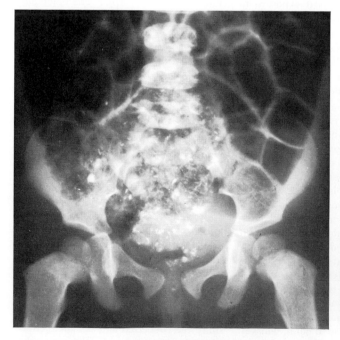

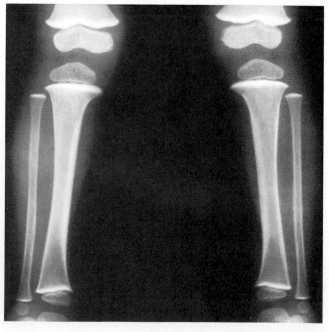

X-RAY PLATES may show evidence of lead ingestion or of an excessive body burden of the metal. The abdominal X ray (left) shows a number of bright opaque particles in the large intestine: bits of lead-containing paint that had been eaten by the 18-month-old subject. The X ray of the same child's legs (right) shows bright "lead lines": excess lead stored at the ends of the long bones.

ciled with postmortem analyses, which indicate that the concentration of lead in bone increases with age, although its concentration in the soft tissues is relatively stable throughout life. The physiological significance of increasing storage in bone is not entirely clear, but it has caused considerable concern. It is quite clear that as the level of intake of lead increases, the rate of absorption may exceed the rate at which lead can be excreted or stored in bone. And when the rates of excretion and storage are exceeded, the levels of lead in the soft tissues rise. Studies in adults indicate that as the sustained daily intake of lead rises above one milligram of lead per day, higher levels of lead in the blood result and metabolic, functional and clinical responses follow [*see illustration on pages 342 and 343*]. The reversible effects abate when the rate and amount of lead absorbed are reduced again to the usual dietary range.

As far as is known, lead is not a trace element essential to nutrition, but this particular question has not been adequately examined. Some of the adverse effects of lead on metabolism have nonetheless been studied in considerable detail. These effects are related to the concentration of lead in the soft tissues. At the level of cellular metabolism, the best-known adverse effect of lead is its inhibition of the activity of enzymes that are dependent on the presence of free sulfhydryl (SH) groups for their activity. Lead interacts with sulfhydryl groups in such a way that they are not available to certain enzymes that require them. In the living organism, under most conditions, this inhibition is apparently partial. Inhibitory effects of lead on other aspects of cellular metabolism have been demonstrated in the test tube. Such studies are preliminary. Most of the effects reported are produced with concentrations of lead considerably higher than are likely to be encountered in the tissues of man, so that speculation about such effects is unwarranted at this point.

The clearest manifestation of the inhibitory effect of lead on the activity of sulfhydryl-dependent enzymes is the disturbance it causes in the biosynthesis of heme. Heme is the iron-containing constituent that combines with protein to form hemoglobin, the oxygen-carrying pigment of the red blood cells. Heme is also an essential constituent of the other respiratory pigments, the cytochromes, which play key roles in energy metabolism. The normal pathway of heme synthesis begins with activated succinate (produced by the Krebs cycle, a major stage in the conversion of food energy to

biological energy) and proceeds through a series of steps [*see illustration below*]. Two of these steps are inhibited by the presence of lead; two others may also be inhibited, but at higher lead concentrations.

Lead is implicated specifically in the metabolism of delta-aminolevulinic acid (ALA) and in the final formation of heme from iron and protoporphyrin. Both of these steps are mediated by enzymes that are dependent on free sulfhydryl groups for their activity and are therefore sensitive to lead. The two steps at which lead may possibly be implicated are the formation of ALA and the conversion of coproporphyrinogen to protoporphyrin. Although the exact mechanism is not known, coproporphyrin (an oxidized product of coproporphyrinogen) accumulates in the urine and the red cells in lead poisoning. Whatever the mechanisms, the increased excretion of ALA and coproporphyrin is almost al-

PATIENTS	LEAD OUTPUT (MILLIGRAMS PER 24 HOURS)		
	MEAN	MEDIAN	RANGE
UNEXPOSED CONTROLS	.132	.157	.012—.175
HOUSEHOLD CONTROLS	.832	.651	.087—1.93
INCREASED LEAD ABSORPTION, NO SYMPTOMS	2.16	1.11	.116—9.60
LEAD POISONING, WITH AND WITHOUT BRAIN DAMAGE DURING EXPOSURE:	44.0	27.0	5.040—104.0
AFTER TREATMENT:	.362	.240	.062—0.850

EXCRETION OF LEAD in feces is an index of exposure to lead. These results of a study by the author and Harold E. Harrison illustrate the massive exposures seen in lead poisoning. Unexposed controls were children with no known exposure to lead. The other groups were children with increased lead absorption (high blood lead), children with lead poisoning and members of their households with neither high blood values nor overt symptoms.

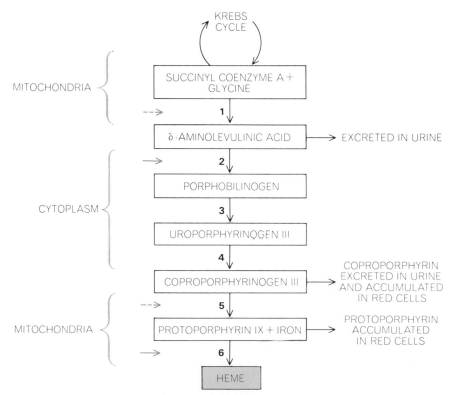

BIOSYNTHESIS OF HEME, a constituent of hemoglobin, is inhibited by lead, resulting in accumulation of intermediates in the synthetic pathway. Of six steps in the pathway, the first and the last two take place in mitochondria, the others elsewhere in the cell cytoplasm. Lead inhibits two steps (*solid colored arrows*) and may inhibit two others (*broken arrows*).

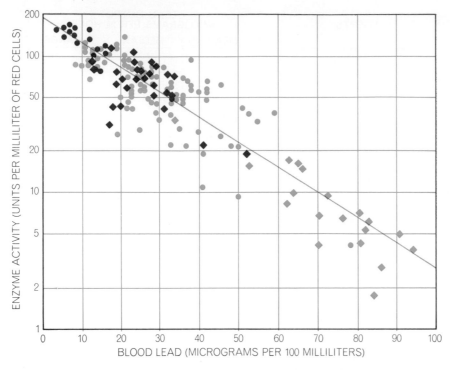

CORRELATION between blood lead and the activity of delta-aminolevulinic acid dehydrase, an enzyme inhibited by lead, was shown by Sven Hernberg and his colleagues at the University of Helsinki. The vertical scale is logarithmic. The values are well correlated, as indicated by the straight regression line, over a wide range of blood-lead levels in groups with different lead exposures: students (*black dots*), automobile repairmen (*black squares*), printshop employees (*colored dots*) and lead smelters and ship scrappers (*colored squares*).

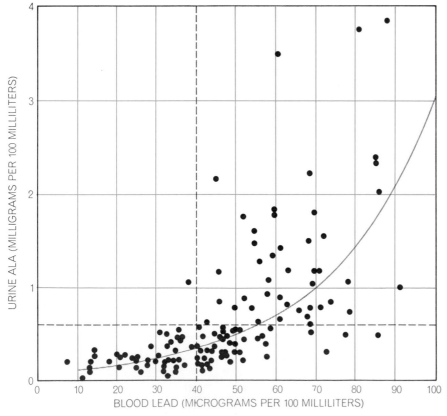

ENZYME SUBSTRATE, delta-aminolevulinic acid (ALA) accumulates in the urine when lead inhibits enzyme activity. Stig Selander and Kim Cramér found that a decrease in lead below about 40 micrograms does not produce a comparable decrease in ALA, suggesting that an enzyme reserve may be involved. Broken lines show presumed normal values.

ways observed before the onset of symptoms of lead poisoning, and the presence of either is therefore important in diagnosis.

The enzyme that catalyzes ALA metabolism is ALA dehydrase. A number of investigators, including Sven Hernberg and his colleagues at the University of Helsinki and Abraham Goldberg's group at the University of Glasgow, have studied the extent to which varying levels of lead in the blood inhibit ALA-dehydrase activity in red blood cell preparations in the laboratory. They have shown a direct relation between the concentration of lead in blood and the activity of the enzyme. Moreover, they find that there seems to be no amount of lead so small that it does not to some extent decrease ALA-dehydrase activity; in other words, there appears to be no threshold for this effect [*see top illustration at left*]. If that is so, however, one would expect to see a progressive increase in the urinary excretion of the enzyme's substrate, ALA, beginning at very low blood-lead levels. This does not seem to be the case. Stig Selander and Kim Cramér in Sweden, correlating blood-lead and urine-ALA values, found that the first measurable increase in urine ALA is observed only after blood lead rises above approximately 30 micrograms of lead per 100 milliliters of whole blood [*see bottom illustration at left*]. The apparent inconsistency between the effect of lead on the activity of an enzyme in the test tube and the accumulation of the enzyme's substrate in the body might be explained by the presence of an enzyme reserve. This hypothesis is consistent with the functional reserve exhibited in many biological systems.

Almost all the information we have on the effect of lead on the synthesis of heme comes from observations of red blood cells. Yet all cells synthesize their own heme-containing enzymes, notably the cytochromes, and ALA dehydrase is also widely distributed in tissues. The observations in red blood cells may therefore serve as a model of lead's probable effects on heme synthesis in other organ systems. Even so, the degree of inhibition in a given tissue may vary and will depend on the concentration of lead within the cell, on its access to the heme synthetic pathway and on other factors. For example, J. A. Millar and his colleagues in Goldberg's group found that ALA-dehydrase activity is inhibited in the brain tissue of heavily lead-poisoned laboratory rats at about the same rate as it is in the blood [*see illustration on opposite page*]. When these workers used amounts of lead that produced an average blood-lead level of 30 micrograms per 100 milliliters of blood, the level of ALA-dehydrase activity in the brain did not differ significantly from the levels found in control rats that had not been given any added lead at all. It is now established experimentally that lead does interfere with heme synthesis in tissue preparations from the kidney, the brain and the liver as well as in red cells but the concentrations of lead that may begin to cause significant inhibition in these organs are not yet known.

Only in the blood is it as yet possible to see a direct cause-and-effect relation between the metabolic disturbance and the functional disturbance in animals or people. In the blood the functional effect is anemia. The decrease in heme synthesis leads at first to a decrease in the life-span of red cells and later to a decrease in the number of red cells and in the amount of hemoglobin per cell. In compensation for the shortage, the blood-forming tissue steps up its production of red cells; immature red cells, reticulocytes and basophilic stippled cells (named for their stippled appearance after absorbing a basic dye) appear in the circulation. The presence of stippled cells is the most characteristic finding in the blood of a patient with lead poisoning. The stippling represents remnants of the cytoplasmic constituents of red cell precursors, including mitochondria. Normal mature red cells do not contain mitochondria. The anemia of lead poisoning is a reversible condition: the metabolism of heme returns to normal, and the anemia improves with removal of the patient from exposure to excessive amounts of lead.

The toxic effect of lead on the kidneys is under intensive investigation but here the story is less clear. In acute lead poisoning there are visible changes in the kidney and kidney function is impaired. Again the mitochondria are implicated: their structure is visibly changed. Much of the excess lead is concentrated in the form of dense inclusions in the nuclei of certain cells, including those lining the proximal renal tubules. Robert A. Goyer of the University of North Carolina School of Medicine isolated and analyzed these inclusions and found that they consist of a complex of protein and lead [see upper illustration, page 336]. He has suggested that the inclusions are a protective device: they tend to keep the lead in the nucleus, away from the vulnerable mitochondria. Involvement of the mitochondria is also suggested by the fact that lead-poisoned kidney cells consume more oxygen than normal cells in laboratory cultures, which indicates

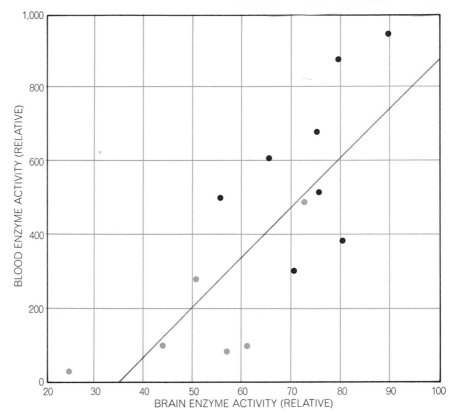

CORRELATION between the activity of ALA dehydrase in the blood and in the brain of normal rats (*black dots*) and lead-poisoned rats (*colored dots*) suggests that the enzyme may be implicated in brain damage, according to J. A. Millar and his colleagues. These data are for severely poisoned rats; in others with blood-lead values of about 30 micrograms per 100 milliliters of blood, brain enzyme activity was not significantly less than in controls.

that their energy metabolism is affected.

Kidney dysfunction, apparently due to this impairment in energy metabolism, is expressed in what is called the Fanconi syndrome: there is an increased loss of amino acids, glucose and phosphate in the urine because the damaged tubular cells fail to reabsorb these substances as completely as normal tubular cells do. The excessive excretion of phosphates is the important factor because it leads to hypophosphatemia, a low level of phosphate in the blood. There is some evidence that, when phosphate is mobilized from bone for the purpose of maintaining an adequate level in body fluids, lead that is stored with relative safety in the bones may be mobilized along with the phosphate and enter the soft tissues where it can do harm. The effect of acute lead poisoning on the kidney can be serious but, like the effect on blood cells, it is reversible with the end of abnormal exposure. Furthermore, the Fanconi syndrome is seen only at very high levels of lead in blood (greater than 150 micrograms of lead per 100 milliliters of blood) and only in patients with severe acute plumbism.

In the central nervous system the toxic effect of lead is least understood. Little

is known at the metabolic level; most of the information comes from clinical observation of patients and from postmortem studies. Two different mechanisms appear to be involved in lead encephalopathy, or brain damage: edema and direct injury to nerve cells. The walls of the blood vessels are somehow affected so that the capillaries become too permeable; they leak, causing edema (swelling of the brain tissue). Since the brain is enclosed in a rigid container, the skull, severe swelling destroys brain tissue. Moreover, it appears that certain brain cells may be directly injured, or their function inhibited, by lead.

The effects I have been discussing are all those of acute lead poisoning, the result of a large accumulation of lead in a relatively short time. There are chronic effects too, either the aftereffects of acute plumbism or the result of a slow buildup of a burden of lead over a period of years. The best-known effect is chronic nephritis, a disease characterized by a scarring and shrinking of kidney tissue. This complication of lead poisoning came to light in Australia in 1929, when L. J. J. Nye became aware of a pattern of chronic nephritis and early death in

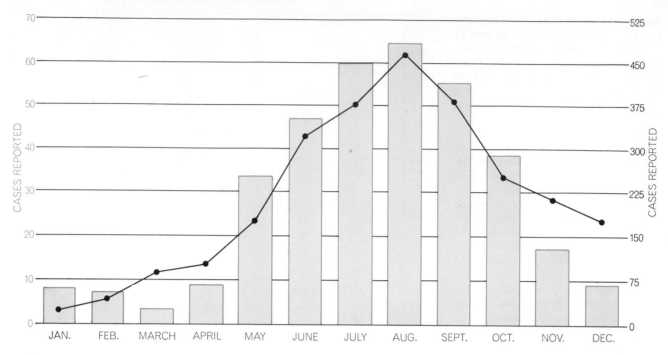

SEASONAL PATTERN of lead-poisoning cases is striking. The bars show the average number of cases reported monthly in Balti- more from 1931 through 1951 (*numbers at left*). Curve shows cases reported monthly in New York City last year (*numbers at right*).

the state of Queensland. Investigation revealed that Queensland children drank quantities of rainwater that was collected by runoff from house roofs sheathed with shingles covered with lead-pigmented paint. In 1954 D. A. Henderson found that of 352 adults in Queensland who had had childhood lead poisoning 15 to 40 years earlier, 165 had died, 94 of chronic nephritis. Chronic lead nephropathy, which is sometimes accompanied by gout, is also seen in persistent, heavy moonshine drinkers and in some people who have had severe industrial exposure. In all these cases, however,

the abnormal intake of lead persists for more than a decade or so before the onset of nephropathy. Most of the patients have a history of reported episodes of acute plumbism, which suggests that they have levels of lead in the tissues far above those found in the general population. Furthermore, there is the suspicion that factors in addition to lead may be involved.

The other known result of chronic overexposure to lead is peripheral nerve disease, affecting primarily the motor nerves of the extremities. Here the tissue damage appears to be to the myelin

sheath of the nerve fiber. Specifically, according to animal studies, the mitochondria of the Schwann cells, which synthesize the sheath, seem to be affected. Various investigators, including Pamela Fullerton of Middlesex Hospital in London, have found that conduction of the nerve impulse may be impaired in the peripheral nerves of industrial workers who have had a long exposure to lead but who have no symptoms of acute lead poisoning.

These findings and others raise serious questions. It is clear that a single attack of acute encephalopathy can cause pro-

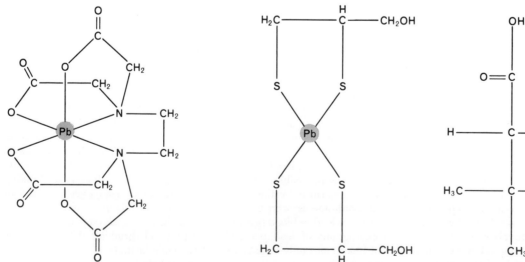

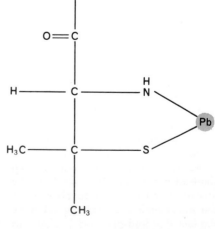

CHELATING AGENTS used in treating lead poisoning bind lead atoms (*Pb*) firmly in one or more five-member chelate rings. Dia- grams show lead chelates formed by EDTA (*left*), BAL (*middle*) and d-penicillamine (*right*). The last structure is still hypothetical.

found mental retardation and other forms of neurological injury that is permanent. Similarly, in young children repeated bouts of symptomatic plumbism can result in permanent brain damage ranging from subtle learning deficits to profound mental incompetence and epilepsy. Can a level of absorption that is insufficient to cause obvious acute symptoms nevertheless cause "silent" brain damage? This question remains unanswered, in part because of the difficulty in recognizing mild symptoms of lead poisoning in children and in part because the experimental studies that might provide some answers have not yet been undertaken.

Classical plumbism—the acute disease—is seen today primarily in children with the pica habit. Before discussing these cases in some detail I shall briefly take up two other current environmental sources of lead: earthenware improperly glazed with lead and lead-contaminated alcoholic beverages.

Michael Klein and his colleagues at McGill University recently reported two cases of childhood lead poisoning, one of which was fatal, that they traced to an earthenware jug in which the children's mother kept a continuously replenished supply of apple juice. The slightly acidic juice was leaching lead out of the glaze, the thin layer of glassy material fused to the ceramic surfaces of the jug. The investigators thereupon tested 117 commercial earthenware food and beverage containers and 147 samples made with 49 different commonly used glazes in the McGill ceramics laboratory. Excessive amounts of lead—more than the U.S. maximum permissible amount for glazes of seven parts per million—were leached out of half the vessels. (The maximum permissible amount should probably be reevaluated, since past methods of testing have not taken account of such variables as the quantity of the food or beverage consumed, its acidity, the length of time it is stored and whether or not it is cooked in the pottery.) As the McGill report points out, the danger of poisoning from lead-glazed pottery has been rediscovered periodically since antiquity. The Greeks knew about the danger but the Romans did not; they made the mistake of storing wine in earthenware. James Lind, who in 1753 recommended lemon or lime juice as a preventive for scurvy, also warned that the juices should not be stored in earthenware jugs. Now the index of suspicion has fallen too low: one physician poisoned himself recently by drinking a cola beverage (and 3.2 milligrams of

POPULATION	EXPOSURE (MICROGRAMS PER CUBIC METER OF AIR)	MEAN BLOOD LEAD (MICROGRAMS PER 100 GRAMS)
RURAL U.S.	0.5	16
URBAN U.S.	1.0	21
DOWNTOWN PHILADELPHIA	2.4	24
CINCINNATI POLICEMEN	2.1	25
CINCINNATI TRAFFIC POLICEMEN	3.8	30
LOS ANGELES TRAFFIC POLICEMEN	5.2	21
BOSTON AUTOMOBILE-TUNNEL EMPLOYEES	6.3	30

RESPIRATORY EXPOSURE to lead is reflected in the mean blood-lead values of various groups, according to John R. Goldsmith and Alfred C. Hexter of the California Department of Public Health. Groups apparently exposed to more lead in the air have generally higher blood-lead values; whether these indicate higher body burdens of lead is not known.

lead) every evening for two years from a mug his son had made for him. Do these cases represent isolated occurrences? How many other people are similarly exposed? Clearly the first step is the testing of earthenware and a reevaluation of its fabrication and use for food and drink.

In the manufacture of moonshine whiskey, lead solder is used in the tubing of distillation units. Moreover, discarded automobile radiators that contain lead often serve as condensers. Lead is therefore found in most samples of confiscated moonshine. Lead encephalopathy, nephritis with gout and other lead-related conditions have been reported in moonshine consumers, largely in the southeastern part of the U.S. The problem of diagnosis is complicated by the fact that the symptoms of acute alcoholism and acute lead poisoning are similar in many ways. (Again there is a historical record. The McGill report noted that the Massachusetts Bay Colony forbade rum distillation in leaded stills in 1723 in an effort to prevent "dry gripes," an intestinal condition. In 1767 Sir George Baker blamed "the endemic colic of Devonshire" on the use of lead-lined troughs in the making of apple cider.)

Childhood lead poisoning in the U.S. is seen almost exclusively in children of preschool age who live in deteriorated housing built before 1940 (when titanium dioxide began to replace lead in the pigment of most interior paints). The causative factors are commonly a triad: a dilapidated old house, a toddler with pica and parents with inadequate resources (emotional, intellectual, informa-

tional and/or economic) to cope with the family's needs. The three factors interact to increase the likelihood that the child will eat chips of leaded paint. A chip of paint about the size of an adult's thumbnail can contain between 50 and 100 milligrams of lead, and so a child eating a few small chips a day easily ingests 100 or more times the tolerable adult intake of the metal! In one study conducted some years ago at the Baltimore City Hospitals and the Johns Hopkins Hospital, Harold E. Harrison and I found that the average daily fecal excretion of lead by children with severe plumbism was 44 milligrams. In a group of normal unexposed children we found a daily fecal lead excretion of less than .2 milligram of lead. In other words, pica for leaded paint results in genuinely massive exposures. And when the abnormal intake ceases, it may be several months or years before blood-lead levels return to normal.

The repeated ingestion of leaded-paint chips for about three months or longer can lead to clinical symptoms and eventually to the absorption of a potentially lethal body burden of lead. During the first four to six weeks of abnormal ingestion there are no symptoms. After a few weeks minor symptoms such as decreased appetite, irritability, clumsiness, unwillingness to play, fatigue, headache, abdominal pain and vomiting begin to appear. These, of course, are all quite nonspecific symptoms, easily ignored as behavior problems or blamed on various childhood diseases. In a few weeks the lassitude may progress to intermittent drowsiness and stupor; the vomiting may become persistent and forceful;

brief convulsions may occur. If the exposure to lead continues, the course of the disease can culminate abruptly in coma, intractable convulsions and sometimes death.

This picture of fulminating encephalopathy is commonest in children between 15 and 30 months of age; older children tend to suffer recurrent but less severe acute episodes and are usually brought to the hospital with a history of sporadic convulsions, behavior problems, hyperactivity or mental retardation. The symptoms tend to wax and wane, usually becoming more severe in summer. (Some 85 percent of all lead-poisoning cases are reported from May through October. This remarkably clear seasonal pattern is still not understood. It may be due at least in part to the fact that the ultraviolet component of sunlight increases the absorption of lead from the intestine.)

The symptoms of even acute encephalopathy are nonspecific, resembling those of brain abscesses and tumors and of viral and bacterial infections of the brain. Diagnosis depends, first of all, on a high level of suspicion. To make a positive diagnosis it is necessary to show high lead absorption as well as the adverse effects of lead. This requires the measurement of lead in blood and other specialized tests. Mild symptoms may be found in the presence of values of between 60 and 80 micrograms of lead per 100 milliliters of blood. As the blood-lead level rises above 80 micrograms the risk of severe symptoms increases sharply. Even in the absence of symptoms, in children blood-lead levels exceeding 80 micrograms call for immediate treatment and separation of the child from the source of lead.

Treatment is with potent compounds known as chelating agents (from the Greek *chēlē*, meaning claw): molecules that tend to bind a metal atom firmly, sequestering it and thus rendering it highly soluble [see the article "Chelation," by Harold F. Walton, beginning on page 344]. Chelating agents remove lead atoms from tissues for excretion through the kidney and through the liver. With chelating agents very high tissue levels of lead can be rapidly reduced to levels approaching normal, and the adverse metabolic effects can be promptly suppressed. Initially two agents are administered by injection: EDTA and BAL. (EDTA, or edathamil, is ethylenediaminetetraacetic acid; BAL is "British Anti-Lewisite," developed during World War II as an antidote for lewisite, an arsenic-containing poison gas.) After the lead level has been reduced another agent, d-penicillamine, may be administered orally as a follow-up therapy.

Before chelating agents were available about two-thirds of all children with lead encephalopathy died. Now the mortality rate is less than 5 percent. Unfortunately the improvement in therapy has not substantially reduced the incidence of brain damage in the survivors. Meyer A. Perlstein and R. Attala of the Northwestern University Medical School found that of 59 children who developed encephalopathy, 82 percent were left with permanent injury: mental retardation, convulsive disorders, cerebral palsy or blindness. This high incidence of permanent damage suggests that some of these children must have had recurrent episodes of plumbism; we have found that if a child who has been treated for acute encephalopathy is returned to the same hazardous environment, the risk of permanent brain damage rises to virtually 100 percent. In Baltimore, with the help of the Health Department and through the efforts of dedicated medical social workers, we are able to make it an absolute rule that no victim of lead poisoning is ever returned to a dangerous environment. The child goes from the hospital to a convalescent home and does not rejoin his family until all hazardous lead

	I NO DEMONSTRABLE EFFECTS	II MINIMAL SUBCLINICAL EFFECTS DETECTABLE	III COMPENSATION	IV FUNCTIONAL INJURY (SHORT, INTENSE EXPOSURE)
METABOLIC EFFECTS	NORMAL	URINARY ALA MAY INCREASE	INCREASE IN SEVERAL METABOLITES IN BLOOD AND URINE	FURTHER INCREASE IN METABOLITES
FUNCTIONAL EFFECTS: BLOOD	NONE	NONE	REDUCED RED CELL LIFE-SPAN. INCREASED PRODUCTION	REDUCED RED CELL LIFE-SPAN WITH OR WITHOUT ANEMIA (REVERSIBLE)
KIDNEY FUNCTION	NORMAL	NORMAL	SOMETIMES MINIMAL DYSFUNCTION	FANCONI SYNDROME (REVERSIBL
CENTRAL NERVOUS SYSTEM	NONE	NONE	?	MINIMAL TO SEVERE BRAIN DAMAGE (PERMANENT)
PERIPHERAL NERVES	NONE	NONE	?	POSSIBLE DAMAGE
SYMPTOMS	NONE	NONE	SOMETIMES MILD, NON-SPECIFIC COMPLAINTS	ANEMIA, COLIC, IRRITABILITY, DROWSINESS; IN SEVERE CASES, MOTOR CLUMSINESS, CONVULSI AND COMA
RESIDUAL EFFECTS	NONE	NONE	NONE KNOWN	RANGE FROM MINIMAL LEARNIN DISABILITY TO PROFOUND MENT. AND BEHAVIORAL DEFICIENCY, CONVULSIVE DISORDERS, BLIND

EFFECTS OF LEAD are associated in a general way with five levels of exposure and rates of absorption of the metal. Level I is associated with blood-lead concentrations of less than 30 micrograms of lead per 100 milliliters and Level II with the 30–50 microgram range. Level III, at which compensatory mechanisms apparently minimize or prevent obvious functional injury, may be associated with concentrations of between 50 and 100 micrograms. Level IV is usually associated with concentrations greater than 80 micrograms but impairment may be evident at lower levels, particularly if compensatory responses are interfered with by some other disease state.

sources have been removed or the family has been helped to find lead-free housing. Cases of permanent brain damage nevertheless persist. It appears that even among children who suffer only one episode, are properly treated and are thereafter kept away from lead, at least 25 percent of the survivors of lead encephalopathy sustain lasting damage.

Clearly, then, treatment is not enough; the disease must be prevented. Children with increased lead absorption must be identified before they become poisoned. Going a step further, the sources of excessive lead exposure must be eliminated.

Baltimore has taken a "case-finding" approach to these tasks. Free diagnostic services were established by the city Health Department in the 1930's. Physicians took advantage of the services, and increasing numbers of cases were discovered. Since 1951 the removal of leaded paint has been required in any dwelling where a child is found with a blood-lead value of more than 60 micrograms. The number of cases reported each year rose for some time as diagnostic methods and awareness improved, but recently it has leveled off. In order to reach children before they are poisoned, however, more is required than

CTIONAL INJURY (CHRONIC
RECURRENT INTENSE EXPOSURE)

REASE ONLY IN CASE OF
ENT EXPOSURE

SIBLE ANEMIA (REVERSIBLE)

RONIC NEPHROPATHY
RMANENT)

ERE BRAIN DAMAGE, PARTICULARLY
HILDREN (PERMANENT)

AIRED CONDUCTION
Y BE CHRONIC)

TAL DETERIORATION.
URES, COMA,
T OR WRIST DROP

TAL DEFICIENCY
EN PROFOUND), KIDNEY
UFFICIENCY, GOUT (UNCOMMON),
T DROP (RARE)

What one can say is that the risk of functional injury increases as the concentration of lead in the blood exceeds 80 micrograms per 100 milliliters. The residual effects persist after blood-lead levels return to normal.

case-finding; what is needed is a screening program that examines entire populations of children in high-risk areas of cities. Chicago undertook that task in the 1960's. Last year New York City inaugurated a new and intensive screening program in which children are being tested for blood lead in hospitals and at a large number of neighborhood health centers; an educational campaign has been launched to bring lead poisoning and the testing facilities to public notice. As in Baltimore, a blood-lead finding of more than 60 micrograms results in an examination of the child's home. If any samples of paint and plaster contain more than 1 percent of lead, the landlord is ordered to correct the condition by covering the walls with wallboard to a height of at least four feet and by removing all leaded paint from wood surfaces; if the landlord does not comply, the city undertakes the work and bills him. Before the new program was begun New York was screening about 175 blood tests a week; by the end of the year it was doing about 2,000 tests a week. Whereas 727 cases of lead poisoning were reported in the city in 1969, last year more than 2,600 were reported. As Evan Charney of the University of Rochester School of Medicine and Dentistry has put it, "the number of cases depends on how hard you look."

Screening is complicated by technical difficulties in testing both children and dwellings. The standard dithizone method of determining blood lead requires between five and 10 cubic centimeters of blood taken from a vein—a difficult procedure in very small children—and the analysis is time-consuming. What is needed is a dependable test that can be carried out on a drop or two of blood from a finger prick. A variety of approaches are now being tried in several laboratories in order to reach this goal; as yet no microtest utilizing a drop or two of blood has been proved practical on the basis of large-scale use in the field. Several appear to be promising in the laboratory, so that field testing in the near future can be anticipated. As for the checking of dwellings, the standard method is laborious primarily because it requires the collection of a large number of samples. Several different portable instruments are under development, including an X-ray fluorescence apparatus that gives a lead-content reading when it is pointed at a surface, but these devices have not yet been proved reliable in the field.

Since World War II the incidence of lead poisoning (usually in the form of

lead palsy) among industrial workers, which was once a serious problem, has been reduced by various control measures. The danger is now limited primarily to small plants that are not well regulated and to home industries.

There is increasing concern over environmental lead pollution. Claire C. Patterson of the California Institute of Technology has shown that the levels of lead in polar ice have risen sharply since the beginning of the Industrial Revolution. Henry A. Schroeder of the Dartmouth Medical School has shown that the burden of lead in the human body rises with age, and that this rise is due almost entirely to the concentration of lead in bone. Although man's exposure to lead in highly industrialized nations may come from a variety of sources, the evidence points to leaded gasoline as the principal source of airborne lead today. These observations have occasioned much speculation. It is nonetheless clear that a further rise in the dissemination of lead wastes into the environment can cause adverse effects on human health; indeed, concerted efforts to lower the current levels of exposure must be made, particularly in congested urban areas.

At the moment there is no evidence that any groups have mean blood levels that approach the dangerous range. Some, however, do have levels at which a minimal increase in urinary ALA, but nothing more, is to be expected. This includes people whose occupation brings them into close and almost daily contact with automotive exhaust. These observations emphasize the need to halt any further rise in the total level of exposure. A margin of safety needs to be defined and maintained. This will require research aimed at elucidating the effects of long-term exposure to levels of lead insufficient to cause symptoms or clear-cut functional injury. With regard to respiratory exposure, it is still not clear what fraction of the inhaled particles reaches the lungs and how much of that fraction is actually absorbed from the lung. Still another important question is the storage of lead in bone. Can any significant fraction of lead in bone be easily and quickly mobilized? If so, under what circumstances is it mobilized? There are more questions than answers to the problems posed by levels of lead only slightly higher than those currently found in urban man. Much research is required.

With regard to childhood lead poisoning, however, we know enough to act. It is impermissible for a humane society to fail to do what is necessary to eliminate a wholly preventable disease.

37

CHELATION

HAROLD F. WALTON
June 1953

It is a chemical process wherein the atoms of a metal in solution are "sequestered" by ring-shaped molecules. Chelating agents have now become important tools of analytical chemistry and technology

THE MOST USEFUL tool of the chemist's trade is the cabalistic scribble of letters and connecting lines with which he pictures a molecule. When, some 90 years ago, the German chemist Friedrich August Kekulé first visualized a molecule as a group of little balls (atoms) joined by sticks, he gave chemistry something much more important than a convenient scheme for writing chemical formulas. Kekulé's visions (one of his discoveries, the benzene ring, actually was suggested to him by a dream of a snake chasing its tail) provided the beginning of our understanding of how a molecule is constructed and of the bonds between atoms. And today the diagram with which a chemist represents a molecule on paper carries a great deal of meaning: it is, in fact, a prediction as to how the substance will behave chemically.

Within the past year or so wide interest has developed in a new branch of chemistry which gets its name from the symbols used to picture its peculiar type of molecule. The chemist's formula for this kind of compound shows rings of atoms in which arrows, representing a special kind of chemical bond, grip a central atom like a claw. The structure is called a chelate ring, from the Greek word *chele*, meaning claw. Chelation is

not a brand-new discovery: there are many chelate compounds in nature, among them the hemoglobin of blood and the chlorophyll of plants. But new ways to use chelation are being found, and there is now rising a flourishing industry which produces made-to-order chelate compounds for many purposes, from softening water to dissolving kidney stones. Chelation also is revolutionizing chemical analysis in the laboratory.

The various uses of the chelate compounds all depend on one fascinating property: the ability of the crablike claw to seize and sequester atoms of metal. A chelate compound will hunt down traces of any given metal in a liquid. It is as if we sent a posse of crabs into a mixed population of flora and fauna with instructions to seize and swallow all the left-handed snails.

TO UNDERSTAND how chelation works we must examine the nature of a chemical bond. According to the modern theory of valence, the atoms in a molecule are bound together by electrons, the charged particles that surround every atom. The bond may be established in one of two ways. An atom may transfer one of its electrons to its neighbor. In that case the atom that loses the electron also loses its electri-

cal neutrality and becomes positively charged, while the atom that receives the electron becomes negatively charged. These two "ions" then are held together by the electrical attraction of their opposite charges. The other way in which two atoms may be bound together is by sharing a pair of electrons—as if two persons were held together by a pair of ropes that belonged not exclusively to either individual but to both together. This is called a covalent bond: the chemist represents it by a single line joining the two atoms. Usually each of the two joined atoms supplies one of the two binding electrons. But sometimes one atom supplies both, and that kind of link is called a coordinate bond. The chemist's symbol for such a bond is an arrow pointing toward the atom which has received the electrons.

Now a chelate ring is simply a group of atoms linked into a ring with one or more coordinate bonds. The atoms that donate the electrons are usually oxygen, nitrogen or sulfur; the acceptor atom, grasped in the claw of arrows, is nearly always a metal. In such a ring the metal atom is gripped more firmly than if it were merely attached to atoms in independent molecules. Another way of saying this is that a metal atom is much more prone to unite with two donor

COORDINATION COMPOUNDS of the cupric ion (Cu^{++}) sequester copper. The arrows denote coordinate bonds. The compound at right is chelated; the one at left is not. The chelated complex is more stable.

atoms in a ring-forming molecule than with the same atoms in two separate molecules. The mechanics of the situation make clear why this is so. To become attached to two separate molecules, the metal atom must capture a donor atom in each molecule separately, and this depends on chance contacts. But when the metal atom becomes attached to one end of a molecule that can form a ring around it, it easily links up with the other end, for the latter is tethered and cannot range far afield.

What kind of molecule do we need to form a chelate ring? In the first place, the molecule must contain at least two atoms that can attach themselves to a metal ion. Secondly, if we are to have a strain-free, stable ring, the atoms forming it must join in such a way that their valence bonds are at their natural angle. The natural angle of the bonds in the carbon atom, which determines the structure of most chelate rings, is slightly more than 108 degrees, and this is the size of the angles of a pentagon. Hence the ideal number of atoms to make a chelate ring is five. There are many known chelate rings with six atoms, but very few have fewer than five or more than six.

One of the best chelating compounds is a chemical with the imposing name of ethylene diamine tetraacetic acid, called EDTA for short. The atoms in this molecule are spaced just the right distance apart to give strain-free chelate rings with five atoms apiece. What is more, the molecule has no fewer than six atoms (two of nitrogen and four of oxygen) that can donate electrons to metals. It will grip an atom of iron with not one but five or six chelate rings: the molecule might fittingly be called an octopus rather than a crab [see bottom diagram on page 346].

It is this kind of molecule, capable of forming several chelate rings at once, that has created the chelation industry. Studies of EDTA and similar chelating agents were begun about eight years ago by the Swiss chemist Gerold Schwarzenbach and have now been taken up in the U. S. by Arthur E. Martell of Clark University and other chemists. Already in commercial production are several chelating compounds, known variously by the trade names Versene, Sequestrene, Nullapon and Trilon.

CHELATION can be applied to any problem in which the presence of metal ions causes trouble. Suppose, to take a common example, that our water supply contains dissolved salts of iron. The iron forms a sediment on standing; it discolors bathtubs and linens; it spoils the taste of tea. On the domestic scale it is very difficult to remove. Thanks to chelation, however, we do not need to remove it to prevent its ill effects. We may,

BEAKER CONTAINING IRON is photographed in the laboratory of the Bersworth Chemical Co. The iron is in the form of an insoluble compound.

BEAKER IS CLEARED by the addition of a chelating agent. The iron is still there, but it is sequestered by the rings of the chelating agent.

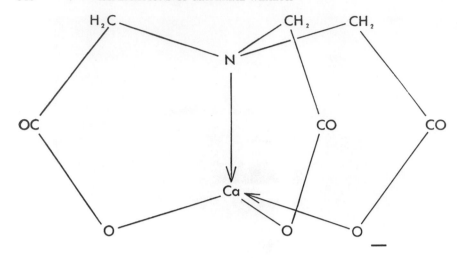

AMMONIATRIACETIC ACID sequesters calcium (Ca) in several chelate rings. The complex is an ion carrying one negative charge (*bottom right*).

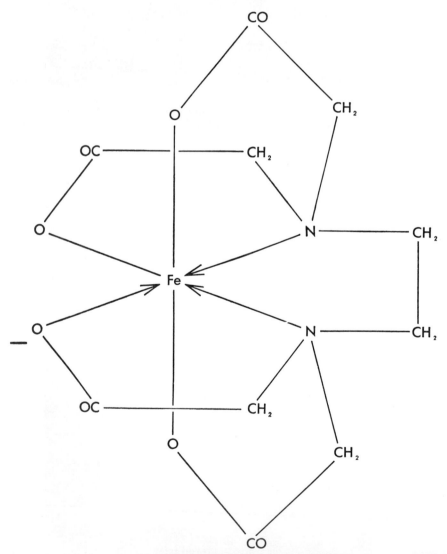

ETHYLENE DIAMINE TETRAACETIC ACID, or EDTA, sequesters iron (Fe) in more rings than ammoniatriacetic acid. It also is negatively charged.

instead, add EDTA to the water. Now the iron will leave no stains, form no sediment: the water becomes sparkling clear. The iron is still there, yet it cannot be detected even by sensitive chemical tests. It is tightly imprisoned and hidden away—"sequestered," in the poetic language of chelation technology—by EDTA's chelate rings.

The softening of water so far has been the largest use of chelation. The chelating compounds combine readily not only with iron but also with the calcium and magnesium ions of hard water. They are too expensive for large-scale use, but are ideal for adding to shampoos, soaps and detergents. Unlike some other water-softeners, chelates of the EDTA type do not lose their strength on prolonged contact with water.

EDTA also is added to dyes to get rid of traces of iron and other metals which may modify or weaken colors or even leave rust stains on the fabric. The metals spoil dyes, incidentally, by forming chelate rings with the dye molecules.

Another field where traces of metal do a lot of damage is in food preservation. Even one part per million of metal ions catalyzes atmospheric oxidation, and this is what makes cut apples turn brown, fats and oils go rancid, orange juice lose its vitamin C and most of its flavor and green vegetables spoil. The addition of one hundredth of 1 per cent of EDTA improves the keeping qualities of such food enormously. EDTA helps preserve rubber latex and high-energy rocket fuels also.

A large dose of EDTA will reduce the concentration of free metal ions in a solution to the vanishing point. With a moderate, calculated dose we can hold the concentration at any level we wish, just as we control the concentration of hydrogen ions, or the acidity, of a solution. EDTA is used in this way to regulate the deposit of metal in electroplating, assuring a smooth, adherent coat.

EDTA can dissolve insoluble salts of metals as well as prevent their formation. This makes it a useful decontaminant for radioactivity: it will wash off invisible films of radioactive metal salts where soap and water will not.

THE SAME property enables EDTA to help dissolve kidney stones, decalcify bone and rid the body of poisonous heavy metals, notably lead and plutonium; EDTA offers almost the only hope of treatment for plutonium poisoning. In such applications it is fed as the calcium salt, so that calcium will not be removed from the blood or bones. On the other hand, EDTA also is useful where we want to take calcium ions out of the blood to prevent clotting. Recent tests have shown that EDTA makes it possible to preserve whole blood nearly twice as long as does the citrate solution which the Red Cross now uses to keep blood from clotting before it is processed.

Citric, malic, lactic and tartaric acids are among nature's chelating agents. They keep metal ions from precipitating in body fluids, much as EDTA prevents precipitation of iron from well waters. The root hairs of plants secrete chelating acids which dissolve such compounds as ferric oxide and calcium carbonate and make the iron and calcium of the soil available to the plant. Humus assists this process, for it, too, contains chelating agents. Soils deficient in humus can be improved by adding EDTA.

Chlorophyll and hemoglobin are chelated compounds of a very special type. They contain the "porphyrin ring": a complicated arrangement of rings within rings which is flat and has four nitrogen atoms placed at the four corners of a square. In the middle of this square is a metal atom gripped by the four nitrogens. In hemoglobin the metal is iron. The iron atom fits very nicely, for it is about the right size, and what is more important, the normal arrangement of the bonds by which it is gripped is flat and not tetrahedral. In chlorophyll the situation is different. Its metal, magnesium, has bonds that normally angle toward the four corners of a tetrahedron. Consequently in the chlorophyll molecule these bonds are strained and the chelate ring is less stable. It is something of a mystery that chlorophyll should contain magnesium rather than some other metal. Once we get chlorophyll out of the living plant cell, its magnesium can be displaced by almost any other metal. To stabilize commercial chlorophyll preparations the magnesium is generally replaced by copper.

FROM ALL THIS it is obvious how chelation becomes a tool of chemical analysis. A chelating agent can be used like a pair of forceps to pluck out a specific kind of atom or ion from a complex mixture. Its selectivity comes from the fact that a given chelating agent grips some metal ions more tenaciously than others. We can apply this selectivity in two ways. Either we hold one metal in solution by chelation while we precipitate or extract another, or we use the chelating agent to make the metal we want insoluble so that we can extract it. The second method is the sharper tool.

Let us examine one particularly vivid example. Every analytical chemist knows that when the chelating agent dimethylglyoxime is introduced into a dilute ammonia solution containing nickel ions, he will get a beautiful scarlet precipitate—so brilliant in color that it has been used in lipsticks. The reaction is an analytical chemist's dream. Of the 98 known elements, only nickel is com-

BLEACHED LEAVES of a citrus plant suffer from chlorosis, a disease caused by a lack of soluble iron in the soil. The condition can be corrected by adding to the soil a soluble chelated complex of the metal.

pletely precipitated by this treatment! No other test is so specific for one metal.

The structure of the chelate compound explains this specificity. The molecules of the dimethylglyoxime chelate that holds the nickel are electrically neutral and completely flat—so flat that they can be stacked up together like a deck of cards to make a compact crystal. Each molecule contains two dimethylglyoxime units, one on each side of the central nickel atom. To form this flat, symmetrical molecule four distinct requirements must be met. First, if the molecule is to be neutral, the metal ion must have a charge of plus two, so that when it combines with the two dimethylglyoxime units no electric charge is left over. Second, it must readily accept electrons from the donor nitrogen atoms. Third, since each dimethylglyoxime contains two nitrogens, the metal ion must join with four nitrogens, and these must lie at the corners of a square, not of a tetrahedron. Finally, the metal ion must be the right size. These requirements are like the tumblers of a lock: to open the lock, the key must fit them all. Nickel is the key that fits this lock. Its ions have a charge of plus two; they coordinate easily with nitrogen; their valence bonds are in one plane and not tetrahedral; the ions are the right size. The ions of copper have almost exactly the same combination of properties as nickel, and they will form a chelate with dimethylglyoxime. But what makes nickel unique here is that the copper chelate is soluble,

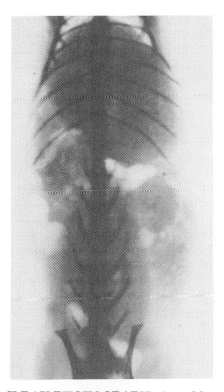

X-RAY PHOTOGRAPH of a rabbit demonstrates the use of a chelating agent to make not only the bones but also the soft tissues visible. The animal was fed chelated lead, sequestered from its usual toxic action. The lead makes the internal organs relatively opaque and is eliminated.

whereas nickel precipitates out of the solution.

DIMETHYLGLYOXIME is one of hundreds of organic reagents for metals used in modern chemical analysis. Nearly all of them form chelate rings. None is quite as selective as dimethylglyoxime and its homologues, or gives quite as beautiful a color as the dimethylglyoxime-nickel reaction. But many do form strongly colored compounds. The combination of color production with selectivity is just what we need for the analysis of complex mixtures. By looking at the colors we can tell what metals are there, and by using a photoelectric colorimeter we can tell how much of each metal is present. One of the best reagents for this kind of analysis is a sulfur-containing chelating agent known as dithizone. With certain metals dithizone forms red or orange chelates which are so intensely colored that as little as a 10-millionth of a gram of the metal can be detected. This method is as sensitive as the spectrograph, and a lot less cumbersome. Dithizone is used to check the content of trace elements in soils and to look for minerals by analyzing the water in streams that flow near the ore beds.

A good way to use chelating agents in analysis is to set two of them against each other in a chemical tug-of-war. The National Bureau of Standards employs this device to measure traces of copper in steel. The elements to be separated are iron and copper; the chelating contestants are EDTA and sodium diethyldithiocarbamate. Neither of these agents is sufficiently selective to separate iron and copper by itself. But EDTA grips iron a trifle more strongly than it does copper, while the other reagent is more partial to copper. When the two reagents are used together, the EDTA gets all the iron (also chromium, cobalt and nickel if they are present), and the sodium diethyldithiocarbamate gets all the copper and nothing else. This copper chelate is intensely colored. It is extracted with an organic solvent—a common practice with metal chelates—and the color is measured with a suitable instrument.

One more example will show the usefulness of these reagents to the analytical chemist. Often he wants to know exactly how much of a dissolved substance is present, by volume, in a liquid; for instance, he may want to measure precisely the total hardness of a sample of water, meaning the sum of its calcium and magnesium ions. He adds to the water a solution of EDTA of exactly known concentration. He drips in the solution a little at a time, as he wishes to shut it off at precisely the moment when the EDTA has swallowed all the calcium and magnesium ions. To determine that point he uses a certain purple-red dye, itself a weakly bound magnesium chelate, which is torn apart by EDTA and changes abruptly from purple-red to sky blue when all the calcium and magnesium ions are gone. Now, from the amount of EDTA solution that was needed to swallow up the ions, he can calculate exactly how much calcium and magnesium was in the water.

From the laundry tub to the analytical laboratory, the versatile chelating agents justify their description as "chemistry's most precise chemicals."

BIBLIOGRAPHIES

I THE STUFF OF LIFE

1. Mineral Cycles

THE YEARLY CIRCULATION OF CHLORIDE AND SULFUR IN NATURE: METEOROLOGICAL, GEOCHEMICAL AND PEDOLOGICAL IMPLICATIONS, PART I. Erik Eriksson in *Tellus*, Vol. 11, No. 4, pages 375–403; November, 1959.

THE YEARLY CIRCULATION OF CHLORIDE AND SULFUR IN NATURE: METEOROLOGICAL, GEOCHEMICAL AND PEDOLOGICAL IMPLICATIONS, PART II. Erik Eriksson in *Tellus*, Vol. 12, No. 1, pages 63–109; February, 1960.

BIOGEOCHEMISTRY OF SULFUR ISOTOPES: PROCEEDINGS OF A NATIONAL SCIENCE FOUNDATION SYMPOSIUM, April 12–14, 1962. Edited by Mead LeRoy Jensen. Yale University, Department of Geology, 1962.

THE CALCIUM, MAGNESIUM, POTASSIUM, AND SODIUM BUDGETS FOR A SMALL FORESTED ECOSYSTEM. G. E. Likens, F. H. Bormann, N. M. Johnson and R. S. Pierce in *Ecology*, Vol. 48, No. 5, pages 772–785; Late Summer, 1967.

THE SULFUR CYCLE IN LAKE WATERS DURING THERMAL STRATIFICATION. M. Stuiver in *Geochimica et Cosmochimica Acta*, Vol. 31, No. 11, pages 2151–2167; November, 1967.

PRODUCTION AND MINERAL CYCLING IN TERRESTRIAL VEGETATION. L. E. Rodin and N. I. Bazilevich. Edited by G. E. Fogg. Oliver & Boyd, Ltd, 1968.

2. The Chemical Elements of Life

TRACE ELEMENTS IN BIOCHEMISTRY. II. J. M. Bowen. Academic Press, 1966.

CONTROL OF ENVIRONMENTAL CONDITIONS IN TRACE ELEMENT RESEARCH: AN EXPERIMENTAL APPROACH TO UNRECOGNIZED TRACE ELEMENT REQUIREMENTS. Klaus Schwarz in *Trace Element Metabolism in Animals*, edited by C. F. Mills, E. & S. Livingstone, 1970.

THE PROTEINS: METALLOPROTEINS, VOL. V. Edited by Bert L. Vallee and Warren E. C. Wacker. Academic Press, 1970.

CERULOPLASMIN: A LINK BETWEEN COPPER AND IRON METABOLISM. Earl Frieden in *Bioinorganic Chemistry*, Advances in Chemistry Series 100.

American Chemical Society, 1971.

TRACE ELEMENTS IN HUMAN AND ANIMAL NUTRITION. E. J. Underwood. Academic Press, 1971.

3. The Water Cycle

WATER: A STUDY OF ITS PROPERTIES, ITS CONSTITUTION, ITS CIRCULATION ON THE EARTH, AND ITS UTILIZATION BY MAN. Cyril S. Fox. Philosophical Library, 1952.

WATER. Arthur M. Buswell and Worth H. Rodebush in *Scientific American*, Vol. 194, No. 4, pages 76–89; April, 1956.

THE ROLE OF SEEPAGE MOISTURE IN SOIL FORMATION, SLOPE DEVELOPMENT, AND STREAM INITIATION. B. T. Bunting in *American Journal of Science*, Vol. 259, No. 7, pages 503–518; Summer, 1961.

4. The Oxygen Cycle

THE ATMOSPHERES OF THE PLANETS. Harold C. Urey in *Handbuch der Physik, Vol. LII, Astrophysics III: The Solar System*, edited by S. Flügge. Springer-Verlag, 1959.

HISTORY OF MAJOR ATMOSPHERIC COMPONENTS. L. V. Berkner and L. C. Marshall in *Proceedings of the National Academy of Sciences*, Vol. 53, No. 6, pages 1215–1226; June, 1965.

ATMOSPHERIC AND HYDROSPHERIC EVOLUTION ON THE PRIMITIVE EARTH. Preston E. Cloud, Jr., in *Science*, Vol. 160, No. 3829, pages 729–736; May 17, 1968.

DISSOCIATION OF WATER VAPOR AND EVOLUTION OF OXYGEN IN THE TERRESTRIAL ATMOSPHERE. R. T. Brinkmann in *Journal of Geophysical Research*, Vol. 74, No. 23, pages 5355–5368; October 20, 1969.

THE EVOLUTION OF PHOTOSYNTHESIS. John M. Olson in *Science*, Vol. 168, No. 3930, pages 438–446; April 24, 1970.

5. The Nitrogen Cycle

AUTOTROPHIC MICRO-ORGANISMS: FOURTH SYM-

POSIUM OF THE SOCIETY FOR GENERAL MICROBI-OLOGY HELD AT THE INSTITUTION OF ELECTRICAL ENGINEERS, LONDON, APRIL, 1954. Cambridge University Press, 1954.

DENTRIFICATION. C. C. Delwiche in *A Symposium on Inorganic Nitrogen Metabolism: Function of Metallo-Flavoproteins*, edited by William D. McElroy and Bentley Glass. The Johns Hopkins Press, 1956.

NITROGEN FIXATION IN PLANTS. W. D. P. Stewart. Athlone Press, 1966.

SYMBIOSIS: ITS PHYSIOLOGICAL AND BIOCHEMICAL SIGNIFICANCE. Edited by S. Mark Henry. Academic Press, 1966.

FIXATION OF NITROGEN BY HIGHER PLANTS OTHER THAN LEGUMES. G. Bond in *Annual Review of Plant Physiology: Vol. XXVIII*, edited by Leonard Machlis, Winslow R. Briggs and Roderic B. Park. Annual Reviews, Inc., 1967.

6. The Carbon Cycle

GEOGRAPHIC VARIATIONS IN PRODUCTIVITY. J. H. Ryther in *The Sea: Ideas and Observations on Progress in the Study of the Seas. Vol. II: The Composition of Sea-Water—Comparative and Descriptive Oceanography*, edited by M. N. Hill. Interscience Publishers, 1963.

THE INFLUENCE OF ORGANISMS ON THE COMPOSITION OF SEA-WATER. A. C. Redfield, B. H. Ketchum and F. A. Richards in *The Sea: Ideas and Observations on Progress in the Study of the Seas. Vol. II: The Composition of Sea-Water—Comparative and Descriptive Oceanography*, edited by M. N. Hill. Interscience Publishers, 1963.

THE ROLE OF VEGETATION IN THE CARBON DIOXIDE CONTENT OF THE ATMOSPHERE. Helmut Lieth in *Journal of Geophysical Research*, Vol. 68, No. 13, pages 3887–3898; July 1, 1963.

GROSS-ATMOSPHERIC CIRCULATION AS DEDUCED FROM RADIOACTIVE TRACERS. Bert Bolin in *Research in Geophysics, Vol. II: Solid Earth and Interface Phenomena*, edited by Hugh Odishaw. The M.I.T. Press, 1964.

IS CARBON DIOXIDE FROM FOSSIL FUEL CHANGING MAN'S ENVIRONMENT? Charles D. Keeling in *Proceedings of the American Philosophical Society*, Vol. 114, No. 1, pages 10–17; February 16, 1970.

PHOTOSYNTHESIS. E. Rabinowitch and Govindjee. John Wiley & Sons, Inc., 1969.

7. The Mechanism of Photosynthesis

PHOTOPHOSPHORYLATION AND THE CHEMI-OSMOTIC HYPOTHESIS. André T. Jagendorf and E. Uribe in *Brookhaven Symposia in Biology*, Vol. 19, pages 215–245; 1966.

ELECTRON TRANSPORT PATHWAYS IN PHOTOSYNTHESIS. Geoffrey Hind and John M. Olson in *Annual Review of Plant Physiology: Vol. XIX*, edited by Leonard Machlis. Annual Reviews, Inc., 1968.

HAEM-PROTEINS IN PHOTOSYNTHESIS. D. S. Bendall and R. Hill in *Annual Review of Plant Physiology: Vol. XIX*, edited by Leonard Machlis. Annual Reviews, Inc., 1968.

II   OUR IMPACT ON THE LAND THAT FEEDS US

8. The Flow of Energy in the Biosphere

FUNDAMENTALS OF ECOLOGY. Eugene P. Odum. W. B. Saunders Company, 1959.

ENERGY EXCHANGE IN THE BIOSPHERE. David M. Gates. Harper & Row, Publishers, 1962.

PHYSICAL CLIMATOLOGY. William D. Sellers. The University of Chicago Press, 1965.

ENERGY FLOW IN BIOLOGY. Harold J. Morowitz. Academic Press, 1968.

CONCEPTS OF ECOLOGY. Edward Kormondy. Prentice-Hall, Inc., 1969.

9. Human Food Production as a Process in the Biosphere

MALNUTRITION AND NATIONAL DEVELOPMENT. Alan D. Berg in *Foreign Affairs*, Vol. 46, No. 1, pages 126–136; October, 1967.

ON THE SHRED OF A CLOUD. Rolf Edberg. Translated by Sven Ahmån. The University of Alabama Press, 1969.

POLITICS AND ENVIRONMENT: A READER IN ECOLOGICAL CRISIS. Edited by Walt Anderson. Goodyear Publishing Company, Inc., 1970.

POPULATION, RESOURCES, ENVIRONMENT: ISSUES IN HUMAN ECOLOGY. Paul R. Ehrlich and Anne H. Ehrlich. W. H. Freeman and Company, 1970.

SEEDS OF CHANGE: THE GREEN REVOLUTION AND DEVELOPMENT IN THE 1970's. Lester R. Brown. Praeger Publishers, 1970.

10. Chemical Fertilizers

EUTROPHICATION: CAUSES, CONSEQUENCES, CORRECTIVES. National Academy of Sciences, Washington, D.C., 1969.

THE CHEMISTRY AND TECHNOLOGY OF FERTILIZERS. Edited by Vincent Sauchelli. Reinhold Publishing Corporation, 1960.

COMMERCIAL FERTILIZERS. G. H. Collings. McGraw-Hill Book Company, Inc., 1955.

FERTILIZER NITROGEN: ITS CHEMISTRY AND TECHNOLOGY. Edited by Vincent Sauchelli. Reinhold Publishing Corporation, 1964.

11. Insects v. Insecticides

PESTICIDES IN THE ENVIRONMENT. In Cleaning our Environment—The Chemical Basis for Action, a report by the Subcommittee on Environmental Improvement, Committee on Chemistry and Public Affairs. American Chemical Society, 1969.

DDT—RESISTANT HOUSEFLIES AND MOSQUITOES. W. V. King in Journal of Economic Entomology, Vol. 43, No. 4, pages 527–532; August, 1950.

12. Pesticides and the Reproduction of Birds

PESTICIDES AND THE LIVING LANDSCAPE. Robert L. Rudd. The University of Wisconsin Press, 1964.

PESTICIDE-INDUCED ENZYME BREAKDOWN OF STEROIDS IN BIRDS. D. B. Peakall in Nature, Vol. 216, No. 5114, pages 505–506; November 4, 1967.

PEREGRINE FALCON POPULATIONS: THEIR BIOLOGY AND DECLINE. Edited by Joseph J. Hickey. The University of Wisconsin Press, 1969.

MARKED DDE IMPAIRMENT OF MALLARD REPRODUCTION IN CONTROLLED STUDIES. Robert G. Heath, James W. Spann and J. F. Kreitzer in Nature, Vol. 224, No. 5214, pages 47–48; October 4, 1969.

13. Third Generation Pesticides

THE EFFECTS OF JUVENILE HORMONE ANALOGUES ON THE EMBRYONIC DEVELOPMENT OF SILKWORMS. Lynn M. Riddiford and Carroll M. Williams in Proceedings of the National Academy of Sciences, Vol. 57, No. 3, pages 595–601; March, 1967.

THE HORMONAL REGULATION OF GROWTH AND REPRODUCTION IN INSECTS. V. B. Wigglesworth in Advances in Insect Physiology: Vol. II, edited by J. W. L. Bement, J. E. Treherne and V. B. Wigglesworth. Academic Press Inc., 1964.

SYNTHESIS OF A MATERIAL WITH HIGH JUVENILE HORMONE ACTIVITY. John H. Law, Ching Yuan and Carroll M. Williams in Proceedings of the National Academy of Sciences, Vol. 55, No. 3, pages 576–578; March, 1966.

14. Pheromones

OLFACTORY STIMULI IN MAMMALIAN REPRODUCTION. A. S. Parkes and H. M. Bruce in Science, Vol. 134, No. 3485, pages 1049–1054; October, 1961.

PHEROMONES (ECTOHORMONES) IN INSECTS. Peter Karlson and Adolf Butenandt in Annual Review of Entomology, Vol. 4, pages 39–58; 1959.

THE SOCIAL BIOLOGY OF ANTS. Edward O. Wilson in Annual Review of Entomology, Vol. 8, pages 345–368; 1963.

III ENERGY IN OUR SOCIETY

15. The Flow of Energy in an Industrial Society

ENERGY IN THE UNITED STATES: SOURCES, USES, AND POLICY ISSUES. Hans H. Landsberg and Sam H. Schurr. Random House, 1968.

AN ENERGY MODEL FOR THE UNITED STATES, FEATURING ENERGY BALANCES FOR THE YEARS 1947 TO 1965 AND PROJECTIONS AND FORECASTS TO THE YEARS 1980 AND 2000. Warren E. Morrison and Charles L. Readling. U.S. Department of the Interior, Bureau of Mines, No. 8384, 1968.

THE ECONOMY, ENERGY, AND THE ENVIRONMENT: A BACKGROUND STUDY PREPARED FOR THE USE OF THE JOINT ECONOMIC COMMITTEE, CONGRESS OF THE UNITED STATES. Environmental Policy Division, Legislative Reference Service, Library of Congress. U.S. Government Printing Office, 1970.

ENERGY CONSUMPTION AND GROSS NATIONAL PRODUCT IN THE UNITED STATES: AN EXAMINATION OF A RECENT CHANGE IN THE RELATIONSHIP. National Economic Research Associates, Inc., 1971.

16. The Energy Resources of the Earth

MAN AND ENERGY. A. R. Ubbelohde. Hutchinson's Scientific and Technical Publications, 1954.

ENERGY FOR MAN: WINDMILLS TO NUCLEAR POWER. Hans Thirring. Indiana University Press, 1958.

ENERGY RESOURCES. M. King Hubbert. National Academy of Sciences–National Research Council, Publication 1000-D, 1962.

RESOURCES AND MAN: A STUDY AND RECOMMENDATIONS. Committee on Resources and Man. W. H. Freeman and Company, 1969.

ENVIRONMENT: RESOURCES, POLLUTION AND SOCIETY. Edited by William W. Murdoch. Sinauer Associates, 1971.

17. Coal

COAL. I.G.C. Dryden in Encyclopedia of Chemical Technology, second edition, Vol. 5, pages 606–678. Interscience Publishers, 1964.

ENERGY SOURCES—The Wealth of the World. Eugene Ayres and Charles A. Scarlott. McGraw-Hill Book Company, Inc., 1952.

EVIDENCE FOR THE CYCLIC STRUCTURE OF BITUMINOUS COALS. H. C. Howard in *Industrial and Engineering Chemistry*, Vol. 44, No. 5, pages 1083–1088; May, 1952.

THE NATURE AND ORIGIN OF COAL AND COAL SEAMS. A. Raistrick and C. E. Marshall. The English Universities Press Ltd., 1939.

18. Tar Sands and Oil Shales

THE K. A. CLARK VOLUME: A COLLECTION OF PAPERS ON THE ATHABASCA OIL SANDS. Edited by M. A. Carrigy. Research Council of Alberta, Information Series No. 45, 1963.

OIL SHALE TECHNOLOGY. H. M. Thorne, K. E. Stanfield, G. U. Dinneen and W. I. R. Murphy. U.S. Bureau of Mines Information Circular 8216, 1964.

OIL SHALES AND SHALE OILS. Harold S. Bell. Van Nostrand Company, Inc., 1948.

19. Clean Power from Dirty Fuels

POWER GAS PRODUCERS: THEIR DESIGN AND APPLICATION. Philip W. Robson. Edward Arnold, 1908.

CLEAN POWER FROM COAL. Arthur M. Squires in *Science*, Vol. 169, No. 3948, pages 821–828; August 28, 1970.

DEALING WITH SULFUR IN RESIDUAL FUEL OIL. S. B. Alpert, R. H. Wolk and A. M. Squires in *Power Generation and Environmental Change*, edited by David A. Berkowitz and A. M. Squires. The M.I.T. Press, 1971.

20. Fuel Cells

FUEL CELL SYSTEMS. Symposia sponsored by the Division of Fuel Chemistry of the American Chemical Society. Advances in *Chemistry Series* No. 47. American Chemical Society, 1965.

RECENT RESEARCH IN GREAT BRITAIN ON FUEL CELLS. F. T. Bacon and J. S. Forrest in *Transactions of the Fifth World Power Conference*; Vienna, 1956, Vol. 15, pages 5, 397–5, 412; 1957.

SYMPOSIUM ON FUEL CELLS. Ernst C. Baars, Chairman, in *Proceedings 12th Annual Battery Research and Development Conference*, pages 2–17; May 21–22, 1958.

TEXTBOOK OF ELECTROCHEMISTRY. G. Kortüm and J. O'M. Bockris. Elsevier Publishing Company, 1951.

21. Fast Breeder Reactors

THE TECHNOLOGY OF NUCLEAR REACTOR SAFETY, VOL. I: REACTOR PHYSICS AND CONTROL. Edited by T. G. Thompson and J. G. Beckerley. The M.I.T. Press, 1964.

FAST REACTOR TECHNOLOGY: PLANT DESIGN. Edited by John G. Yevick. The M.I.T. Press, 1966.

AEC AUTHORIZING LEGISLATION, FISCAL YEAR 1971, HEARINGS BEFORE THE JOINT COMMITTEE ON ATOMIC ENERGY, PART 3. U.S. Government Printing Office, 1970.

22. The Prospects of Fusion Power

PROGRESS IN CONTROLLED THERMONUCLEAR RESEARCH. R. W. Gould, H. P. Furth, R. F. Post and F. L. Ribe in *Presentation Made before the President's Science Advisory Committee, December 15, 1970, and AEC's General Advisory Committee, December 16, 1970.*

WORLD SURVEY OF MAJOR FACILITIES IN CONTROLLED FUSION. *Nuclear Fusion*, Special Supplement 1970, STI/Pub/23. International Atomic Energy Agency, 1970.

WHY FUSION? William C. Gough in *Proceedings of the Fusion Reactor Design Symposium, Held at Texas Tech University, Lubbock, Texas, on June 2–5, 1970.*

23. The Hydrogen Economy

HYDROGEN FROM OFF-PEAK POWER: A POSSIBLE COMMERCIAL FUEL. R. A. Erren and W. Hastings Campbell in *The Chemical Trade Journal and Chemical Engineer*, Vol. 92, No. 2392, pages 238–239; March 24, 1933.

HYDROGEN: KEY TO THE ENERGY MARKET. G. De Beni and C. Marchetti in *Euro Spectra*, Vol. 9, No. 2, pages 46–50; June, 1970.

HYDROGEN, MASTER-KEY TO THE ENERGY MARKET. Cesare Marchetti in *Euro Spectra*, Vol. 10, No. 4, pages 117–130; December, 1971.

THE HYDROGEN ECONOMY. D. P. Gregory, D. Y. C. Ng and G. M. Long in *The Electrochemistry of Cleaner Environments*, edited by J. O'M Bockris. Plenum Press, 1972.

IV THE LEGACY OF ENERGY USE

24. Flame

COMBUSTION, FLAMES AND EXPLOSIONS OF GASES. Bernard Lewis and Guenther von Elbe. Academic Press, Inc., 1961.

EXPLOSION AND COMBUSTION PROCESSES IN GASES. Wilhelm Jost. McGraw-Hill Book Company, Inc., 1946.

FLAMES: THEIR STRUCTURE, RADIATION, AND TEMPERATURE. A. G. Gaydon and W. G. Wolfhard. Chapman and Hall, Ltd., London, 1953.

25. The Chemical Effects of Light

LIGHT-SENSITIVE SYSTEMS: CHEMISTRY AND APPLICATION OF NONSILVER HALIDE PHOTOGRAPHIC PROCESSES. Jaromir Kosar. John Wiley & Sons, 1965.

THE MIDDLE ULTRAVIOLET: ITS SCIENCE AND TECHNOLOGY. Edited by A. E. S. Green. John Wiley & Sons, 1966.

ENERGY TRANSFER FROM HIGH-LYING EXCITED STATES. Gisela K. Oster and H. Kallmann in *Journal de Chimie Physique et de Physico-Chimic Biologique*, Vol. 64, No. 1, pages 28–32; January, 1967.

PHOTOPOLYMERIZATION OF VINYL MONOMERS. Gerald Oster and Nan-Loh Yang in *Chemical Reviews*, Vol. 68, No. 2, pages 125–151; March 25, 1968.

FLASH PHOTOLYSIS AND SOME OF ITS APPLICATIONS. George Porter in *Science*, Vol. 160, No. 3834, pages 1299–1307; June 21, 1968.

26. The Control of Air Pollution

SMOG: A REPORT TO THE PEOPLE. Lester Lees, Mark Braly, Mahlon Easterling, Robert Fisher, Kenneth Heitner, James Henry, Patricia J. Horne, Burton Klein, James Krier, W. David Montgomery, Guy Pauker, Gary Rubenstein, and John Trijonis. The Ward Ritchie Press, 1972.

AIR POLLUTION: VOLS. I AND II, edited by Arthur C. Stern. Academic Press, 1962.

AIR POLLUTION CONTROL. William L. Faith. John Wiley and Sons, Inc., 1959.

PHOTOCHEMISTRY OF AIR POLLUTION. Philip A. Leighton. Academic Press, 1961.

WEATHER MODIFICATION AND SMOG. M. Neiburger in *Science*, Vol. 126, No. 3275, pages 637–645; October, 1957.

27. The Global Circulation of Atmospheric Pollutants

INTERNATIONAL SYMPOSIUM ON TRACE GASES AND NATURAL AND ARTIFICIAL RADIOACTIVITY IN THE ATMOSPHERE. *Journal of Geophysical Research*, Vol. 68, No. 13, pages 3745–4016; July 1, 1963.

PROCEEDINGS OF THE CACR SYMPOSIUM: ATMOSPHERIC CHEMISTRY, CIRCULATION AND AEROSOLS, AUGUST 15–25, 1965, VISBY, SWEDEN. *Tellus*, Vol. 18, No. 2–3, pages 149–684; 1966.

MAN'S IMPACT ON THE GLOBAL ENVIRONMENT: ASSESSMENT AND RECOMMENDATIONS FOR ACTION. Report of the Study of Critical Environment Problems (SCEP). The M.I.T. Press, 1970.

28. Radioactive Poisons

IONIZING RADIATION. Earl Cook in *Environment: Resources, Pollution and Society*, edited by William M. Murdoch. Sinauer Associates, Inc., 1971.

ESTIMATING RADIOELEMENTS IN EXPOSED INDIVIDUALS. Jack Schubert in *Nucleonics*, Vol. 8, Nos. 2, 3, and 4, pages 13–28, 66–78 and 59–67; February, March and April, 1951.

MAXIMUM PERMISSIBLE AMOUNTS OF RADIOISOTOPES IN THE HUMAN BODY AND MAXIMUM PERMISSIBLE CONCENTRATIONS IN AIR AND WATER. National Bureau of Standards Handbook 52. National Bureau of Standards, 1953.

THE LATE EFFECTS OF INTERNALLY-DEPOSITED RADIOACTIVE MATERIALS IN MAN. Joseph C. Aub, Robley D. Evans, Louis H. Hemplemann, and Harrison S. Martland in *Medicine*, Vol. 31, No. 3, pages 221–329; September, 1952.

29. The Circulation of Radioactive Isotopes

RADIOACTIVITY IN GEOLOGY AND COSMOLOGY. T. P. Kohman and N. Saito in *Annual Review of Nuclear Science*, Vol. 4, pages 401–446; 1954.

THE RADIOACTIVITY OF THE ATMOSPHERE AND HYDROSPHERE. Hans E. Suess in *Annual Reviews of Nuclear Science*, Vol. 8, pages 243–256; 1958.

V IMPLICATIONS OF MATERIAL WEALTH

30. Food Additives

FOOD STANDARDS COMMITTEE REPORTS. Great Britain Ministry of Agriculture, Fisheries and Food. Her Majesty's Stationery Office, London.

REPORTS OF THE CODEX COMMITTEE ON FOOD ADDITIVES. Food and Agriculture Organization of the United Nations, Rome.

REPORTS OF THE JOINT FAO/IAEA/WHO EXPERT COMMITTEE ON FOOD IRRADIATION. Food and Agriculture Organization of the United Nations, Rome.

REPORTS OF THE JOINT FAO/WHO EXPERT COMMITEE ON FOOD ADDITIVES. Food and Agriculture Organization of the United Nations, Rome.

HANDBOOK OF FOOD ADDITIVES. Edited by Thomas E. Furia. The Chemical Rubber Co., 1968.

31. Poisons

CLINICAL TOXICOLOGY OF COMMERCIAL PRODUCTS. Marion N. Gleason, Robert E. Gosselin and Harold C. Hodge. Williams & Wilkins Company, 1957.

MECHANISM OF THE TOXICITY OF THE ACTIVE CONSTITUENT OF DICHAPETALUM CYMOSUM AND RE-

LATED COMPOUNDS. Rudolph A. Peters in *Advances in Enzymology*, Vol. 18, pages 113–160; 1957.

LA SCIENCE EXPÉRIMENTALE. Claude Bernard. Librairie J.-B. Baillière & Fils, 1878.

32. Radiation-Imitating Chemicals

THE BIOLOGICAL ACTIONS AND THERAPEUTIC APPLICATIONS OF THE B-CHLOROETHYLAMINES AND SULFIDES. Alfred Gilman and Frederick S. Philips in *Science,* Vol. 103, No. 2, 675, pages 409–415; April 5, 1946.

THE CEHMICAL AND GENETIC MECHANISMS OF CARCINOGENESIS. I. NATURE AND MODE OF ACTION. A. Haddow in *The Physiopathology of Cancer,* edited by Freddy Homburger, pages 565–601. Hoeber-Harper, 1959.

CHEMICALS WHICH PRODUCE THE SAME BIOLOGICAL EFFECTS AS ATOMIC RADIATIONS. Peter Alexander in *Atomic Radiation and Life,* pages 220–228. Pelican Book, 1957.

33. Free Radicals in Biological Systems

FREE RADICALS IN BIOLOGICAL SYSTEMS: PROCEEDINGS OF A SYMPOSIUM HELD AT STANFORD UNIVERSITY, MARCH, 1960. Edited by M. S. Blois, Jr., H. W. Brown, R. M. Lemmon, R. O. Lindblom and M. Weissbluth. Academic Press, 1961.

FREE RADICALS IN TISSUE. Irvin Isenberg in *Physiological Reviews*, Vol. 44, No. 3, pages 487–513; July, 1964.

FREE RADICALS. William A. Pryor. McGraw-Hill Book Company, 1966.

RADIATION & AGEING: PROCEEDINGS OF A COLLOQUIUM HELD IN SEMMERING, AUSTRIA, JUNE 23–24, 1966. Edited by Patricia J. Lindop and G. A. Sacher. Taylor & Francis Ltd, 1966.

34. Beryllium and Berylliosis

THE METAL BERYLLIUM. Edited by D. W. White, Jr., and J. E. Burke. The American Society for Metals, 1955.

PNEUMOCONIOSIS: BERYLLIUM, BAUXITE FUMES, COMPENSATION. Edited by Arthur J. Vorwald. Paul B. Hoeber, Inc., 1950.

STUDIES ON THE MECHANISM OF PROTECTION BY AURINTRICARBOXYLIC ACID IN BERYLLIUM POISONING. I. Jack Schubert, Marcia R. White and Arthur Lindenbaum in *Journal of Biological Chemistry*, Vol. 196, No. 1, pages 279–288; May, 1952.

STUDIES ON THE MECHANISM OF PROTECTION BY AURINTRICARBOXYLIC ACID IN BERYLLIUM POISONING. II: EQUILIBRIA INVOLVING ALKALINE PHOSPHATASE. Jack Schubert and Arthur Lindenbaum in *Journal of Biological Chemistry*, Vol. 208, No. 1, pages 359–368; May, 1954.

STUDIES ON THE MECHANISM OF PROTECTION BY AURINTRICARBOXYLIC ACID IN BERYLLIUM POISONING. III: CORRELATION OF MOLECULAR STRUCTURE WITH REVERSAL OF BIOLOGIC EFFECTS OF BERYLLIUM. Arthur Lindenbaum, Marcia R. White and Jack Schubert in *Archives of Biochemistry and Biophysics*, Vol. 52, No. 1, pages 110–147; September, 1954.

35. Mercury in the Environment

THE GENERAL PHARMACOLOGY OF THE HEAVY METALS. H. Passow, A. Rothstein and T. W. Clarkson in *Pharmacological Reviews*, Vol. 13, No. 2, pages 185–224; June, 1961.

ABSORPTION AND EXCRETION OF MERCURY IN MAN: I-XIV. Leonard J. Goldwater *et al.* Papers published in *Archives of Environmental Health*, 1962–1968.

CHEMICAL FALLOUT: CURRENT RESEARCH ON PERSISTENT PESTICIDES. Edited by Morton W. Miller and George C. Berg. Charles C Thomas, Publisher, 1969.

MERCURY IN THE ENVIRONMENT. Geological Survey Professional Paper 713. United States Government Printing Office, 1970.

36. Lead Poisoning

THE EXPOSURE OF CHILDREN TO LEAD. J. Julian Chisolm, Jr., and Harold E. Harrison in *Pediatrics,* Vol. 18, No. 6, pages 943–958; December, 1956.

THE ANAEMIA OF LEAD POISONING: A REVIEW. H. A. Waldron in *British Journal of Industrial Medicine,* Vol. 23, No. 2, pages 83–100; April, 1966.

THE RENAL TUBULE IN LEAD POISONING, I: MITOCHONDRIAL SWELLING AND AMINOACIDURIA. Robert A. Goyer in *Laboratory Investigation*, Vol. 19, No. 1, pages 71–77; July, 1968.

THE RENAL TUBULE IN LEAD POISONING, II: IN VITRO STUDIES OF MITOCHONDRIAL STRUCTURE AND FUNCTION. Robert A. Goyer, Albert Krall and John P. Kimball in *Laboratory Investigation*, Vol. 19, No. 1, pages 78–83; July, 1968.

THE USE OF CHELATING AGENTS IN THE TREATMENT OF ACUTE AND CHRONIC LEAD INTOXICATION IN CHILDHOOD. J. Julian Chisolm in *The Journal of Pediatrics,* Vol. 73, No. 1, pages 1–38; July, 1968.

LEAD POISONING IN CHILDHOOD — COMPREHENSIVE MANAGEMENT AND PREVENTION. J. Julian Chisolm, Jr., and Eugene Kaplan in *The Journal of Pediatrics*, Vol. 73, No.6, pages 942–950; December, 1968.

37. Chelation

CHELATING AGENTS AND METAL CHELATES. Edited by F. P. Dwyer and D. P. Mellor. Academic Press, 1964.

CHEMISTRY OF THE METAL CHELATE COMPOUNDS. Arthur E. Martell and Melvin Calvin. Printice-Hall, Inc., 1951.

INDEX